Contents

Introduction   iv

Biology
B5a In good shape   1
B5b The vital pump   10
B5c Running repairs   18
B5d Breath of life   26
B5e Waste disposal   33
B5f Life goes on   41
B5g New for old   51
B5h Size matters   57

B6a Understanding bacteria   62
B6b Harmful microorganisms   69
B6c Microorganisms – factories for the future?   77
B6d Biofuels   82
B6e Life in soil   87
B6f Microscopic life in water   94
B6g Enzymes in action   102
B6h Genetic engineering   106

Chemistry
C5a Moles and empirical formula   111
C5b Electrolysis   121
C5c Quantitative analysis   128
C5d Titrations   135
C5e Gas volumes   141
C5f Equilibria   148
C5g Strong and weak acids   155
C5h Ionic equations   161

C6a Energy transfers – fuel cells   167
C6b Redox reactions   172
C6c Alcohols   178
C6d Chemistry of sodium chloride (NaCl)   183
C6e Depletion of the ozone layer   189
C6f Hardness of water   195
C6g Natural fats and oils   202
C6h Analgesics   207

Physics
P5a Satellites, gravity and circular motion   211
P5b Vectors and equations of motion   220
P5c Projectile motion   227
P5d Momentum   232
P5e Satellite communication   241
P5f The nature of waves   248
P5g Refraction of waves   256
P5h Optics   267

P6a Resisting   273
P6b Sharing   280
P6c Motoring   286
P6d Generating   293
P6e Transforming   299
P6f Charging   306
P6g It's logical   314
P6h Even more logical   321

Answers to SAQs
Biology   326
Chemistry   331
Physics   340

Glossaries
Biology   346
Chemistry   352
Physics   356

Periodic Table   359

Physics formulae   360

Index   361

Introduction

To the pupil
This book is divided into three sections – Biology, Chemistry and Physics. Each of these sections is then arranged by Item, as in the exam specification. Each Item has the following features.
- **Self-Assessment Questions (SAQs)** are placed within the text and refer to the material that has gone before. Answers to these are provided at the back of the book.
- **End-of-chapter questions**, including exam-style questions. Answers to these are not provided in this book.
- **Summaries** at the end of the chapter that show the information you need in order to do well in the exam.
- **Higher-level** text, questions and summaries are shown by a side bar marked with the letter 'H'.
- **Context boxes** that give you the opportunity to read about the history behind the discoveries and to learn about real-world applications.
- **Worked example boxes** (where relevant).

The book also contains glossaries, a Periodic Table, a list of physics formulae and an index at the back of the book.

To the teacher
The *Cambridge Gateway Sciences* series has been written to cover the new Gateway Specification (B) developed by OCR.

This text contains materials for the Separate Science specification.

For further information on all accompanying materials, visit the dedicated website at www.cambridge.org/gatewaysciences

5a In good shape

Skeletons

What keeps your body in shape? The main structure that does this is your **skeleton**. Without it, you would be a shapeless heap on the ground.

Humans have an **internal skeleton**. This means that the skeleton is *inside* the body. Figure 5a.1 shows the main bones in your skeleton. It is mostly made of a material called **bone**. Some parts – for example, the nose, the pinna (outside 'flap') of the ear and the ends of leg bones – are made of a more flexible material called **cartilage**. Birds, reptiles, amphibians and most fish also have internal skeletons made of bone and cartilage. But the internal skeleton of sharks is made of cartilage only.

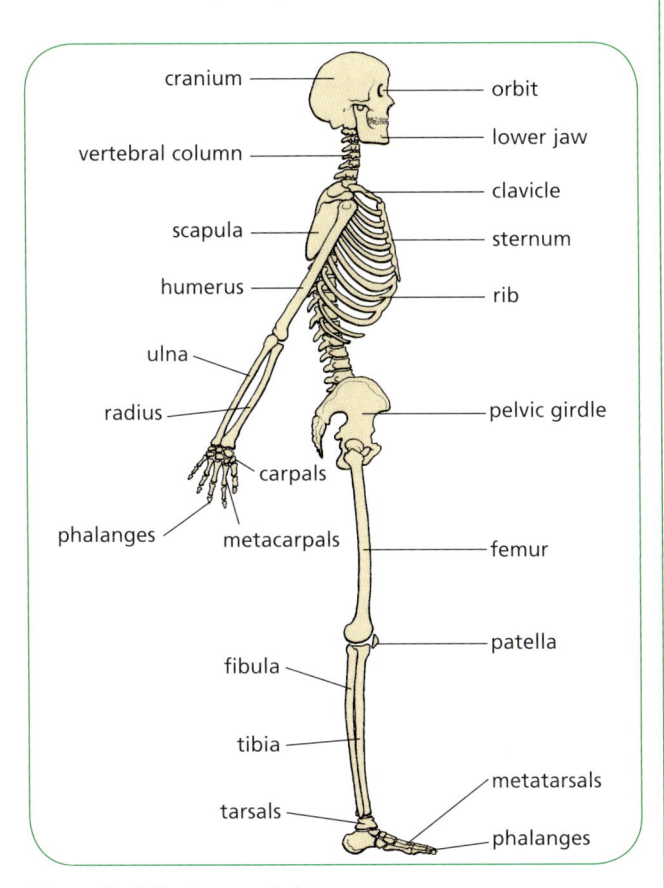

Figure 5a.1 The human skeleton.

Insects and some other invertebrates have a completely different kind of skeleton. It is on the *outside* of the body, and so it is called an **external skeleton**. It is made of a tough, strong material called **chitin** (pronounced ky-tin). Part of an insect's skeleton is shown in Figure 5a.2.

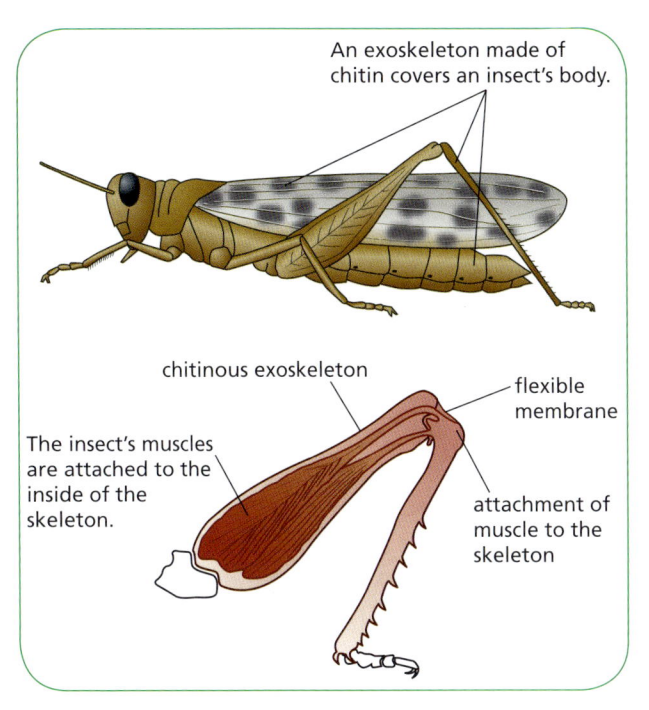

Figure 5a.2 An insect's exoskeleton.

Earthworms have a completely different kind of skeleton. Rather than being made of something hard, it is made of liquid. Liquids are incompressible, so they keep the same volume all the time. The liquid in the earthworm pushes out against the body wall and keeps it firm and in shape. This is called a **hydrostatic skeleton**.

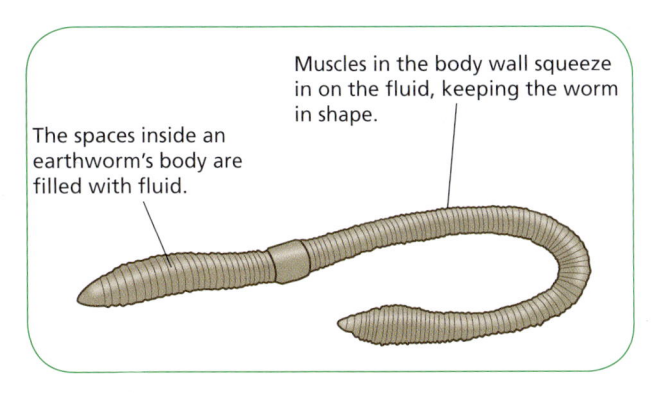

Figure 5a.3 An earthworm's hydrostatic skeleton.

SAQ

1. Is an earthworm's skeleton an external skeleton or an internal skeleton?
2. Suggest a definition of the term *skeleton*. Try to make sure it fits all the examples of skeletons described above.

B5a In good shape

Which is the best kind of skeleton?

We, like all vertebrates, have an internal skeleton. Insects have an external one. Which is better?

This is a difficult question! There are many times more insects on Earth than there are vertebrates, and many different kinds of insects. So, clearly, an external skeleton works perfectly well. If it did not, then insects would not be so successful.

All the same, there are a few problems with external skeletons that don't occur with internal ones.

- An internal skeleton provides a **framework** around which the rest of the body can be supported. External skeletons do this too but, because the skeleton is around the *outside*, it is more difficult to support parts of the body that are deep *inside* it. With an internal skeleton, no part of the body is far from a bone. This means that exoskeletons cannot support such large bodies as internal ones can. This is one of the main reasons why there are no giant insects as big as ourselves. Giant insects would need massive exoskeletons, and these would need to be thick and heavy – so heavy that the insects would probably be unable to support the weight or move easily.
- An internal skeleton can **grow** with the body. As you grow, your bones also grow. But an insect's exoskeleton cannot do this. To be able to grow, the insect has to shed its old exoskeleton. Underneath, a new one has been growing, but it is still soft. As soon as it sheds the old one, the insect puffs itself up, stretching the new skeleton to a larger size. The new skeleton then hardens. So, for a short time, the insect is soft and vulnerable to predators while its new skeleton is slowly getting harder. Many insects shed their skeletons four or five times as they grow into adults.
- One of the functions of a skeleton is to provide a firm **attachment for muscles**. An internal skeleton can have a much larger surface area than an external one, and so it can provide more places where muscles can be attached.
- External skeletons provide a protective layer around the animal's body. But, to allow movement, there must be some flexible places, which are weak spots in the insect's armour. In general, internal skeletons can have more joints than external ones, so a body with an internal skeleton can sometimes be more **flexible**.

SAQ

3 Suggest a *disadvantage* of internal skeletons compared with external ones.

4 Figure 5a.5 shows how the leg of a locust changes in length during its lifetime.

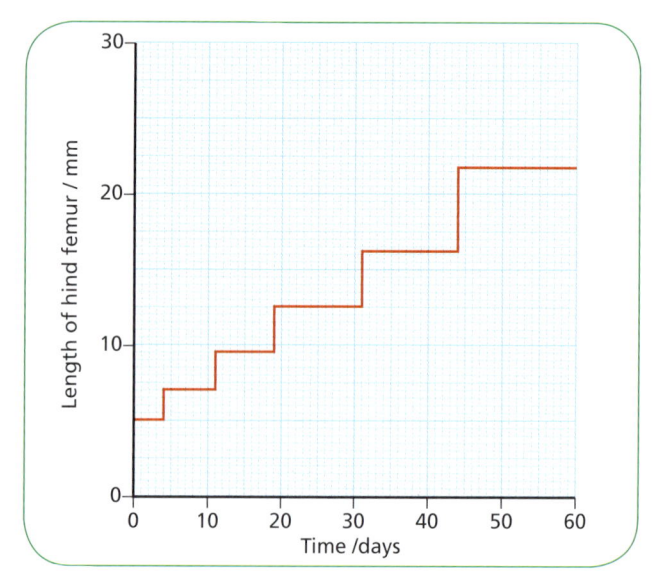

Figure 5a.5 Changes in the length of a locust's leg as it grows to an adult.

a Explain why the length of the leg increases in steps, not smoothly.

b Use the information in the graph to work out how many times the locust moulted its skeleton.

Figure 5a.4 This dragonfly was a water-living nymph less than an hour before this photo was taken. Now it has to wait until its new exoskeleton has hardened before it can fly.

Bone and cartilage

Figure 5a.6 shows the structure of the top of a person's **femur**. The femur is a **long bone**. Long bones include the ones in the limbs, as opposed to bones such as the pelvis (hip bones) or cranium (skull bones).

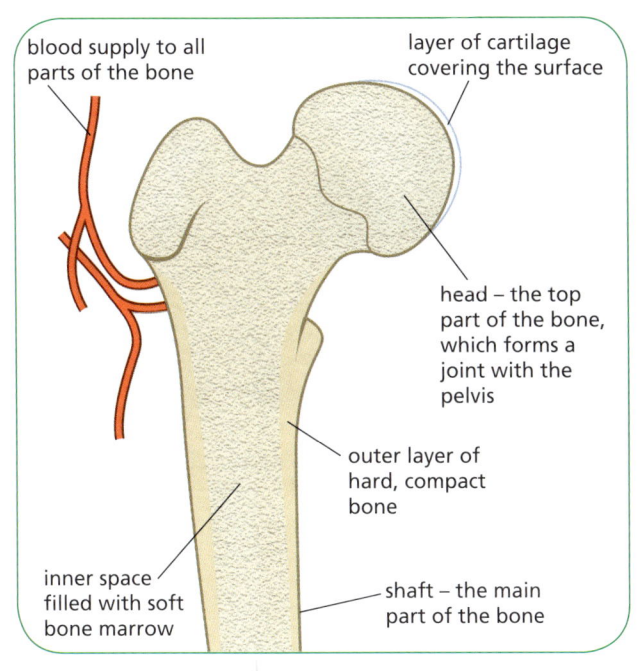

Figure 5a.6 The structure of a long bone.

You can see that the bone is made of three main tissues – bone, cartilage and marrow. All of these contain living cells. The bone is supplied by blood vessels, which branch out and carry supplies of oxygen, glucose and other nutrients to the cells inside the bone. If you break or cut a living bone, it will bleed.

The end of the femur is covered with a layer of smooth, slippery cartilage. This makes it easy for the head of the femur to slide over the bone in the socket of the pelvis when you walk. As well as being more slippery, cartilage is a more flexible material than bone. Bone tissue contains **calcium phosphate** and other salts, which make it very hard. Cartilage does not contain calcium phosphate.

The middle of the shaft contains a material called **bone marrow**. This is much softer than bone and cartilage. Having a hollow centre actually makes the bone stronger than if it was solid all the way through, and less likely to break. (The bone marrow is where red and white blood cells are made.)

Bone growth

Your bones all began as cartilage. As you grew, calcium phosphate was gradually deposited into some of the cartilage, so that it changed into bone. This process is called **ossification**. Forensic scientists can use this to determine the age of a dead body – the younger the person, the more cartilage is present compared to bone.

A long bone, such as the femur, grows by adding more and more cells at regions called **growth plates** (Figure 5a.7). Here, cartilage cells divide repeatedly, producing new cartilage tissue and lengthening the bone. The new tissue gradually turns to bone. When a person is fully grown, the growth plates disappear and are replaced by bone.

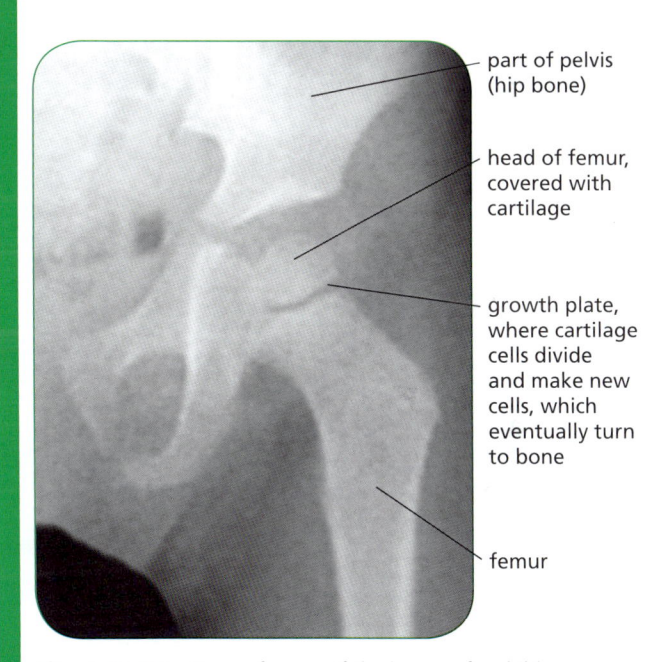

Figure 5a.7 An X-ray of some of the bones of a child.

Even when a bone is fully grown, it does not just stay the same. Cells in the bone constantly break parts of it down and replace them with new bone. This helps the bone to adjust to the way the body is being used. For example, if a person trains to be a weightlifter, a sprinter or a dancer, the bones will grow stronger in the places where most forces are put onto them.

This ability to grow and change throughout life can help a bone to fight off infection. Because they are made of living tissue, bone and cartilage can become infected just like any other part of the body. One kind of bone infection is called **osteomyelitis**. This is caused by bacteria infecting

B5a In good shape

the bone, often following some damage to the bone, such as a fracture (a break). Osteomyelitis is quite a rare disease. It can usually be cured using antibiotics, especially if the infection is detected and treated at an early stage.

Broken bones

Despite being made of a tough, slightly flexible material, bones can break. A fracture is often the result of a large force being applied to a bone, but sometimes a sharp knock will break a bone. Broken bones usually heal, given time.

Types of fracture

Doctors classify fractures into different types. Three types of fracture are shown in Figure 5a.8.

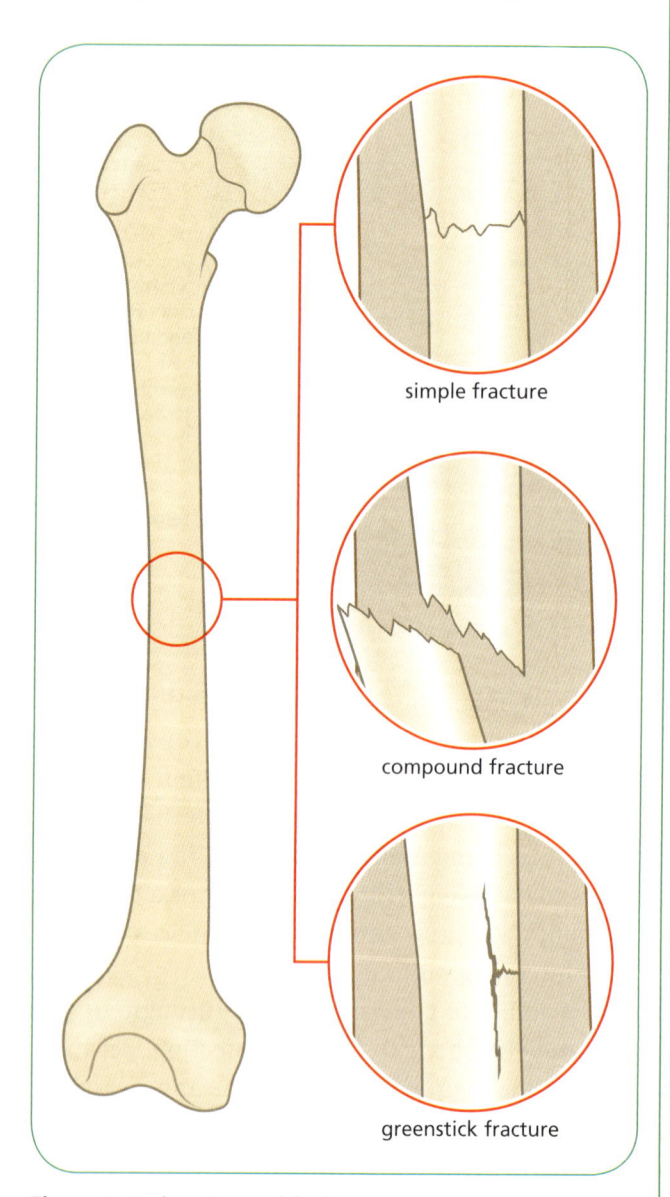

Figure 5a.8 Three types of fracture.

Simple fractures involve a simple, clean break in the bone. The break goes all the way through the bone but the bone does not move out of place, and no tissues around the bone are damaged. Simple fractures usually heal quickly if the bone is held in place using a plaster.

Figure 5a.9 A hard plaster holds a limb firmly in place so that the broken bone cannot move and gradually heals in the correct position.

Compound fractures happen when the bone breaks apart completely and part of it pushes through the skin. Apart from being especially painful, this kind of break is dangerous because it could let pathogens (such as bacteria) into the wound so that the bone or other tissue gets infected. A compound fracture often needs surgery to move the bone back into place before the limb is put in plaster and the bone left to heal gradually.

Greenstick fractures usually happen in children, because the bones have not finished growing and are more flexible than in an adult. Only one side of the bone is broken, so the bone is still held in place. The limb is usually held firmly in position using a hard plaster, and the break normally heals very quickly.

SAQ

5 A person with a suspected fracture in the vertebral column should not be moved except by a paramedic or other medically qualified person. Using what you know about the structure of the nervous system, explain why moving the person could be very dangerous.

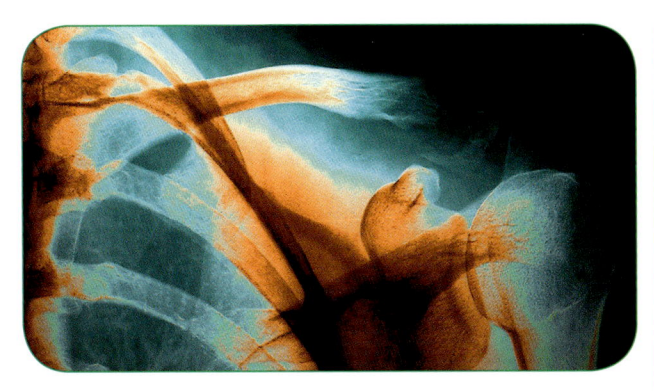

Figure 5a.10 This X-ray shows a broken collar bone. You can feel where your collar bone should be by feeling just below your shoulder. A broken collar bone is a common injury caused by a fall from a horse or a bicycle.

Osteoporosis

As someone gets older, their bones gradually become less dense and lose their strength. Sometimes, the density of the bone drops so low that it becomes really easy to break it. A person with bones like this has **osteoporosis**.

Osteoporosis is more common in women than in men. It usually develops after the menopause (when periods stop), because the body secretes less of the hormone oestrogen. Oestrogen helps to keep bones strong. It is estimated that half of all women over 75 years of age have fractures in their vertebrae (backbones) that they don't know about, because of osteoporosis. This is what causes some older people to have a 'hunched' posture – their vertebrae are not strong enough to hold the body up straight.

You can do a lot to help to reduce the risk of developing osteoporosis when you are older. Eating a diet with plenty of calcium in it helps bones to become strong and dense, and the stronger they are to start with, the less likely they are to become weak in later life. Taking exercise that requires your bones to carry your weight, such as walking or dancing, is also good because the bones respond by growing stronger.

Bone loss in space

Our bones constantly reshape themselves according to the forces we exert on them. The bones of an astronaut in space don't have anywhere near as much force applied to them as on Earth. The lack of gravity means that the lower vertebrae, the hip bones and the leg bones don't have to bear weight. Over time, they gradually lose mass, as the bone cells break down the unwanted bone so that the chemicals in it can be used elsewhere in the body. On average, between 1% and 2% of bone mass is lost during one month in space.

That is fine while the astronaut is in space, but it is dangerous when they come back to Earth. Their bones may have become so weak that they can no longer stand up, and run the risk of bone fractures.

NASA does a lot of research on preventing and treating this problem. The answers they find can also be used to reduce the risk of Earth-bound people getting osteoporosis.

Astronauts in a spacecraft do exercises each day to make sure their bones have forces applied to them. The force of gravity is replaced by forces from springs and elastic straps, which the astronauts push against to move their legs and arms. The astronauts may also spend time standing on a vibrating plate with their legs held down by elastic straps. This pushes and pulls on their leg bones, helping to keep them strong.

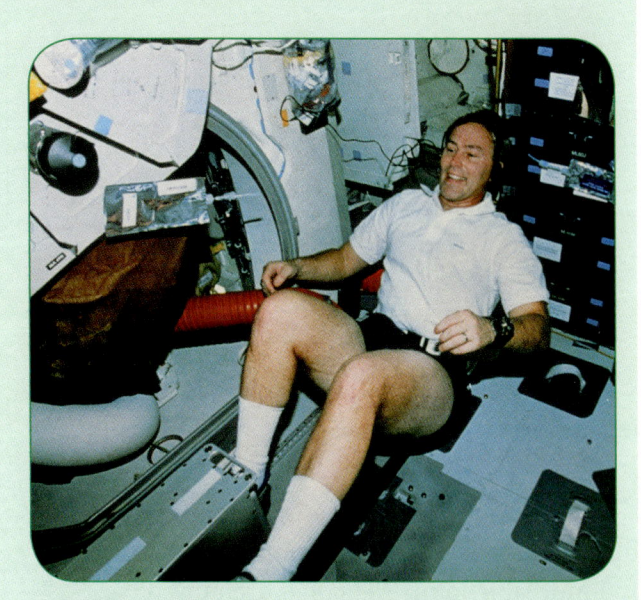

Figure 5a.11 Working leg muscles against a machine that provides mechanical resistance can avoid some of the worst effects of weightlessness.

Joints

A **joint** is a place where two bones meet. The bones are usually held together by strong, slightly stretchy cords called **ligaments**.

Types of joint

At a **fixed joint**, the bones are held tightly together and no movement can take place between them. Examples of a fixed joint include the **sutures** that hold the bones of the cranium together.

A **hinge joint** allows the bones to move in one plane, like a door opening and closing. The elbow, knee and finger joints are hinge joints (Figure 5a.12).

A **ball and socket joint** allows a bone to move in a circular manner, using a ball on one bone sitting in a rounded socket in the other bone. The hip joint and shoulder joint are examples. You can clearly see the ball at the head of the femur and the socket in the pelvis in the X-ray in Figure 5a.7.

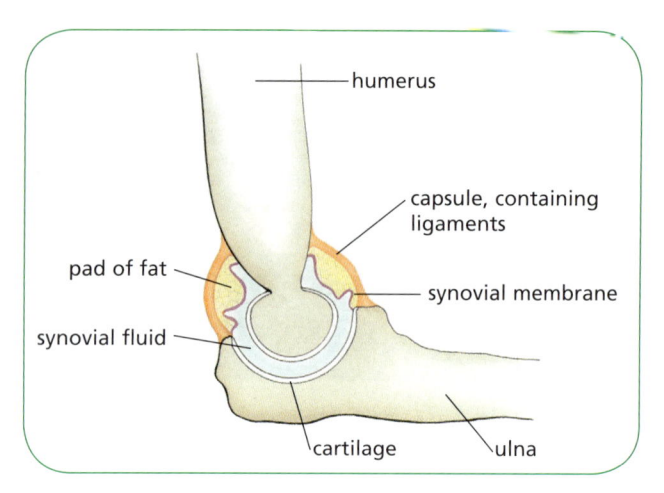

Figure 5a.13 Section through the elbow joint.

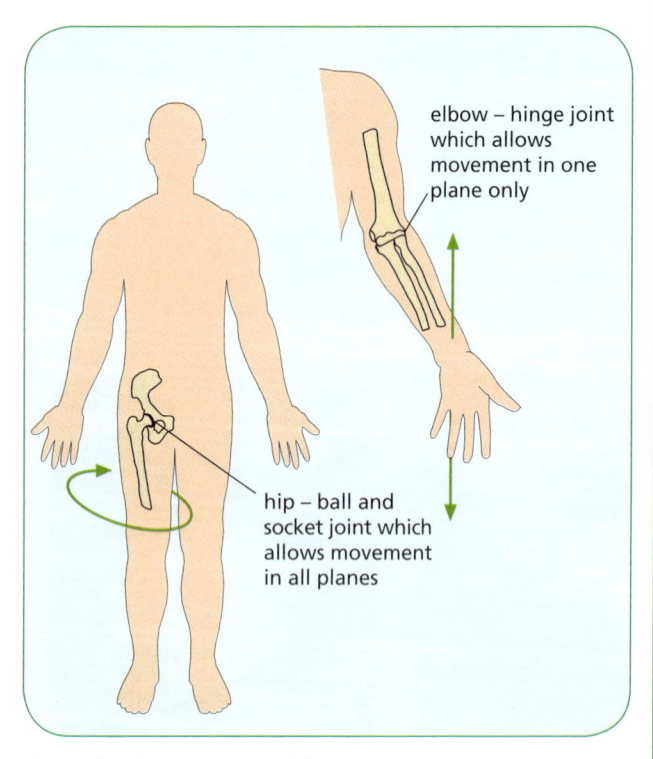

Figure 5a.12 Movement at joints.

Hinge joints and ball and socket joints are different types of **synovial joint**. A synovial joint is a moveable joint. The bones are moved by **muscles** attached to the bones by strong cords called **tendons**. Figure 5a.13 shows the structure of the elbow joint.

How a synovial joint works

The ligaments in a synovial joint hold the bones tightly together but they can stretch a little, which allows the bones to move relative to each other. The ligaments help to form the **capsule** of the joint, which encloses it and keeps the **synovial fluid** in place.

This fluid is secreted by the **synovial membrane**. The fluid helps to **lubricate** the joint, allowing the bones to move without rubbing against each other. The lubrication reduces friction. Friction is also reduced by the glassy smooth cartilage that covers the ends of the bones.

Joint replacements

Joints wear out as we get older. The cartilage wears away, allowing the ends of the bones to grate against each other. This can become extremely painful. In some people, the wear and tear builds up so much that they can scarcely move some of their joints. You may have read about this in Item C1f *Designer polymers*, p.185 in *Gateway Science*.

Some joints can be replaced with artificial ones. You may know someone who has had a hip replacement or a knee replacement. These operations are major ones but they are usually very successful.

In a hip replacement, the socket in the pelvis is often lined with a very smooth, slippery kind of hard 'cement' made of polythene. In other cases, a metal socket may be used. The top of the femur is replaced using a metal shaft (which fixes into the person's own femur) and a metal ball, again covered with a very slippery kind of cement.

B5a In good shape   7

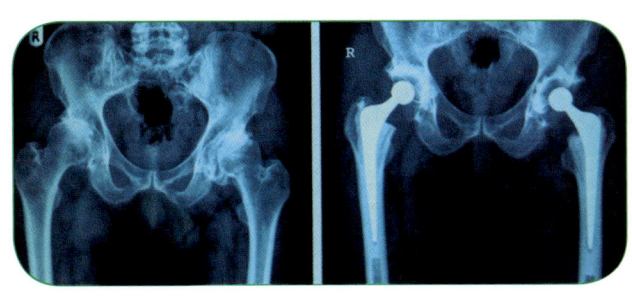

Figure 5a.14 These are X-rays of a person's hips before and after having two artificial hips fitted.

Having a hip replacement means that the person can move easily again – they can walk without constantly battling with pain. There are some risks, as there always are with a major operation. For example, bacteria could get into the wound and cause infection. If the operation is not done well, the length of the leg could be altered, making the person walk with a limp. And, because the person has to keep the limb fairly immobile for some time after the operation, a blood clot might develop and cause a pulmonary thrombosis. You can read more about this on pages 23–24.

SAQ

6 Use a series of bullet points to summarise the advantages and disadvantages of joint replacement surgery. You may be able to think of others, as well as the ones described on this page.

Movement at the elbow joint

Figure 5a.15 shows how the arm can bend and straighten at the elbow joint.

The **biceps** and **triceps** are muscles. Muscles have cells that can use energy to get shorter. This is called **contraction**. However, muscles *cannot* make themselves get longer again. They only get longer when they are pulled by something else.

This is why there are *two* muscles at the joint. When the biceps contracts, it pulls on the tendons attaching it to the bones, pulling the radius closer to the scapula. The arm bends or **flexes**.

To straighten the arm, the triceps contracts, pulling on the tendons attached to the ulna and the scapula. This pulls the end of the ulna closer to the scapula, extending the arm. While this is happening, the biceps stops contracting – in other

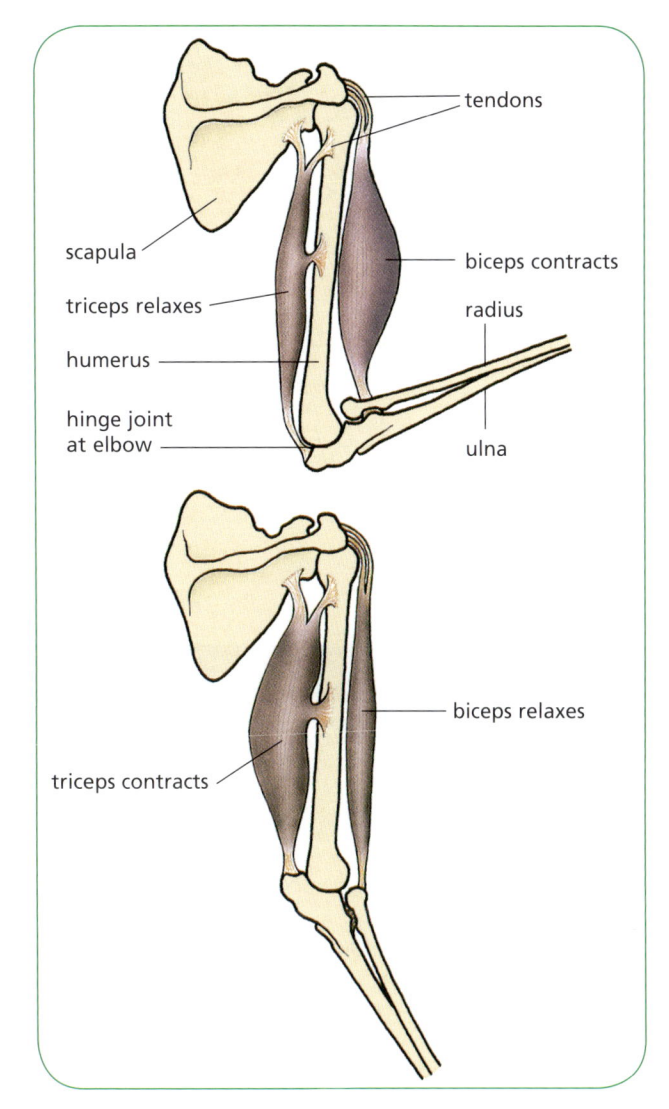

Figure 5a.15 Movement of the forearm.

words, it **relaxes**. This allows it to be pulled long and thin again.

When two muscles act together like this, one moving the bones at a joint in one direction, and the other moving them the opposite way, they are said to be **antagonistic muscles**.

SAQ

7 What must the triceps do while the biceps is contracting, in order to allow the arm to be straightened?

The arm as a lever

The arm moves like a lever. The elbow joint is the pivot (fulcrum), and forces are applied by the muscles, the weight of the arm and the weight of anything the arm is lifting.

8 B5a In good shape

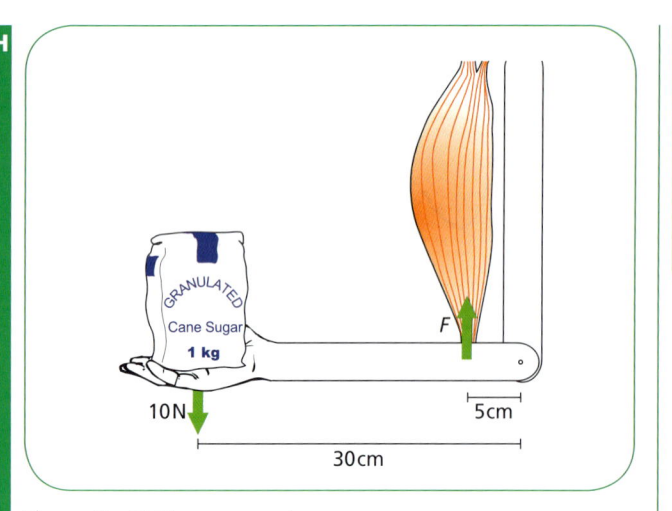

Figure 5a.16 The arm as a lever.

SAQ

8 Calculate the force that the biceps muscle must exert in order to lift the bag of sugar in Figure 5a.16. Use the formula:

anticlockwise force × distance from pivot
= clockwise force × distance from pivot

Summary

You should be able to:

- give examples of animals with internal skeletons, external skeletons and hydrostatic skeletons
- describe the occurrence of chitin, bone and cartilage in skeletons
- describe advantages of internal skeletons over external skeletons
- describe the structure of a long bone
- **H** explain that bone and cartilage can repair themselves, for example if they are infected
- describe how the skeleton ossifies as a person grows up
- describe different kinds of bone fractures and state that X-rays can detect them
- **H** explain why a person with a fracture (especially a spinal fracture) should not be moved
- outline the effects of osteoporosis
- explain what is meant by the term *joint* and describe examples of different kinds of joints
- describe the structure of a synovial joint
- **H** explain the functions of the different parts of a synovial joint
- understand that some joints can be replaced
- **H** discuss advantages and disadvantages of joint replacements
- identify the main bones in the arm
- describe how the biceps and triceps act as antagonistic muscles to move the arm
- **H** explain how the arm acts as a lever

B5a In good shape

Questions

1   a   Describe *three* differences between bone and cartilage.

    b   Name *one* kind of vertebrate that has a skeleton made entirely of cartilage.

    c   Name the material that an insect's external skeleton is made of.

    d   Describe *three* advantages of an internal skeleton over an external skeleton.

2   For each of the following structures, describe *one* part of the body where you would find them. Be as precise as you can.

    a   fixed joint

    b   tendon

    c   ligament

    d   cartilage

    e   ball and socket joint

3   Gymnasts put great demands on their bones and joints.

Figure 5a.17

   a   Which bones are bearing the weight of the gymnast in the photograph?

   b   How might this affect the development of these bones?

Gymnasts train to allow some of their joints more movement than usual.

   c   Name *one* joint in the gymnast in the photograph that is letting the bones move more than usual.

   d   Suggest the part of the joint that changes to allow this greater movement.

   e   Suggest how the position of the gymnast in the photograph could damage her elbow joints.

B5b The vital pump

Circulatory systems

You already know that many different substances – including oxygen and nutrients – are transported around the human body in the **blood**. The blood flows inside **blood vessels** and is pumped around the body by the **heart**. The heart and blood vessels make up the **circulatory system**.

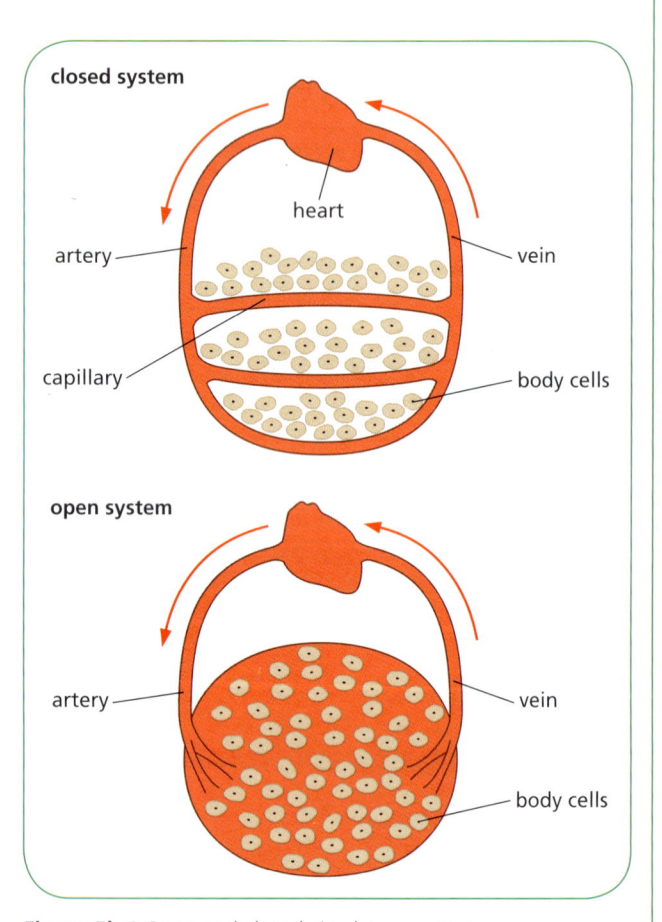

Open and closed circulatory systems

Some animals don't have a circulatory system at all. Very small organisms, such as the single-celled protist *Amoeba*, are so tiny that all the substances the cell requires can get quickly to every part of it just by diffusion.

But most multicellular (many-celled) animals do have some kind of circulatory system. Many have a system like ours, in which blood is carried around the body inside blood vessels. This is called a **closed circulatory system** (Figure 5b.1). All vertebrates and many invertebrates, such as earthworms, have a closed circulatory system.

Arthropods (invertebrates with jointed legs, including insects) do something different. Their blood simply fills all the spaces in their body. This is called an **open circulatory system**.

SAQ

1. Describe *three* similarities and *one* difference between an open circulatory system and a closed circulatory system.
2. Suggest *one* advantage of a closed circulatory system, and *one* advantage of an open circulatory system.

Double and single circulatory systems

You may already know that the kind of blood system that we have is called a **double circulatory system** (Figure 5b.2). There are two separate circuits around which the blood travels. The circuit that takes blood to the lungs and back is called the **pulmonary circulation**. The circuit that takes blood to the rest of the body and back is called the **systemic circulation**.

Fish, however, have a **single circulatory system**. Their blood is pumped from the heart to the gills, and it then carries on around the body before going back to the heart. There is only one circuit.

H SAQ

3. On the diagram of the single circulatory system of a fish in Figure 5b.2b, identify the ventricle and the atrium of the heart. Make a simple diagram of the heart and label it to show where these are.
4. Most animals that have a double circulatory system have a four-chambered heart. Explain why these two features are usually found together.
5. Some animals with a double circulatory system – for example, frogs – have only a three-chambered heart. Do you think that is any better than having a single circulatory system with a two-chambered heart, like that of a fish? Explain your answer.

Figure 5b.1 Open and closed circulatory systems.

B5b The vital pump   11

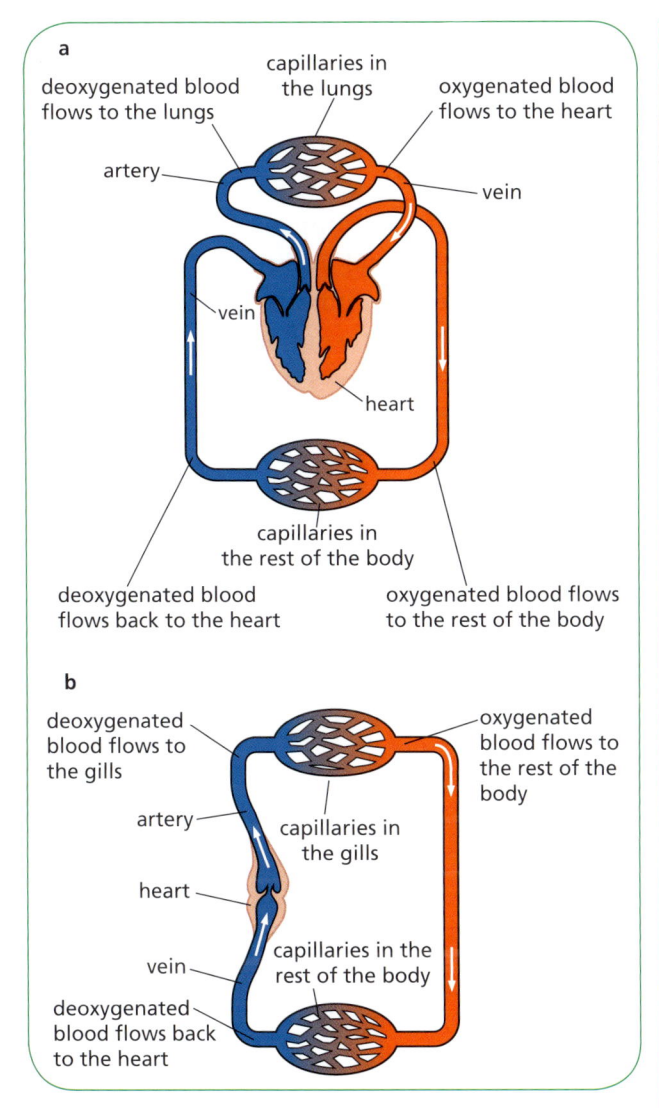

Figure 5b.2 Humans have a double circulatory system (**a**) while fish have a single circulatory system (**b**).

Sharing out the blood

The oxygenated blood that is pumped out of the heart is shared out between all the different organs in your body. The quantity of blood that flows to each organ is affected by the number and size of the blood vessels that supply it. The **arterioles** that supply the tissues have muscles in their walls that can contract and make the lumen (the space inside them) smaller. So, if some of the organs need extra blood for a while, the arterioles supplying less needy organs can close down a little, while the ones supplying the needy organs can remain fully open.

Discovering circulation

It is strange to think that people did not always realise that blood flowed around the body. It wasn't until around AD 150 that anyone did careful dissections and looked at the parts of the circulatory system. The person who first did this was a physician called Galen, who lived in Greece. He could see that the heart was made of two sides, completely separated from each other by a septum down the middle. He couldn't work out how blood could possibly get from one side to the other, so he suggested that maybe there were tiny holes in the septum that it could seep through.

Galen made many discoveries about the human body, and he wrote documents and made drawings suggesting how doctors should treat patients with different conditions. For hundreds of years afterwards, doctors all did what Galen said. No-one ever really challenged his ideas. Dissecting human bodies was thought to be completely wrong – the dead should be respected. No-one did experiments to find out how the human body really worked, or to find out which treatments worked best for different diseases.

The person who first discovered how the blood travelled around the body, and how it got from one side of the heart to the other, was William Harvey. He was a physician and scientist who lived in London between 1578 and 1657. He did many experiments on animals (including some on living dogs, which we would find completely horrific today). He had a very difficult time convincing doctors that he was right and that Galen was wrong. Questioning Galen's ideas was considered to be almost heresy. It took many years before doctors accepted the truth of Harvey's careful descriptions of the blood circulation, and adjusted their treatments accordingly.

The heart

You have probably already learned about the structure of the heart. It is shown in Figure 5b.3.

Most of the heart is made of a special kind of very powerful muscle, called **cardiac muscle**. The wall of the heart is so thick that, even though the heart is full of blood, the cells in its wall need their own blood vessels to supply them with the oxygen and food they need to provide energy for them to contract. The blood is carried to the heart muscle by the **coronary artery**, which branches off from the aorta.

12 B5b The vital pump

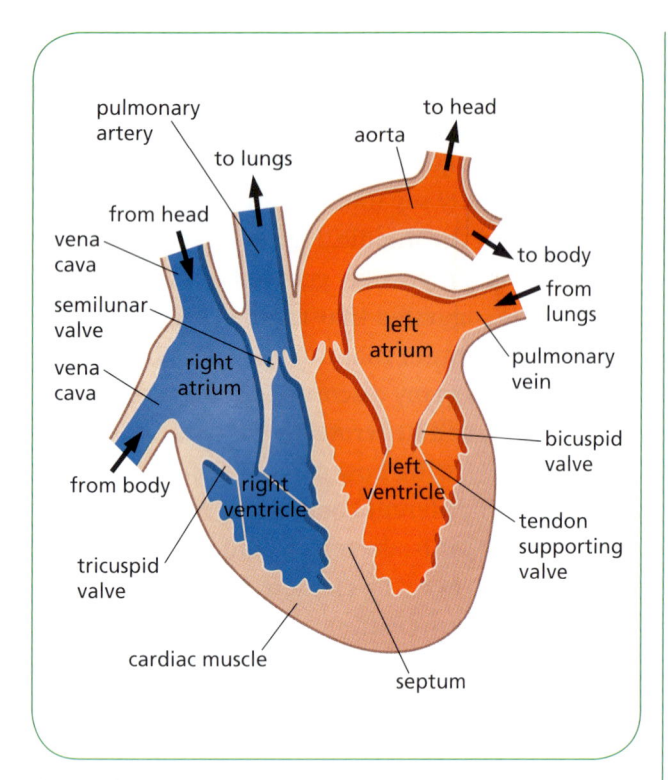

Figure 5b.3 The human heart.

Blood pressure

About 65 times a minute, the muscle in the walls of your heart contracts and squeezes inwards on the blood. This increases the pressure of the blood, and forces it up and out through the pulmonary artery and the aorta. You have already found out how we can use this to measure a person's **pulse rate**. Each surge of pressurised blood from the heart pushes outwards on the walls of the arteries, and we can feel this by placing fingers over an artery that is near the surface – for example, in the wrist, the neck or the temple (Figure 5b.4). The pulse rate tells us how fast the heart is beating.

The pressure of blood in the blood vessels is called **blood pressure**. It is highest in the blood vessels that lead directly out of the heart – the arteries. As the blood flows into the smaller arterioles and then the capillaries, getting further and further from the heart, its pressure drops. By the time it reaches the veins, the pressure is much lower, and the blood no longer pulses.

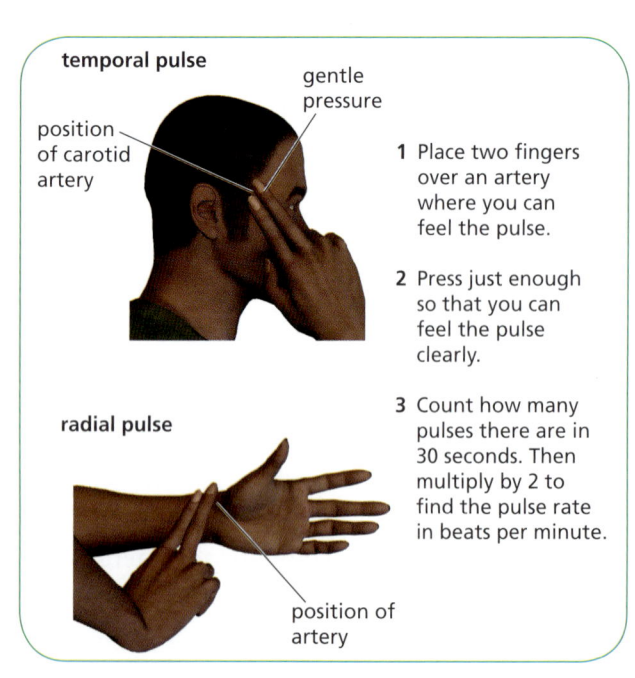

Figure 5b.4 How to measure a person's pulse rate.

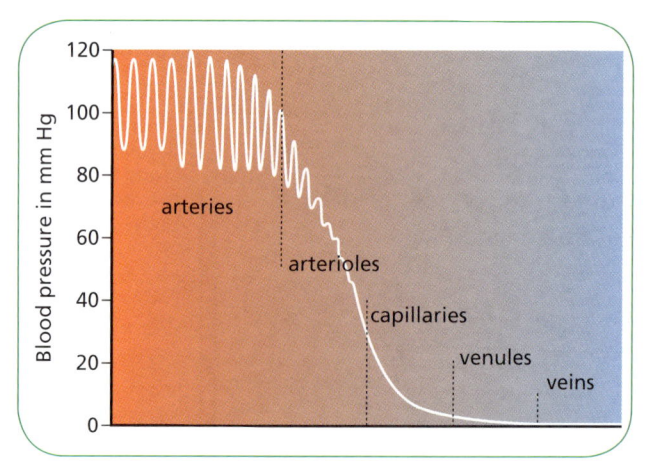

Figure 5b.5 Blood pressure in the circulatory system.

SAQ

6 Figure 5b.5 shows how blood pressure changes as the blood travels once around the body.

(Notice that the blood pressure is measured in 'mmHg'. This stands for 'millimetres of mercury', because doctors measure it using an instrument in which the pressure pushes a column of mercury up a tube.)

  a What is the maximum pressure of the blood as it passes around the circulatory system?

  b Explain why the pressure of blood in the arteries oscillates (goes up and down).

  c Even though it is at a very low pressure, blood in the veins flows back towards the heart. Explain why it does this. (If you have forgotten, look back at *Gateway Additional Science*, pages 19–20.)

Diastole All muscles are relaxed. Blood flows into heart.
- semilunar valves shut, preventing blood from flowing into ventricles
- atrio-ventricular valves open
- muscles relax, allowing blood to flow into heart from veins

Atrial systole Muscles of atria contract. Muscles of ventricles remain relaxed. Blood forced from atria into ventricles.
- semilunar valves remain shut
- valves in veins are forced shut by pressure of blood, stopping blood from flowing back into the veins
- muscles of atria contract, squeezing blood into the ventricles

Ventricular systole Muscles of atria relax. Muscles of ventricles contract. Blood forced out of ventricles into arteries.
- semilunar valves are forced open by pressure of blood
- atrio-ventricular valves are forced shut by pressure of blood
- muscles of ventricles contract, forcing blood out of ventricles

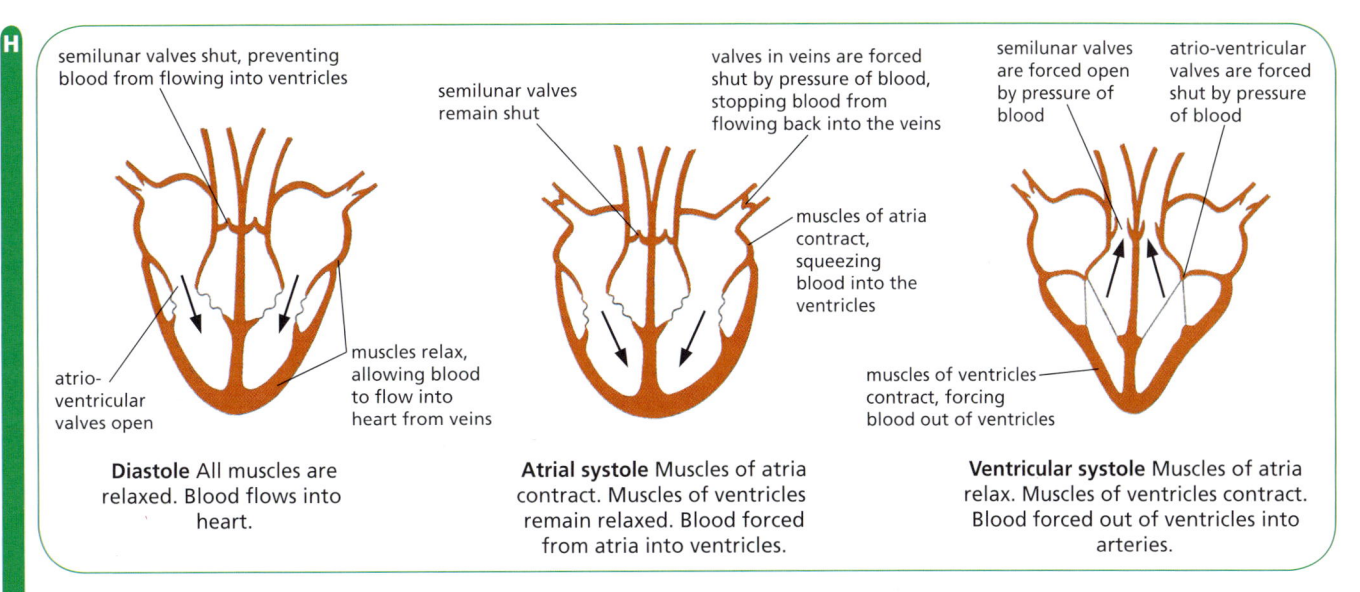

Figure 5b.6 The cardiac cycle.

The cardiac cycle

The heart beats by contracting and relaxing the muscles in its walls. When they contract, the volume inside the heart becomes smaller, raising the pressure of the blood and squeezing it out through the arteries. When the muscles relax, the volume inside the heart becomes larger, reducing the pressure and allowing blood to flow in.

Figure 5b.6 shows how this happens. The sequence of events is called the **cardiac cycle**.

Changing heart rate

What makes your heart beat faster? It could be just thinking about something that makes you feel excited or very emotional. Doing exercise always causes the heart rate to increase. Feeling frightened has the same effect.

Exercise increases heart rate because the muscles need faster deliveries of oxygen, to allow them to respire faster and release energy so that they can contract. The faster movement of the blood also helps to carry carbon dioxide away more quickly. The increase in the heart rate is caused by nerve impulses travelling from the brain to the heart, making it beat faster and more strongly.

Feeling nervous or excited causes the adrenal glands to release a hormone called **adrenaline**. Adrenaline is sometimes called the 'fight or flight' hormone, because it helps to prepare the body for action. This is fine if you really do need extra energy for your muscles – for example, if you are about to run a race, or are being attacked by a lion – but in most situations where you feel nervous it doesn't help at all.

The pacemaker

Nerves and hormones both affect heart rate by stimulating the **pacemaker** (Figure 5b.7, page 14). This is a little patch of muscle in the wall of the right atrium. The cells in it contract and relax rhythmically, and they set the pace of contraction for the whole heart. Each time they contract, they send out a surge of electrical impulse (a very tiny one) which spreads over the heart surface. As the impulse reaches the other cells in the heart, they contract.

The pacemaker is also known as the **SAN**, which stands for **sino-atrial node**. As the electrical impulse spreads out from the SAN across the walls of the right and left atria, they cause the muscles in the walls of the atria to contract. However, the impulse can't easily get down into the walls of the ventricles. The only conducting pathway for it is down through the septum to the bottom of the ventricles, and then up through their muscular walls.

There is a short delay before this happens, as the impulse is held up for a fraction of a second at a patch of muscle in the septum called the **AVN**, or **atrio-ventricular node**. This means that the atria contract before the ventricles.

14 **B5b** The vital pump

1 Muscle cells in the SAN contract rhythmically, sending out electrical impulses.

2 The impulse spreads rapidly through the walls of the atria, making the muscles contract.

3 The impulse is held up for a split second at the AVN.

4 The impulse travels down through the septum and then sweeps up the walls of the ventricles, making them contract.

Figure 5b.7 The pacemaker sets the rhythm of the heart beat.

SAQ

7 **a** Suggest what would happen if the electrical impulse from the pacemaker reached *all* the parts of the heart muscle at the same time.

b Can you think why it is useful for the impulse to go down through the septum and then sweep upwards through the ventricle walls?

It is not uncommon for a person's pacemaker to go wrong, especially as they get older and if they have heart disease. Many people with faulty pacemakers go back to living an almost normal life when they are given an artificial pacemaker. You can read about this on pages 22–23 in *Gateway Additional Science*.

ECGs

ECG stands for **electrocardiogram**. It is a recording of the electrical activity taking place in the heart. It can give doctors a lot of information about the way the heart is working. For example, it can show whether the electrical impulse produced by the pacemaker is spreading over the heart properly, and whether the heart muscles are responding to it.

A heart-stopping experience

Each year, on average, 49 people in England and Wales are struck by lightning. That's a risk of about one person in 1.2 million being struck. Of these, three people (on average) die each year – which means that around one person in 19 million is likely to be killed by lightning. For reasons that no-one has quite worked out, four out of every five deaths caused by lightning are of men rather than women.

The deaths are usually caused by cardiac arrest – a heart attack. The electrical current that surges through the body during a lightning strike can stop the pacemaker contracting, and completely confuse the electrical activity in the heart. The heart muscle cells all start doing their own thing, all contracting individually and not in time with each other.

Figure 5b.8 Troy Trice was hit by lightning while he was playing football. The lightning burnt through his helmet and blew his shoes off, but he has made a full recovery.

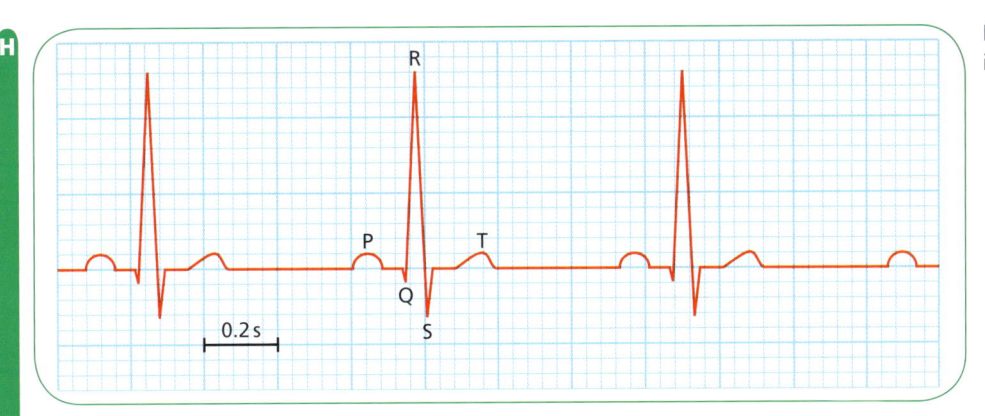

Figure 5b.9 An ECG for a heart that is working normally.

Figure 5b.9 shows an ECG for a person whose heart is working normally. It is a kind of graph of electrical activity. The *x*-axis is time (you can see that 1 cm represents 0.2 seconds). The *y*-axis is the magnitude of the electrical impulse. The different parts of the ECG are given different letters:
- P is the wave of electrical activity just before the atria contract.
- Q, R and S represent the wave of electrical activity just before the ventricles contract.
- T is the relaxation of the ventricles.

Figure 5b.10 shows how doctors record a person's ECG.

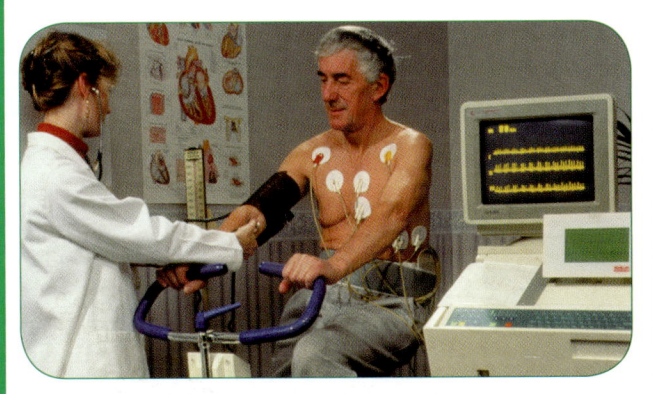

Figure 5b.10 This patient is having an ECG following diagnosis of problems with his heart. The electrodes stuck onto his chest record the electrical activity in his heart, which is recorded on the screen behind him.

SAQ

8 a How many heart beats are represented by the ECG in Figure 5b.9?

   b Calculate the person's heart rate.

Echocardiograms

An **echocardiogram** is another way in which doctors can collect information about how well a person's heart is working. It is usually done just after or just before an ECG is made.

The echocardiogram is produced using ultrasound, in a similar way to an ultrasound scan that a pregnant woman is given to check that her baby is developing properly. The way in which the sound waves are reflected from the heart allows images to be constructed showing the size and shape of the heart chambers. A computer can analyse the reflected sound waves to produce 2D and 3D images of the heart at various stages of contraction. This can show whether the muscles in the heart walls are contracting and relaxing normally, whether the volume of the ventricles and atria is within a normal range, and how strongly the heart is pumping. The doctor can watch the image changing as the heart beats. (If you type 'echocardiogram' into an internet search engine, you should be able to find some video clips of this.)

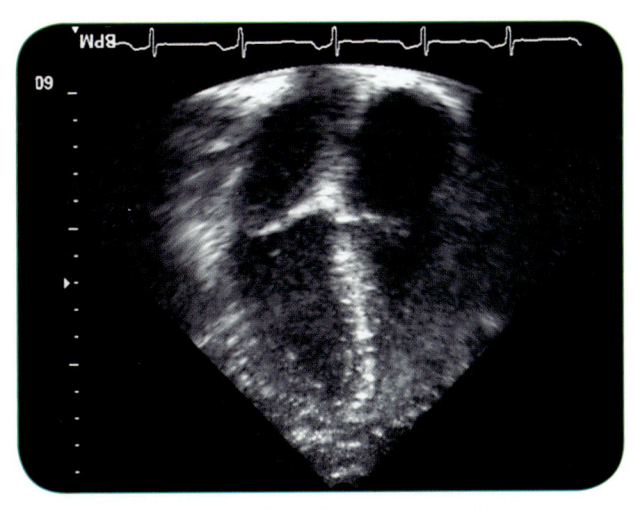

Figure 5b.11 This is an echocardiogram image of a person's heart beating. You can see the atria at the top and the ventricles at the bottom. There is also an ECG trace to show doctors the electrical activity of the heart.

16 B5b The vital pump

Summary

You should be able to:

- give examples of animals with no circulatory system, and with closed and open systems
- describe what is meant by a closed circulatory system
- explain the difference between a single and a double circulatory system
- [H] explain the link between types of circulatory system and having a two- or four-chambered heart
- describe how Galen and Harvey helped our knowledge of the circulatory system
- describe the structure of the heart
- describe what causes the pulse, and how to measure pulse rate
- interpret data on the pressure changes in arteries, capillaries and veins
- [H] describe the cardiac cycle and interpret graphs relating to it
- describe how physical activity affects heart rate
- [H] describe how adrenaline affects heart rate
- describe the role of the pacemaker
- [H] explain how the pacemaker coordinates heart activity
- outline the use of ECGs and echocardiograms
- [H] interpret data from ECGs

Questions

1. **a** For each of these organisms, state:
    - whether it has a circulatory system
    - if so, whether it is open or closed
    - whether it is single or double.
      i human
      ii grasshopper
      iii fish

   b Explain the difference between a single circulatory system and a double circulatory system.

   continued on next page

Questions - *continued*

2 Table 5b.1 shows the volume of blood supplied to different organs in the body when at rest and when doing light exercise and vigorous exercise.

	Volume of blood in dm³ per minute			
	Skin	Muscle	Brain	Kidneys and digestive system
At rest	1	2	1	3
Light exercise	3	7	1	2
Vigorous exercise	2	25	1	1

Table 5b.1

a Explain why the volume of blood supplied to the muscles needs to be greater when a person is carrying out vigorous exercise. (Try to think of *four* reasons.)

b Calculate the total volumes of the blood supplied each minute to these four organs at rest, during light exercise and during vigorous exercise.

c Explain what happens in the body to increase this total volume per minute.

d Suggest why the volume of blood supplied to the brain during exercise does not change, whereas the volume of blood supplied to the skin increases.

H 3 Figure 5b.12 shows the pressure changes in the left ventricle and left atrium during one heart beat. You'll also find it helpful to look at Figure 5b.5 and perhaps a diagram showing the structure of the heart while you are thinking about the answers to these questions.

a What is the maximum pressure that is produced inside the left ventricle?

b Explain how this pressure is produced.

c What is the maximum pressure that is produced inside the left atrium?

d Explain why the pressure produced in the atrium does not need to be as great as the pressure produced inside the ventricle.

e Think back to what you know about the structure of the heart. If another line was drawn on the graph showing the pressure in the *right* ventricle, how would it differ from the line for the left ventricle?

f The graph shows one complete heart beat. Calculate the heart rate in beats per minute, showing your working.

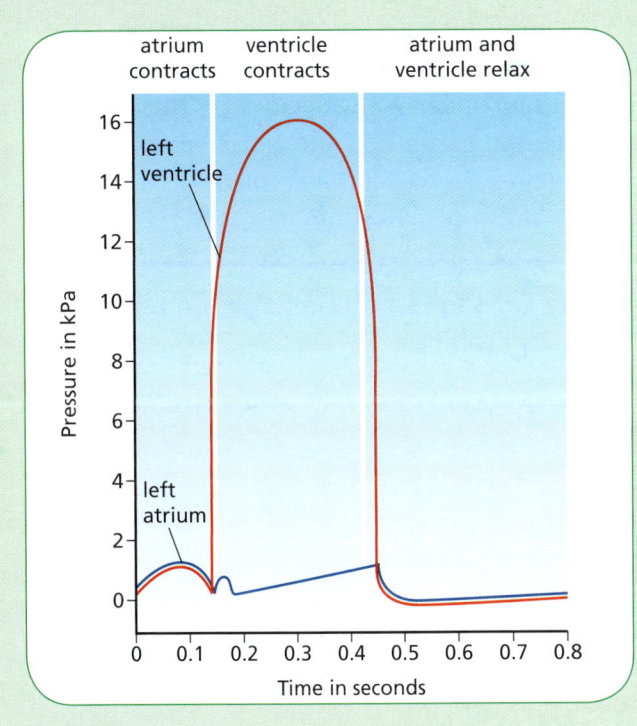

Figure 5b.12

B5c Running repairs

Heart problems

Many people develop some kind of problem with their heart as they get older. Some children are born with an abnormal heart. Today, many heart problems can be successfully treated, usually with surgery.

Some of the commonest heart problems are:
- an irregular heart beat
- a hole in the heart
- damaged or weak valves
- coronary heart disease.

Irregular heart beat

There are many different reasons why a person might have an irregular heart beat. They are usually to do with a faulty pacemaker, or a problem with the conducting tissues in the walls of the heart, so that the electrical impulses from the pacemaker cannot spread through it correctly. Many people have slightly irregular heart beats which cause them no problems. But sometimes this can be an indication of a serious problem that needs treatment.

A doctor using a stethoscope will probably be the first person to realise that a patient's heart is not beating regularly. The doctor may then book the patient in for an ECG and an echocardiogram, to find out exactly what the problem is.

SAQ

1 Figure 5c.1 shows an ECG of a person with an irregular heart beat.

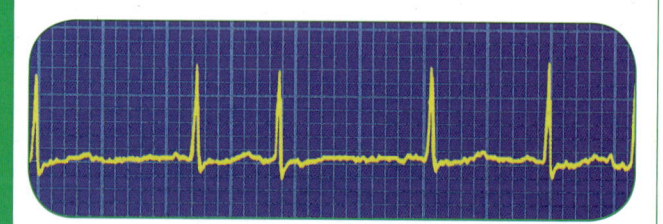

Figure 5c.1 An ECG of an irregular heart beat. Each large square across the graph represents 0.2 s, as in Figure 5b.9.

  a Describe *two* ways in which this ECG differs from the normal ECG shown in Figure 5b.9.

  b Does the problem appear to be in the atria or the ventricles? Explain your answer.

  c Suggest how an artificial pacemaker might help this person.

Hole in the heart

While a fetus is developing in the uterus, it gets its oxygen from its mother's blood, via the placenta and umbilical cord. The oxygenated blood flows from the placenta back to the fetus's heart in the vena cava. So the blood enters the *right* side of the heart, not the left side as it will once the baby's lungs begin to work.

In a developing fetus, there is a hole in the septum that allows this oxygenated blood to pass into the left atrium. This ensures that some of the oxygenated blood passes into the aorta and travels all around the body.

Usually, when a baby takes its first breath, this hole closes up. But in some babies it does not. The hole remains open, allowing blood to move between the two atria. Oxygenated blood gets mixed up with deoxygenated blood (Figure 5c.2). In severe cases, this means that the body tissues get so little oxygen delivered to them that there is a threat to the baby's life.

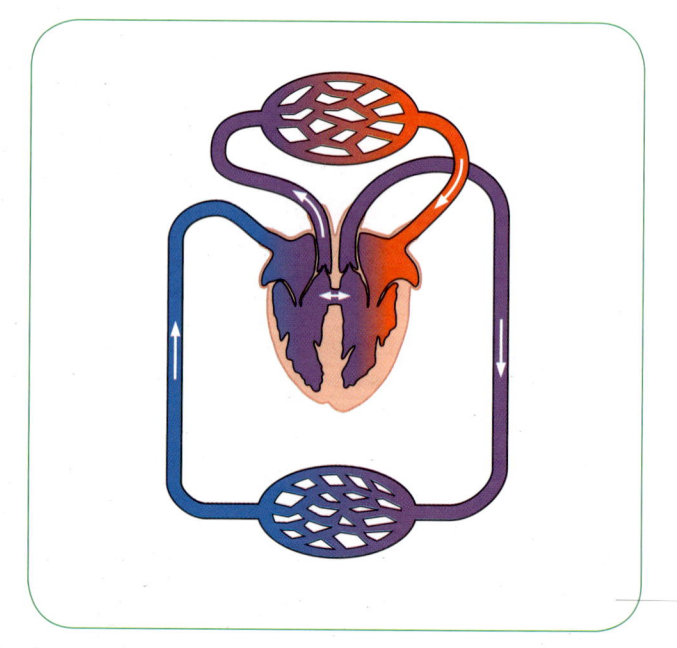

Figure 5c.2 In babies with a hole in the heart, blood can move between the two sides of the heart.

SAQ

2 Describe the differences between the blood circulation shown in Figure 5c.2, and that shown in Figure 5b.2a on page 10.

Surgery to correct a hole in the heart used to be done by open heart surgery, which was a big operation. Now it is usually done by threading a tube containing a little umbrella-shaped device into a vein in the leg. The device is gently pushed along the vein, into the vena cava and then into the heart. The surgeon then manoeuvres it into the hole and 'opens' the umbrella, blocking the hole. The surgeon will generally use ultrasound images to show exactly where the device is throughout the procedure.

Faulty valves

Sometimes a baby is born with faulty heart valves. Instead of the blood flowing in one direction through the heart, it can move back and forth as the heart beats. Heart valves can also become damaged later in life, sometimes as a result of an infection.

Faulty valves can be replaced with artificial valves, made from metal or plastic. You may have read about this in *Gateway Additional Science*, page 22. These work well, but they have a tendency to cause blood to clot when it is in contact with them. So the patient may need to take anti-clotting drugs (page 24) for the rest of their life.

Coronary heart disease

This disease is caused by parts of the coronary artery becoming partly blocked and stiffened. You may have read about it in *Gateway Additional Science*, pages 21–23.

The blockage prevents enough oxygen getting to the heart muscle. The muscle may stop contracting, or it may die. The first sign a person may have that the coronary artery is becoming blocked is a pain in the chest, especially when the person has been doing something energetic. This is called **angina**.

The standard treatment for a blocked or badly damaged coronary artery is **bypass surgery**. Figure 5c.3 shows how this is done.

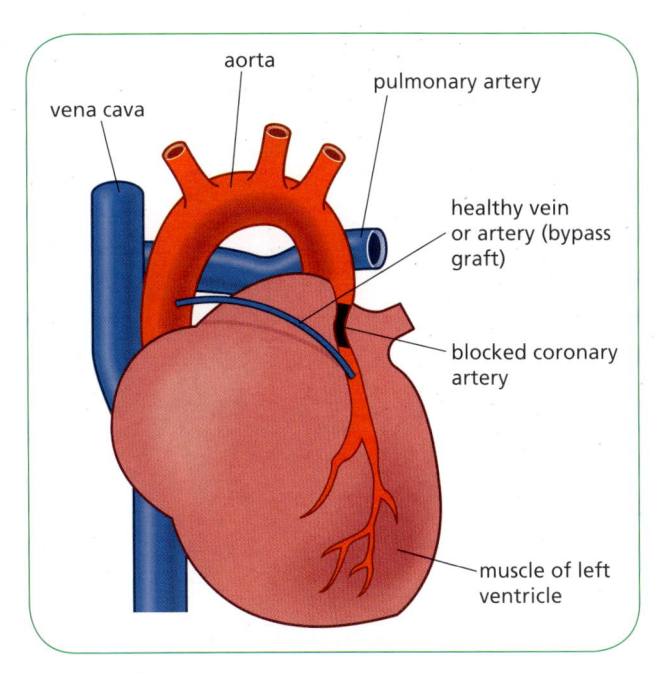

Figure 5c.3 In bypass surgery, a piece of healthy blood vessel (from the patient's leg, for example) is connected to let blood flow around the blocked part of the coronary artery. Blood reaches the heart muscle, and the heart can beat normally.

SAQ

3 Explain how bypass surgery will relieve the symptoms of coronary heart disease.

Transplants and heart assist devices

If a heart is very badly damaged, it may be impossible to repair it. If that is the case, then a surgeon may carry out a **heart transplant**. You may have read about this on page 23 in *Gateway Additional Science*.

SAQ

4 Using what you already know about heart transplants, make a list of the advantages of this operation, and a list of its disadvantages. (If you have forgotten, then look up this information on page 23 in *Gateway Additional Science*.)

It often takes a very long time to find a heart that is a suitable match for someone who needs a transplant. If that is the case, it may be possible to use a **heart assist device** to keep the patient's own heart going while they wait for a suitable donor heart to be found. Figure 5c.4 shows one kind of heart assist device. It consists of a pump that is connected between the left ventricle and

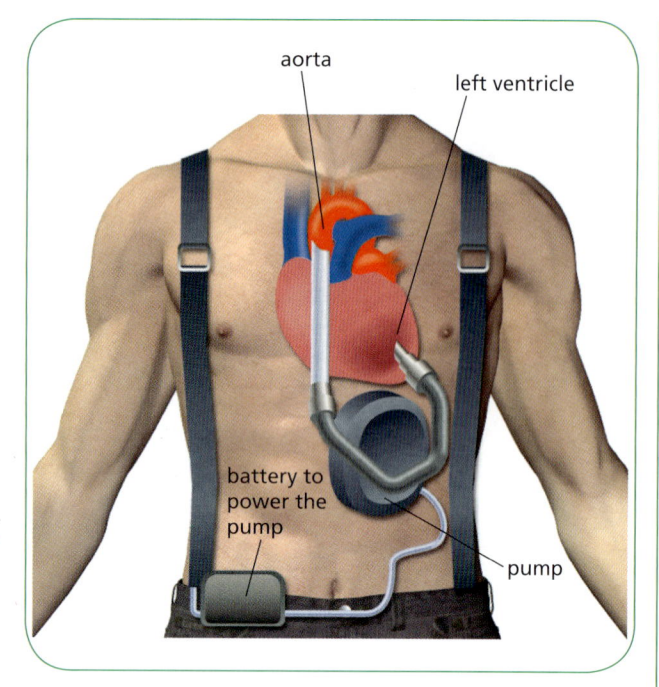

Figure 5c.4 A heart assist device helps to keep a patient's heart going until a donor heart can be found for a transplant.

the aorta. The pump takes over the work of the left ventricle, allowing it to rest and perhaps to heal itself. The pump is powered by a battery that the person carries strapped to their body.

Lifestyle and the circulatory system

As we age, wear and tear on the heart and blood vessels cause gradual changes in them, so that the circulatory system gradually becomes less effective at moving blood around the body. The rate at which this deterioration happens is greatly affected by a person's lifestyle. There is a lot that we can do to slow down the ageing process in the heart, and to reduce the risk of developing coronary heart disease.

Diet is always in the news. It has been really difficult to collect reliable and unambiguous evidence about how what we eat affects the heart. This is partly because we are all genetically different from each other, and our genes have a big effect on the risk of developing heart disease. So any study looking at the effects of different diets always has some uncontrolled variables in it. All the same, it makes sense to avoid too much of foods that are thought to be harmful. In particular, it seems that a diet high in **animal fats** and **cholesterol** can increase the risk of **plaques** (*Gateway Additional Science*, page 22) building up in the coronary arteries. There are also links between a high **salt** content in the diet and high blood pressure, and this is a known risk factor for developing coronary heart disease.

Drinking too much **alcohol** is also a risk factor, because it appears to increase the level of harmful fats in the circulatory system. However, there is some evidence that small amounts of alcohol can actually have a protective effect. But the jury is still out on this.

Smoking also increases the risk of heart disease. A smoker is much more likely to develop blood clots in the arteries than a non-smoker. For a smoker, stopping smoking is the lifestyle change that will have the biggest effect on his or her risk of developing coronary heart disease.

Stress is another thing to try to reduce, and there is a link between high levels of uncontrollable stress (for example, being unhappy in your job) and the development of coronary heart disease. Once again, though, it has been difficult to find clear evidence for this, and there are many people who thrive on stress and don't appear to suffer from it. Again, genes are probably involved, and also the kind of stress – stress that you are in control of is very different from stress that is imposed on you and that you can't escape from. We cannot avoid all stress, and life would be very dull without at least some controllable stress.

Some **drugs** can cause damage to the heart. **Cocaine**, for example, has been known to cause a person to have a heart attack and die on the very first occasion that they used it. **Heroin** and other opiates (such as morphine) also cause damage to the heart, especially if the drug is injected into a blood vessel. The injections often introduce pathogens that can infect the blood system and the heart, causing irreparable damage. One person writes:

'In case anyone thinks this is a joke, it's how this writer's best friend died. He (my mate Keith) got septicaemia from an unsterile needle. This damaged the valves of his heart and he nearly died shortly after. He had to have operations to put in plastic heart valves and a pacemaker. He lived for a few years after that but then his heart gave out.

That's reality.'

Blood

The components of blood

You probably already know that blood has four main components – **red cells**, **white cells**, **plasma** and **platelets**. They are described on pages 17–18 in *Gateway Additional Science*, where you can also see a photograph of blood taken through a microscope. Table 5c.1 summarises their functions.

Component	Function
red cells	contain the pigment haemoglobin, which transports oxygen from the lungs to the tissues
white cells	defence against pathogens (disease-causing organisms); phagocytes engulf and digest them, while lymphocytes produce specific antibodies to help to destroy them
plasma	the liquid in which the other blood components float; it contains many substances in solution that are being transported around the body – for example, glucose, amino acids and hormones
platelets	help with blood clotting

Table 5c.1

SAQ

H

5 This question tests how much you can remember about the relationship between structure and function of blood cells, described on pages 17–18 in *Gateway Additional Science*.

   a Make a diagram of a red blood cell. Add annotations to explain how its structure is adapted to its function.

   b Describe *one* way in which the structure of a phagocytic white blood cell is adapted to its function.

Giving blood

When you arrive at the blood donation centre, you will be asked some questions about yourself, even if you are a regular donor. You will be asked if you are HIV positive, if you might ever have had hepatitis B or C, or if you have ever injected non-prescribed drugs. If you answer 'yes' to any of these, you won't be allowed to give blood.

SAQ

6 Explain why a person saying 'yes' to any of these questions should not be allowed to give blood.

Next, a nurse will use a sterile lancet to take a tiny drop of blood from your thumb. This will be tested for haemoglobin levels. If your haemoglobin level is too low, then you might become anaemic if you give blood.

To collect the blood, a needle will be inserted into a vein in your arm, and blood allowed to flow out through a tube into a sterile bag (Figure 5c.5). It takes about 10 minutes to collect approximately 500 cm^3, which is the quantity that most people give.

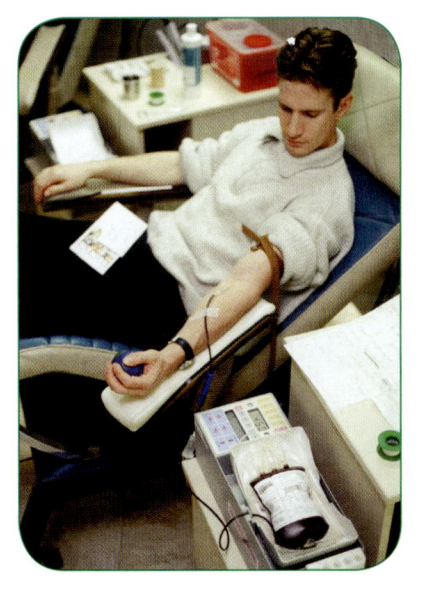

Figure 5c.5 Giving blood at a donor clinic. The blood is taken from a vein in the forearm, and collected in the bag at the front of the picture.

Receiving blood

No-one is given a blood transfusion unless it is really necessary. Not only would this waste precious supplies of blood, but there is always a small risk associated with receiving blood that is not your own.

The main reasons why a person might be given blood are:

- because they are having surgery and losing a lot of blood
- because they have lost a lot of blood in an accident
- because they have anaemia or another problem with their own blood.

The blood will normally be given by a drip, attached to a tube leading into a vein in their arm.

Blood groups

Do you know your blood group? There are four different blood groups – A, B, O and AB. Table 5c.2 shows how common each of these blood groups is amongst people living in the United Kingdom.

Blood group	Percentage of population in the UK
A	42
B	9
AB	3
O	46

Table 5c.2

You will also be either **Rhesus positive** (Rh⁺) or **Rhesus negative** (Rh⁻). Around 85% of people in the UK are Rhesus positive. The Rhesus system of classifying blood is a different method from the ABO system.

Most of the time, you don't need to know your blood group. It only becomes important if you need a blood transfusion, or if you want to give blood.

Your blood group actually refers to some molecules called **agglutinins** that you have on your red blood cells. If you have A type agglutinins, then your blood group is A. If you have B agglutinins, you are blood group B. If you have both kinds, you are blood group AB. If you have neither, you are blood group O.

Your blood plasma may contain **antibodies** against some of the agglutinins. There are two kinds – anti-A and anti-B antibodies. Which ones you have depends on the agglutinins on your red blood cells, as shown in Table 5c.3.

Blood group	Agglutinins on red cells	Antibodies in plasma
A	A	anti-B
B	B	anti-A
AB	A and B	none
O	none	anti-A and anti-B

Table 5c.3

Problems come when someone is given blood whose cells carry agglutinins that will be attacked by the antibodies in their plasma. Imagine, for example, that a group A person is given group B blood. Her own plasma contains anti-B antibodies. These will lock on to the B agglutinins on the donated red blood cells, making the cells clump together. This is called **agglutination**. Blood vessels become blocked by the clumps of cells, reducing the supply of oxygen to tissues. Organs all over the body may be damaged and stop working – a condition called 'multiple organ failure'. It can be fatal.

The importance of blood group

On 1 October 2004, an elderly woman went into a private hospital in Leeds for an operation that would help her to walk more freely and in less pain. She had gone in for pre-operative tests on 27 September, when a blood sample had been taken.

Her blood was tested to find out her blood group, in case she needed a transfusion during her operation. The biomedical scientist who did the tests was fully qualified, and had been doing tests like this for 25 years. But this time he got it wrong. He mistakenly identified the woman's blood group as AB Rh⁻. Her actual blood group was O Rh⁻.

According to correct procedure, someone else should have checked this result. But this did not happen.

During her operation, the woman was given group A Rh⁻ blood. This would have been fine if she had really been AB Rh⁻. But group A blood is incompatible with group O blood. Her condition deteriorated rapidly. Despite being given transfusions of the correct type of blood over the next few days, she died of multiple organ failure on 5 October.

The biomedic went on trial for manslaughter. But the case collapsed, as the court found that it was the procedures in place at the laboratory that were at fault, rather than his negligence.

SAQ

7 Read the box on page 22. Explain why the woman in Leeds died after she was given the wrong type of blood.

8 Copy and complete Table 5c.4, to show which type(s) of blood can safely be given to a person with each blood group.

Group of donor	Group or groups of recipients who can safely receive the blood
A	
B	
AB	
O	

Table 5c.4

9 Sometimes, a person needing a blood transfusion is given **serum** (blood plasma with no cells or platelets in it) rather than whole blood. Suggest an advantage of this.

Blood clotting

When you cut yourself, your blood should form a **clot** to seal the wound. This not only stops blood escaping, but it prevents pathogens from getting into your body and causing infection.

Figure 5c.6 shows how a blood clot forms.

The blood needs to contain many different chemicals in order for it to clot properly. Perhaps the most important is a protein called **fibrin**. Clotting also needs plenty of calcium ions, and a number of different substances called **clotting factors**. Vitamin K, which is found in green leafy vegetables such as broccoli and spinach, is especially important. Alcohol, however, slows down blood clotting.

Some people have an inherited disease called **haemophilia**. They are unable to make another clotting chemical called **factor 8**, so their blood does not clot when it should. You might imagine that this would cause them to bleed to death, but in fact one of the biggest problems is when bumps and knocks cause blood vessels to bleed into the person's joints. This is very painful, and it can do serious harm.

Although it is obviously important that blood clots quickly at a wound, we *don't* want it clotting

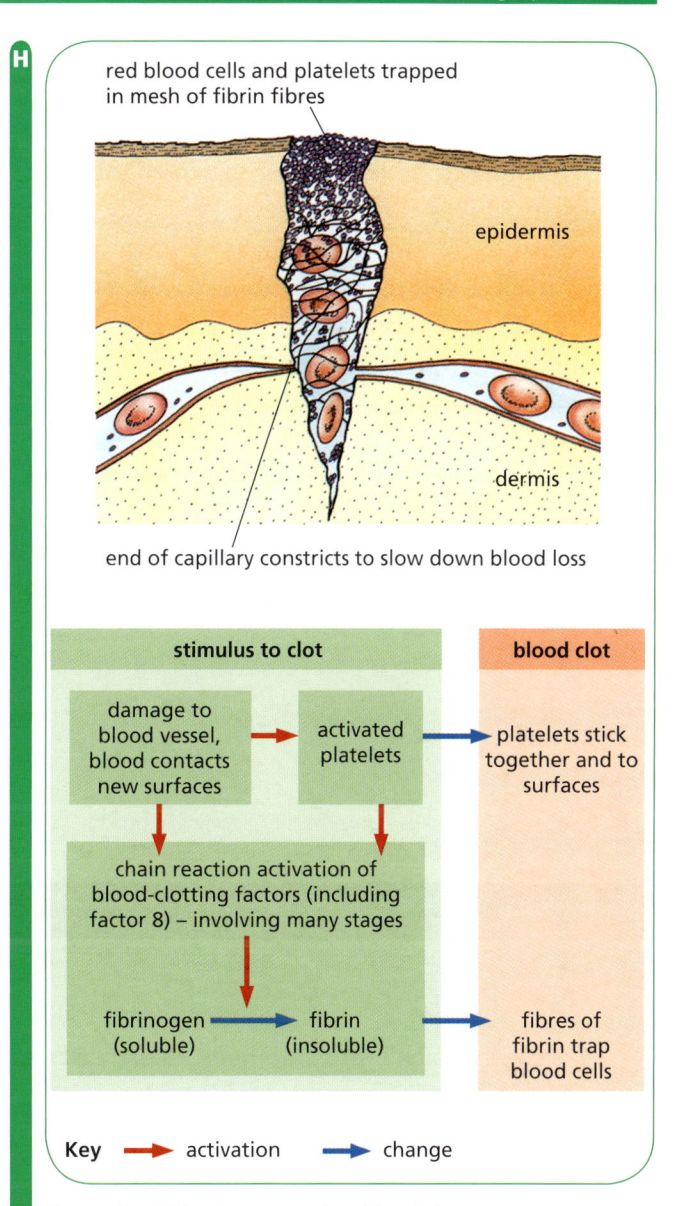

Figure 5c.6 What happens when blood clots.

inside blood vessels. This is called a **thrombosis**. It can be dangerous, particularly if part of the clot breaks off and gets swept away in the blood to one of the small arterioles supplying the lungs or the heart. It could form a blockage there, and this could be fatal. A thrombosis is especially likely to occur if a person has been sitting still for a long time – for example, during a long journey in a car or aeroplane – so that their blood has not been flowing freely back from the legs to the heart. If a clot forms inside one of the veins deep inside the legs, it is called **deep vein thrombosis** or **DVT**.

Passengers on long-haul flights are sometimes advised to take half an **aspirin** before they fly. Many doctors say that this should only be done after first seeking medical advice. Aspirin works as an **anti-coagulant** – a substance that slows down blood clotting (Figure 5c.7). Cranberry juice also has anti-coagulant properties. A person who has suffered a thrombosis, either in a deep vein or in their coronary arteries, may be prescribed strong anti-coagulant drugs such as **heparin** or **warfarin**. These drugs are sometimes given to patients who have recently had surgery. This is so that they do not form unnecessary and possibly dangerous clots in the body, where the tissues have been damaged.

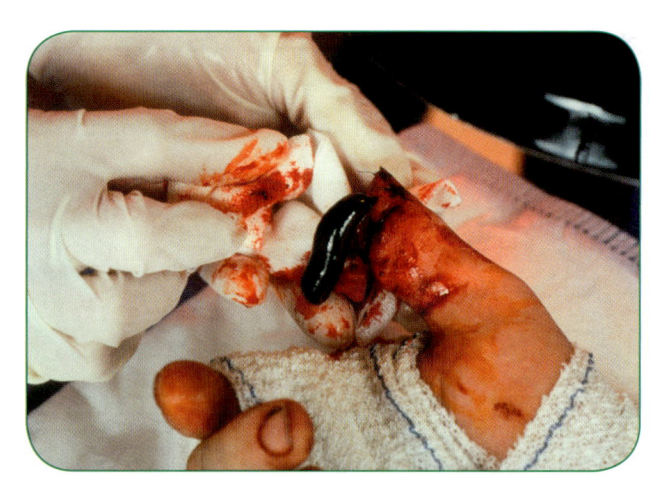

Figure 5c.7 An accident has severed this man's thumb. It is quite easy to connect the arteries when a part of the body is reattached (because they have thick walls) but much more difficult to connect the veins. This means that blood may not be drained away. Leeches, which feed on blood, can help out. They inject an anti-clotting substance called hirudin, which allows the blood to flow freely and helps the wound to heal.

Summary

You should be able to:

- describe some heart conditions, their causes and effects, and their treatment
- describe how lifestyle can affect the circulatory system
- identify the components of blood and outline their functions
- list the different blood groups, and describe the processes of blood donation and transfusion
- explain the interactions between blood of different groups, and explain which blood groups can be safely donated to which recipients
- outline the process of blood clotting
- list some substances that are required for blood to clot, and some which reduce blood clotting
- describe the uses of anti-coagulants

Questions

1. a What is meant by a *hole in the heart*?

   b Explain why a hole in the heart can mean that body cells do not get enough oxygen delivered to them.

   c Describe how a hole in the heart can be treated.

2. Rita is 63. She has been getting pain in her chest when she goes shopping. Her doctor says she has angina, and wants her to have an ECG to find out exactly what is causing it.

   a What is an ECG?

   b The ECG suggests that there is a blockage in one of Rita's coronary arteries. She is booked in for bypass surgery. Explain what this is, and how it might help Rita.

3. a Which of the following substances help blood to clot, and which ones slow down blood clotting?

   vitamin K   aspirin   heparin   alcohol   warfarin

   b Explain why someone might be prescribed an anti-coagulant drug.

   c People who travel on long-haul flights may be advised to take half an aspirin just before they travel. Suggest why this is a good idea.

4. Table 5c.5 shows part of a chart that doctors in some countries use to estimate the chances of a woman having a heart attack.

	\multicolumn{8}{c}{Percentage of women who are expected to have a heart attack within 5 years}
	\multicolumn{2}{c}{Age 40–49}	\multicolumn{2}{c}{Age 50–59}	\multicolumn{2}{c}{Age 60–69}	\multicolumn{2}{c}{Age 70–79}
	no diabetes	with diabetes	no diabetes	with diabetes	no diabetes	with diabetes	no diabetes	with diabetes
non-smokers	1	3	3	7	5	12	7	23
smokers	4	7	6	13	12	22	15	33

Table 5c.5

   a Use Table 5c.5 to identify *three* factors that increase the risk of a woman having a heart attack.

   b Which *one* of these factors can a person completely eliminate from their lifestyle?

   c Imagine you are a GP. You have a female patient who has diabetes and smokes. What will you say to her to convince her to stop smoking? (Use figures from the table to support your argument.)

B5d Breath of life

Gaseous exchange

All living cells respire. Respiration is a metabolic reaction in which glucose is oxidised, releasing energy that can be used by the cell.

Most of the time, our cells respire aerobically. In aerobic respiration, glucose reacts with oxygen, producing carbon dioxide and water. We therefore need to take oxygen into the body, and remove carbon dioxide from it. This is called **gaseous exchange**.

Gaseous exchange surfaces

In humans, the gaseous exchange surface is the **alveoli** in the **lungs**. But many living organisms do not have lungs.

If you look back at page 27 in *Gateway Additional Science*, you can remind yourself why a single-celled organism such as an amoeba does not need lungs. It is so small that it can manage just by oxygen and carbon dioxide diffusing across its cell membrane.

Earthworms are much bigger than an amoeba, but they can still manage by using their body surface to exchange respiratory gases. For this to work, their skin has to be quite thin, without any protective covering on it (Figure 5d.1). So they have to keep it moist, to stop the cells in it drying out. They do this by secreting watery mucus onto it. This is the main reason why earthworms cannot live in dry places, but spend most of their lives in the soil where it is damp.

Larger organisms need to have specially developed gaseous exchange surfaces. These provide a larger surface area than their skin would, so more oxygen and carbon dioxide can move in and out at the same time.

We have lungs to do this. They are tucked away inside the chest cavity. Here, they can keep moist. If they were hanging outside our bodies, they would quickly dry out. Having them inside the body means that we can live in dry places. Inside the body they are also protected from physical damage, and there is less chance of infection.

SAQ

1 Explain how each of these features of our lungs helps gaseous exchange to take place. (You may like to look back at page 11 in *Gateway Additional Science*, if you have forgotten this.)
   a large surface area of alveoli
   b moist surface
   c thin lining
   d good blood supply

Amphibians, such as frogs, also have lungs as adults. This allows them to live on land. Their lungs are not full of tiny alveoli like ours. This means their surface area to volume ratio is much smaller, so frogs cannot rely only on their lungs for gaseous exchange. They also have to have thin skin, like an earthworm (Figure 5d.2). And, just like earthworms, this restricts the places that they can live. Many terrestrial (land-living) amphibians are only found in places where the air stays fairly moist.

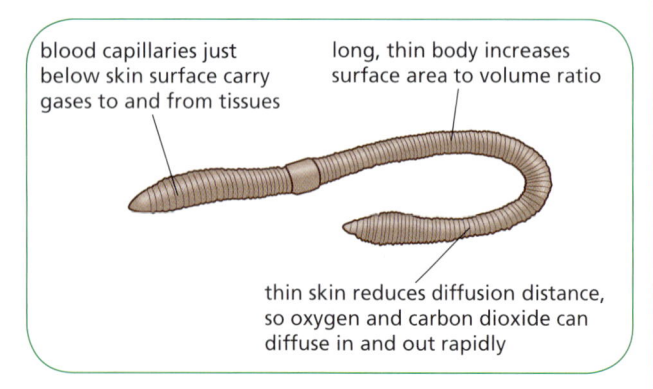

Figure 5d.1 Gaseous exchange in an earthworm.
- blood capillaries just below skin surface carry gases to and from tissues
- long, thin body increases surface area to volume ratio
- thin skin reduces diffusion distance, so oxygen and carbon dioxide can diffuse in and out rapidly

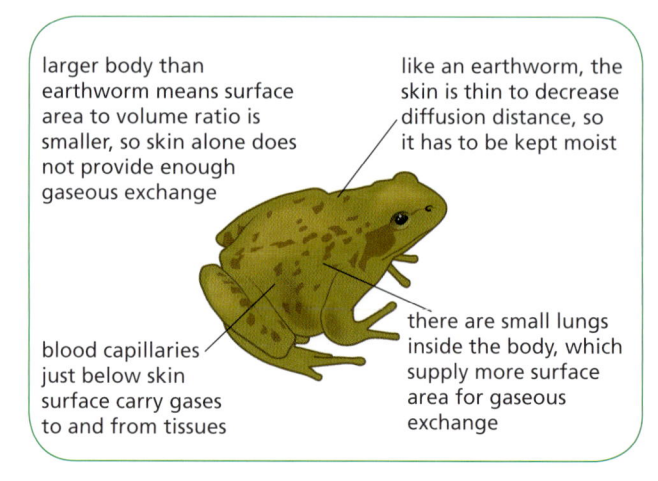

Figure 5d.2 Gaseous exchange in a frog.
- larger body than earthworm means surface area to volume ratio is smaller, so skin alone does not provide enough gaseous exchange
- like an earthworm, the skin is thin to decrease diffusion distance, so it has to be kept moist
- blood capillaries just below skin surface carry gases to and from tissues
- there are small lungs inside the body, which supply more surface area for gaseous exchange

Fish only have **gills**. Fish gills rapidly dry out once the fish are on land. What's more, if you take a fish out of water the gill filaments tend to stick together, because of the surface tension of the water on them. This reduces the surface area that is exposed to the air around them, so not enough gaseous exchange can take place. Fish are therefore restricted to living in water.

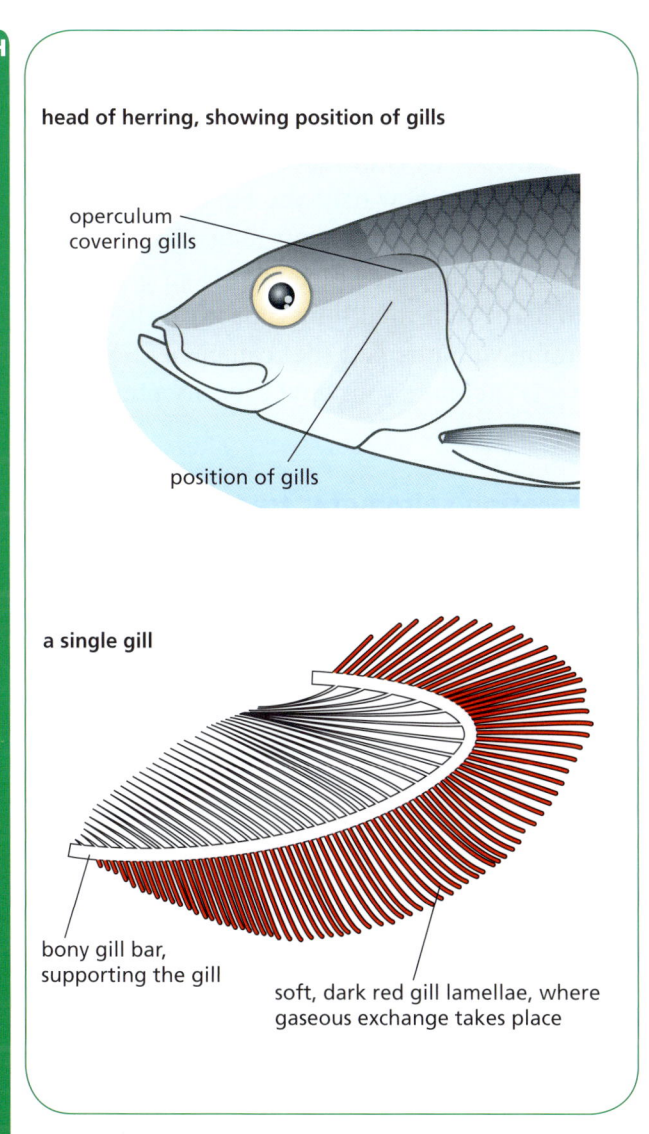

Figure 5d.3 Gaseous exchange in a fish.

SAQ

2 Explain how these features of fish gills help gaseous exchange to take place.
   a being made up of many gill filaments
   b gill filaments being very thin
   c having many tiny blood capillaries, with blood that contains haemoglobin

Breathing

Because our lungs are tucked away from the air, inside our body, we have to work to bring air into contact with them. This is called **breathing**.

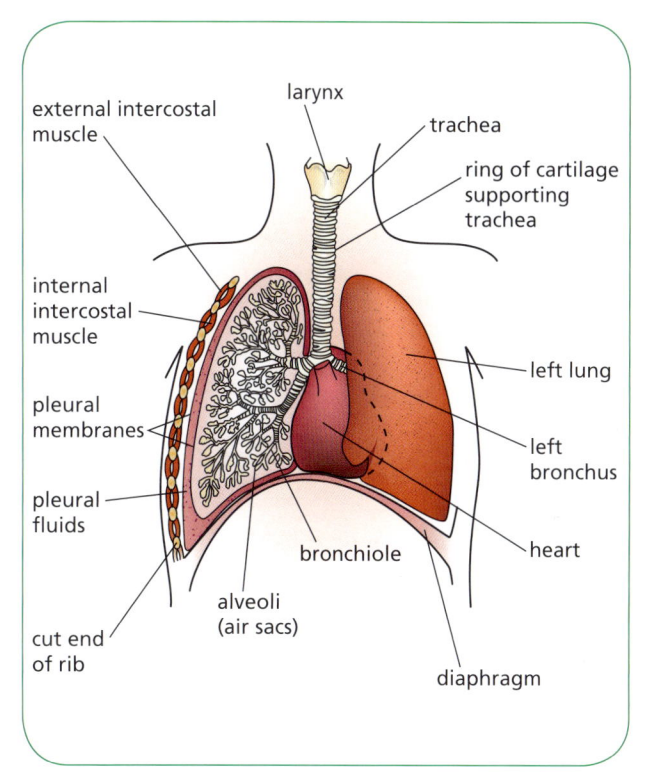

Figure 5d.4 The gaseous exchange system in humans.

Inspiration and expiration

Inspiration means breathing air into the lungs. **Expiration** means breathing air out of the lungs. Figure 5d.5 (page 28) shows how this happens.

When you breathe in, your external intercostal muscles contract. This pulls the ribcage upwards and outwards, which increases the volume inside your chest cavity. At the same time, the diaphragm muscle also contracts, pulling the diaphragm downwards and increasing the volume in the chest cavity even more. This means that the pressure inside the chest cavity decreases. The air outside your body is at a higher pressure, so it flows into your lungs down its pressure gradient.

When you breathe out, both sets of muscles relax. The ribs drop downwards into their resting position, and the diaphragm springs back upwards into its normal domed shape. This decreases the volume inside the chest cavity, increasing the pressure and squeezing air out of the lungs.

28 B5d Breath of life

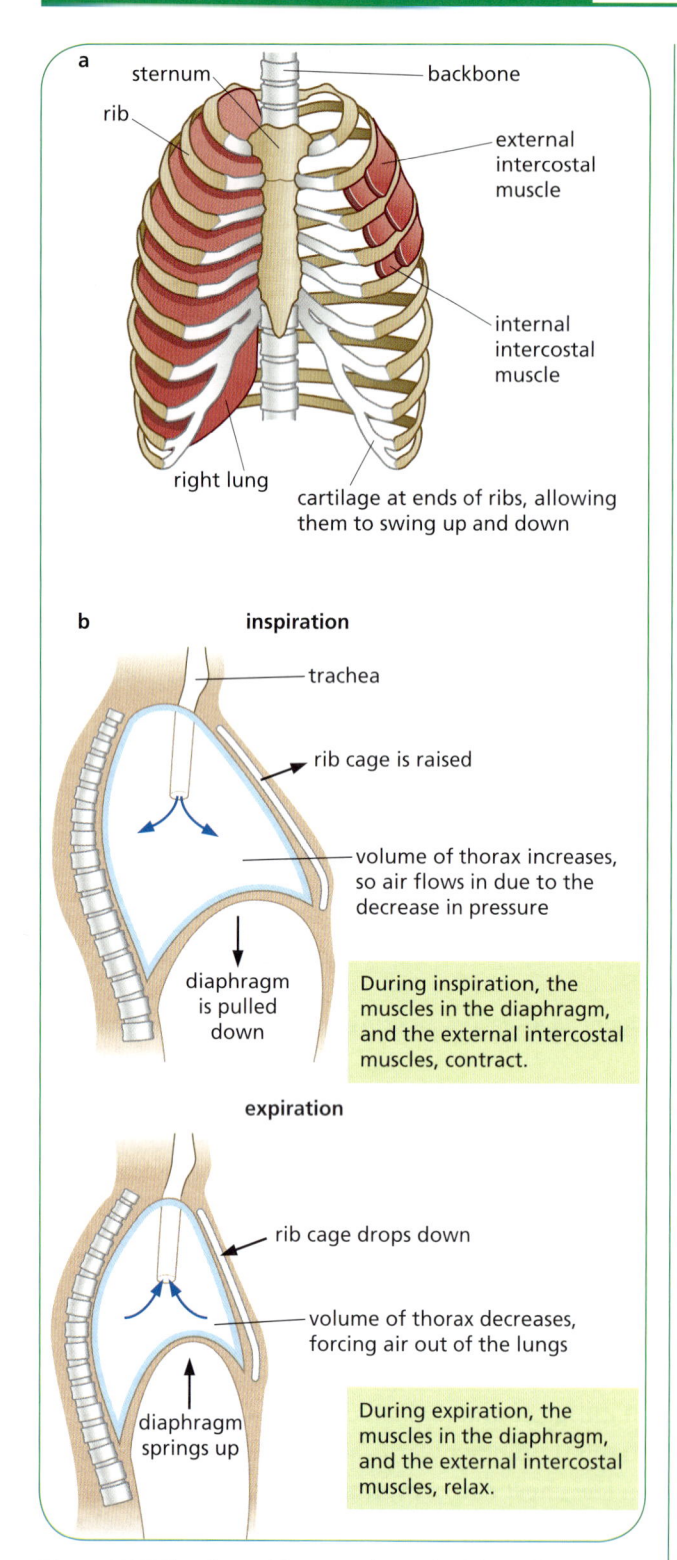

Figure 5d.5 The ribs and intercostal muscles (**a**), and how the intercostal muscles and diaphragm produce breathing movements (**b**).

SAQ

3 Explain the difference between:

  a *inspiration* and *expiration*

  b *breathing* and *respiration*.

Lung capacities

You may be able to use a spirometer or other measuring instrument, to find out how much air you move in and out of your lungs when you breathe.

During normal breathing, most people move about 0.5 dm^3 of air in and out of their lungs with each breath. This is called your **tidal air**. But your lungs have a much greater capacity than this. If you take the very deepest breath in that you can manage, and then the very largest breath out, you may be able to move as much as 3.5 dm^3 of air. This is called your **vital capacity**.

But no matter how hard you breathe out, there is always some air left in your lungs, because they never collapse completely. This air is called **residual air**. Your total **lung capacity** – the maximum volume of air that can be held in your lungs – is therefore your vital capacity plus your residual capacity.

Interpreting spirometer traces

A spirometer is an instrument that makes a record of the volumes of air that you breathe in and out with each breath. There are several different kinds, and one model is shown in Figure 5d.6.

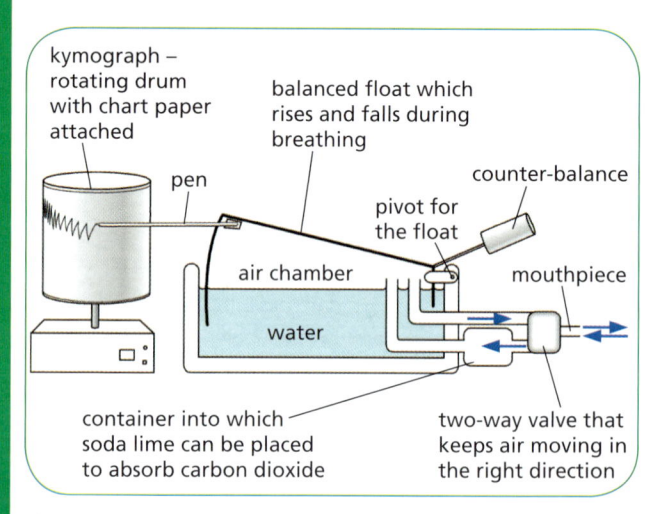

Figure 5d.6 A spirometer.

The person whose breathing is to be measured breathes in and out through the mouthpiece. The air they breathe in comes directly from the air chamber. The air they breathe out goes through some soda lime, to remove carbon dioxide from it. If this didn't happen, then the concentration of carbon dioxide in the air in the chamber would increase, which could be dangerous for the person who is breathing it.

As the air goes in and out of the chamber, the chamber moves up and down. These movements are recorded by a pen writing on a revolving chart. Figure 5d.7 shows the kind of record that you might get.

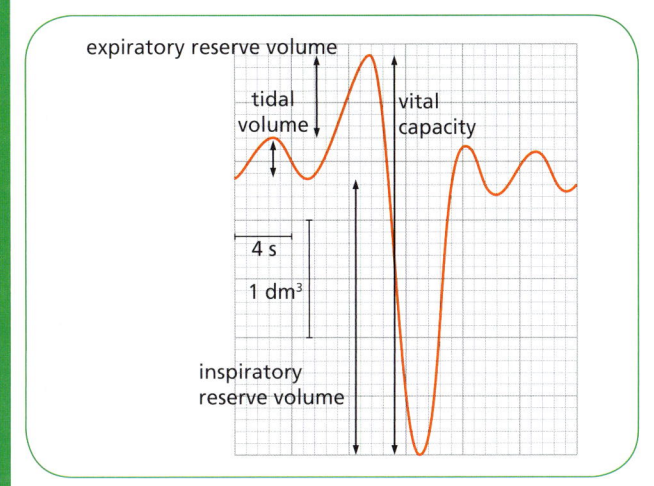

Figure 5d.7 A chart made using a spirometer.

SAQ

4 A doctor used a spirometer to investigate a patient's breathing. The chart in Figure 5d.7 was produced.
   a When the patient breathed out, did the chamber of the spirometer go up or down?
   b Use the chart to find the patient's tidal volume.
   c Use the chart to find the patient's vital capacity.
   d Describe what the doctor asked the patient to do to find his vital capacity.
   e Explain why a spirometer cannot measure a person's total lung capacity.

Respiratory diseases

Each time you breathe in, millions of tiny particles are carried into your nose and trachea, along with the air. Some of these might be harmful. For example:
- tar and other substances in cigarette smoke can cause **lung cancer**
- particles in the exhaust gases of vehicles, especially from badly maintained diesel engines, and also cigarette smoke, contain **particulates** which can irritate the lungs and cause damage to them
- pathogens can cause **bronchitis** or **pneumonia**.

The lungs are effectively a dead end – the only way out is the same as the way in. So anything that gets down into the lungs tends to stay there. It is therefore very important to stop too many of these harmful things getting in. Our main defences against the entry of harmful substances into the lungs are the **mucus** and **ciliated cells** that line the inside of the trachea and the bronchi (Figure 5d.8).

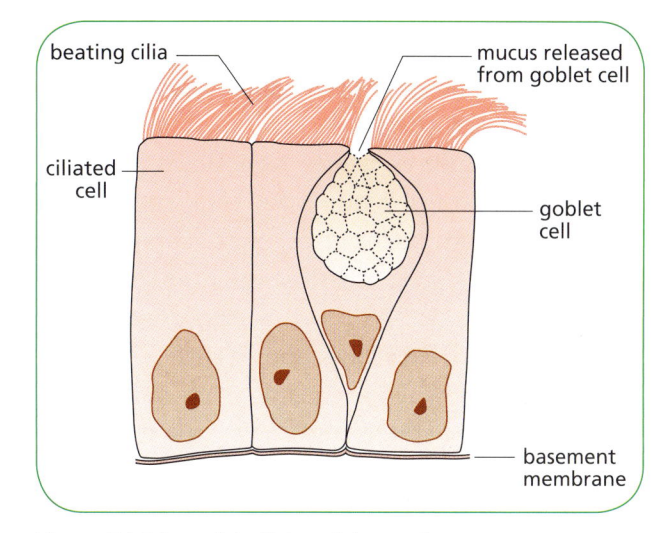

Figure 5d.8 Part of the lining of the respiratory passages.

SAQ

5 Explain how the goblet cells and ciliated cells in the lining of the trachea and bronchi help to prevent harmful particles getting into the lungs.

Bronchitis and pneumonia

Bronchitis means 'inflammation of the bronchi'. It is caused by bacteria that infect the lining of the bronchi. Bronchitis is very common in people who smoke, and may be a chronic disease – that is, one that they have almost all the time. If a non-smoker gets bronchitis, it usually lasts for a few weeks at most.

If you have bronchitis, your goblet cells tend to make extra mucus. You will probably have a cough, as your body tries to remove the infection from your bronchi.

Bronchitis can sometimes lead to **pneumonia**. This happens when bacteria get right inside the lungs and breed there, causing painful inflammation. Pneumonia is a dangerous disease and it can kill, but some types are easy to treat with antibiotics.

Asbestosis

Asbestosis is a disease of the lungs that is caused by breathing in fibres of asbestos. You may have read it about on pages 266–267 in *Gateway Science*. It is an **industrial disease** – one that people get because of the work that they do. Asbestos used to be widely used in buildings, because it does not burn. People working in factories that made asbestos products, or who worked with it in buildings, may have been exposed to asbestos fibres over many years.

The asbestos fibres can get right down into the alveoli in the lungs. White blood cells attempt to destroy them by phagocytosis, but of course their enzymes don't digest asbestos. The lungs become badly inflamed.

A person with asbestosis finds it difficult to get their breath when exercising. They may have a severe, chronic cough, and chest pain. It isn't possible to repair the damage that has been caused, so the best that can be done for a patient is to give them oxygen to breathe.

Many people have died because of asbestosis. Having this disease also increases the risk of developing a type of lung cancer, mesothelioma, which is not usually treatable.

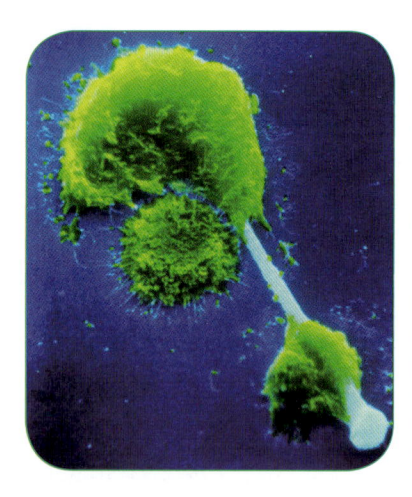

Figure 5d.9 Scanning electron micrograph (SEM) showing two white blood cells impaled on an asbestos fibre. Normally, foreign particles in the lungs are engulfed by these cells. Asbestos fibres, however, are virtually indestructible and puncture the cells, destroying them and releasing their contents into the lung tissue.

Lung cancer

Lung cancer is a disease where the genes that control cell division become altered. Cells begin to divide out of control, forming a **tumour**.

Lung cancer is almost always caused by something in a person's environment. Almost everyone who gets lung cancer has been a cigarette smoker, and their disease is caused by the tars and other carcinogens in cigarette smoke. Another cause, as we have seen, is breathing in asbestos fibres, and some people develop lung cancer after prolonged exposure to other people's cigarette smoke. Lung cancer is entirely a lifestyle disease, avoidable if people take care and don't smoke.

Lung cancer is very difficult to treat, and most cases are fatal. It causes about 35 000 deaths each year in England and Wales. Lung cancer is uncommon in countries where the air is relatively clean and few people smoke.

Vicious fibres

James Allput was 32 when he died, on 27 April 2005. He had never worked anywhere that exposed him to asbestos. But his father had done – he worked as a scaffolder. For a while, James's father had worked on the scaffolding at a power station, where asbestos was present.

James had been unwell for some time, before he was diagnosed with lung cancer in September 2004. The cancer seems to have developed because he had inhaled asbestos fibres that his dad had inadvertently brought home on his clothes.

If James had been working with asbestos himself, and if it could be proved that the company he worked for had been negligent in not giving him sufficient protection, his family could have claimed compensation for this death. But he wasn't working for anyone at the time he was exposed to the asbestos fibres. It's going to be a long battle, trying to get compensation from James's father's employer.

Cystic fibrosis

Not all diseases of the lungs are caused by a person's environment. You may remember that cystic fibrosis is a disease caused by a person's genes (*Gateway Science*, pages 60–63).

A person with cystic fibrosis produces very sticky mucus, which accumulates in the lungs and provides a breeding ground for bacteria. Infections can seriously damage the lungs. At present, there is no cure for cystic fibrosis, although a lot of research is being done into ways in which gene therapy might be able to get correct copies of the faulty genes into the cells in the lungs and elsewhere in the body.

Asthma

Asthma is a condition in which the airways leading to the lungs can sometimes become inflamed. It is a very common condition in the UK, and one adult in 13 is regularly treated for asthma.

An asthma attack happens because the white blood cells of the immune system overreact to some stimulus. The trigger for an attack can be different for different people, but many people with asthma find that car exhaust fumes, pet hairs or dust mite faeces act as triggers. Most people with asthma find their symptoms are worse at night than in the daytime.

When a person has an asthma attack, they cannot breathe. The muscles in the walls of their airways contract, making the airways partly close up. However hard the person tries, they cannot get enough air into their lungs. Extra mucus may be made, which makes it even more difficult for oxygen to get into the blood. It is very frightening.

Someone having an asthma attack can find relief by using an inhaler. This often contains a drug called salbutamol. Salbutamol belongs to a class of drugs called bronchodilators. As their name suggests, these drugs make the muscles in the walls of the airways relax, opening them up and allowing air to get through.

Over the long term, another kind of drug can be helpful. Many people regularly take steroids – again, using an inhaler – which reduce the inflammation in the airways.

Summary

You should be able to:

- outline the methods of gaseous exchange in an amoeba, earthworm, fish and human, and relate these to their size and habitats
- describe how fish gills allow gaseous exchange in water
- identify the main parts of the human respiratory system
- explain how the structure of the human respiratory system allows efficient gaseous exchange
- describe how breathing movements are brought about
- understand the terms tidal air, vital capacity and residual air
- interpret data on lung capacities
- describe the causes and symptoms of asthma, bronchitis, pneumonia, lung cancer, cystic fibrosis and asbestosis

B5d Breath of life

Questions

1. a Copy and complete Table 5d.1.

Organism	Gaseous exchange surface	Where does it live?
amoeba		
earthworm		
fish		
frog		

    Table 5d.1

    b Explain why an amoeba does not need a specialised gaseous exchange surface.

    c Explain how the gaseous exchange surface of each of the following organisms affects where it is able to live.
       i   fish
       ii frog

2. Explain the difference between:

    a *tidal air* and *vital capacity*

    b *residual air* and *total lung capacity*.

3. Table 5d.2 shows the percentages of men and women in Britain who had asthma between 1994 and 1998.

Year	1994	1995	1996	1997	1998
Percentage of men with asthma	6.75	6.93	7.06	7.23	7.32
Percentage of women with asthma	6.71	7.09	7.27	7.55	7.65

    Table 5d.2

    a Draw a line graph to display these figures. Put both lines on the same pair of axes.

    b Describe the overall trend in the percentage of people with asthma between 1994 and 1998.

    c Compare the figures for men and women.

5e Waste disposal

Defecation and excretion

Your body is constantly producing waste material. We have several different organs whose function is to get rid of it, before it does any harm.

Some of the waste has never actually been truly *inside* the body at all. When food passes through the digestive system, it is travelling along a space that connects directly with the outside world. The alimentary canal is a tube running continuously from the mouth to the anus. If you were small enough, you could go all the way along someone's alimentary canal, and out at the other end, without crossing any barriers at all.

Much of the food that we eat is digested – broken down into small molecules that seep through the wall of the intestine. They are then carried away in the blood or lymph. But some of it is not digested. This waste food continues along the canal and eventually reaches the rectum. Here – together with old cells that have rubbed off from the lining of the alimentary canal – it is collected into **faeces**. These are eventually passed out of the body through the anus. This is **defecation**.

But there are other kinds of waste that we produce. You are getting rid of some it now, as you are reading this.

This waste is produced by some of the metabolic reactions that take place in our cells. The waste that you are getting rid of now is **carbon dioxide**, which is a waste product of respiration. All our cells produce it, all of the time, and it is carried in the blood to the lungs. Here it is lost from the body in our expired air.

Removing waste products of metabolism, such as carbon dioxide, is called **excretion**. These waste products are often **toxic** – they would harm the body if they were not removed. There are several different excretory products. For example, we excrete **urea**, which is removed from the body in the urine that is produced in the **kidneys**.

Figure 5e.1 shows the main excretory products of the human body, and the organs that are responsible for excreting them.

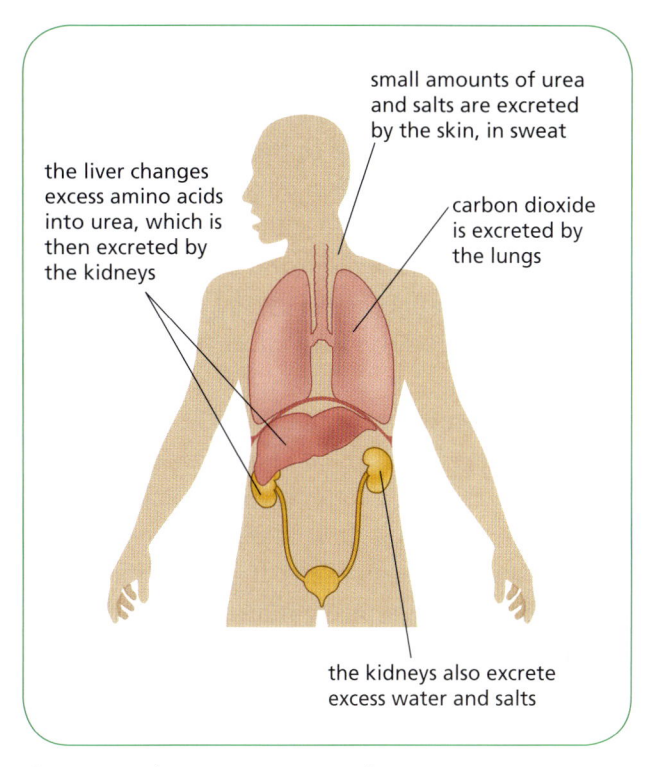

Figure 5e.1 The excretory organs in humans.

SAQ

1 **a** Copy and complete Table 5e.1. Use the information in Figure 5e.1, and also in the paragraphs above.

Waste substance	Where is it produced?	Which organ removes it from the body?
undigested food		
carbon dioxide		
urea		

Table 5e.1

b Use a highlighter or coloured pencil on your table to indicate which of these substances is lost from the body by defecation.

c Use a different colour to indicate which of these substances are lost from the body by excretion.

Excreting carbon dioxide

Carbon dioxide is produced continuously in the body, by respiration. It diffuses out of cells as they respire, and into the blood plasma. It dissolves

34 B5e Waste disposal

in the plasma and is carried to the lungs in the blood. It then diffuses across the walls of the blood capillaries and alveoli, and is breathed out in expired air.

It is essential that this happens all the time. If carbon dioxide builds up in the blood, it can do considerable harm. Carbon dioxide dissolves in water to form a weak acid, called **carbonic acid**, so high levels of carbon dioxide make the blood more acidic. Changes in pH can be very damaging to cells, partly because they affect the activity of enzymes.

Your brain constantly monitors the pH of the blood. If the pH of the blood falls, this usually means that there is a lot of carbon dioxide dissolved in it. This is most likely to happen because you have been exercising hard, and your muscle cells were respiring more rapidly than usual. Your brain responds by sending nerve impulses to the diaphragm and the intercostal muscles. These nerve impulses make the muscles contract more often and more strongly, which means that you breathe more quickly and more deeply. This helps to remove the extra carbon dioxide from the blood. This, of course, also helps to get more oxygen into the body, which can be used by the respiring muscle cells.

SAQ

2 Figure 5e.2 shows a spirometer trace made by a student. She took several breaths while resting, and then ran on the spot while still breathing through the spirometer.

a The **tidal volume** is the volume of air moved in or out of the lungs in one breath. Use the graph to find the student's mean tidal volume before she began to exercise.

b Calculate her average breathing rate, in breaths per minute, before she began to exercise.

c Describe how her tidal volume changed when she exercised.

d Describe how her breathing rate changed when she exercised.

e Explain how these changes were achieved.

f Explain why breathing faster and more deeply while she was exercising would help to avoid possible damage to her body cells.

g Suggest what will happen to her tidal volume when she stops exercising.

Excreting urea

When you eat foods containing protein, the protein molecules are broken down to amino acids in the alimentary canal. The amino acids are absorbed into the blood and transported around the body dissolved in the blood plasma. They are used by cells to make their own proteins.

If you eat more amino acids than you need, your body is not able to store the excess. Instead, they are broken down in the **liver**. This is called **deamination**. It produces a waste product called **urea**.

Urea contains nitrogen, which has come from the amino acids, so it is sometimes known as a **nitrogenous excretory product**.

The urea diffuses out of the liver cells and dissolves in the blood plasma. Urea is a toxic substance, so it must not be allowed to build up in the blood. The blood stream delivers it to the **kidneys**, where it is excreted in urine. A small amount of urea is also lost through the skin, dissolved in sweat.

The kidneys

The kidneys are our most important excretory organs. They produce a liquid, **urine**, which contains several different waste products. The main one is urea, but urine also contains waste water and salt.

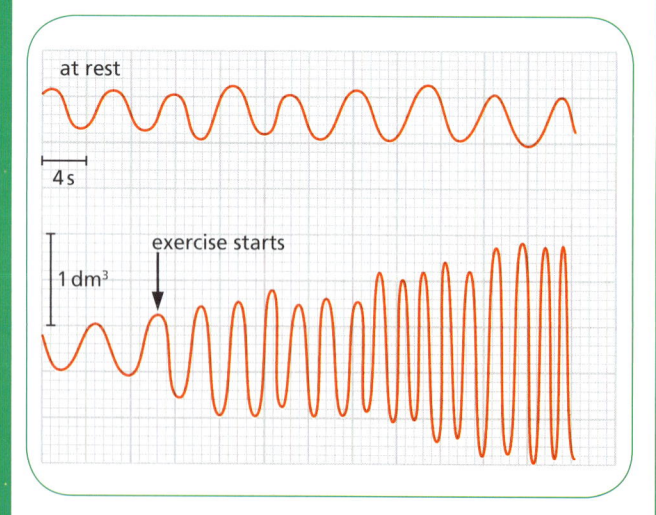

Figure 5e.2

Structure of the excretory system

The kidneys are part of the excretory system. This system also includes the ureters, bladder and urethra. Figure 5e.3 shows the positions of these structures in the body. Figure 5e.4 shows the structure of a kidney. Blood is brought to the kidneys in the **renal arteries** and taken away in the **renal veins**.

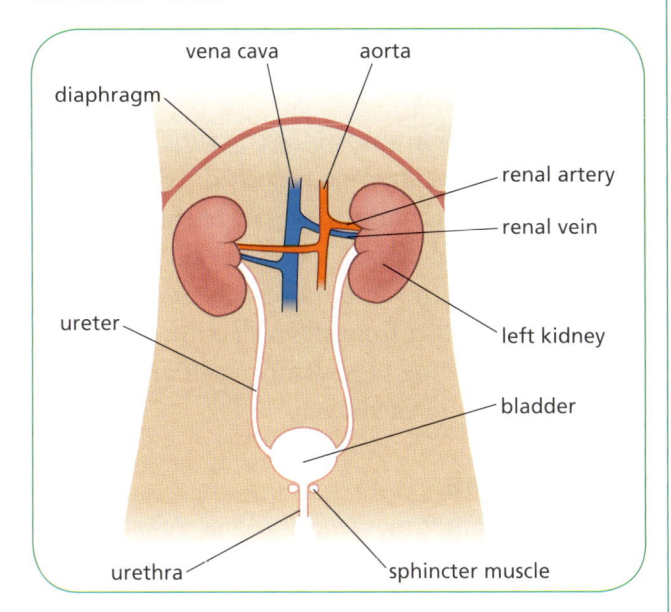

Figure 5e.3 The human excretory system.

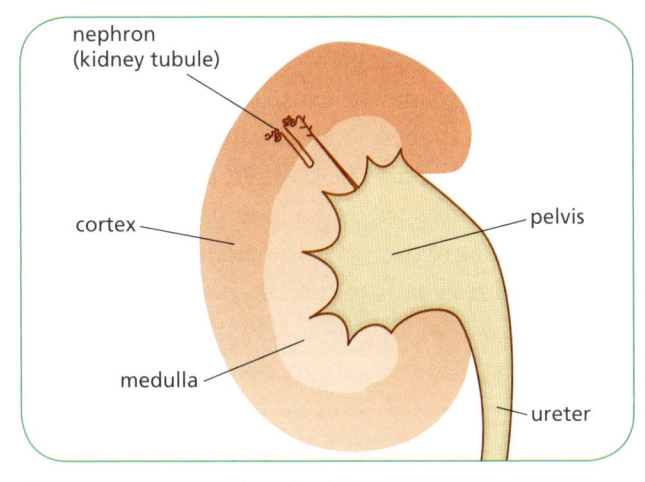

Figure 5e.4 A section though a kidney.

How kidneys make urine

Each kidney contains thousands of tiny tubes called **tubules** or **nephrons** (Figure 5e.5). These remove waste products from the blood. This process happens in two stages.

1 First, the kidney tubules **filter** the blood. To help the filtration to happen, the blood that arrives at the kidney tubule is kept at a slightly higher pressure than in most places in the body. The blood capillary that brings blood to the kidney tubule is quite wide, but the one taking blood away is narrow. This means that the blood cannot get away easily, so quite a high pressure builds up, squeezing the blood through the walls of the capillary and into the kidney tubule.

The walls of the capillary and the kidney tubules have small holes in them. All the small, soluble molecules in the blood pass through these tiny holes and into the kidney tubules. Water, urea, glucose and salts can all go through. Large molecules, like proteins, cannot get through the holes and stay in the blood. The red and white blood cells also stay in the blood capillaries.

2 Next, the kidney tubules **reabsorb** some of the substances that have filtered through into them. This is because some of the substances that have been squeezed into the kidney tubules are needed by the body. All of the glucose, a lot of the water and some of the salts need to be kept in the blood.

There are blood capillaries wrapped around each kidney tubule. As the fluid flows along

filtration – small molecules, such as water, glucose, salts and urea are squeezed out of the blood into the start of the kidney tubule

reabsorption – useful substances, especially glucose, some water and some salts, are taken back into the blood

the remaining liquid, called urine, flows into the ureter

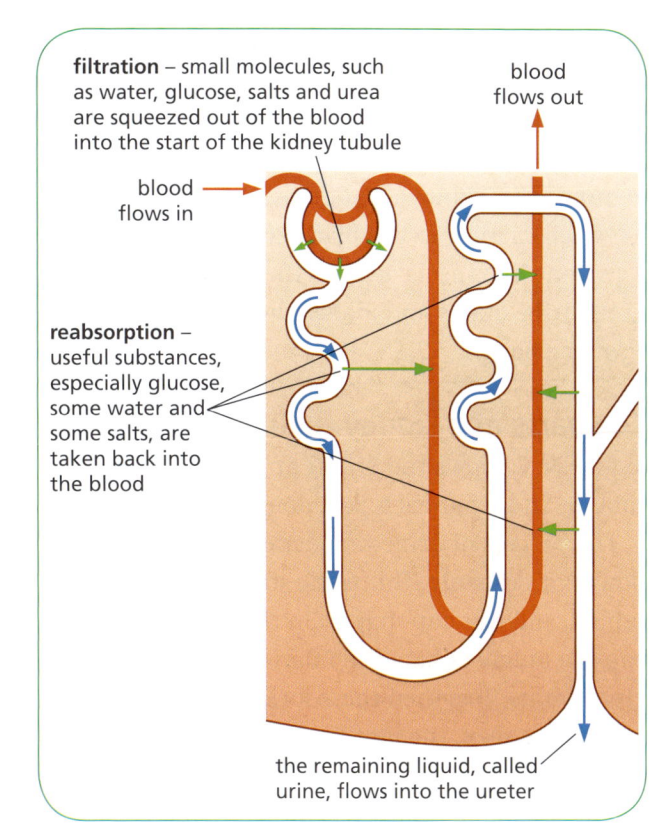

Figure 5e.5 How urine is made in a kidney tubule.

inside the tubule, these useful substances are taken back into the blood.

The remaining fluid is made up of urea and salts, dissolved in water. This is urine. It flows out of the kidney tubules and into the ureters, which carry it to the **bladder**. It is stored here. When the muscle around the **urethra** relaxes, the tube opens and allows the urine to flow out of the body.

The parts of a kidney tubule

Figure 5e.6 shows the different parts of a kidney tubule, and what each part does.

The process of urine production begins at a **Bowman's capsule**. This is the start of the kidney tubule, and it is cup-shaped. In the centre of the cup is a tangle of blood capillaries, called the **glomerulus**. Together, these form a 'filtration unit'. As we have seen, the liquid part of the blood is squeezed out of the capillary and into the capsule, leaving behind all the cells and larger molecules.

Next, the fluid flows down into the first coiled part of the tubule, which has blood capillaries close to it. This is where all the glucose and some of the water in the fluid are taken back into the blood. This process is called **selective reabsorption**. To help it happen quickly, the membranes of the cells lining the tubule are folded, providing a large surface area. Some substances are moved against their concentration gradient, which requires the cells to use energy. These cells therefore contain a lot of mitochondria.

As the fluid continues along the tubule, more salts and more water are reabsorbed. This mostly happens in the second coiled part of the tubule, and also in the final part, which is called the **collecting duct**. The amount of salts and water that are reabsorbed from the collecting duct can be varied, according to how much salt and water there is already in the blood.

Kidney failure

Kidneys can sometimes stop working. People can manage perfectly with only one kidney, but if both of them fail then the person may die. Without working kidneys, urea builds up in the blood. It is toxic in large concentrations, and the body organs gradually become unable to function normally.

Kidney failure may be caused by an infection. A person with diabetes may develop kidney failure, and it can also be brought about by having a high blood pressure over a long period of time.

There are only two ways in which a person with complete kidney failure can be effectively treated. The best way is to give them a kidney transplant (pages 52–54). However, there are not enough kidneys available for transplant, and it is usually not possible to find a suitable kidney straight away. So the person is treated using a **dialysis** machine.

Figure 5e.7 shows how one type of dialysis machine works. The person's blood is made to flow through many tiny tubes made of a partially permeable membrane (like Visking tubing). These are surrounded by **dialysis fluid**. The dialysis fluid is mostly water, but it also contains dissolved glucose and salts, including sodium ions, at just the right concentration for the body. There is no urea in dialysis fluid. As the blood flows through

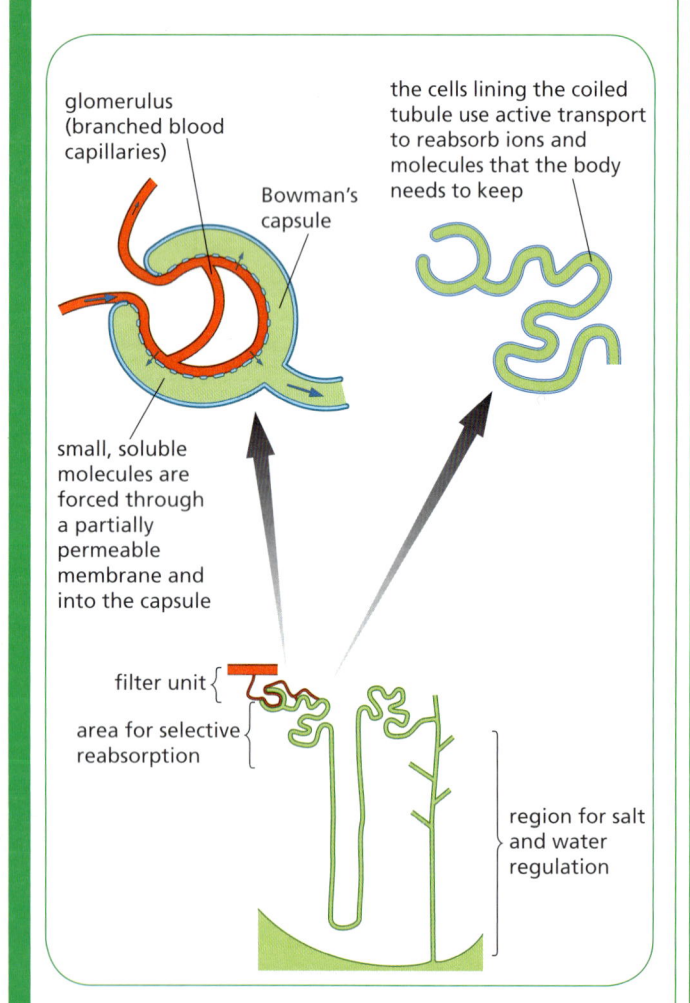

Figure 5e.6 The parts of a kidney tubule.

the tubes, small molecules that can pass through the tiny holes in the tubing diffuse through it.
- Urea diffuses down its concentration gradient, from the blood into the fluid.
- Water, glucose and sodium ions diffuse down their concentration gradients, which may be in either direction, depending on how concentrated they are in the person's blood.
- Blood cells and large molecules stay in the blood.

SAQ

3   a Explain why the machine is designed so that blood flows through many narrow tubes, instead of one larger one.

   b In Figure 5e.7, you can see that the blood is kept at a higher pressure than the dialysis fluid. Suggest how this helps to make the process of dialysis work effectively.

   c In terms of the pressure of the blood, how is dialysis similar to the way kidneys work?

   d Describe *one* more similarity between the working of a dialysis machine and kidneys.

   e Describe *two* differences between the working of a dialysis machine and the kidneys.

Dialysis is a life-saver for thousands of people, but few of them enjoy it. It takes several hours, at least twice a week, and the person has to be connected up to the machine all that time. In between times, the patient has to be very careful what they eat – they must not eat too much protein or salty foods like crisps.

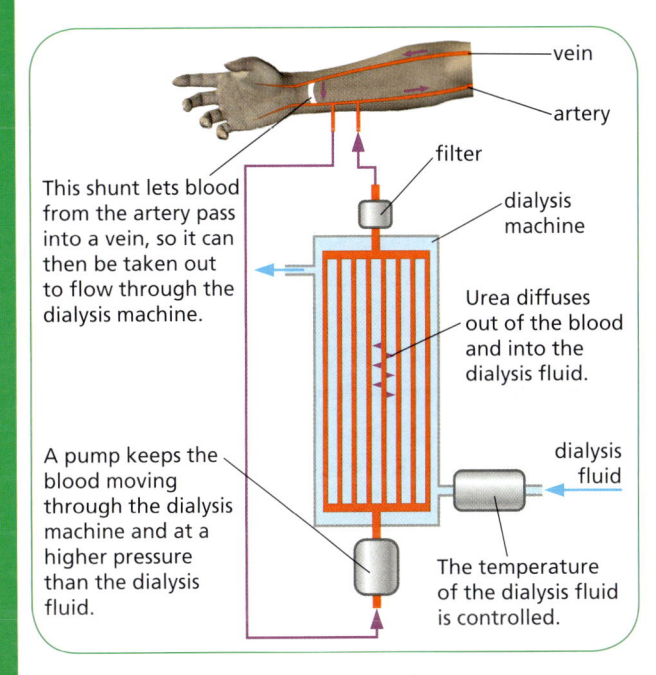

Figure 5e.7 How a dialysis machine works.

Bird pee, kidney stones and gout

Have you ever seen a bird peeing? Birds don't produce liquid urine as we do. Instead, they produce a white paste, rich in uric acid.

Uric acid, like urea, is a nitrogenous excretory product made in the liver from excess amino acids. It's an alternative way to get rid of nitrogenous waste. Birds use it because it is not as soluble as urea, and can be excreted as a paste, rather than as a watery liquid. It's thought that they have evolved this way because bird embryos develop inside a hard-shelled egg. The uric acid that the embryo produces is tucked away inside a membrane in the egg, keeping it away from the embryo. If birds excreted urea, the embryo would have to float in a sea of its own liquid urine.

Humans, too, produce uric acid, but normally in much lower amounts than urea. Uric acid is not very soluble, and if a person has abnormally high amounts of it in their blood, then it can crystallise out of solution and form kidney stones inside the nephrons. Sometimes, these stones get washed down into the ureters and block them. This is very painful. A person with a painful kidney stone may be able to remove it by drinking lots of water, so it gets flushed out. In some cases, ultrasound treatment is given to break up the stones.

Uric acid can also crystallise out of the blood in other parts of the body. If this happens inside a joint, it can cause a very painful condition called gout. For some reason, this is most common in the big toe. It is difficult to predict who is at risk from gout, but there seems to be a link with high blood pressure and obesity, and with drinking a lot of alcohol.

Regulating the amount of urine

You don't produce the same amount of urine each day. On some days, you might produce only a small volume of urine, which is quite concentrated (dark in colour). On other days, you might produce a lot more, which is less concentrated.

The quantity and concentration of the urine a person produces depends on how much water they have in their blood. If the blood has more water than the body requires, then the excess water is removed from the body in the urine. So there will be a lot of urine, and it will be dilute. If, on the other hand, the blood is low in water, then only a small amount of concentrated urine will be produced, to conserve as much water as possible in the blood.

We gain water in the body by drinking, or by eating food that contains water (such as fresh fruit). We lose water not only in our urine, but also as **sweat** from the skin, and in the moist air that we breathe out. So if you do a lot of vigorous exercise, especially on a hot day, you lose more water than usual. If you don't drink extra fluids, the kidneys will conserve water in the body by excreting only small volumes of urine.

H The concentration of the urine that is produced by the kidneys is controlled by a hormone called **antidiuretic hormone**, ADH. This hormone is produced in the **pituitary gland**, right in the middle of the head, just beneath the brain.

The concentration of the blood – which depends on how much water there is in it – is sensed by the **hypothalamus**, which has a direct connection to the pituitary gland. When the water content of the blood is low, the hypothalamus senses this and sends nerve impulses to the pituitary gland. The pituitary responds by secreting ADH (Figure 5e.8).

The ADH dissolves in the blood plasma and is transported around the body. Its target organs are the kidneys. The ADH affects the walls of the collecting ducts. It makes them very permeable to water. This means that water in the urine can escape from the collecting duct, seeping through the walls and into the blood. So the urine that is left is more concentrated than usual, and there is not very much of it.

H If, however, there is too much water in the blood, then no ADH is secreted. The walls of the collecting ducts become impermeable to water, so as the urine flows along the collecting duct it all has to stay inside. This means that a lot of urine is made, and it is dilute.

1 The blood water content is monitored by the hypothalamus.

2a If there is a lot of water in the blood, very little ADH is secreted.

2b If there is too little water in the blood, the hypothalamus causes the pituitary gland to secrete ADH.

3a The kidneys do not reabsorb much water.

3b The ADH makes the kidneys reabsorb a lot of water from the urine. The water goes back into the blood.

4a A lot of dilute urine is produced.

4b Only a small amount of concentrated urine is produced.

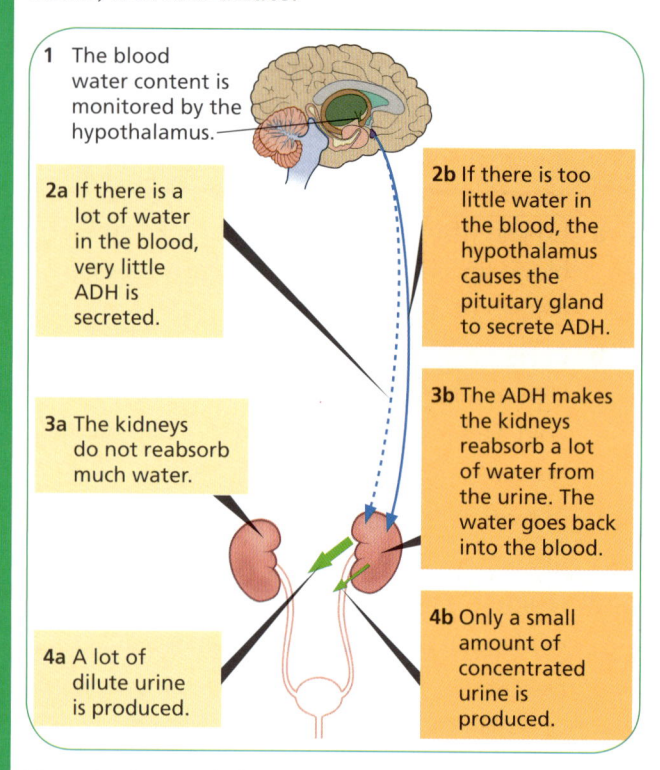

Figure 5e.8 Control of water level in the blood by negative feedback.

The skin

The skin is our largest organ. It is the part of our body that is most in touch with our surroundings. It forms a protective layer, preventing harmful substances or bacteria getting into the body. It is waterproof, so the water inside our cells and tissues cannot escape through it. It helps us to keep track of what is going on around us, because it contains sense organs for temperature, pressure and touch. Figure 5e.9 shows some of the features that allow skin to carry out all these jobs.

Skin also has a very important role in regulating temperature. One of the ways in which it does this is by producing sweat when the body temperature becomes too high. Sweat contains a lot of water. The water takes heat from the hot tissues in the skin, and uses the heat to make the liquid water turn into water vapour.

Sweat is produced from blood plasma. As the blood flows through the tangle of capillaries

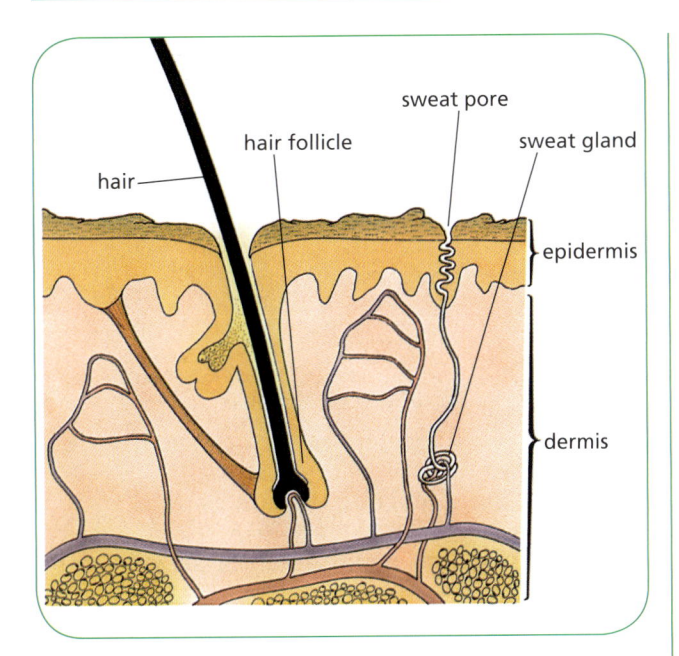

Figure 5e.9 A section through human skin.

around a sweat gland (Figure 5e.10), liquid seeps out through the capillary walls. The gland adjusts the composition of the liquid a little, before allowing it to flow upward through the sweat pore onto the skin surface.

As sweat is made from blood plasma, it contains not only water, but also many of the substances that are transported round the body in solution in the blood. These include salts and urea. There isn't all that much urea, compared with the quantity that the kidneys excrete, but all the same we can think of the skin as a supplementary excretory organ.

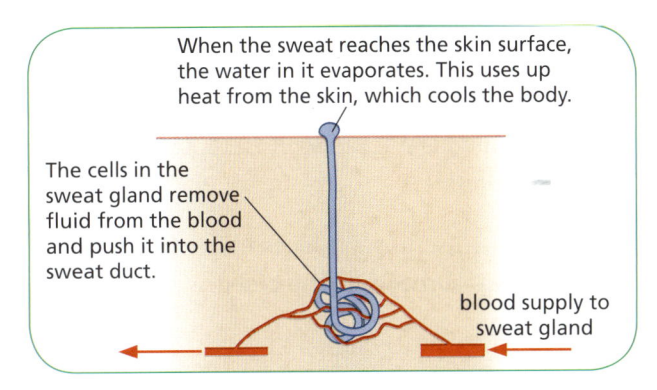

When the sweat reaches the skin surface, the water in it evaporates. This uses up heat from the skin, which cools the body.

The cells in the sweat gland remove fluid from the blood and push it into the sweat duct.

Figure 5e.10 How sweat cools the skin.

Summary

You should be able to:

- define excretion, and explain how it is different from defecation
- describe the positions of the skin, lungs, liver and kidneys, and name the excretory products that they remove from the body
- describe how carbon dioxide is excreted
- **H** explain how breathing rate is adjusted when carbon dioxide levels in the blood rise
- state that urea is produced in the liver from excess amino acids
- describe the structure of a kidney
- **H** describe the structure of a kidney tubule (nephron)
- describe how kidneys work, in terms of filtration and reabsorption, and list the components of urine
- **H** explain how filtration and selective reabsorption take place
- explain how a kidney dialysis machine works
- describe how the amount of urine produced is affected by water intake, temperature and exercise
- **H** explain how ADH produced by the pituitary gland helps to keep the water content of the blood steady, using a negative feedback mechanism
- describe the structure of the skin, and state that urea and salts are excreted in sweat
- explain how the evaporation of the water in sweat helps to cool the body

B5e Waste disposal

Questions

1. **a** Urea is a *nitrogenous excretory product*. Explain what this term means.

   b Name the organ in which urea is made.

   c Name *two* organs that excrete urea. Which one of these excretes the most urea?

   d Describe the difference between urea and urine.

2. These four sentences about how kidneys work are in the wrong order, and they are also incomplete. Rewrite them in the correct order, completing them with suitable words.

   - The remaining fluid flows along the collecting _____ and into the ureters.

   - Blood is filtered at _____ pressure, and the filtrate flows into the start of the kidney tubule.

   - The useful substances in the filtrate, including _____ and _____, are reabsorbed into the blood.

   - The urine is stored in the _____ before being removed from the body.

3. **H** The water content and the glucose content of the blood are controlled using negative feedback mechanisms. If you have forgotten about the control of blood glucose, look back at pages 49–50 in *Gateway Science*.

   Copy and complete Table 5e.2 to compare these two control mechanisms.

	Control of blood water content	Control of blood glucose content
Which organ monitors this factor, and senses when it changes from normal?		
Which hormone is secreted when the level gets too high (glucose) or too low (water)?		
Which endocrine gland secretes this hormone?		
What are the target organs for the hormone?		
What effect does the hormone have on its target organ?		
How does this help to bring the level back to normal?		

   Table 5e.2

5f Life goes on

Human reproduction

Humans reproduce by **sexual reproduction**. Maybe one day we will be able to produce new people using asexual methods – **cloning** – but so far this hasn't happened, and it is very debatable whether we should allow it to be done, even if it became possible. For the moment, all new humans will continue to begin as a single-celled **zygote**, formed by the fusion of a male gamete (sperm) and a female gamete (egg).

The male reproductive organs

Figure 5f.1 shows the structure of the male reproductive organs.

The male gametes – sperm cells – are made in the two **testes**. These are outside the body, in two sacs of skin called the **scrotum**. The structure of a sperm cell is described on page 30 in *Gateway Additional Science*.

The sperm cells are made inside lots of tiny tubes inside each testis, and then stored in the **epididymis**. They can be carried away from the testes along tubes called **sperm ducts**. The two sperm ducts join up with the urethra just below the bladder. The urethra continues through the penis and opens at its tip. The urethra can carry both urine and sperm, usually at different times.

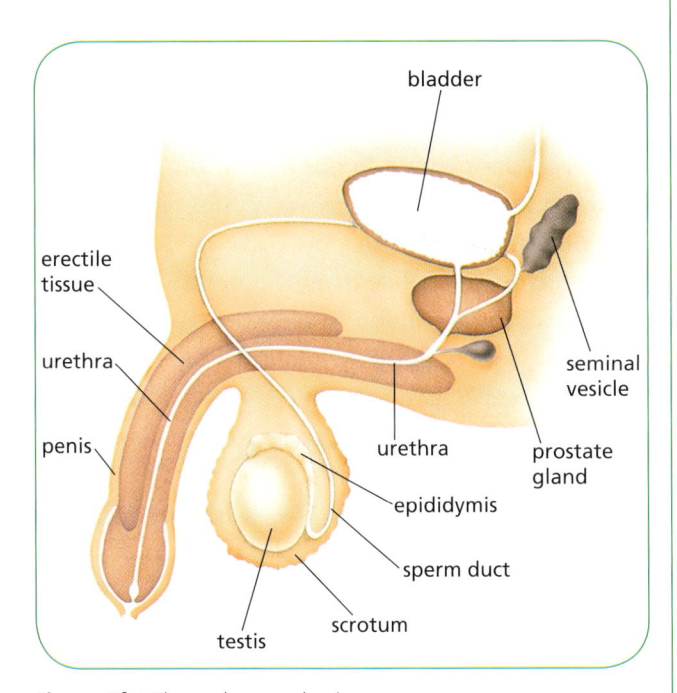

Where the sperm ducts join the urethra, there is a gland called the **prostate gland**. This makes a fluid that the sperm cells swim in. Just behind the prostate gland are the two **seminal vesicles**, which also secrete fluid. The sperm cells plus these fluids are called **semen**.

SAQ

1 Write down, in order, the structures through which a sperm cell passes as it moves from where it is made to the outside world.

The female reproductive organs

Figure 5f.2 shows the structure of the female reproductive organs.

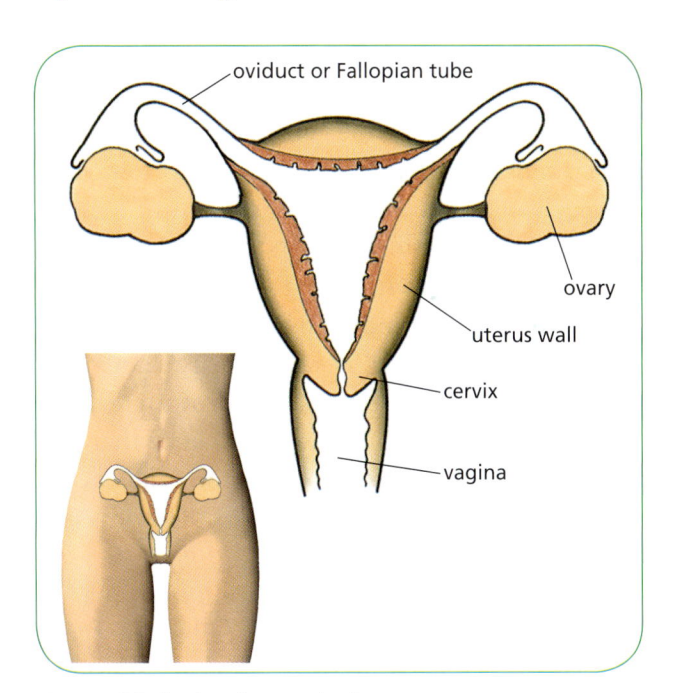

Figure 5f.2 The female reproductive organs.

The female gametes – eggs, egg cells or ova – are made in the two **ovaries**. The structure of an egg (ovum) is described on page 31 in *Gateway Additional Science*.

Leading away from each ovary are the **oviducts** (often called the Fallopian tubes). These tubes don't connect directly with the ovaries, but each one ends in a funnel that 'catches' the eggs when they are released from the ovary. The two oviducts lead to the womb or **uterus**. This has thick outer walls

Figure 5f.1 The male reproductive organs.

made of strong muscles, and a softer lining. The uterus is about the size of a clenched fist, but it stretches enormously when a woman is pregnant.

At the base of the uterus is a narrow opening, guarded by muscles, called the **cervix**. This leads into the **vagina**, which opens to the outside.

In women, the urethra does not join up with any other tubes, as it does in a man. It has its own separate opening to the outside.

Fertilisation

Approximately once a month, an egg leaves one of a woman's ovaries and enters one of her oviducts. It takes it a couple of days to move slowly down the oviduct and reach the uterus. For the first 8 to 24 hours of this journey, it is able to be fertilised by a sperm cell, but after that it is too late.

When a man is sexually excited, blood is pumped into spaces inside the penis, making it firm and erect. The penis is now able to be placed inside a woman's vagina, and semen can be ejaculated close to the cervix. After this, the sperm cells have to make their own way towards the egg.

A single ejaculation deposits about one million sperm cells. They swim, using their tails, up through the cervix and then through the thin layer of moisture that lines the uterus. They can swim about 4 mm per minute. Most of them will never reach an egg, but there is a chance that a few of them will make it as far as the oviducts. If a sperm cell meets an egg in one of the oviducts, then fertilisation can occur.

Figure 5f.3 shows how fertilisation takes place. The head of the successful sperm cell enters the egg. Then the nucleus of the sperm cell and the egg nucleus fuse together. A zygote has been formed.

SAQ

2 Explain why fertilisation happens in the oviducts, never in the uterus.

3 Which of these are haploid cells, and which are diploid? (Look back at *Gateway Additional Science* page 29 if you have forgotten these terms.)
   - sperm
   - egg
   - zygote

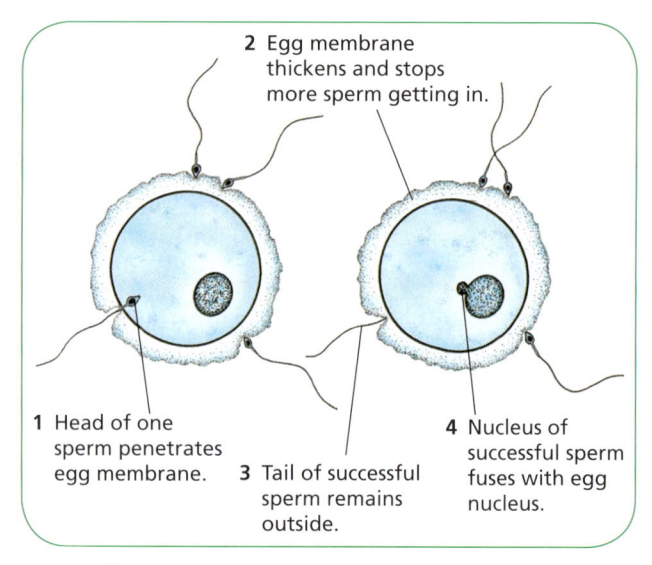

Figure 5f.3 Only one sperm cell fertilises an egg cell.

1 Head of one sperm penetrates egg membrane.
2 Egg membrane thickens and stops more sperm getting in.
3 Tail of successful sperm remains outside.
4 Nucleus of successful sperm fuses with egg nucleus.

Implantation and development

After fertilisation, the zygote continues to move slowly along the oviduct. As it goes, it divides by mitosis. After several hours, it has formed a ball of cells. This is called an **embryo**.

By the time the embryo reaches the uterus, it is a ball of 16 or 32 cells. It sinks into the spongy lining of the uterus, a process called **implantation** (Figure 5f.4 on page 43).

The cells of the embryo continue to divide by mitosis. As the embryo grows larger, a **placenta** is formed, which connects the embryo to the wall of the uterus. The placenta has many finger-like projections which fit closely into the uterus wall. The mother's blood in the uterus wall, and the embryo's blood in the projections are brought close together – but they are kept entirely separate and do not mix. Their closeness, however, allows substances such as oxygen and soluble nutrients (for example, glucose and amino acids) to diffuse from the mother's blood into the embryo's blood. Waste substances from the embryo, especially carbon dioxide and urea, diffuse the other way. You can remind yourself about how the placenta is adapted for this, on page 13 in *Gateway Additional Science*.

SAQ

4 Using your knowledge of blood groups, explain why it is important that the mother's blood does not mix with her embryo's blood.

B5f Life goes on   43

4 After several hours, a ball of cells is formed.

3 The zygote divides by mitosis.

2 Fertilisation. A sperm nucleus fuses with the egg nucleus, forming a zygote.

5 The cells in the ball keep dividing as it moves down the oviduct. It is now called an embryo.

1 Ovulation. A mature follicle bursts, and releases an egg into the oviduct.

6 Implantation. The embryo sinks into the soft lining of the uterus.

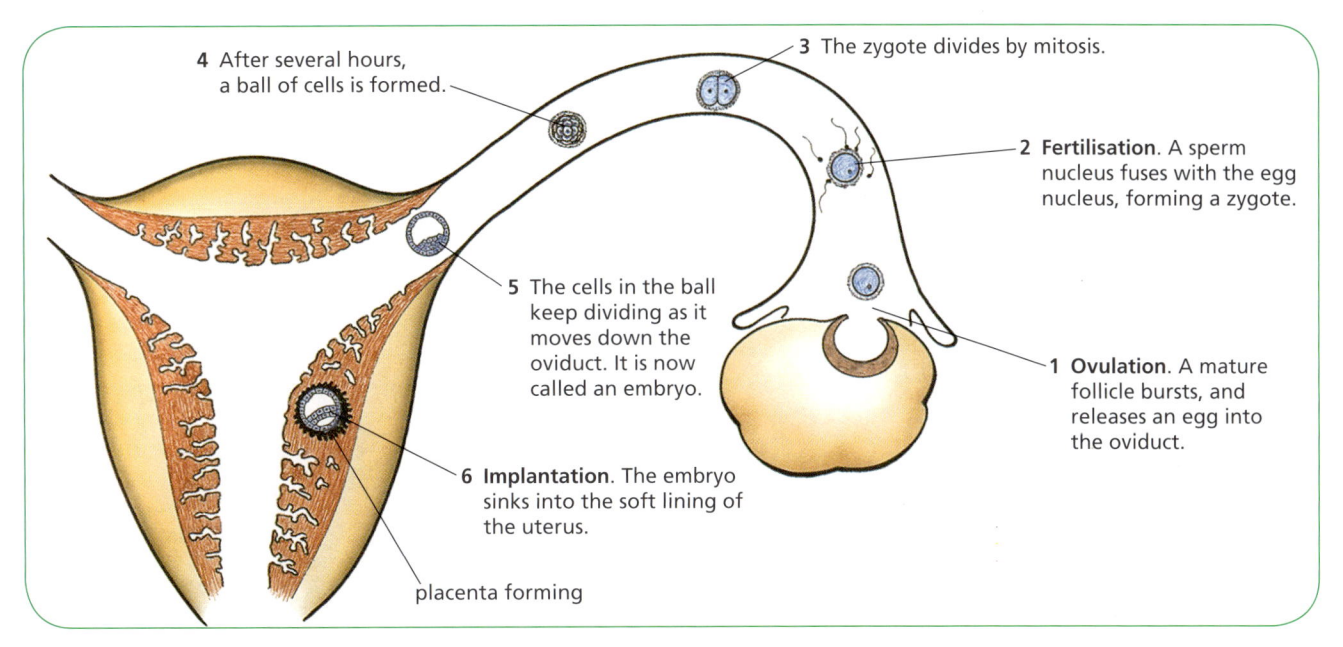

placenta forming

Figure 5f.4 Stages leading to implantation.

The embryo is connected to the placenta by the **umbilical cord**, which contains two arteries and a vein. The embryo floats in a liquid called **amniotic fluid**, which is made by a strong membrane, called the **amnion**, surrounding the embryo.

By the eleventh week after fertilisation, all of the embryo's body organs have been formed, and it is now called a **fetus**. It will remain in the uterus for a total of about nine months, before it is ready to be born (Figure 5f.5). Throughout this time, the placenta must remain in place, and the lining of the uterus must remain thick and spongy, with a good blood supply. This is ensured by a hormone called **progesterone**, which is secreted by the placenta throughout pregnancy.

The menstrual cycle

We have seen that, on average, one egg is released into the oviduct every month in an adult woman. Before the egg is released, the lining of the uterus becomes thick and spongy, to prepare for the arrival of an embryo. It is full of tiny blood vessels, ready to supply the embryo with nutrients and oxygen.

If the egg is *not* fertilised, however, it is dead by the time it reaches the uterus. It does not sink into the uterus lining, but passes out through the vagina. The spongy lining is not needed now, so it gradually disintegrates. It is slowly lost through the vagina. This is called **menstruation**, or a period. It usually lasts for about five days.

After menstruation, the lining of the uterus builds up again, so that it will be ready to receive an embryo if the next egg is fertilised. The cycle begins all over again (Figure 5f.6 on page 44).

This cycle is called the **menstrual cycle**. 'Menstrual' means 'monthly', and in many women the cycle does last around a month. But it can be shorter than this or longer than this. Some women have very regular and predictable cycles, while for others the cycles can last for different lengths of time from month to month.

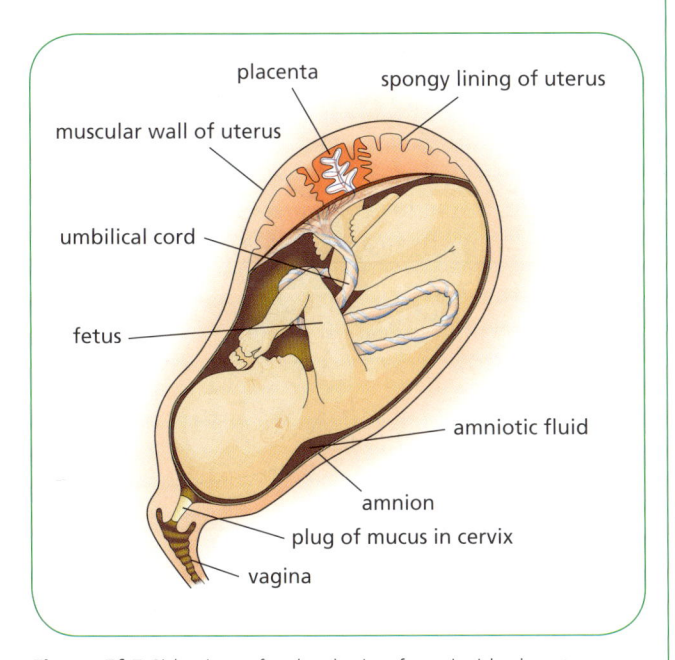

placenta
spongy lining of uterus
muscular wall of uterus
umbilical cord
fetus
amniotic fluid
amnion
plug of mucus in cervix
vagina

Figure 5f.5 Side view of a developing fetus inside the uterus.

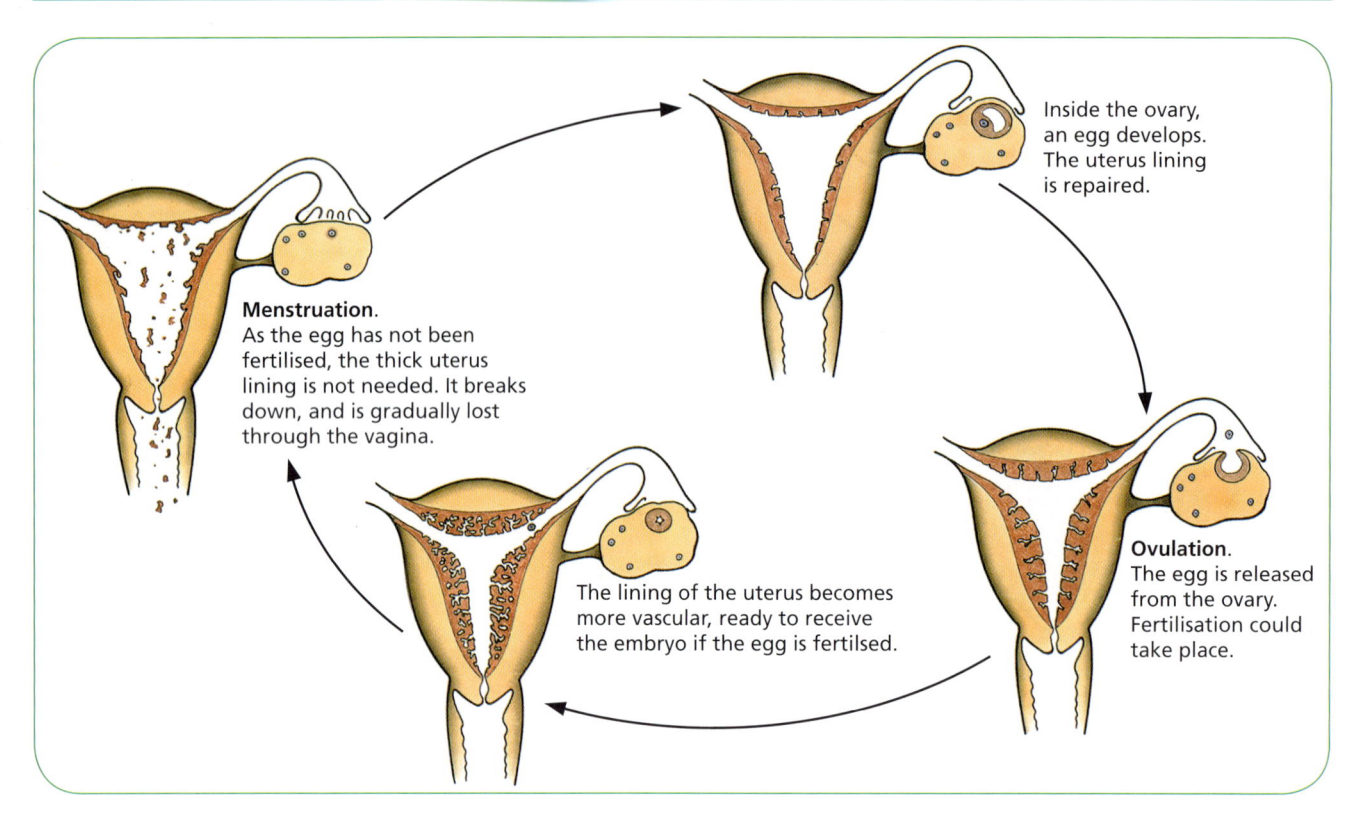

Figure 5f.6 The menstrual cycle.

Hormonal control of the menstrual cycle

A woman's menstrual cycle is controlled by hormones secreted by her ovaries and pituitary gland. Figure 5f.7 shows how the amounts of these hormones change during a 28-day cycle.

During menstruation, the pituitary gland secretes two hormones called **FSH** (follicle stimulating hormone) and **LH** (luteinising hormone). These hormones stimulate the ovary to secrete other hormones – in particular, **oestrogen**.

As the oestrogen level increases, it has a negative feedback effect on the pituitary gland, making it reduce the secretion of FSH and LH. The oestrogen has another effect, causing the lining of the uterus to thicken and get ready for the possible arrival of an embryo.

When the oestrogen level gets really high, the pituitary gland reacts by suddenly secreting a lot more LH. This surge of LH stimulates the ovary to release an egg into the oviduct. This usually happens in the middle of the cycle, round about day 14.

The part of the ovary that the egg came from now secretes **progesterone**. Progesterone adds to the effect of oestrogen on the uterus lining, maintaining it in its thick, spongy state.

If the egg is not fertilised, then the secretion of progesterone and oestrogen gradually slows down, so their levels in the blood fall. The lack of progesterone and oestrogen means that the lining of the uterus is no longer maintained, so it breaks down and is lost.

Now that progesterone and oestrogen are only

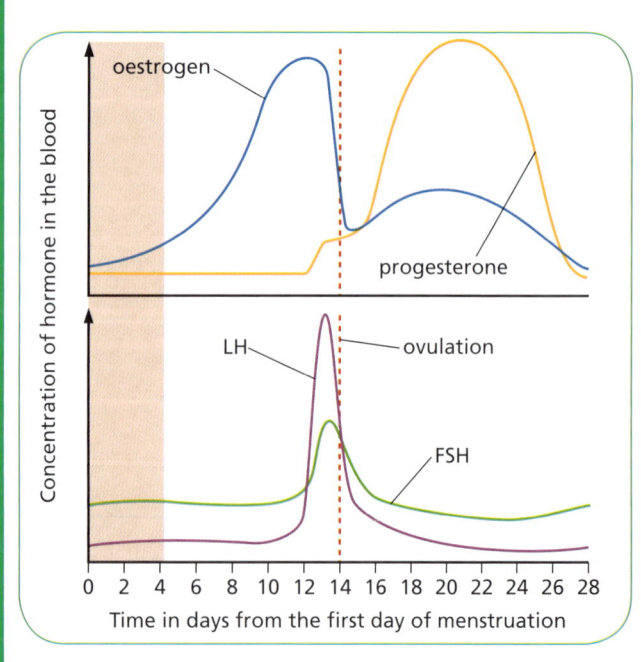

Figure 5f.7 Changes in hormone levels during the menstrual cycle.

present in low amounts, the pituitary gland is no longer inhibited, and it starts to secrete FSH and LH again. The cycle has gone full circle.

SAQ

5 Contraceptive pills contain progesterone and oestrogen. Using the information about the interactions between progesterone, oestrogen, LH and FSH, suggest how the pill works.

Fertility treatment

Infertility

Not every couple who would like to have children is able to do so. In Britain, a couple is said to be infertile if they have been trying for a baby for 12 months and haven't been successful.

There are many different causes of infertility. About 50% of them are caused by a problem in the woman's reproductive system, 35% are related to the man and the remaining 15% don't have any identifiable cause. Some of these couples can be helped by fertility treatment.

If an infertile couple ask a doctor for help, then tests will be done to try to find out the cause of the problem. Until this is known, it isn't really worth trying out any treatments. Fertility treatment is expensive. Hospitals cannot usually afford to spend as much on fertility treatments as they would like, so not every couple is able to get all the help they would like to have.

Artificial insemination

Sometimes, the problem might be that the man's sperm are not very active. They don't seem to be able to swim through the cervix and find an egg. If the woman is ovulating normally, then it is worth trying collecting some of the man's sperm and inserting them closer to the oviducts, actually inside the uterus. This is called **artificial insemination**. This treatment is also used if the reason for failure is that the man's sperm ducts are blocked.

If the man's sperm are unable to fertilise an egg, then the couple may decide to use sperm from a donor.

Using FSH

Another common problem can be that the woman is not ovulating. This is the cause of around 20% of all infertility cases.

One way of trying to get her ovaries to produce and release eggs is to give her FSH, or a mixture of FSH and LH. This helps to make ovulation happen. Sexual intercourse with her partner then stands a good chance of resulting in fertilisation.

It can be tricky getting the amount of FSH just right. Too little, and the woman will not ovulate. Too much, and she may produce several eggs all at once. If two or three of them are fertilised, then the couple may find they have twins or triplets. They may be delighted by this, but, in general, doctors would prefer just one fetus to develop in the uterus. Multiple pregnancies (where there are twins, triplets or more) are riskier than a single pregnancy, both in terms of the mother's health and the health of her babies.

'In vitro' fertilisation

'In vitro' means 'in glass'. *In vitro* fertilisation, usually known as **IVF**, involves fertilising an egg outside a woman's body, in a laboratory. This is how 'test tube babies' are made. The fertilisation, though, is done in a Petri dish, not a test tube.

IVF is more expensive, and less likely to be successful, than either artificial insemination or FSH treatment. It may be used when these have failed, or when the infertility is caused by blockage in the woman's oviducts, preventing her eggs from meeting a sperm.

The treatment starts by giving the woman FSH or other hormones to induce ovulation (Figure 5f.8, page 46). This is because several eggs are needed at once, to give the best chance of success. The eggs are usually collected using a thin, flexible tube that is inserted through the vagina and cervix and into the oviducts. The doctor often uses an ultrasound scan to help to guide the tube into the right place.

The eggs are placed in fluid in a Petri dish, and sperm from her partner are added. They are left in a warm liquid, with plenty of oxygen available, for about 3 days, and then inspected to see if any of the eggs have been fertilised. If they have, they may have divided to form a little ball of cells – an

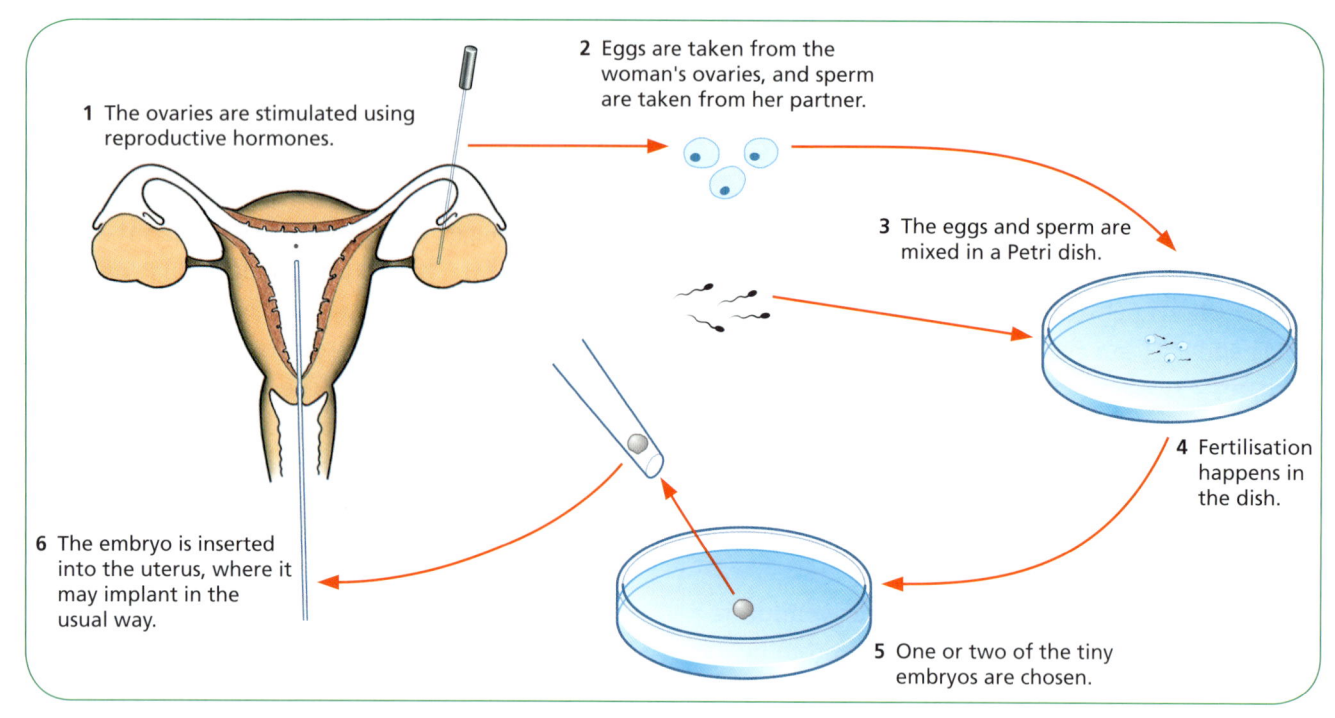

Figure 5f.8 The IVF treatment cycle.

1 The ovaries are stimulated using reproductive hormones.
2 Eggs are taken from the woman's ovaries, and sperm are taken from her partner.
3 The eggs and sperm are mixed in a Petri dish.
4 Fertilisation happens in the dish.
5 One or two of the tiny embryos are chosen.
6 The embryo is inserted into the uterus, where it may implant in the usual way.

embryo. One or two of these will be selected, and inserted into the woman's uterus. With a little luck, at least one of them will implant, and then develop in the usual way.

If IVF fails, it may be because the sperm cells are not able to get into an egg. The doctors might then try actually injecting a sperm nucleus into an egg, while they are in the Petri dish. This is called **ICSI**, which stands for intra-cytoplasmic sperm injection (Figure 5f.9).

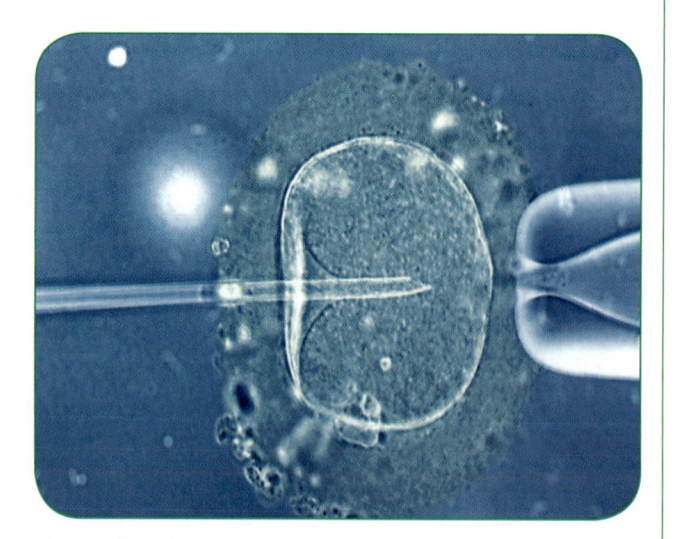

Figure 5f.9 A human egg being injected with a micro-needle containing a single sperm. This type of IVF is known as intra-cytoplasmic sperm injection (ICSI). The injected sperm fertilises the egg, which develops into an embryo and is then implanted into the woman's uterus.

Egg donation

In some cases of infertility, the woman is not able to produce eggs, even with hormonal treatment. For example, she might have had her ovaries removed because of cancer, or they might simply not work properly. In this case, it will never be possible for her to have a child from her own eggs. But, so long as her uterus is healthy, she could carry a fetus and give birth to a baby.

In this case, the couple might decide to use an egg donated by someone else. Quite often, the egg is given by another member of the family – the woman's sister, perhaps. But some women are prepared to donate eggs even for someone they do not know. In Britain, women are not allowed to do this for payment, but they will be paid all their expenses.

Surrogacy

Egg donation is used when a woman can't produce eggs but can carry a child. But sometimes the opposite is true. The woman can produce eggs, but she cannot carry a child, perhaps because there is a problem with her uterus.

In that case, she may be able to find another woman to carry her child for her. She will go through an IVF cycle, and then one or two of the

Donating an egg

Why would a woman decide to donate her eggs to someone that she does not even know? Here is what one young woman said about it.

'I'm young, fit and healthy, and I just felt that this was something I could easily do that would make a huge difference to someone else. I'd heard that couples were finding it really difficult to find an egg that they could use to start a baby. I'd seen a notice up in the health centre, so I thought I would give it a go.

They give you a really thorough health check. They tell you all about what will happen, and give you plenty of chances to back out if you change your mind. But I stuck with it. I really wanted to help someone out – I could imagine someone feeling desperate to have a child and thought how I could make her dream a reality.

They give you hormones to make you ovulate, which wasn't much fun. They made me feel pretty ill for a couple of days. They do the procedure for collecting the eggs in hospital. It only takes a short while. They give you a sedative (which felt really nice – I felt sleepy and floating and not worried about anything at all) and then get the eggs using a tube that goes into the vagina. There's a really, really fine needle on the end that they use to suck the eggs out of your ovaries.

It only took about 15 minutes. Afterwards, I was sleepy and groggy for a bit, as the sedative wore off. Once I'd got myself back together, the doctor talked me through the side-effects I might get. There's something called OHSS, which is something to do with overstimulation by the hormones, which can make you pretty ill. But I didn't get that. By the next day or so, I was absolutely back to normal.

They don't tell you who has your eggs. But they do keep a record. One day, a child might want to know who her biological mother is. It's a bit odd. I'm not sure how I'd feel if a person I didn't know came up to me and said they'd come from my egg.'

embryos will be inserted into the other woman's uterus. This woman will act as a **surrogate** mother. Once the child is born, she will give it immediately to the woman who provided the egg – the child's biological mother.

Why would anyone want to be a surrogate mother? As for egg donation, in Britain you cannot be paid for doing this. Sometimes, a close relative might act as a surrogate mother – for example, a sister or even mother. It's a really difficult thing to do. Most women could not contemplate having a baby developing in their uterus for nine months, and then giving it away when it is born.

Ovary transplants

Recently, several women have been able to become pregnant after having an ovary transplant. This might be done after a woman's ovaries have stopped working for some reason, or after they have been removed, perhaps because of cancer. It looks as though even just transplanting some ovary tissue can work – there's no need to transplant a whole ovary.

Like all transplants, a good match has to be found between donor (the woman who gives the ovary tissue) and recipient (the woman who receives it). One of the first successful transplants took place between twin sisters. As they were genetically identical, there was no risk of rejection. One of the sisters had three children, while the other was unable to have children because her ovaries were not working. A small amount of tissue was taken from the fertile twin, and attached to her sister's ovaries. Soon afterwards, she began to produce healthy eggs, and quickly became pregnant. Her baby girl, born in June 2005, was the first child to be born after an ovary tissue transplant.

SAQ

6 Explain why transplants between identical twins are very unlikely to be rejected.

7 The sisters did not know if the baby girl had developed from an egg from the donor or from the recipient. Explain why it would be impossible for them to find out.

Issues relating to fertility treatment

Our increasing ability to help women to become pregnant creates a lot of questions which we have to try to answer. Some of them are to do with cost, while others are more to do with ethics and morals. Here are just some of them. There are no correct answers. You might like to debate some of these with your friends, and come to your own decision.

- Fertility treatment is expensive – wouldn't it be better for the National Health Service to spend money on really ill people, like cancer patients?
- Does an infertile couple have a right to have a child, or is their infertility just something they have to accept and live with?
- The older a woman, the less likely it is that she will become pregnant after fertility treatment. So should there be an upper age limit on who can have treatment? Recently, a 66-year-old woman gave birth to a baby after fertility treatment. Might there be problems for a child born to a mother in her late sixties?
- Should fertility treatment be given to a woman who does not have a partner with whom she intends to spend her life?
- During IVF, should a couple have a right to choose an embryo that is of a particular sex, or that contains or does not contain particular genes? For example, if there is a risk of their child inheriting cystic fibrosis, should they be able to choose an embryo that does not have the cystic fibrosis alleles? Should they be able to choose one that will have blonde hair and blue eyes?
- How many embryos should be inserted into the uterus after IVF? The more there are, the more likely it is that the treatment will succeed. But this also increases the chances of multiple pregnancy, with all its attendant risks.
- Should women who want to donate eggs, or to be surrogate mothers, be paid for it?
- Should a child born to a couple who are not his biological parents have the right to find out who his biological parents are? The law in Britain says that he does have this right. Might this mean that fewer men are likely to donate sperm, or fewer women to donate eggs?
- During IVF, many more embryos are usually produced than can be used. What should happen to them? Can they be discarded, or is that destroying life? They can survive for many years frozen at very low temperatures – should they be frozen, ready to be used later if need be? Could they be used for research – for example, for the extraction of stem cells?

In the UK, these questions and others are debated by the Human Fertilisation and Embryology Authority, HFEA for short, and they make rulings about what can and cannot be allowed, particularly in relation to IVF treatment. After you've thought about what answers you would give to the questions, you could look at the HFEA website, at www.hfea.gov.uk, to see if they have made any recommendations on a particular question, and if they match yours.

Fetal investigations

Most pregnant women in Britain have an **ultrasound scan** during their pregnancy. The scan can show the sex of the fetus, whether there are any obvious abnormalities, and whether the woman is carrying twins. Ultrasounds are cheap and very safe.

If there is reason to suspect that there may be problems that won't show up on an ultrasound scan, then doctors may suggest that the mother has other tests done. These include amniocentesis and chorionic villus sampling.

Amniocentesis

Amniocentesis can be carried out 15 to 16 weeks into the pregnancy. A fine needle is inserted through the mother's abdomen wall and into the amniotic fluid that surrounds her fetus. A little of the fluid is withdrawn, and taken away for testing.

Who might have amniocentesis? It is usually offered to any pregnant woman over the age of 35. As a woman gets older, the chance that her baby might have Down's syndrome increases. This happens when an egg with an extra chromosome 21 is fertilised by a normal sperm, so the zygote has three copies of chromosome 21 instead of two. (You can read about Down's syndrome in *Gateway Science*, page 56.) The amniotic fluid will contain some loose cells that have come from the fetus, and these can be checked to see if they contain an extra chromosome.

Many different genetic abnormalities can be checked out in a similar way. For example, if two parents know that they are both carriers for cystic fibrosis, the fetal cells can be screened to see if the fetus has two copies of the cystic fibrosis allele, which would mean that he will have the disease.

Chorionic villus sampling

This test is carried out earlier than amniocentesis, at between 8 and 10 weeks. It is a slightly more risky procedure, increasing the chance of miscarriage by about 1%. Like amniocentesis, it involves inserting a fine needle through the mother's abdomen wall, but this time a sample is taken from the tiny folds in the placenta. The cells can then be tested to check for genetic disorders.

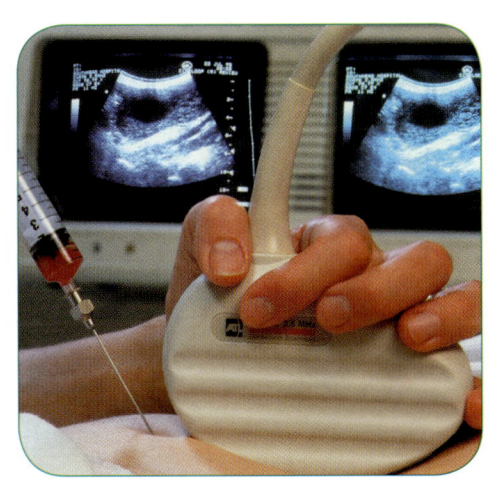

Figure 5f.10 This woman is having a chorionic villus sampling procedure. You can see the fine needle that will be used to take a sample of the villi from her placenta. The doctor is using an ultrasound scan to help guide the needle into exactly the right position. Amniocentesis is done in a similar way.

Ethical issues related to fetal screening

What happens if amniocentesis or chorionic villus sampling shows that the fetus does have an abnormality? Say, for example, that it is found to have Down's syndrome. Should the mother have an abortion, or should she keep her baby?

The detection of fetal abnormalities always raises difficult issues. Who can decide if the life of a child with a disability will be worth living or not? Most disabled people say they are glad to be alive. Yet some women who find they are carrying a fetus with an abnormality feel that they want an abortion. It is an enormously difficult decision to make.

Neither chorionic villus sampling nor amniocentesis gives quick results. By the time a woman gets the results of an amniocentesis test, she will have been pregnant for almost five months. It is really difficult to decide to abort a child after carrying it for that long.

So what is the point of the mother having these tests done, if she will keep the child anyway? It may be a good idea to find out any problems early on, so that the parents are mentally prepared for the birth of their child, knowing what the difficulties may be and how they are going to deal with them.

If fertilisation happens through IVF, then of course tests can be done on the tiny embryo, to check if it has an extra chromosome 21, or the alleles for cystic fibrosis. If so, then that embryo could be discarded and another one used instead. This avoids the difficult issue of deciding whether to have an abortion or not. But is this ethically acceptable? If it is OK to choose an embryo that does *not* have a feature like cystic fibrosis, is that any different from choosing one that will have blue eyes, for example? Or from choosing one that will be a boy, rather than a girl?

There are no easy answers. There's no doubt that, as we become more able to control what happens during reproduction and to find out information about the unborn child, we will come to use our knowledge more and more widely. Perhaps one day things that seem ethically unacceptable now will come to feel normal and right. It's important that a whole range of people debate these issues and help to make decisions.

50 B5f Life goes on

Summary

You should be able to:

- give a definition of fertilisation
- describe the structure of the male and female reproductive organs, and state the functions of each
- describe what happens during the menstrual cycle
- [H] explain how LH, FSH, oestrogen and progesterone control the menstrual cycle
- describe the following treatments for infertility: artificial insemination, use of FSH, *in vitro* fertilisation, egg donation, surrogacy and ovary transplants
- [H] discuss the issues and arguments relating to fertility treatments
- describe amniocentesis and chorionic villus sampling, and explain how they can be used to identify conditions such as Down's syndrome in the fetus
- [H] discuss the ethical issues raised by fetal screening

Questions

1. Give the name of the organ that matches each description.

    a. where eggs are made _____

    b. where sperm are made _____

    c. where fertilisation happens _____

    d. where the fetus develops _____

    e. where exchange of substances between mother and fetus takes place _____

2. Which of these fertility treatments might help a couple who cannot have a child because the woman's oviducts are blocked? Explain your answer, including an explanation of why the other methods would not work.
    - artificial insemination
    - IVF
    - ovary transplant
    - injections of FSH

3. The average success rate for IVF treatment using fresh eggs in the UK is:
    - 28.2% for women under 35
    - 18.3% for women aged 38–39
    - 23.6% for women aged 35–37
    - 10.6% for women aged 40–42.

    The average success rate for donor insemination (artificial insemination) treatment in the UK is:
    - 14.1% for women under 35
    - 4.9% for women aged 40–42.
    - 8.3% for women aged 35–39

    a. Present these data as a chart.

    [H] b. With reference to the data, discuss whether either of these two fertility treatments should be offered to women over the age of 40.

5g New for old

We are not built to last forever. Over time, parts of the body gradually wear out. This happens to all of us as we age.

For some people, however, one body part may deteriorate, or be damaged, relatively early on in their lives, when the rest of their body is still quite healthy. If that one part could be replaced, then they could carry on living an active life for many more years.

This is now possible for several different organs. Sometimes, a machine can take over the function of the organ. In other cases, a transplant of an organ from someone else provides the best hope of success.

Mechanical replacements

There are now quite a few body organs that can be replaced by machines. Some of these are shown in Figure 5g.1.

Mechanical replacements inside the body

We can use metal and plastic to replace joints. People with cataracts can be given plastic eye lenses. We can even give someone a plastic and metal heart, to pump their blood around the body when their own heart has failed. Usually, this is done just to keep someone going until a real heart can be found, so they can have a heart transplant. The 'old' heart may be left in place, and an artificial pump is attached to it to take on some of the workload, rather than completely replace it. This can sometimes even allow the person's own heart to partly repair itself. (Figure 5c.4 on page 20 shows a 'heart assist device' of this kind.)

These mechanical replacements need to be made of materials that won't be too easily damaged by body fluids, and that won't be attacked by the person's immune system. They are called **bio-inert** materials. Many plastics are bio-inert, and so are some metals such as gold and titanium.

There is an extra problem if these organs are inserted into a growing child. The mechanical organ does not grow. Doctors have a choice of inserting an organ that is too large for the child, which they will 'grow into', or putting in one that is the right size now but that will have to be replaced in a couple of years' time.

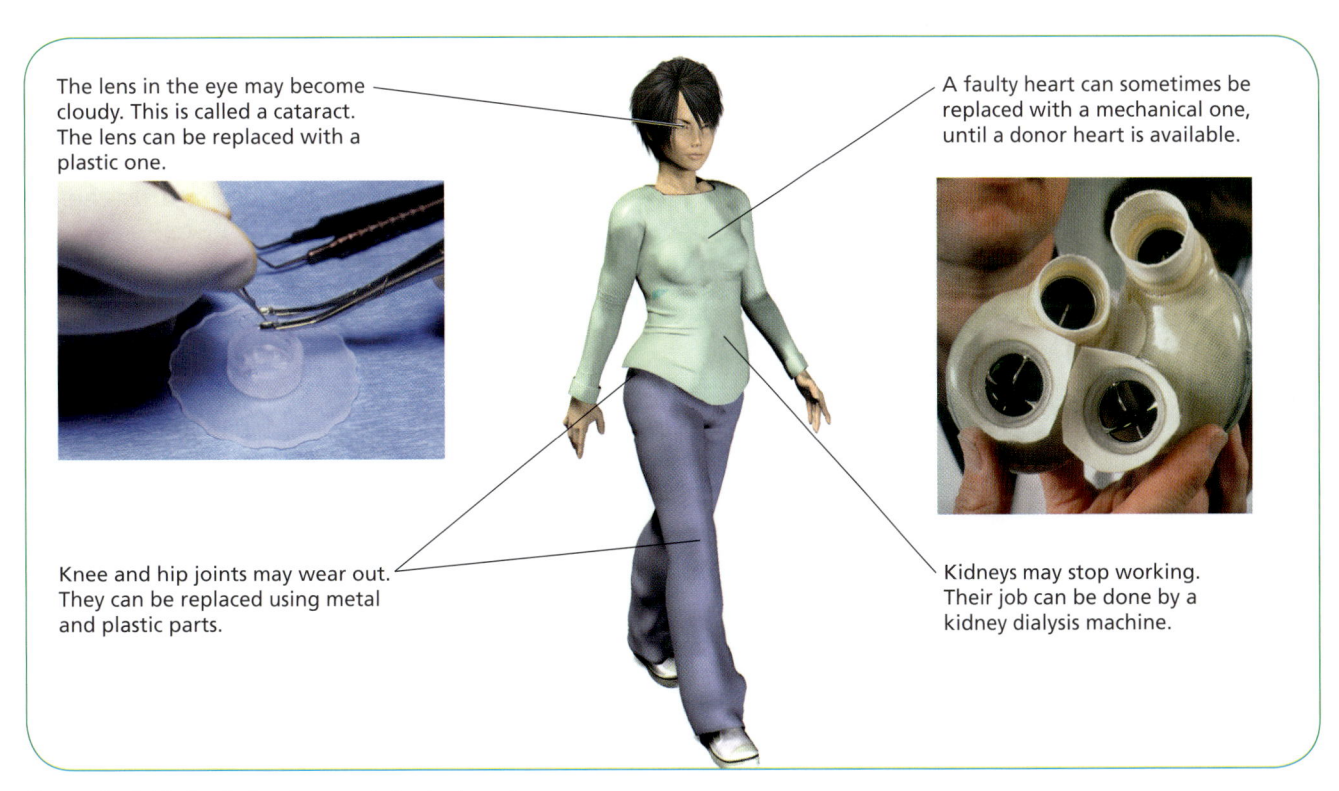

The lens in the eye may become cloudy. This is called a cataract. The lens can be replaced with a plastic one.

A faulty heart can sometimes be replaced with a mechanical one, until a donor heart is available.

Knee and hip joints may wear out. They can be replaced using metal and plastic parts.

Kidneys may stop working. Their job can be done by a kidney dialysis machine.

Figure 5g.1 Mechanical replacements for body parts.

Yet another problem is that some replacement organs need a power supply. They can't run on the body's fuels. A mechanical heart, for example, needs a constant supply of electricity. The power supply is attached to the outside of the person's body, so that it can easily be replaced when it runs out.

SAQ

1 Sometimes, just *parts* of a heart can be replaced, rather than the entire organ. Name *two* parts of the heart that can be replaced with mechanical parts.

Mechanical replacements outside the body

So far, no-one has designed an artificial kidney that can be put inside someone's body. Instead, people whose kidneys have stopped working often use a dialysis machine. (Have a look at pages 36–37 to remind yourself about dialysis.)

During heart surgery, the heart is usually stopped – it is difficult for a surgeon to perform a delicate operation on something that is constantly moving. The person's blood, however, still needs to be pushed around the body, and be able to pick up oxygen and deliver it to the body cells. The blood is passed through a **heart–lung machine**, which carries out these functions all the time the operation is taking place (Figure 5g.2).

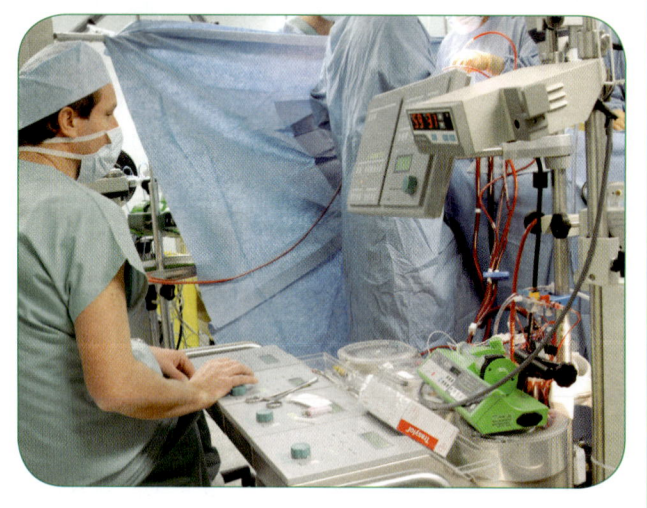

Figure 5g.2 During a heart operation, a doctor constantly monitors the heart–lung machine, which adds oxygen to the blood and removes carbon dioxide, as well as keeping the blood at the right temperature.

Another 'outside the body' machine is the **iron lung**. This machine helps a person who cannot breathe for themselves, usually because they are paralysed from the neck down. Their whole body, except the head, is enclosed in the machine, with airtight seals. A pump regularly increases and decreases the pressure inside the machine. When the pressure inside is raised, air is pushed out of the person's lungs. When the pressure is lowered, air flows into the lungs. Iron lungs were first used in the 1920s, to help people affected by poliomyelitis. This is an infectious disease, which causes partial paralysis, and it was not uncommon then. There are still some iron lungs in use today.

Organ transplants

We have already seen that a badly damaged heart (*Gateway Additional Science*, page 23) can sometimes be replaced with a heart from a donor, and that ovary transplants are sometimes done (page 47). Figure 5g.3 shows several more organs that can be transplanted – that is, taken from a donor and put into a recipient.

An organ will only be used if the donor:
- does not have a transmissible disease, especially HIV or hepatitis
- has the same blood group as the recipient
- is approximately the same age and body weight as the recipient
- has agreed that his or her organs can be used, or the donor's family agrees to this after the donor's death.

Organs from dead donors

Most organs for transplant are taken from someone who has recently died, following a road traffic accident, for example. The organs need to be removed quickly from the body, so that they do not have time to deteriorate before they are placed in the recipient's body. Sometimes this can cause difficulties in deciding exactly when a person is 'dead', especially if they are being kept alive on a life support machine in hospital. Is it when their heart stops beating? Or is it when their brain stops showing any signs of activity? Usually, death is determined by 'brain death' – when no signs of activity can be found in the brain – and

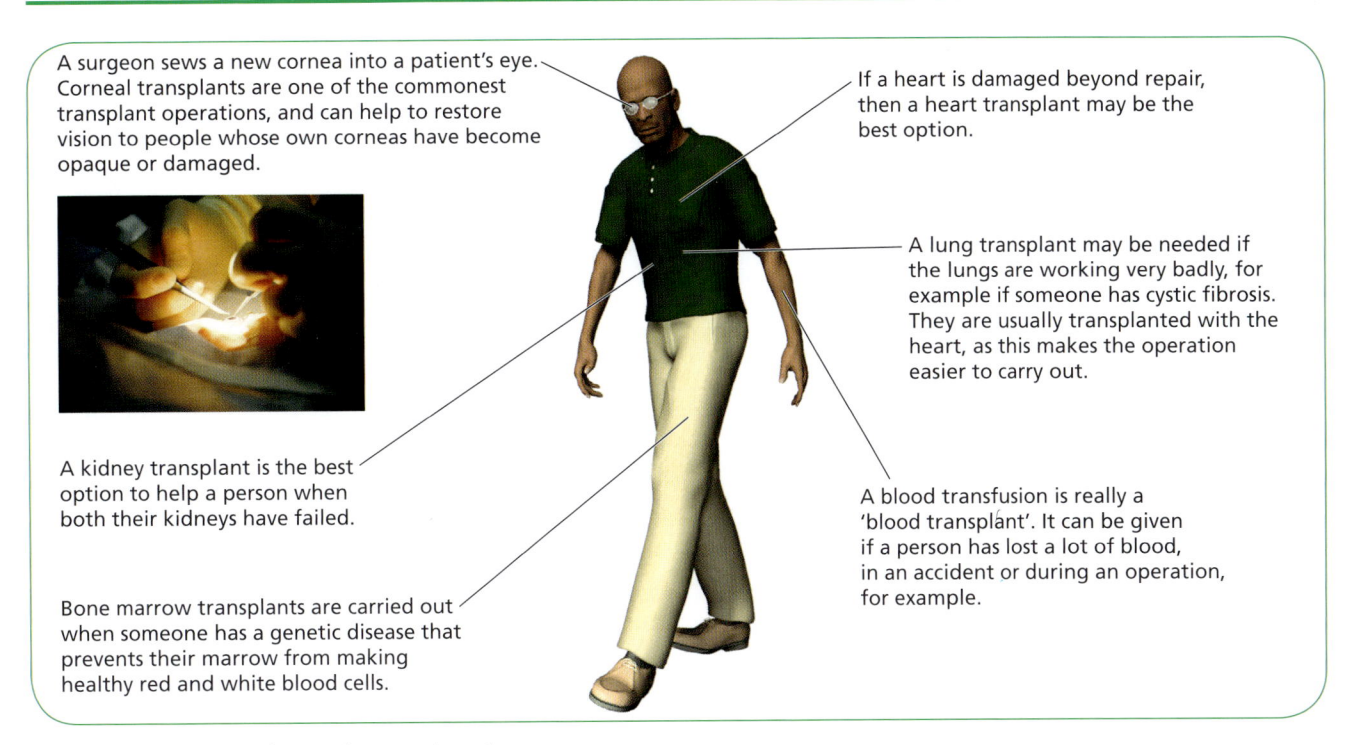

A surgeon sews a new cornea into a patient's eye. Corneal transplants are one of the commonest transplant operations, and can help to restore vision to people whose own corneas have become opaque or damaged.

If a heart is damaged beyond repair, then a heart transplant may be the best option.

A lung transplant may be needed if the lungs are working very badly, for example if someone has cystic fibrosis. They are usually transplanted with the heart, as this makes the operation easier to carry out.

A kidney transplant is the best option to help a person when both their kidneys have failed.

A blood transfusion is really a 'blood transplant'. It can be given if a person has lost a lot of blood, in an accident or during an operation, for example.

Bone marrow transplants are carried out when someone has a genetic disease that prevents their marrow from making healthy red and white blood cells.

Figure 5g.3 Body parts that can be transplanted.

most organs for transplants are taken from such patients. In some countries, including Canada and Australia, doctors are beginning to think about taking organs from patients whose hearts have stopped beating but who have not been declared brain dead. This will probably increase the number of organs available for donation, but is it right? The debate is still wide open.

Even if there is no doubt at all that an accident victim is dead, there can still be problems in using their organs for transplant. Doctors need to be sure that the person would have been willing to donate their organs. This is easy if they registered with the National Health Service Organ Donor Register, but most people don't even think about this while they are alive. Close relatives may be able to give permission for organs to be removed for transplant if they know that the dead person would have wanted this to be done, but this is a difficult thing to ask someone who has just lost a loved one. However, quite often the relatives say that knowing that an organ could help someone else to survive has helped them to deal with the death of their child or other close relative. All the same, there is a severe shortage of donor organs available, and many people needing a transplant have to wait years before a suitable organ can be provided for them.

Figure 5g.4 An organ to be used for transplant is packed in ice and transported in a box that keeps it cool.

Another issue that reduces the number of suitable organs for transplant is that the donor and the recipient need to be approximately the same size. For example, you could not transplant the heart of a 25-year-old man into a 3-year-old child.

Our cells have molecules in their membranes that are unique to each of us, called our **tissue type**. The main difficulty lies in finding a donor with a similar tissue type to the recipient. If cells with different molecules are put into the body, then the lymphocytes recognise them as foreign, and launch an attack. The 'invading' cells are killed.

This almost always happens if an organ from one person is put into the body of another. The transplanted organ is attacked. Its cells are destroyed. The organ is rejected by the recipient's body.

It is not usually possible to find a perfect tissue match, but doctors like to find one that is as close as possible. Even then, the recipient will need to be given **immunosuppressant** drugs. These reduce the activity of the immune system, giving the transplanted organ a better chance of avoiding rejection. These drugs will need to be taken for the rest of the patient's life.

SAQ

2 Suggest the possible side-effects that there could be from immunosuppressant drugs.

Organs from living donors

Sometimes, it is possible to take an organ from a living donor, so long as this still leaves the donor with their own working organ. A person can manage perfectly well with just one kidney, for example.

Living donors are usually close relatives of the recipient. A mother might donate a kidney to her daughter, or bone marrow might be taken from a brother to give to his sister. A transplant has a much better chance of success if the organ is taken from a living donor, for a number of reasons.

- If the living donor is a relative, their organ may be a very close tissue match for the recipient. If they are identical twins, then the match is perfect.
- The organ can be removed from the donor and given to the recipient in the same hospital, so there is no need to transport the organ any distance.
- The operation can be planned and carried out at a time when the recipient is still healthy enough to have a good chance of survival.

Organs for cash?

It is illegal in Britain for anyone to buy or sell an organ to be used in a transplant. Yet there is a huge shortage of organs for people who desperately need transplants. Why shouldn't people sell one of their kidneys?

In fact, several cases have been discovered when exactly this has happened. At least two British doctors have been found guilty of buying kidneys from living donors in other countries, and some patients have sourced their own transplant in this way. The donors are usually living in developing countries where the payment that they can get for one of their organs can provide them with more money than they could earn in a year. There is even some evidence that organs are being obtained from people without their knowledge or permission.

SAQ

3 Outline *two* reasons why there are risks in being a living donor.

People can have very strong opinions about whether or not this should be allowed. Some think that there is no problem with it. After all, someone with one kidney is just as healthy as someone with two, so if someone wants to sell one of their kidneys – and knows what the risks are – then there is no reason why they should not. Others think that it is ethically wrong. They feel strongly that giving an organ for transplant out of generosity is fine, but the donor should not earn any money for it.

Another controversial source of organs for transplants is from executed prisoners. Organs are certainly obtained in this way in some countries, and then sold to people who need transplants. In one country in the Far East, the official line of the government is that the prisoners voluntarily donate their organs, but it is difficult to be sure that this is true. Many people feel that these sales are an abuse of the prisoners' human rights.

As long as it remains illegal to buy or sell organs in the UK, and as long as there is a shortage of organs for transplants, a black market in body parts could thrive.

Face transplant

In 2005, 36-year-old Isabelle Dinoire was bitten very badly by her own dog, and lost her nose, chin and lips. She survived and was in no danger of dying, but her life without her face would have been almost intolerable. She was offered the chance of being the world's first face transplant patient, and she accepted.

There were very big risks, not just because this was a new and untried technique. The arteries, veins and nerves in what remained of the woman's face needed to be painstakingly joined to those in the transplant. The danger of rejection was high. The doctors could not predict how much facial expression the woman might have – it would depend very much on how well her nervous system was able to link up with the muscles in her 'new' face.

Many people find the idea of a face transplant absolutely horrifying. It seems almost as though you are transplanting one 'person' onto another person's body. But is it really any different from transplanting a kidney or a cornea?

By February 2006, Mme Dinoire felt brave enough to appear before the world's media. She was absolutely sure that she had made the right decision. She told reporters: *'I now have a face like everyone else. A door to the future is opening.'*

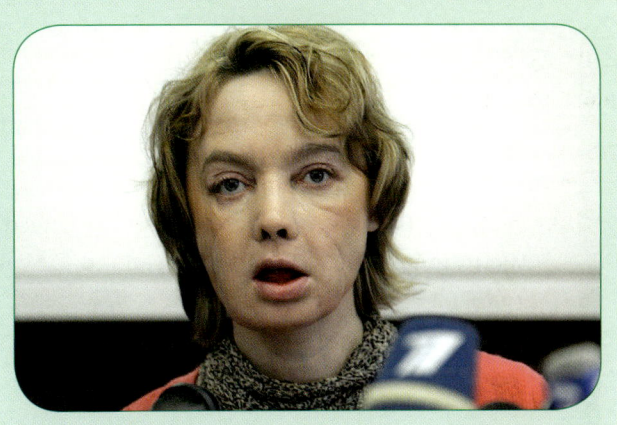

Figure 5g.5 Isabelle Dinoir, some months after the 15-hour operation that gave her a new face.

Summary

You should be able to:

- list some body parts that can be replaced by mechanical organs, both inside and outside the body
- describe the problems involved in using mechanical replacements
- list some body parts that can be transplanted from one person to another
- describe the criteria that must be met for a transplant from a dead or living donor to be successful
- explain why it is possible for living people to donate an organ
- explain why there is a shortage of organs for transplant
- **H** explain why transplant patients need to take immunosupressant drugs
- discuss the ethical issues relating to transplants

56 B5g New for old

Questions

1   a   List *four* organs that can be mechanically replaced.

    b   Name *one* of these mechanical replacements that is used outside the body.

    c   List *three* organs that can be transplanted from one person to another.

2   a   Outline the criteria that must be met for a dead person to be a suitable donor.

    b   Discuss the reasons why there are not enough donor organs to go around.

    c   It has been suggested that one way of increasing the supply of organs would be to assume that everyone is willing to donate organs after their death unless they have specifically said that they do not want to do this. Do you consider that this would work better than having a national donor register? Explain your reasons.

3   Figure 5g.6 shows the number of people needing a kidney transplant, the number of donors and the number who received a transplant, for each year between 1996 and 2005.

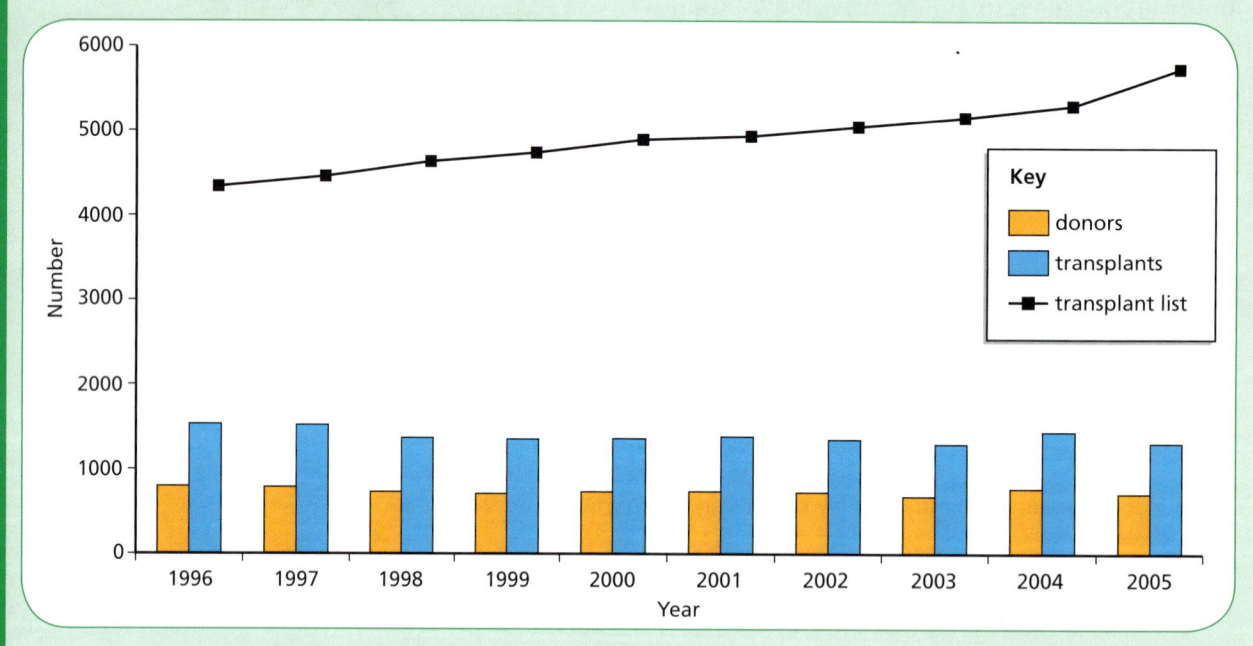

Figure 5g.6

a   Explain why the number of transplants each year is greater than the number of donors.

b   Compare the trend in the number of people needing a kidney transplant with the trend in the number of kidneys available.

c   Discuss how these trends might increase public pressure to allow the sale of organs for cash.

d   The success rate for a kidney transplant from a living donor is about 90%, while the success rate for one from a dead donor is about 80%. Discuss the reasons for this difference.

5h Size matters

What is growth?

As we have seen (*Gateway Additional Science*, page 32), growth can be defined as a permanent increase in size. We grow as our cells divide repeatedly, by mitosis, ending up with a body containing millions of cells. These have all been produced by the one original cell that began our life – the zygote. These cells are all genetically identical.

This happens in both animals and plants, but there are a few differences between them. In most animals, growth takes place only in early life, whereas most plants grow continuously. Also, in animals, most parts of the body can grow. In plants, however, growth can only take place at the tips of shoots and roots, or sideways in stems to form trunks and branches of trees (*Gateway Additional Science*, page 33). The growing regions of a plant are called **meristems** (Figure 5h.1).

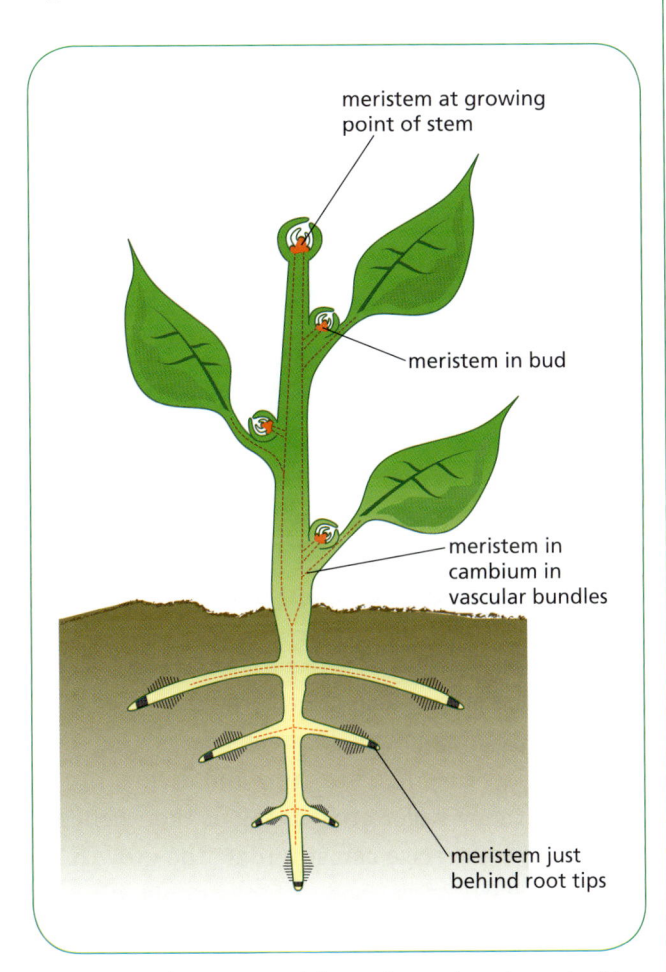

Figure 5h.1 The positions of the meristems (growing points) in a plant.

Human growth

Humans start to grow as soon as the egg is fertilised. While a fetus develops in the uterus, and for up to 20 years after birth, a person grows in height. Most of their growth in body mass also takes place during these years.

Measuring growth

Humans change a lot as they grow older. Although the changes are gradual, they can be divided into five main stages – infancy, childhood, puberty and adolescence, adulthood or maturity, and old age. These are described on page 34 in *Gateway Additional Science*.

Children are usually measured several times during the first few months after their birth, to make sure that they are growing normally. Charts have been drawn showing the average height, body mass and head circumference of children at each age, and the child's measurements can be compared with these (Figure 5h.2). You can remind yourself about this using pages 36–37 in *Gateway Additional Science*.

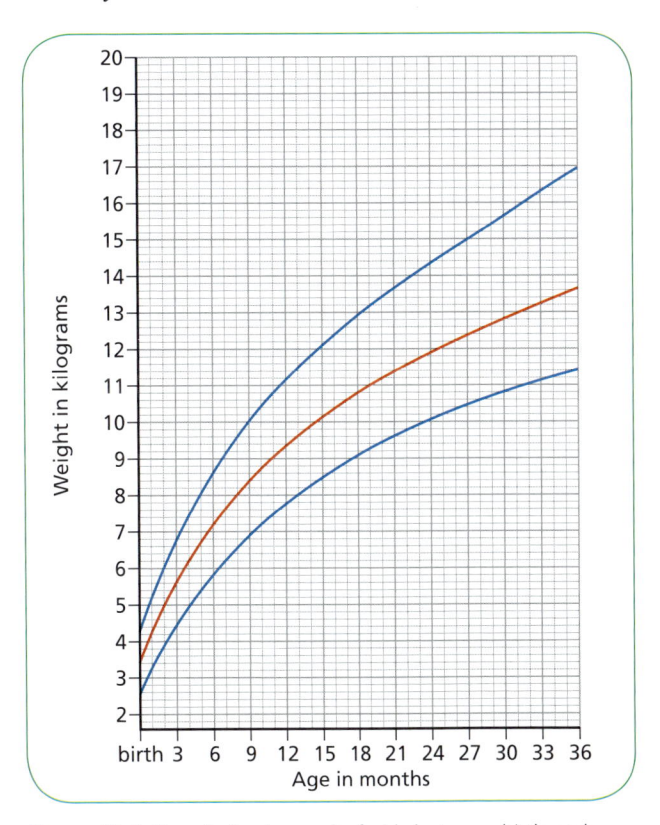

Figure 5h.2 Growth (body mass) of girls between birth and three years old.

SAQ

1 Figure 5h.2 shows a growth chart for girls between birth and three years old. The red line shows the average body mass for all girls. The two blue lines show the upper and lower weight limits between which 90% of children fit.

   a A two-year-old girl has a body mass of 16 kg. Should her parents be worried about this? Explain your answer.

   b During which time in the first three years after birth do girls grow most rapidly?

   c Calculate the average rate of growth of a girl, in kg per year, between birth and three years old.

Factors affecting growth

What determines how tall you will grow, or what your body mass will be when you are an adult? These are affected both by your **genes** and your **environment**.

There are many different genes that can affect a person's potential to grow tall. If you have tall parents, then you might have inherited a set of alleles that will enable you to grow tall. On the other hand, you might have inherited a different set, which could increase the chance that you will be shorter than either of your parents. It is because there are a lot of different genes involved, each of them with several alleles, that humans can be any height in between the smallest and tallest extremes.

Some of these genes work by affecting the **hormones** produced in your body. One of the most important hormones in determining a person's growth pattern is called **human growth hormone**, or **hGH**. (It is also sometimes just called growth hormone, or GH.)

Human growth hormone is produced by the pituitary gland, in the middle of the head. It is secreted throughout our lives, not only when we are growing. It stimulates growth in children, including the growth of the long bones in the arms and legs. In an adult, it helps to maintain the strength of bones and muscles.

If a child's pituitary gland does not secrete enough hGH, then they will remain small unless given treatment. Reduced production of hGH is usually the result of faulty genes. The child can be given regular doses of hGH through their childhood, which will help them to grow to a normal size.

At the other end of the scale, the pituitary gland may sometimes secrete too much hGH. This can produce an extremely tall person, a 'pituitary giant'. If it is detected early enough, the person's growth can be slowed down by destroying part of the pituitary gland. This can be done by irradiating it.

Figure 5h.3 The man on the left is a pituitary dwarf, resulting from being born with a thyroid gland that did not secrete enough hGH. The man on the right is a pituitary giant, who has grown much more than average because his pituitary gland is overactive.

Diet and exercise also affect growth. A diet that is low in proteins means that cells cannot get all the materials they need to grow and divide, so the person may never grow as tall as their genes would allow. Exercise can stimulate the growth of certain parts of the body – for example, if a person trains as a discus thrower, the muscles, bones and tendons in their upper arm may grow larger and stronger.

hGH cheats

Human growth hormone is used by some people to increase the growth of their muscles and bones in order to help them to do better in their sport. This is not allowed by the organisations that control sporting competition. Apart from being unfair, this abuse of hGH can cause long-term health problems. It can cause abnormal growth of some parts of the body, and imbalances in the secretion of other hormones.

It is very difficult to detect abuse of hGH, because this hormone is already present in the body. The hGH that people can buy is produced by genetically modified bacteria, which follow the 'instructions' on the human gene that has been inserted into them. So the hGH that is abused is identical with the hGH that is naturally present in the body. Until recently, the only effective way that an illegal user could be caught was if he or she was found with hGH in their possession. That did actually happen at the world swimming championships in Australia in 1998, when customs officials found 13 vials of hGH in the luggage of a swimmer.

Recently, Southampton University has developed a test for hGH that might make it possible to test athletes to find out if they are abusing hGH. Instead of testing for hGH itself, the test will measure the amounts of two hGH 'markers' in the body – substances whose concentration increases when excessive hGH is present. The test should be ready to be used at the London Olympics in 2012.

How long will you live?

Life expectancy in Britain has been steadily increasing. Life expectancy is the average age to which people live. Figure 5h.4 shows how long a baby born in a particular year could be expected to live, on average. In just 20 years, life expectancy has increased by four years. Can it go on increasing? Might a baby born in 2040 be expected to live to about 100 years old?

To answer that question, we have to know something about why life expectancy has been increasing. There seem to be quite a number of reasons, all of which have added together to produce this large effect. Here are some of them.

- Medicine has made it possible to keep many people healthy who otherwise would have died. For example, we can now cure most bacterial infections using antibiotics. Surgery can cure people with damaged hearts or other conditions that might have been fatal only 20 or 30 years ago.
- Childbirth is now much safer for both mother and baby than it was in the past. In the late 19th century, in some parts of the country, more than one in ten babies died before their first birthday. Now the rate is about five deaths per thousand.
- Working conditions are much better than they used to be. We now understand how some substances, such as asbestos and coal dust, can cause long-term health problems and early death. Workers are now protected from these substances, so industrial diseases such as asbestosis are becoming much less common.
- Housing conditions have improved. Almost everyone in Britain now has electricity and safe drinking water supplied to their homes. There are sewage pipes to take waste away to treatment works, so there is no danger

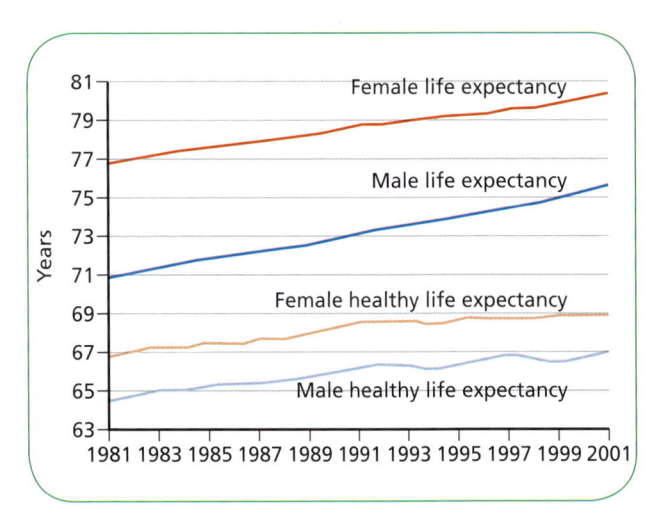

Figure 5h.4 Life expectancy at birth, in Britain, between 1981 and 2001.

of diseases being spread by water supplies becoming contaminated.

- We have a better diet than in the past. A wide range of foods is now available to everyone, whereas in the 18th century many people could not afford to buy enough to eat, let alone have a balanced diet.

It is possible that life expectancy might go on and on increasing, but there will be an upper limit on how long a human body can last. Over time, cells lose their ability to divide, so organs gradually wear out and cannot be repaired by new growth. More and more damage to body organs builds up as we age, and cannot be repaired.

SAQ

2   Figure 5h.5 shows life expectancy in various countries.

   a   In which part of the world is life expectancy lowest?

   b   Suggest reasons for this.

Problems of increased longevity

You might think it will be good to have people living longer and longer. Indeed, it may be – but there is also a downside.

Even though people are living longer, they may not always be really healthy during the later stages of their lives. When we plot graphs for *healthy* life expectancy, we get lines between 67 and 69 years for women and 65 and 67 years for men (Figure 5h.4). This means that we can expect to spend the last eight to ten years of our lives in poor health.

Another problem is providing support for all these extra old people. Older people are more likely to become ill and need care, perhaps in hospital or at home. They are not working, so their income will usually come from their pensions – paid either by the government, or by a company with whom they have been saving during their working years. At the moment, most people can retire at 65 years old, but in future this age will be raised so that there is enough money to pay pensions to all the people who are no longer working.

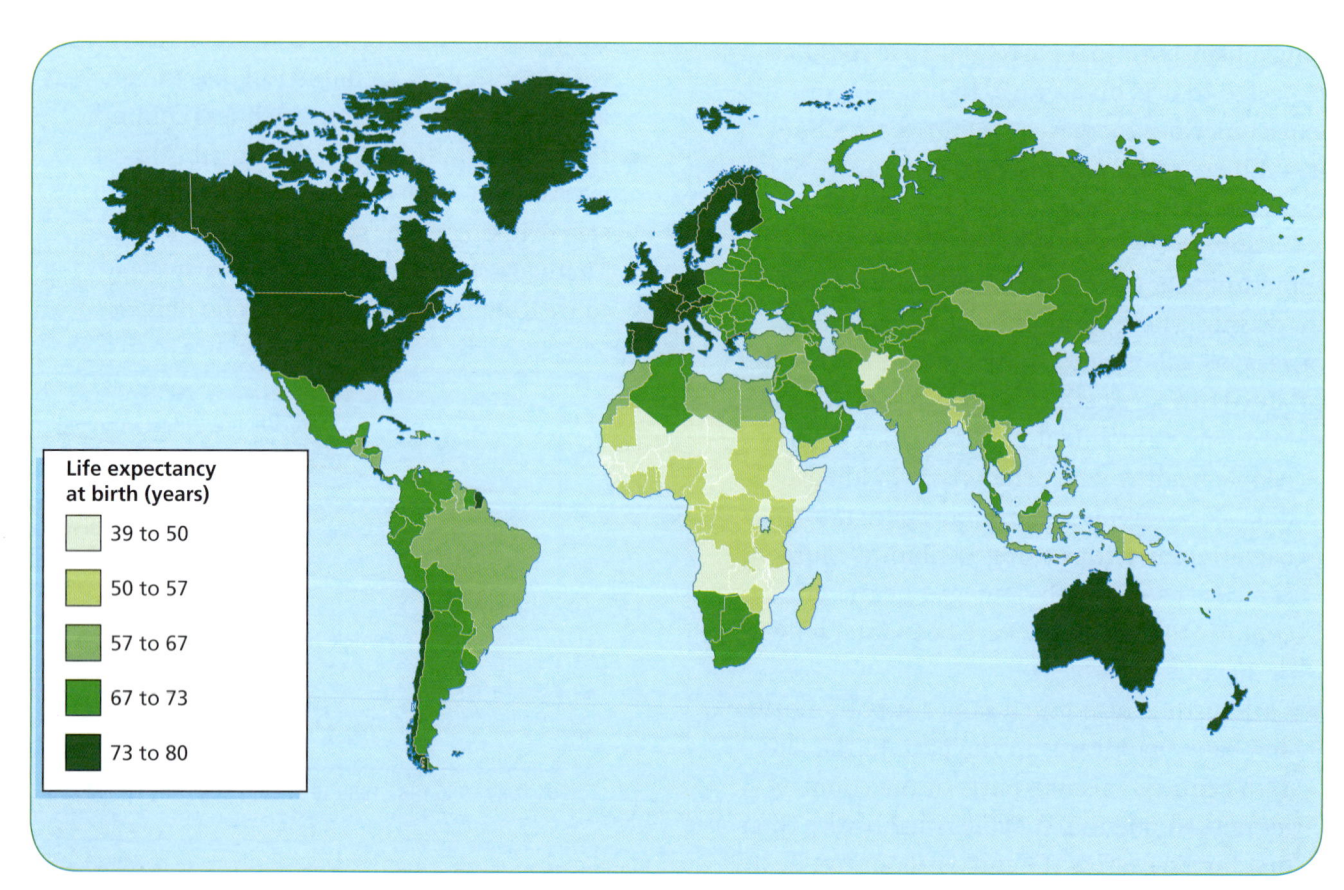

Figure 5h.5 Life expectancy in different countries around the world.

Summary

You should be able to:

- describe the differences between growth in animals and in plants
- state that growth takes place by mitosis, which produces identical cells
- describe the five stages in a human's life
- describe the factors that determine how much a person grows
- [H] explain how the pituitary gland affects growth
- describe how babies' growth is measured and compared to average growth charts
- interpret data on human growth
- suggest possible causes for the increase in life expectancy
- [H] discuss possible problems that the increase in life expectancy may cause

Questions

1. a What is a meristem?

    b Describe *two* places in a plant where you would find a meristem.

    c Animals do not have meristems. Describe one other way in which the growth of an animal differs from the growth of a plant.

2. If a child is growing more slowly than normal, tests will be done to make sure that there is enough growth hormone, hGH, in his blood.

    a Explain how a doctor would use growth charts to decide whether a child was growing normally.

    [H] b Name the gland that secretes hGH.

    c Describe the role of hGH in the body.

    d Human growth hormone used to be obtained from the pituitary glands of people who had died. Now it is made by genetically modified bacteria. Suggest *two* advantages of this.

[H] 3. a Using the information in Figure 5h.5, summarise the differences in life expectancy for people living in Africa, Asia, Europe, North America and South America.

    b Explain why life expectancy in Britain is increasing. Would you expect to see similar increases in other parts of the world? Explain your answer.

    c Discuss the possible personal and national problems that could arise as a result of our increasing life expectancy.

B6a Understanding bacteria

Bacteria (singular: bacterium) are the most abundant living things on Earth. Too small to see, they are all around us – on our bodies and inside them, in the soil and water, in the food that we eat and in the air that we breathe. Luckily, most of these bacteria are completely harmless, and some of them are very useful to us.

The structure of a bacterium

The structure of a common bacterium called *Escherichia coli* is shown in Figure 6a.1. Bacteria are single-celled organisms. Their cells are much smaller than ours, usually no more than a few microns (thousandths of a millimetre) long. They differ from animal cells and plant cells in not having a nucleus. Instead, their single, circular molecule of DNA simply floats freely in the cytoplasm. Just as in our own cells, the DNA controls the activities of the cell, including its replication.

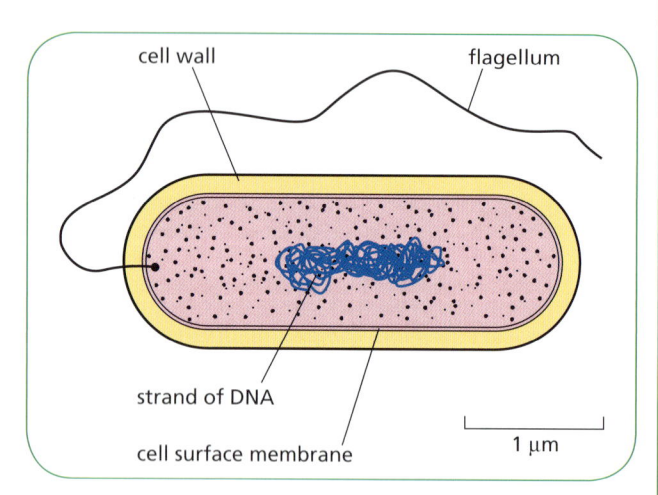

Figure 6a.1 The structure of a common bacterium, *E. coli*.

Bacteria are always surrounded by a cell wall. The cell wall is not exactly the same in all bacteria, but it *never* contains cellulose. The cell wall maintains the shape of the cell. As in plant cells, the cell wall stops the cell from bursting if it takes up a lot of water from its surroundings.

Bacterial cells do not contain mitochondria. Nor do they contain vacuoles like the ones that are found in plant cells. Some of them are able to photosynthesise, but they do not have chloroplasts.

E. coli has a long, whip-like **flagellum**, which it uses to swim. Not all bacteria have a flagellum. Many bacteria are not able to move around under their own propulsion system, but rely instead on movements of water or air to carry them from one place to another.

SAQ

1. Suggest how bacteria can photosynthesise, if they do not have chloroplasts.

H 2. The symbol 'μm' stands for 'micrometre'. One micrometre is one millionth of a metre.

   Measure the bacterium in Figure 6a.1, and then calculate the magnification, using the formula:

   $$\text{magnification} = \frac{\text{length in diagram}}{\text{actual length}}$$

Classifying bacteria

Bacteria can be classified according to the relationships they have with each other, just as we classify other living things (*Gateway Science*, page 75). However, it is sometimes useful to classify them according to their shape. This can be helpful if you are trying to identify bacteria that you can see using a microscope. Once you know the shape of the bacterium, this reduces the possibilities to a relatively small number, making it easier to work towards a final identification. Figure 6a.2 shows the main structural groups into which bacteria can be classified.

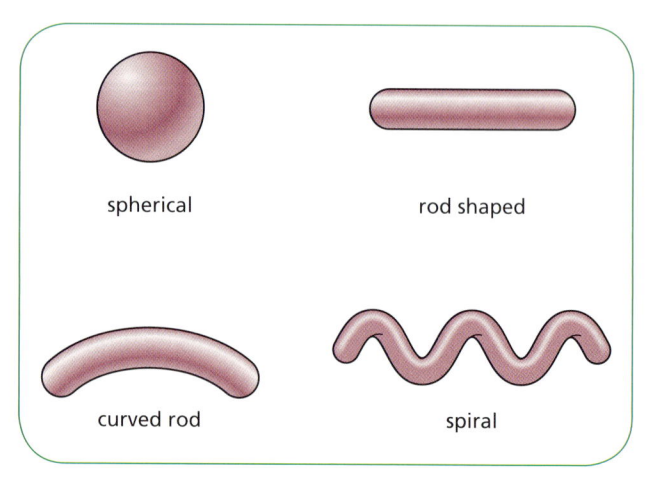

Figure 6a.2 Bacteria come in four main shapes.

SAQ

3 State the shape group that *E. coli* belongs to.

4 Figure 6a.3 shows some *Staphylococcus* bacteria (yellow) in the lining of the trachea.

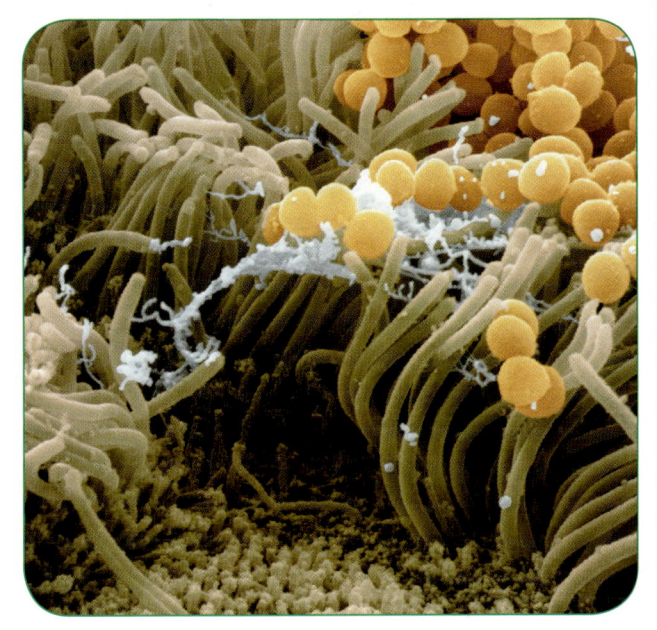

Figure 6a.3 This photograph (8000×), taken with a powerful electron microscope, shows the surface of the trachea (the windpipe).

  a Which shape group do these *Staphylococcus* bacteria belong to?
  b Name the long, hair-like structures that you can see in the photograph.
  c Suggest what the white material in the centre of the photograph might be.
  d Suggest what is happening in the photograph.

How bacteria feed

Bacteria have been on Earth much longer than any plant or animal. A huge range of different methods of feeding has evolved in bacteria.

The ultimate energy source for plants and animals is energy in sunlight. Plants use this energy, in photosynthesis, to make glucose and other organic substances. These contain energy, which is passed on to other organisms along a food chain.

Some species of bacteria are able to photosynthesise in a similar way to plants. Some feed on organic nutrients in living plants or animals. Many others feed on organic nutrients in dead bodies or excretory products of plants and animals. These are decomposers, helping waste substances to decay and releasing the nutrients they contain into the environment so that they can be re-used (*Gateway Additional Science*, pages 96–100).

There is a huge range of substances that different species of bacteria can use as an energy source. There are bacteria that can digest oil spilled from oil tankers, several kinds of plastics or the waste materials in our sewage. There seems to be almost nowhere on Earth where some kind of bacterium is not able to make a living.

Some bacteria do not rely on energy that originally came from the Sun. They are able to extract energy from simple inorganic chemicals. This is a completely different way of feeding from that carried out by plants and animals. These **chemotrophic** bacteria are found in many different places, such as around the 'black smokers' in the depths of the oceans. Here, vents in the sea floor constantly spew out water at temperatures of up to 370 °C (Figure 6a.4). The water contains a wide range of different dissolved substances, including sulfur and iron. 'Black smokers' are black because of the iron sulfide that forms. The bacteria that live around these vents, in complete darkness and in the almost total absence of oxygen, make a living by using the energy in the iron sulfide and other minerals.

Figure 6a.4 An artist's impression of a deep sea vent, or 'black smoker'. At the base of the food chain are bacteria that get their energy from iron sulfide and other chemicals. They supply energy to other organisms including crabs and giant tube worms.

While these bacteria survive at what seem like impossibly high temperatures, other bacteria are adapted to live at especially low temperatures. Some of them have enzymes that can work at 0 °C – these bacteria can continue to decay food even when you have stored it away in the fridge. There are also bacteria that can live at a very high or very low pH (Figure 6a.5).

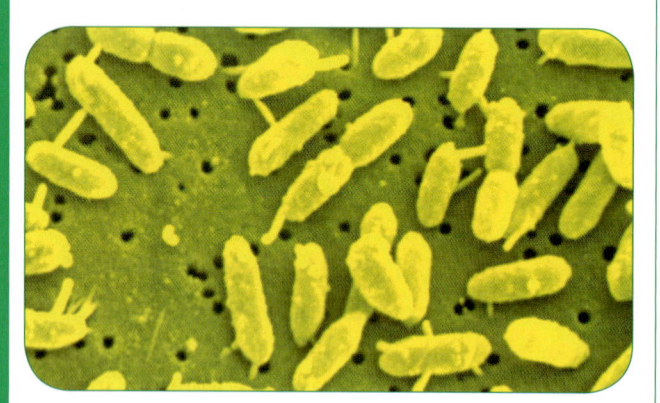

Figure 6a.5 This bacterium, called *Acidophilus*, lives and breeds in very acidic waste from mines.

How bacteria reproduce

Bacteria reproduce by splitting in two. This is called **binary fission**. You can see some bacteria doing this in Figure 6a.5. Binary fission is a kind of asexual reproduction. The two new bacteria that are formed are genetically identical.

If they have the right conditions – an ideal temperature, plenty of food and plenty of oxygen (if they need to respire aerobically) – bacteria can reproduce at an astonishing rate. *E. coli*, for example, can undergo binary fission about every 20 minutes. So, if you begin with only one bacterium, you will have eight after one hour, 64 after two hours, 512 after three hours and so on. (Can you work out how many there will be after 24 hours?)

We can make use of the speed of reproduction of bacteria when we use them to make products that we want (Figure 6a.6). For example, some *E. coli* bacteria have been genetically modified so that they contain the gene for human insulin. Some of these bacteria are put into a large container, called a fermenter, where they are given all the conditions that they need – a perfect temperature, plenty of oxygen and a good food supply. The bacteria multiply rapidly to produce a large population, all of which make insulin in their cells. The insulin can be collected and used by people with diabetes.

Figure 6a.6 These fermenters contain *E. coli* bacteria that are being grown to produce proteins used in drugs. The tubes going in and out of the big steel vessels are supplying the bacteria with everything they need, so they can reproduce rapidly.

This very rapid speed of reproduction helps to explain why a bacterial infection can take over your body so quickly. Perhaps just two or three bacteria get into a person's body through a cut in their skin. Within one day, there is a huge population of the bacteria, spreading through the body in the bloodstream. They have a chance to build up into huge numbers before the immune system is able to launch an effective defence.

This also explains why food can go bad so quickly if it is left in a warm place. No-one is going to eat food that is covered with fungi or colonies of bacteria. But often, you cannot actually see any effects of the bacteria, and might not be aware that there are dangerous populations of them in the food. For example, a joint of meat might be roasted in a restaurant kitchen, and then just put on one side ready to be served in sandwiches later on. Somehow, bacteria get onto the cooling meat – perhaps from a dirty knife, from dirty hands, or just a few that drop onto the meat from the air. If it is warm in the kitchen, there could be an enormous population of bacteria in the meat by the time it is served to the unsuspecting guests. Careful food hygiene, which is described on pages 96–97 in *Gateway Additional Science*, can make sure that this does not happen.

B6a Understanding bacteria   65

Making use of bacteria

We have many different ways of making use of bacteria (Figure 6a.7). We have already seen that we can genetically modify them so that they produce human proteins. That is a very recent technology, but there are some that have been around for many years, and some that people were using thousands of years ago.

Yoghurt is made using two different species of bacteria – *Streptococcus thermophilus* and *Lactobacillus bulgaricus*. These bacteria are added to warm milk. Usually, the milk is **pasteurised** – heated to a high temperature for a very short time, to kill most of the bacteria in it. (If it is heated for too long, it loses some of its flavour.)

SAQ

5 Explain why the milk must be pasteurised *before* the bacteria are added, not afterwards.

If you want to make yoghurt at home, the easiest way to add the bacteria to the milk is to use some 'live' yoghurt. This still contains the living bacteria. The live yoghurt is a **starter culture** – it is a culture of bacteria that are used to start off the yoghurt-making process.

Once added to the warm milk, the bacteria ferment the sugar in it, changing the sugar to lactic acid. This is what gives the yoghurt its distinctive sour taste. The lactic acid lowers the pH of the milk, which makes the protein

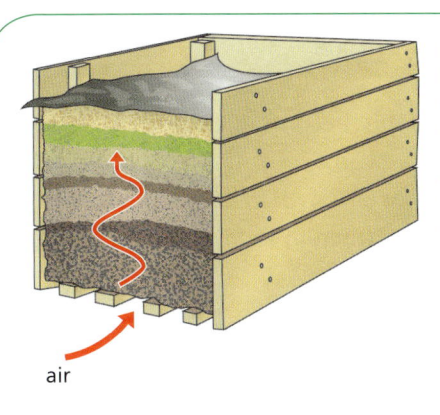

Bacteria help to break down lawn mowings and other garden rubbish, turning it into compost, which can be added to the soil to provide nutrients for plants.

These bales of newly cut grass will be wrapped in plastic so that no air can get in. Bacteria inside the bale will rot it down to make a tasty, nutritious substance called silage, which will be fed to cattle during the winter when no grass is growing.

Bacteria help to break down the proteins and fats in milk and turn it into solids (curds) and a watery liquid (whey). The curds shown here will be pressed and made into cheese. Bacteria carry on working in the cheese while it is stored, producing its characteristic flavour.

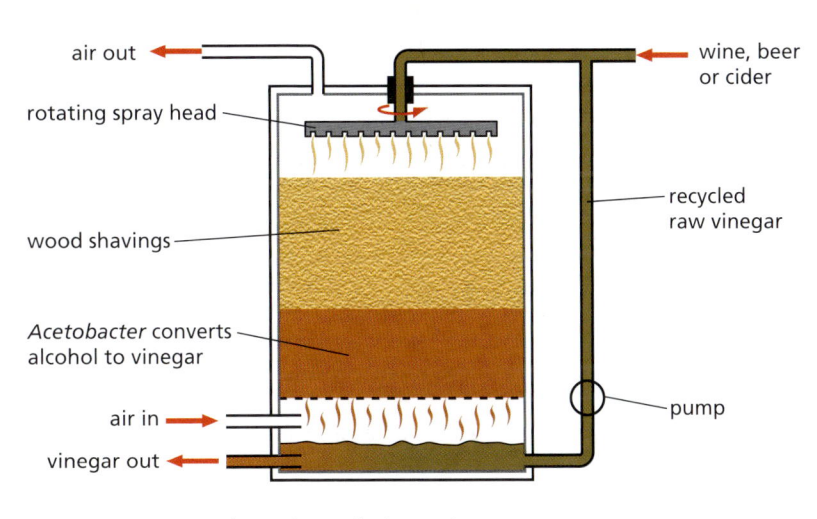

A bacterium called *Acetobacter* converts wine, beer or cider to vinegar.

Figure 6a.7 Some of the ways humans use bacteria.

66 B6a Understanding bacteria

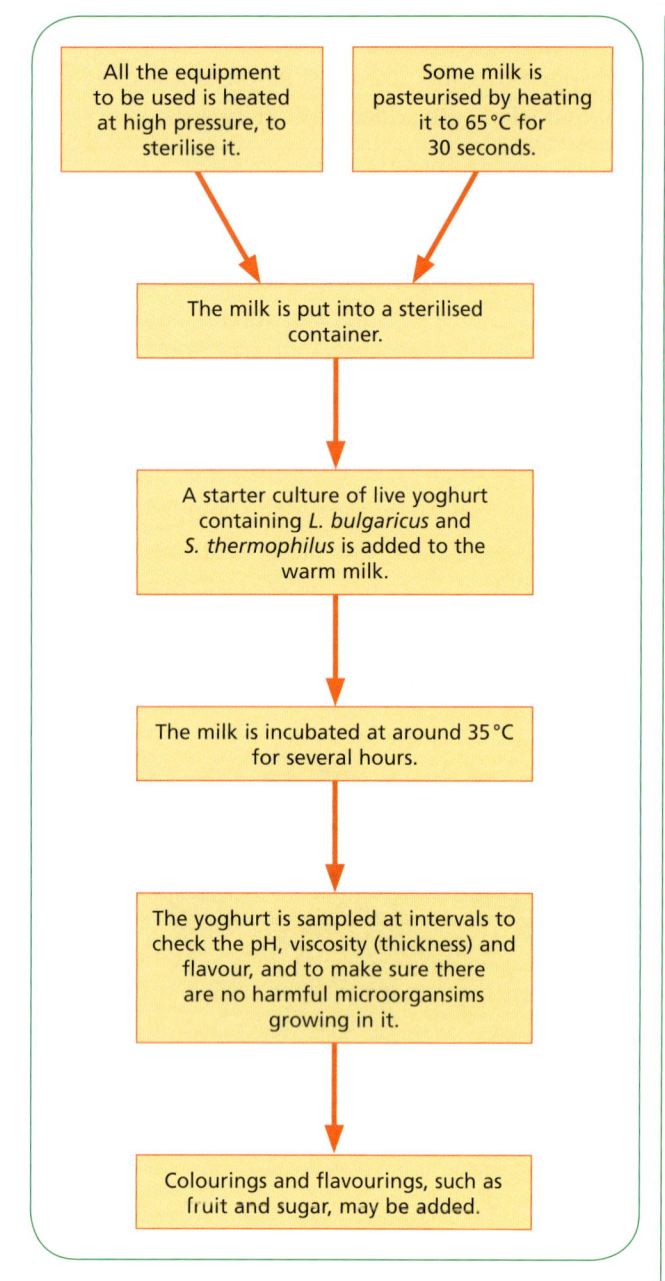

Figure 6a.8 How to make yoghurt.

Flowchart:
- All the equipment to be used is heated at high pressure, to sterilise it.
- Some milk is pasteurised by heating it to 65 °C for 30 seconds.
- The milk is put into a sterilised container.
- A starter culture of live yoghurt containing *L. bulgaricus* and *S. thermophilus* is added to the warm milk.
- The milk is incubated at around 35 °C for several hours.
- The yoghurt is sampled at intervals to check the pH, viscosity (thickness) and flavour, and to make sure there are no harmful microorgansims growing in it.
- Colourings and flavourings, such as fruit and sugar, may be added.

molecules in it change shape. This makes the yoghurt become thicker. Sugar and other flavourings can be added later, just before the yoghurt is packaged ready for sale.

Handling bacteria safely

If you are going to make your own yoghurt, it is very important that you understand how to ensure that it is safe to eat. The bacteria that you add in the starter culture are not at all harmful – in fact, there is some evidence that they can actually improve the health of the digestive system. However, there is always the possibility that other bacteria might get into the yoghurt, and some of these – as you will see on pages 69 to 70 – can be very harmful.

This is the main reason why all the equipment that is to be used for making the yoghurt is sterilised before use. Imagine that it was just washed out, and that whoever did the washing did not do a very good job, so there were a few bacteria lurking in a corner of the equipment. When the warm milk is added, they are provided with a source of nutrients. As the milk is incubated, these bacteria breed rapidly, making use of these nutrients and the ideal temperature for their growth. Anyone who eats the yoghurt is eating thousands of these bacteria. Serious food poisoning could result.

Death by yoghurt

In 1989, 27 people became seriously ill, and one died, after eating commercially produced hazelnut yoghurt. They had contracted a rare and very serious form of food poisoning called botulism.

This food poisoning is caused by a bacterium called *Clostridium botulinum*. The bacterium is very common. It survives as tough, heat-resistant spores that live in the soil. We often eat small quantities of these spores, but they do us no harm.

The danger comes when the spores have been kept in conditions where they can germinate and become active bacteria. This happens when the spores are in a source of nutrients and in conditions where there is no oxygen. These bacteria only grow and reproduce in anaerobic conditions.

As their population increases, they produce chemicals called toxins. Some of these are highly dangerous, affecting the nervous system. (Botox, injected to reduce wrinkles in people's faces, comes from this bacterium.) When food is canned, it is important to heat it to a high enough temperature to destroy any *C. botulinum* spores that might be present in it.

continued on next page

Death by yoghurt - *continued*

In the 1989 outbreak, the manufacturer making the hazelnut yoghurt had bought in some canned hazelnut conserve. The manufacturer did not know that the canning process had not reached a high enough temperature. Unknown to everyone, *C. botulinum* had been breeding in a can of conserve, producing its potentially fatal toxin. When the hazelnut conserve was added to the yoghurt before packaging, the toxin went in as well.

People who ate the contaminated yoghurt experienced severe symptoms. It began with their vision becoming blurred, followed by vomiting and abdominal pain. If they had not been treated, this could have led to paralysis of the breathing muscles and death. This did happen to one person, but for the others rapid treatment with antitoxin, which neutralised the effects of the toxin, allowed them to recover.

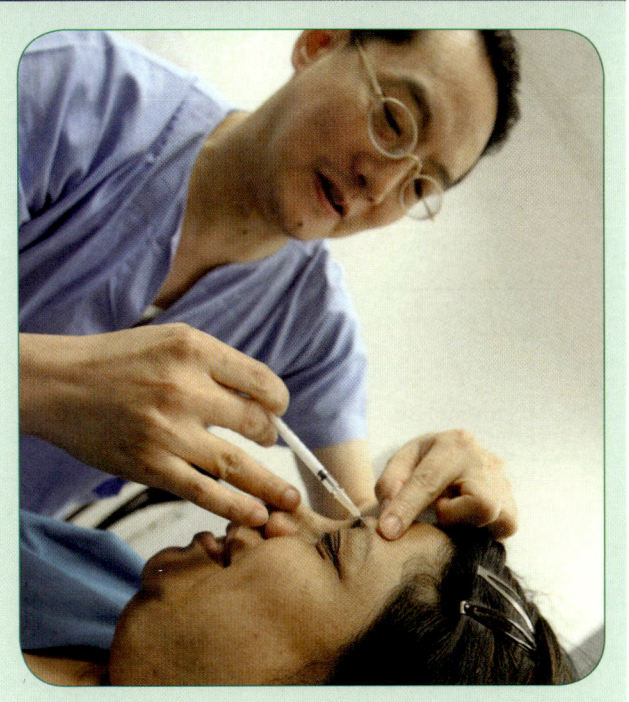

Figure 6a.9 The deadly botulinum toxin can be safely used in tiny quantities to relax muscles and reduce wrinkles.

Summary

You should be able to:

- describe the structure of a bacterial cell, and label a diagram of a bacterium
- describe the functions of the parts of a bacterial cell
- compare the structure of a bacterium with a plant cell and an animal cell
- describe the four main shapes of bacteria
- (H) explain why bacteria can live in a very wide range of habitats
- describe how bacteria reproduce
- (H) explain how the rapid reproduction of bacteria is linked to food spoilage and infectious disease
- list some uses of bacteria in making food and compost
- describe the main stages in making yoghurt
- (H) explain why bacteria must be handled safely

68 B6a Understanding bacteria

Questions

1. Copy and complete Table 6a.1, comparing the structures of bacterial, animal and plant cells.

Feature	Bacterial cell	Animal cell	Plant cell
cell membrane	yes	yes	yes
cell wall	yes, but not made of cellulose	no	
cytoplasm			
nucleus			
DNA			
mitochondria			
chloroplasts			
vacuole			

Table 6a.1

2. Bacteria are involved in making several products that we eat, or which we feed to animals.

   Copy Figure 6a.10, and then complete it by linking each raw material with the product or products that bacteria can produce from it.

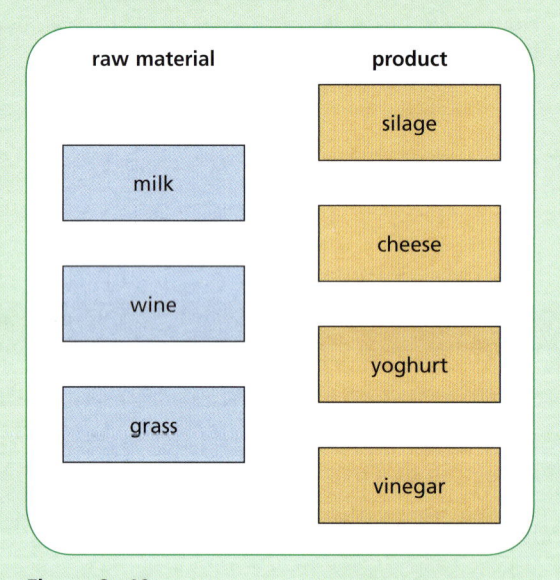

   raw material: milk, wine, grass
   product: silage, cheese, yoghurt, vinegar

   Figure 6a.10

3. Explain *why* each of these steps is carried out when making yoghurt.

   H

   a Sterilising all of the equipment.

   b Pasteurising the milk.

   c Adding the starter culture and incubating.

   d Sampling.

6b Harmful microorganisms

Infectious diseases

Infectious diseases are caused by **pathogens** – microorganisms that enter the body, breed and make us ill. Four kinds of organisms can be pathogens – viruses, bacteria, fungi and protozoa (*Gateway Science*, pages 18–19). Table 6b.1 lists some diseases caused by each of these.

Type of microorganism	Examples of diseases they cause
viruses	influenza, chickenpox, smallpox, AIDS
bacteria	tuberculosis, food poisoning, septic wounds
fungi	athlete's foot
protozoa	malaria, dysentery

Table 6b.1 Diseases caused by the four types of microorganisms.

Figure 6b.1 shows some of the ways in which pathogens can get into the body.

The stages of an infectious disease

You probably know that, when you pick up an infectious disease such as food poisoning, you sometimes don't have any symptoms for a few days. The entry of the pathogen into your body is called **infection**. The time between infection and having symptoms of disease is called the **incubation period**.

This happens because you have probably only taken a few hundred or so bacteria into your body. It takes time for them to begin breeding inside you and building up their population. As these bacteria feed inside your body, they produce chemicals called **toxins**. Toxins are poisons, and they make you feel ill. They cause the **symptoms** of the disease – such as feeling nauseous and perhaps having a high temperature.

In most cases, our immune system is able to destroy the pathogens. This takes a little time, so you may feel ill for a few days before the symptoms begin to subside. You can remind yourself about how the immune system fights against pathogens by looking back at *Gateway Science*, pages 21–22.

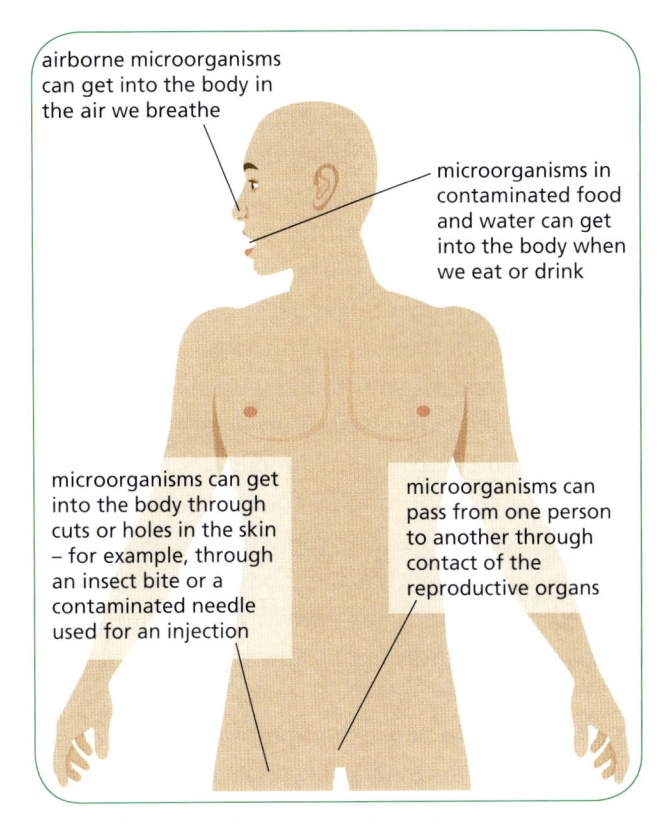

Figure 6b.1 How pathogens enter the body.

- airborne microorganisms can get into the body in the air we breathe
- microorganisms in contaminated food and water can get into the body when we eat or drink
- microorganisms can get into the body through cuts or holes in the skin – for example, through an insect bite or a contaminated needle used for an injection
- microorganisms can pass from one person to another through contact of the reproductive organs

Some examples of infectious diseases

Food poisoning

Several different kinds of bacteria can cause food poisoning. Three of the commonest are *Salmonella*, *Escherichia coli* and *Campylobacter*.

We get food poisoning when we eat food that contains large numbers of these bacteria. This is most likely to happen if the food has been left in a warm place for some time, giving an opportunity for any bacteria that were in the food to breed and form a large population.

70 B6b Harmful microorganisms

How do bacteria get into food in the first place? Often, they come from someone's dirty hands, especially if they were not washed after using the toilet. They can just fall in from the air. Raw vegetables and meat often contain bacteria that were present in soil, or on an animal's skin or inside its gut before it was slaughtered. Flies walking on food may transfer pathogens from their feet or saliva. Flies often visit rubbish heaps to find places to lay their eggs, so they can bring pathogens from there onto your food. What's more, they feed on your food by vomiting saliva onto it, then sucking it back up together with the softened and partly digested food.

When food is thoroughly cooked, most bacteria will be killed. This is why you will see labels on ready meals saying that the food should be really hot before serving it, and why it is important to cook a chicken thoroughly. Raw meat, including chicken, often contains *Salmonella* or other bacteria that can cause food poisoning, but if the meat is thoroughly cooked then these will be killed and the meat will be absolutely safe to eat. Keeping raw foods away from cooked ones will ensure that new bacteria don't get back onto the food after the original ones have all been killed. This is why it is good hygiene to wrap food kept in a fridge, and also to keep cooked food on higher shelves than raw food.

SAQ

1 Figure 6b.2 shows the number of reported cases of food poisoning in the UK between 1986 and 2001.

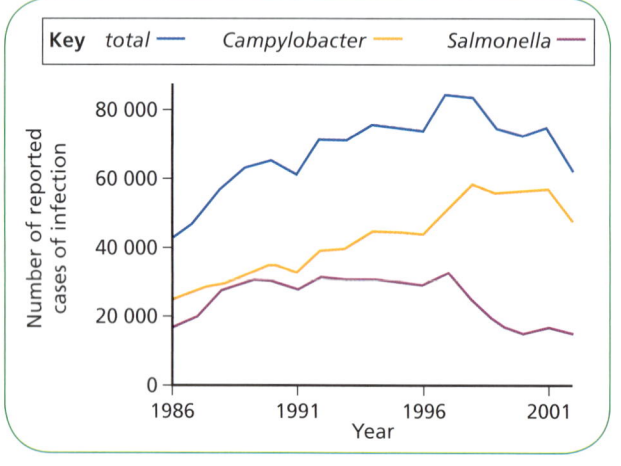

Figure 6b.2

a Compare the trends for food poisoning by *Campylobacter* and *Salmonella* shown in the graph.

b It has been suggested that the increase in cases of food poisoning in the 1980s and 1990s could have been caused by:
- people shopping less frequently and so storing food for longer
- people eating more ready meals.

Explain how these might have caused the food poisoning increase.

c In 1989, it was found that many eggs and chickens sold in shops were contaminated with *Salmonella*. All flocks of chickens were tested for this bacterium, and any that contained it were slaughtered. Now, many chickens are vaccinated against *Salmonella*. Using the information in the graph, discuss whether there is any evidence that these measures have affected the number of cases of food poisoning.

d After a natural disaster, such as an earthquake, electricity supplies may be disrupted. Explain why this could cause an increase in cases of food poisoning.

Cholera

Cholera is caused by a bacterium called *Vibrio cholerae* (Figure 6b.3). This bacterium can get into water supplies that are contaminated by sewage. It lives and breeds in the alimentary canal, causing serious diarrhoea. The faeces of an infected person carry millions of the bacteria, so it is very important to ensure that they cannot get into water that will be used for drinking, washing or cooking.

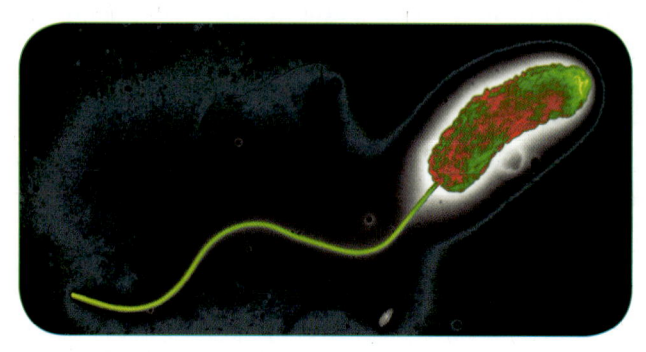

Figure 6b.3 The cholera bacterium has a flagellum that it can use to swim through water.

SAQ

2 Suggest what you could do to make drinking water safe, if you were living in a town where there was a cholera outbreak.

Cholera is usually only found in countries where the sewage system is inadequate, and it used to be common in the UK before the 20th century. Now it is rare in the UK. An outbreak often begins after a natural disaster, such as flooding, an earthquake or a volcanic eruption. These catastrophes can damage the sewage systems, allowing sewage and drinking water to mix. The cholera bacterium can then spread quickly through a population, causing an **epidemic** of cholera.

Without treatment, cholera can be fatal. Death is caused by rapid dehydration, because the person loses so much fluid through diarrhoea.

Figure 6b.4 In 1994, organised murder of one tribe by another in the African country of Rwanda caused a refugee crisis. Thousands of people in this refugee camp died of cholera.

Dysentery

Dysentery, like cholera, is an illness that affects the digestive system and causes diarrhoea. Many cases of dysentery are caused by a protozoan called *Entamoeba*. This is a single-celled organism with cells similar to animal cells. Figure 6b.5 shows some of these protozoa in someone's colon.

The protozoa can get into the digestive system in contaminated food or drink. Most people just get diarrhoea for a few days and then get over it. However, if you travel abroad, especially in tropical countries, you could come into contact with more dangerous forms of dysentery, so it is important to take care over what you eat and drink. Like cholera, epidemics of dysentery often occur following natural disasters, when sewage systems and health services break down.

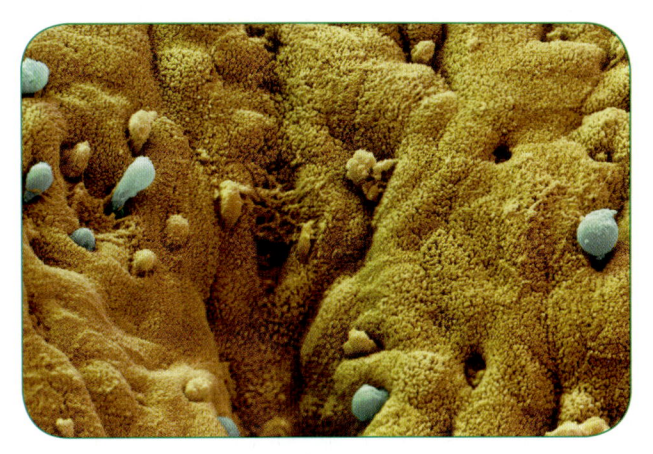

Figure 6b.5 This is a magnified view of the inside of the colon (1200×). The blue blobs are protozoa that cause dysentery.

SAQ

3 What precautions should you take to avoid a gut infection while you are on holiday in a tropical country?

Influenza

Influenza (most people call it flu) is caused by a virus. The virus gets into the body in droplets of moisture that you breathe in. These droplets can be produced by someone's saliva when they are talking, or when they sneeze. The virus can also be present on moist surfaces, and it can get into you if you touch something with the virus on it, and then put your fingers into your mouth or nose.

Flu gives you a cough, aches and pains and a high temperature. Most people get over it after a week or so, but you may still not feel quite right for another week or more after that.

There are several different types of the flu virus, and it keeps changing all the time. When a new form appears, it can spread rapidly around the world, because people have not yet had a chance to become immune to it. A worldwide spread of infection like this is called a **pandemic**. Millions of people died in a flu pandemic in the winter from 1918 to 1919 and then another in 1968.

Figure 6b.6 This poster was used in Chicago during the 1918–1919 flu pandemic, which killed millions of people.

We are due for another pandemic any time now. Some scientists think that it is likely to come from Asia, where a virus that causes bird flu may mutate into a form that is highly infectious to people.

Septicaemia

If bacteria get into a wound, they may be able to enter the blood system and breed there. This is called **blood poisoning**, or **septicaemia**.

The bacteria may release large quantities of toxins, which can damage many different body organs and cause a condition called **septic shock**. This can be fatal if not quickly treated. The patient is usually given large doses of antibiotics, often through a drip that delivers them directly into the blood.

SAQ

4 Suggest why septicaemia can result in damage to many different body organs.

Should we destroy the smallpox virus?

Smallpox is a dangerous, disfiguring and often fatal disease caused by a virus. It can kill more than a quarter of people who are infected, and even the survivors are often badly scarred or blind. A worldwide vaccination campaign, coordinated by the World Health Organization, appeared to have eradicated smallpox from the whole world by 1977.

You might think that would be the end of it. No more smallpox. But governments in several countries want to keep small amounts of the smallpox virus, safely locked away in laboratories. They are worried that other countries might have some, and could use it as a weapon of bioterrorism. If they keep some of the virus, then they can develop vaccines against it (Figure 6b.7).

Ever since 1977, there has been heated debate about this. In 2005, the World Health Assembly voted not to destroy the two remaining (known) stocks of the virus, which are kept in the USA and in Russia.

Figure 6b.7 These scientists are working in a laboratory in Maryland, in the USA. They are culturing the smallpox virus which they will then use to produce new vaccines against it.

The fight against pathogens

Our own immune systems put up a very good fight against most infections. But they are not always successful. Until the middle of the 20th century, infectious diseases were a major cause of death in the UK, as they still are in some parts of the world today.

Louis Pasteur

Until the 19th century, no-one knew that microorganisms existed, let alone that some of them could cause disease. Louis Pasteur was one of the world's first microbiologists, carrying out research in his laboratory in France which made enormous differences to the death toll of hospital patients from infections.

Pasteur's early work, in the 1850s, showed that microorganisms, which he could see with a microscope, were present in fermenting wine. He began to suspect that some diseases might also be caused by microorganisms. This is called the 'germ theory of disease'.

Pasteur built on his theory to suggest that infections might be carried from one person to another on the dirty hands and instruments of surgeons in hospitals. He persuaded them to disinfect metal instruments using boiling water or steam, and to wash their hands regularly. He identified several kinds of microorganism that cause disease, including the bacterium that causes cholera and another that causes anthrax.

Joseph Lister

Lister lived in Scotland around the same time as Louis Pasteur was carrying out his investigations in France. Unlike Pasteur, Lister was a practising surgeon; he worked in Edinburgh Royal Infirmary. At that time, many patients survived an operation but then died later, as a result of septicaemia. Even before he heard about Pasteur's work, Lister had recognised the connection between dirty hands and instruments and the spread of disease to patients, and made sure that everything was really clean in the operating rooms where he worked.

Once Lister heard about Pasteur's discovery of microorganisms and their links to disease, he was able to take his ideas further. He sprayed carbolic acid, which destroys microorganisms, onto wounds, and found that this greatly reduced the chance of septicaemia developing. Carbolic acid was one of the first **antiseptics** – something that acts against sepsis (infection). Lister advocated using it for any open wound, not just ones made during surgery.

Figure 6b.8 Louis Pasteur was one of the very first people to study microorganisms. His laboratory was 'state of the art' in the 19th century.

Figure 6b.9 Joseph Lister.

He explained:

'In conducting the treatment, the first object must be the destruction of any septic germs which may have been introduced into the wounds, either at the moment of the accident or during the time which has since elapsed. This is done by introducing the acid of full strength into all accessible recesses of the wound by means of a piece of rag held in dressing forceps and dipped into the liquid.'

This cannot have been very pleasant for his patients, especially as anaesthetics had not yet been developed, but it was found to greatly improve their chances of survival without needing to have a limb amputated. Lister also introduced the idea of doctors and nurses washing their hands in antiseptic before and after touching patients.

Today, antiseptics are used by almost everyone, for cleaning wounds to stop them getting infected. You may have had antiseptic cream or iodine (which works as an antiseptic) put onto a cut or graze. Notice that antiseptics are not the same as disinfectants. Antiseptics are something that we can use to destroy microorganisms on the body, while disinfectants are used for cleaning non-living objects, such as floors, work-surfaces or surgical instruments. Some disinfectants, like bleach, are very powerful cell killers and can damage human tissue.

Alexander Fleming

Like Lister, Alexander Fleming was born in Scotland, but he worked as a microbiologist in a London hospital. His most famous discovery came in 1928. Fleming had been growing a type of bacterium on agar in a Petri dish. One dish was left lying around for a while, and when Fleming went back to it he noticed that there was a fungus growing on the agar as well. Around the patch of fungus, the agar was clear – there were no bacteria growing on it.

Fleming realised that the fungus must be producing something that killed bacteria – an **antibiotic**. He wrote up his findings in a research paper, which was read by two scientists working in Oxford – Howard Florey and Ernst Chain. They

Figure 6b.10 Alexander Fleming in his laboratory. A chance entry of fungal spores through a window and onto the agar in a Petri dish led to his discovery of penicillin.

worked on trying to produce large quantities of the substance, which they called 'penicillin', because it was made by a fungus called *Penicillium*.

All three scientists received the Nobel Prize for medicine in 1945. The discovery of antibiotics has probably been the single most important development in medicine in the last century. It is difficult to imagine how many people died of bacterial infections before we had antibiotics available to kill the bacteria. Today, there are many different antibiotics which, between them, are able to kill most kinds of pathogenic bacteria without damaging us. However, the more we use antibiotics the more likely it is that bacteria will evolve to become resistant to them. There is a constant effort to find new antibiotics, to keep ahead in the 'arms race' with bacteria.

SAQ

5 Look back at Table 6b.1 on page 69. List three diseases from the table that could be treated with antibiotics, and three that could not.

B6b Harmful microorganisms

Summary

You should be able to:

- describe how disease-causing (pathogenic) organisms enter the body
- name the types of microorganisms causing different named diseases
- describe the causes and transmission of food poisoning, cholera, dysentery and influenza
- **H** name the organisms that cause food poisoning, cholera and dysentery
- interpret data relating to the incidence of infectious diseases
- describe the stages of an infectious disease
- explain why natural disasters can cause a spread of dysentery, cholera and food poisoning
- describe the work of Pasteur, Lister and Fleming
- describe how antiseptics and antibiotics can be used in the control of disease

Questions

1 Copy and complete Table 6b.2.

Disease	Pathogen that causes it	How it is transmitted
food poisoning		
cholera		
dysentery		
influenza		

Table 6b.2

2 Explain the difference between each of the following pairs.

   a antibiotics and antiseptics

   b bacteria and viruses

   c infection and incubation

   d food poisoning and septicaemia (blood poisoning)

continued on next page

Questions - *continued*

H 3 The map in Figure 6b.11 shows what happened after cholera broke out in a shanty town in Peru in 1991. It is thought that the cholera bacterium was accidentally brought to Peru in the bilge water of a Chinese ship.

a Officials in Peru suggested that the outbreak began because of overcrowding and poor hygiene in the shanty town. Greenpeace, however, stated that it was caused because the government had stopped ensuring that the water supply was treated with chlorine. Suggest how both of these circumstances might help cholera to spread.

b Using the information on the map, describe what happened after the initial outbreak.

c Suggest what could have been done to stop the cholera epidemic spreading.

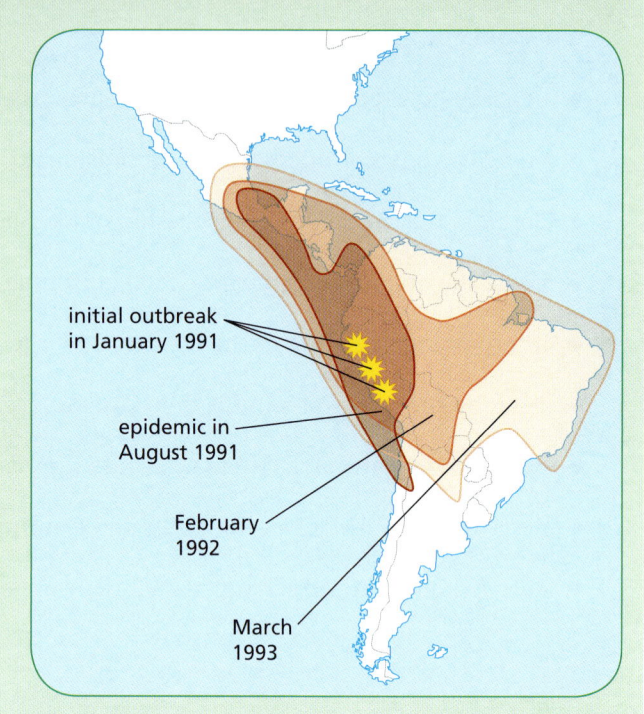

Figure 6b.11

4 One form of septicaemia is caused by a bacterium that also causes meningitis – inflammation of the membranes that surround and protect the brain. Septicaemia or meningitis caused by this bacterium is often fatal.

There are several types of this bacterium, called 'serogroups', and vaccines have been developed to immunise people against serogroup C. The graph in Figure 6b.12 shows the numbers of people who suffered from infections by two types of the bacterium between 1997 and 2003.

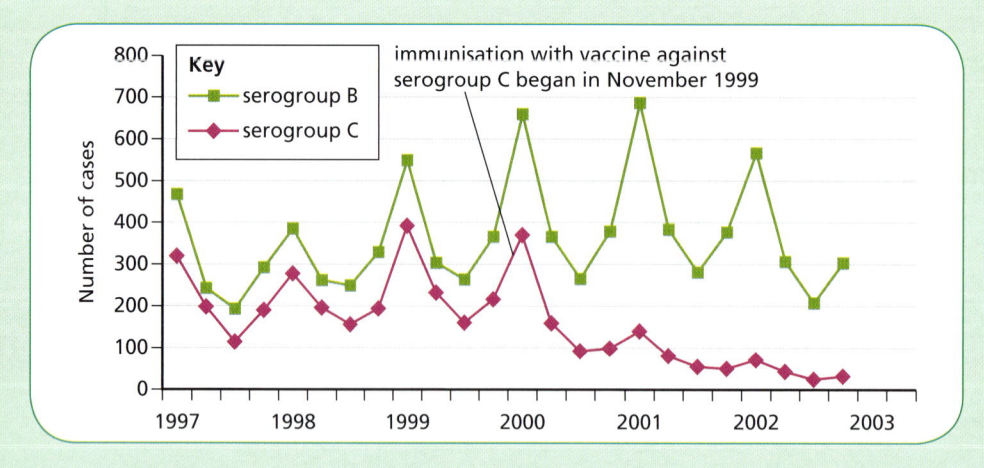

Figure 6b.12

a Suggest why the curves for infection by both types of bacterium go up and down.

b Compare the patterns shown by the two serogroups.

c Immunisation of young people against serogroup C began in November 1999. Do these data suggest that the immunisation campaign has been successful? Explain your answer.

6c Microorganisms – factories for the future?

Yeast

Yeast is a fungus. Yeast cells are very small, so you need a microscope to see them. Figure 6c.1 shows the structure of a yeast cell.

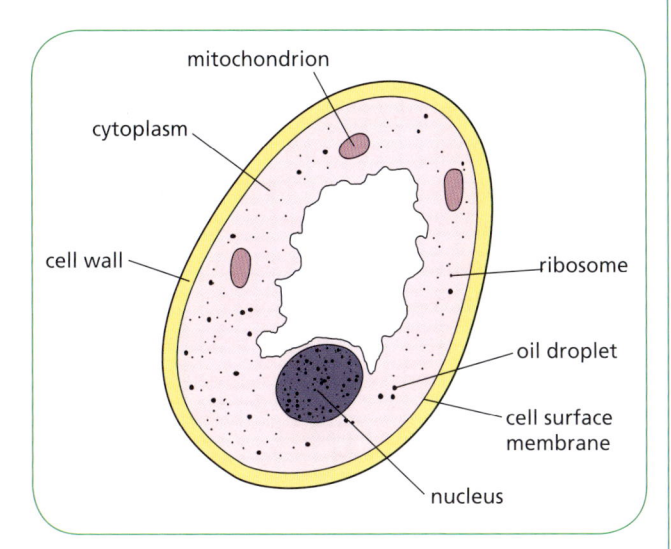

Figure 6c.1 The structure of a yeast cell.

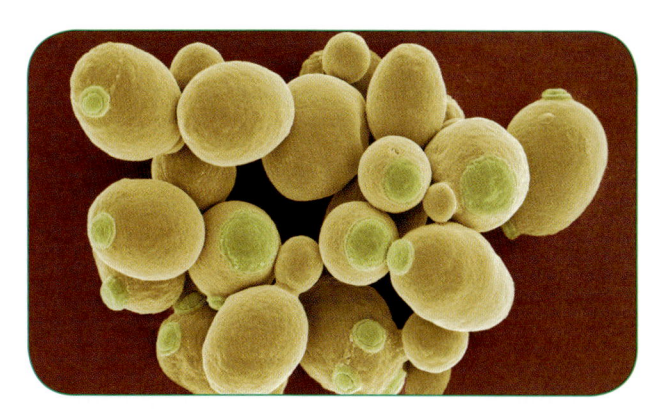

Figure 6c.2 This photograph was taken using an electron microscope. Each yellow-brown blob is a yeast cell. You can see that many of them are growing buds, which will split away from the parent cell.

SAQ

1 A yeast cell is about 10 μm long at its widest part. There are 1000 μm in 1 mm. Calculate how many times the yeast cell is magnified in Figure 6c.1.

'Wild' yeasts are found on the surface of fruit, such as grapes. You can sometimes see the yeast – it looks like a powdery covering on the fruit, and you can rub it off with your fingers. The yeast feeds on the sugars in the fruit.

Yeast can survive for a long time as dry **spores**. The spores start to become active when they come into contact with water, especially if there is a little bit of sugar in the water that they can use as food. The spores produce new yeast cells, and once a cell has grown to a certain size, it can split into two, by 'budding' off a smaller cell (Figure 6c.2). This is a kind of asexual reproduction. The new cell is genetically identical to the original cell.

The rate at which yeast reproduces depends on several factors:
- food availability – yeast needs sugars such as glucose in order to grow and divide
- temperature – the optimum temperature for most kinds of yeast is around 30–40 °C

- pH – yeasts grow best at a neutral to slightly acidic pH
- presence of other substances – yeast stops growing if the alcohol concentration around it builds up too high.

At temperatures below its optimum, a 10 °C rise in temperature approximately doubles the rate of growth of a yeast population. So, for example, yeast will grow twice as quickly at 15 °C as at 5 °C.

SAQ

2 a Explain why yeast is *not* able to grow at temperatures above about 40 °C.
  b Explain why the growth of a yeast population increases when temperature increases, so long as the temperature stays below its optimum.

How yeast respires

Yeast cells are able to respire either aerobically or anaerobically. If plenty of oxygen is available, then they respire just like our cells, using glucose and oxygen and producing carbon dioxide and water.

However, if oxygen is in short supply, they respire anaerobically. In this process, they break down glucose and produce carbon dioxide and alcohol. This is also known as **fermentation**.

glucose → carbon dioxide + alcohol

$C_6H_{12}O_6 \rightarrow 2CO_2 + 2C_2H_5OH$

SAQ

3 How does anaerobic respiration in yeast differ from anaerobic respiration in human cells? (If you have forgotten, look back at page 2 in *Gateway Science*.)

We use fermentation by yeast for many different purposes. These include:
- breaking down sugar in waste water from food processing factories, so that it will not cause harmful pollution when it flows into rivers
- making bubbles of carbon dioxide that help bread to rise
- making alcoholic drinks, including beer, wine and cider.

Making alcoholic drinks

For thousands of years, humans have used yeast to make drinks containing alcohol. Alcohol is a drug, and it affects the activity of the brain and the speed at which nerve impulses travel. Some cultures and religions ban the drinking of alcohol. However, making beer, wine and cider have become important industrial activities all over the world.

All three of these drinks rely on fermentation by yeast, using sugars that are naturally present in fruits and seeds. Beer is made from barley grains. Wine is made from grapes. Cider is made from apples.

As well as alcohol, yeast makes carbon dioxide when it ferments sugars. This makes some drinks 'fizzy'. However, most alcoholic drinks are bottled after the fermentation has finished and most of the carbon dioxide has escaped. There are some beers where the yeast is left in and which still contain carbon dioxide that the yeast has made, and these are called 'naturally conditioned' beers. But in most beers, the yeast has been filtered out, and carbon dioxide is added afterwards.

Figure 6c.3 shows how beer is brewed.

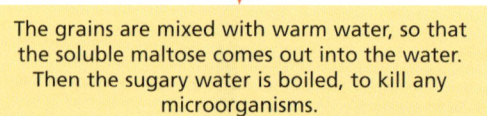

Figure 6c.3 How beer is made.

SAQ

4 State the two stages in the beer brewing process during which microorganisms are killed.

5 Explain why less alcohol would be made if air was allowed to mix with the mixture of yeast and sugars.

6 When wine is made, the sugars come from crushed grapes. There is no need to boil the grape juice before the yeast is added, and no hops are used. Write a revised version of the first four boxes in Figure 6c.3, to describe how wine is made.

Pasteurisation

In the mid 19th century, Louis Pasteur was asked to help out wine makers who were having trouble with their wine going sour. The alcohol in the wine was turning to vinegar, and they did not know why. Pasteur thought this was being caused by unwanted microorganisms getting into the wine. He suggested that the wine makers should try heating their wine for a short time – long enough to kill these unwanted microorganisms but not so long that it spoiled the flavour of the wine. The process became known as **pasteurisation**.

Pasteurisation is used today to help to preserve many different kinds of drinks, including milk, fruit juices, beer and wine. There are several different techniques that can be used, each involving heating to different temperatures for different times. For example, the liquid can be heated to 63 °C for up to 30 minutes, or to 71 °C for 15 seconds. The different methods of pasteurisation have different effects on the flavour of the drink. Pasteurisation does not kill all of the microorganisms in the liquid, but it does kill the yeast and most kinds of bacteria that could have contaminated it. If living yeast was left in the beer, it might continue to ferment and produce more carbon dioxide, which could burst the bottle.

However, it also possible to leave yeast in the bottle and allow it to produce more carbon dioxide, so long as you make sure there is not enough sugar to allow it to make too much. This kind of beer is called 'bottle conditioned'. Many brewers of 'real ale' do not filter or pasteurise their beer, as they believe that this spoils its flavour.

Small beer

In the past, houses did not have water brought to them in pipes, and there were no water treatment works producing water that was safe to drink. Households would collect their water from a spring or a well. There were still many houses in the UK for which this was necessary right up to the middle of the 20th century.

That water could contain dangerous microorganisms – for example, the cholera bacterium or the polio virus. The water was therefore often boiled before drinking.

It was much safer to drink beer rather than water. The fermentation process killed most microorganisms other than the yeast, because they could not survive in the alcoholic liquid. So people tended to drink beer rather than water. Early in the day, for breakfast, they would normally drink 'small beer'. This was a kind of beer containing only a little alcohol, and even children drank it. The beer was often not filtered, so the yeast remained in it, making a porridge-like substance that was very nutritious.

Figure 6c.4 This engraving shows activity at a 17th century brewery.

Making spirits

Spirits are drinks with a much higher alcohol content than beer, wine or cider. They include:
- whisky, made from malted barley
- rum, made from sugar cane
- vodka, made from potatoes.

[H] When beer is made, yeast ferments sugars from malted barley. This produces alcohol. However, most strains (varieties) of yeast cannot survive once the alcohol content increases above about 14%, although there are some strains that can continue to survive in up to 18% alcohol. Once the yeast has made this much alcohol, the fermentation stops.

To get a drink containing more alcohol than this, the liquid has to be **distilled**. In the UK, companies are not allowed to do this without a special licence. They have to pay tax on the high-alcohol drinks that they produce, called excise duty. The premises are kept secure and inspected regularly by customs officials, to make sure that all the alcoholic drinks that are made are fully accounted for and that tax is paid on them.

The distillation process is very similar to the method you may have used to get pure water from sea water (*Gateway Additional Science*, page 210). Figure 6c.5 shows some of the huge copper flasks used in a distillery for making whisky. The alcoholic liquid that has been produced by the yeast is heated inside the flasks. Alcohol has a lower boiling point than water, so the alcohol, and also many other chemicals that are present, turn to vapour and go up into the tall pipe leading out of the flask. Towards its end, the pipe is surrounded by cold water, and this makes the vapours condense to form a liquid containing a high proportion of alcohol. This will become whisky. Whisky can have an alcohol content of up to 40 or 50%.

Most of the water in the fermented liquid, and of course also the yeast, is left behind in the flask. This mixture can be used to make animal feeds.

Figure 6c.5 Distillation flasks in a whisky distillery.

Summary

You should be able to:

- describe the structure of yeast, and how it reproduces
- list factors that affect the rate of growth of a yeast population
- [H] explain how temperature affects the rate of growth of yeast
- state the word equation for fermentation (anaerobic respiration by yeast)
- [H] state the balanced symbol equation for fermentation

continued on next page

Summary - continued

- state the sources from which beer, wine and cider are made
- describe how beer or wine is made
- [H] describe how pasteurisation is carried out, and why it is done
- describe how spirits such as whisky, rum and vodka are made by fermentation followed by distillation
- [H] explain how increasing alcohol content stops fermentation, depending on the strain of yeast involved

Questions

1. This list of statements describes how rum is made from sugar cane. Using what you know about how whisky is made, write them down in the correct order.
   A The yeast ferments the sugars in the molasses.
   B The molasses is mixed with water and then pasteurised.
   C The alcoholic liquid is distilled, by boiling it and collecting the liquid formed from the cooled vapours.
   D Sugar cane is cut and processed to make a sticky, sugary substance called molasses.
   E Yeast is added to the warm, pasteurised molasses solution.

2. a Explain why spirits such as vodka or whisky cannot be made by fermentation alone.

   b Food processing factories may produce waste water that contains a lot of sugar. Explain how this water could affect the life in a river if it was allowed to flow into the river untreated.

   c Suggest how yeast could be used to change the waste water so that it will not damage the environment.

3. [H] An investigation was carried out to determine how temperature affects the rate of respiration of yeast cells. Some yeast was added to glucose solution in five different tubes, which were kept at five different temperatures. As carbon dioxide was produced by the yeast, it was collected in a gas syringe. The time taken to collect 10 cm^3 of carbon dioxide was recorded.

   The results are shown in Table 6c.1.

   a List *two* variables that should have been kept the same in the experiment.

   b Display these results as a graph.

   c Describe the effect of temperature on the rate of carbon dioxide production between 0 °C and 40 °C.

   d Suggest an explanation for the effect you have described.

   e Explain the results obtained at 50 °C.

Temperature in °C	Time taken to collect 10 cm^3 of carbon dioxide, in minutes
0	took too long to measure
10	62
20	29
30	15
40	8
50	12

Table 6c.1

B6d Biofuels

A **biofuel** is a fuel that is made from biological material. Biofuels include:
- **biogas**, which is made from rotting organic waste
- **bioethanol**, which is made by the action of yeast on sugars.

These two fuels, and the ways in which they are made, are described in *Gateway Additional Science*, pages 82–83.

Biogas

Biogas is formed when bacteria rot down **biodegradable** organic material such as dead plants and animal waste products. It contains mainly methane, some carbon dioxide, water vapour and small traces of hydrogen, nitrogen and hydrogen sulfide.

SAQ

1 Which two of these six gases can burn and release energy?

As biogas is formed by the activity of microorganisms, it happens most quickly at temperatures at which their enzymes work best. In general, temperatures between 32 °C and 39 °C are ideal. Above or below these temperatures, the biogas contains less methane and more carbon dioxide.

Sources of biogas

The production of biogas can happen naturally – for example, in marshy places. The bacteria that produce biogas live in places where there is not much oxygen – and these are the conditions found in marshes, where there are dead plants beneath the water. The gas is sometimes called 'marsh gas'. It can occasionally ignite spontaneously, forming 'will-o'-the-wisps' – dancing flames that flitter across the surface of a marsh (Figure 6d.1).

A similar mix of gases is produced inside the digestive systems of ruminant animals – that is, animals such as cattle and sheep that 'chew the cud'. These animals have bacteria inside a special

Figure 6d.1 People used to think the 'will-o'-the-wisp' flames were carried by goblins, who lured people into dangerous places in the marsh where they were drowned.

stomach (the rumen), which digest the cellulose in their diet of grass. The bacteria produce methane, which the animals belch out into the atmosphere. There is a story that has been around for a long time about a cow exploding when a vet opened up its stomach in an operation and lit a cigarette – but, disappointingly, it does not seem to be true.

Too much methane in the atmosphere is not good, because it contributes to the greenhouse effect and therefore to global warming. Methane is a more effective greenhouse gas than carbon dioxide – one molecule of methane has a much greater effect than one molecule of carbon dioxide. If we collect and use the methane in biogas as a fuel, we could help to slow down the rate of climate change. Moreover, it might help to reduce our dependence on other, non-renewable fuels such as oil.

Special **biogas digesters** (see Figure 4e.7 in *Gateway Additional Science*, page 83) have been used for a long time in countries where there are many places with no mains electricity supply or mains sewage system. Digesters can operate on a **continuous flow** system, where there is a steady supply of organic waste at one end, and a continuous removal of methane and waste sludge at the other.

In the UK, homes that cannot be linked to the sewage system may have a **septic tank**. This is a bit like a biogas digester – the sewage runs into it and is broken down by bacteria to form a safe liquid and sludge (Figure 6d.2). However, people don't usually collect methane from a septic tank. The sludge from the tank is collected about once a year and taken away to be safely disposed of.

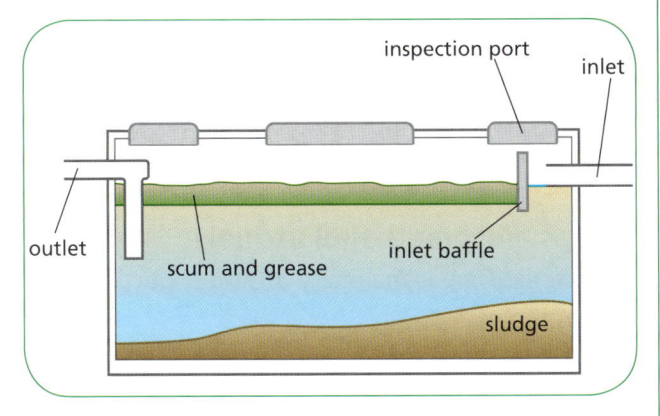

Figure 6d.2 A septic tank.

How biogas is formed

When biodegradable waste is added to a biogas digester, or buried underground at a landfill site, it begins to rot almost immediately. The process happens in three main stages, involving three groups of bacteria.

- Firstly, if there is any oxygen in the waste material, **aerobic bacteria** begin to break it down. This can only happen for a short while, until the oxygen is used up.

- Next, **anaerobic bacteria** (ones that can survive without oxygen) act on the waste. They produce a lot of ammonia, which smells unpleasant and is damaging to the environment, and also ethanoic acid. The bacteria also release many other potentially harmful substances from the waste, such as toxic metal ions, which can leach out if water is allowed to flow through it. A well-designed landfill site will not allow this to happen (Figures 6d.3 and 6d.4).

- Lastly, slower-acting anaerobic bacteria break down the remaining substances in the waste, producing methane.

Figure 6d.4 This pipe is coming up from rubbish buried beneath the ground at a landfill site. The pipe allows methane to escape into the air. The engineer tests it to keep track of how the rubbish is breaking down, and to make sure it is safe.

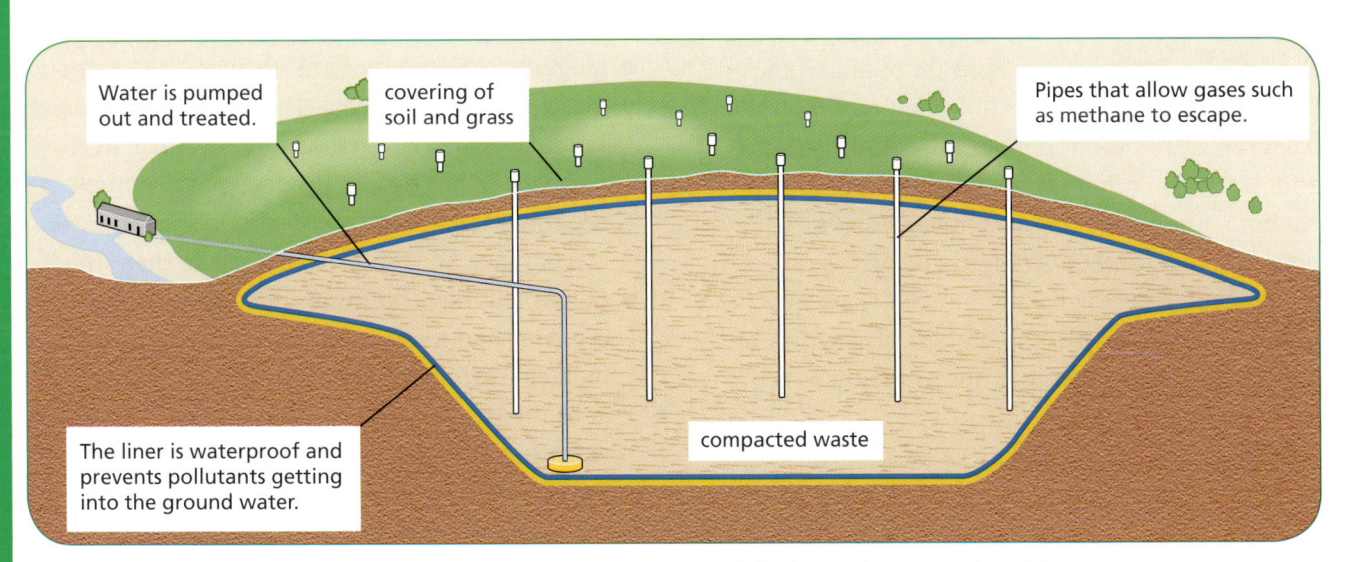

Figure 6d.3 A landfill site is not just a hole in the ground – it must be carefully designed to prevent harmful substances damaging the surrounding environment.

Depending on the kind of waste, and the conditions in which it is being broken down, biogas can contain very different proportions of the different gases. In general, the more methane the better, because this is the useful part of the gas mixture. It also makes the gas safer. Biogas containing between 5% and 15% of methane is explosive when it mixes with air. However, if there is more methane than this, it cannot explode. The gas produced at the site shown in Figure 6d.5 contains about 50% methane.

Figure 6d.5 These pipes are carrying methane, collected from a landfill site in Cheshire. The huge landfill site – about 2 square kilometres – accepts one million tonnes of waste each year, and produces 9000 m³ of biogas, which is used to generate electricity.

The temperature at which the biogas is produced affects the proportion of methane to carbon dioxide in it, because it affects the activities of the micro-organisms that are rotting the rubbish. There are two groups of anaerobic bacteria involved. One group, called mesophilic bacteria, work best at temperatures between 29 °C and 41 °C. The other group, called thermophilic bacteria, are most active at temperatures between 49 °C and 60 °C (Figure 6d.6). Overall, temperatures between 32 °C and 39 °C are best, producing biogas with the highest proportion of methane, and producing it quite quickly.

SAQ

2 Predict how biogas production will differ in winter compared with summer. Explain your prediction.

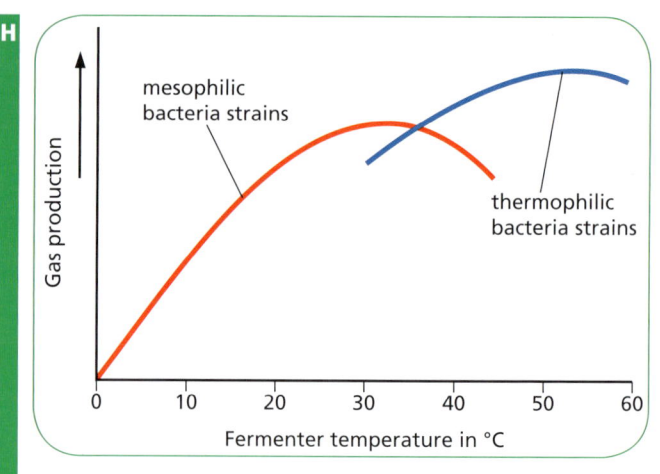

Figure 6d.6 The effect of temperature on biogas production.

Biogas and bioethanol as fuels

We can use biogas for many different purposes. For example:
- it can be burned to provide heat, which can heat water for houses
- it can be burned to generate electricity
- it can be used as a fuel for vehicles, such as buses.

In each case, methane burns in air, releasing heat energy:

methane + oxygen → carbon dioxide + water
CH_4 + $2O_2$ → CO_2 + $2H_2O$

The main use of bioethanol is in vehicle engines, and an increasing number of vehicles in Brazil now use this fuel.

What are the advantages of using biofuels, rather than petrol or diesel in car engines, or rather than gas or coal for electricity generation?
- Biogas and bioethanol are 'cleaner' fuels than diesel, petrol, oil or coal. This means that they produce less pollutants when they burn. In particular, they do not produce **particulates** – the tiny particles of carbon and other solids that can damage our lungs if we breathe them in.
- Biogas and bioethanol produce less carbon dioxide than fossil fuels. When fossil fuels are burned, carbon dioxide is produced which goes into the air and can contribute to global warming. Ethanol does this too – but it is a 'carbon-neutral' fuel. This means that the carbon in the ethanol was taken out of the air by photosynthesising plants, shortly before the ethanol was made. So the carbon dioxide being

put back into the air only matches the carbon dioxide that was recently taken out. When we burn fossil fuels, we are releasing carbon dioxide that has been locked up in the Earth for millions of years.
- Biogas and bioethanol reduce our use of fossil fuels, which are a non-renewable resource that is rapidly running out.

SAQ

3 Think back to what you know about the effects of cigarette smoke. Explain how particulates affect the lungs.

H Overall, biofuels help us to conserve scarce resources. Biogas actually makes some use of rubbish that would otherwise just get thrown away, causing its own pollution problems. As biogas is a mixture of gases, not just methane, it contains less energy for a given volume than natural gas – but that is easily outweighed by its much cheaper production and the fact that it is renewable. Using biofuels increases the sustainability of our energy usage, providing a resource that will continue to be available for as long as we keep producing biodegradable rubbish. And, of course, burning biogas gets rid of methane, a gas which otherwise would increase the greenhouse effect and increase the problem of global warming.

Explosion at Loscoe

There had been a landfill site at Loscoe, in Derbyshire, ever since 1971, making use of a deep quarry that had once been used as a brickworks. Up until 1977, it had been used only to dispose of 'inert' waste – that is, non-biodegradable rubbish such as concrete rubble, plastics and metals. But then a licence was granted to the company that owned the tip, allowing them to include some domestic waste as well.

There were houses all around the site. People living there kept on complaining about the houseflies and smells from the tip, but no-one took much notice. Then, in 1984, people started to complain that the plants in their gardens were dying. Some of them even noticed that the soil was getting quite hot in places. But still no-one realised just what was happening.

Then, in March 1986, there was a large explosion inside one of the houses. It was completely destroyed – amazingly, no-one was killed.

Only then did the council begin to monitor methane levels in the soil and air around the site. They found that there had been dangerous quantities of methane in several of the houses, so it was just luck that nothing worse had happened. Belatedly, pipes were inserted into the underground waste, allowing methane to flow upwards, where it was safely burned.

Summary

You should be able to:

- describe how bacteria rot organic material to produce methane and carbon dioxide (biogas)
- describe how biogas is produced in digesters, septic tanks, marshes, animal digestive systems and landfill sites, including continuous flow systems
- state the composition of biogas, and how this is affected by temperature
- **H** explain how the proportion of methane affects the danger of explosion of biogas
- explain why different kinds of bacteria are needed to produce biogas
- explain how biogas production is affected by temperature
- describe some uses of biogas
- describe the advantages of using biofuels such as bioethanol and biogas
- **H** explain these advantages

Questions

1. a Name the *three* main gases that are present in biogas.

   b Explain why biogas is more useful when it contains a high proportion of methane.

2. a What is bioethanol and how is it produced?

   b Explain why bioethanol is said to be 'carbon-neutral'.

   H c Do you consider biogas to be carbon-neutral? Explain your answer.

3. Discuss each of the following.

   a Cattle contribute significantly to the greenhouse effect.

   b Burning biogas can reduce the rate of global warming.

   c Biogas is a 'cleaner' fuel than diesel.

   d Using biofuels is more sustainable than using fossil fuels.

6e Life in soil

What is soil?

If you put some soil into water, shake it up and let it settle, you will see something like the diagram in Figure 6e.1. This gives an idea of what soil contains.

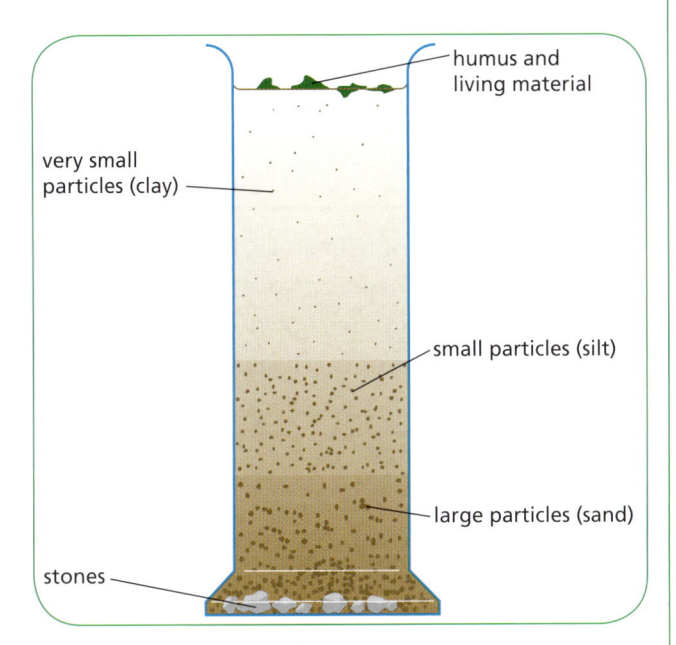

Figure 6e.1 The contents of soil.

Most of what you will see is soil particles. These are little grains of minerals, which were produced by the weathering and erosion of rocks. They are **rock particles**. The particles can be of several different sizes. The larger ones are **sand** particles, and the smaller ones are **silt** and **clay** particles. There may also be some really big particles, so big that we can call them gravel, stones or pebbles.

You may also see some things floating on the top of the water. These are probably the remains of living organisms that have started to decay, such as partly rotted dead leaves. This decaying material is **humus**. Humus is a dark, soft material.

You might see some **living organisms** in the soil, such as earthworms. There will be millions of living organisms that you *cannot* see, because they are so small. These are the **microorganisms**, including bacteria and fungi. Soil is teeming with them.

Soil also contains air and water, although of course you won't see these if you have shaken the soil up with water. Figure 6e.2 shows what tiny specks of soil would look like if you magnified them.

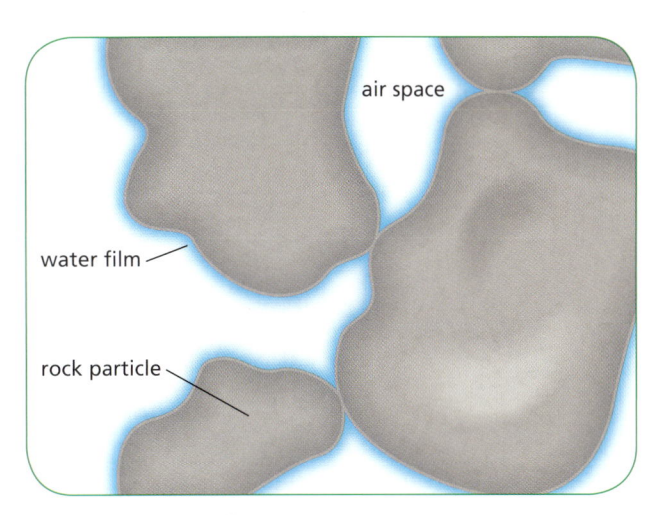

Figure 6e.2 Grains of soil, close up.

SAQ
1 List the main components of soil.

Organisms in soil

There are many different species of living organisms that spend most, if not all, of their lives in the soil. Many of these are **aerobic** organisms, and they rely on the **oxygen** that is present in the air spaces between the soil particles. If the soil gets too wet, then these spaces may fill up with water instead of air, and we say the soil is **waterlogged**. This considerably reduces the amount of oxygen in the soil. Although soil organisms do need some water, too much of it can make it very difficult for them to live there. If you look on a lawn after there has been really heavy rain for several days, and the soil is waterlogged, you may see earthworms that have crawled out of the soil, needing to find more oxygen.

Figure 6e.3 (overleaf) shows some of the different kinds of organisms that live in soil.

There are also many plant roots in the soil. Plants rely on soil for:
- anchoring their roots firmly, so that the plant cannot be uprooted by grazing animals, or blown away by the wind
- a source of minerals – for example, inorganic ions like nitrate, which they need to make proteins and other substances
- a source of water.

88 B6e Life in soil

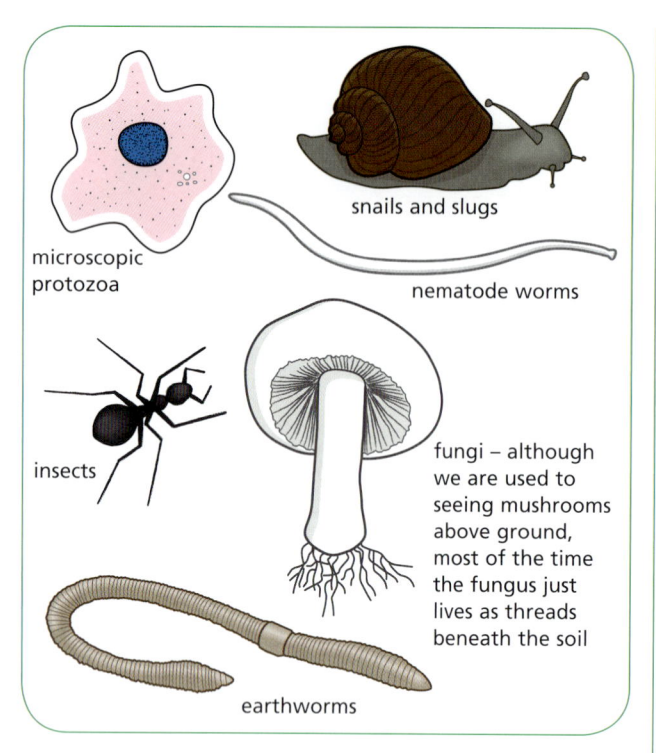

Figure 6e.3 Some soil organisms (not drawn to scale).

Soil food webs

Figure 6e.4 shows how some of the different kinds of soil organisms are linked together in a **food web**.

Just as in any food web, there are producers and consumers in this web. However, the web differs from 'above-ground' webs because it contains huge numbers of **detritivores**. These are organisms that feed on dead parts of plants and animal waste products.

SAQ

2 Different species of nematodes feed in different ways, and so they appear twice in the food web in Figure 6e.4. They are shown as herbivores. Do you think it is correct to classify them all as herbivores? Explain your answer.

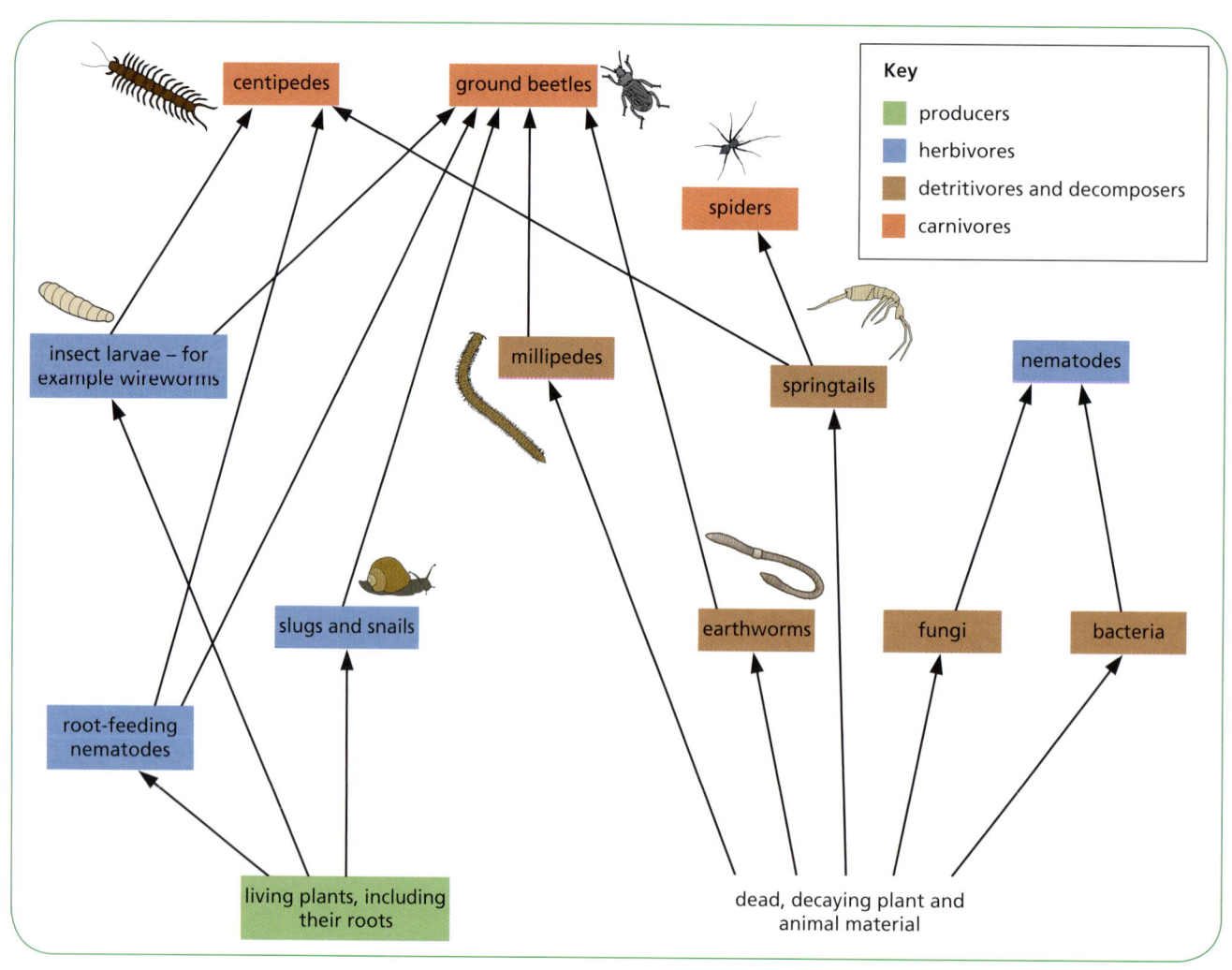

Figure 6e.4 A soil food web.

Improving the soil

Gardeners often talk about 'improving' their soil. They mean making it better for plants to grow in, so that they get strong, healthy plants with green leaves, colourful flowers or high yields of vegetables.

As we have seen, plants need minerals, oxygen and water from soil, as well as good support for their roots.

Minerals are provided in the soil particles themselves, and also from the activities of detritivores and decomposers, which produce nitrates and other ions that plants can absorb and use. Detritivores and decomposers are most active in soils with a high humus content, so adding compost or other organic material – for example, horse manure – helps to provide a good source of minerals for plants.

In order to be able to absorb the minerals they need, most plants need a soil with a neutral or slightly alkaline pH (Figure 6e.5). If a gardener or farmer has a soil with an acidic pH, it can be improved by adding lime. The lime neutralises the acidity in the soil, as well as providing an extra source of calcium ions, which plants need for forming strong connections between their cells.

Oxygen is needed by plant roots, so that their cells can respire and produce the energy that they need for carrying out active transport. Roots get their oxygen by diffusion from the air spaces in the soil. If the soil is compacted – for example, by people walking over it a lot, or by heavy machinery moving across it – then this can squeeze the soil particles very close together, so

Figure 6e.6 Spiking a lawn lets air into the soil, helping the grass to grow better.

there is little air present. Plants don't grow well in compacted soil. Gardeners sometimes aerate lawns by spiking holes all over them (Figure 6e.6).

Plants get their water from the soil. The water clings to the surfaces of the soil particles by surface tension. The smaller the soil particles, the larger the surface area to volume ratio, and the more water the soil will hold. Soil with big particles, such as sandy soil, doesn't hold much water and can rapidly dry out. Adding humus, which is very good at absorbing and holding the water, can help to make sure water is available to plant roots even if it has not rained for a while.

Digging soil helps to turn it over and aerate it. Digging mixes up different layers of soil. Humus tends to form mostly near the soil surface, so digging can help to move some of this humus lower down. It also helps to move other nutrients upwards – soluble mineral ions may get washed

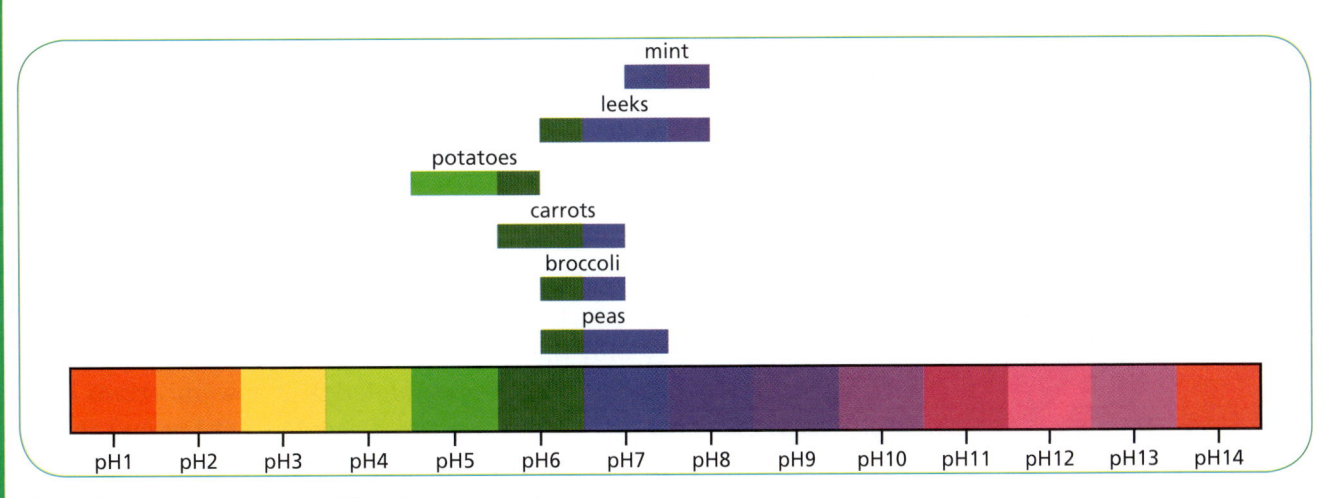

Figure 6e.5 pH ranges in which different plants grow best.

90 B6e Life in soil

down into the deeper layers of the soil when it rains, perhaps too deep for plant roots to reach them. Turning the soil over makes these minerals more accessible to the roots.

SAQ

3 Explain why plants cannot take up many of the mineral ions in the soil unless they have plenty of oxygen.

Earthworms

Of all the animals that live in the soil, earthworms probably do the most to help make the soil fertile and good for plant growth.

- Earthworms make burrows through the soil. The burrows help to let rainwater get right down into the soil, and also help it to drain away without causing waterlogging. The burrows allow air to penetrate the soil, providing oxygen for plant roots and other organisms.
- Some earthworms feed on the humus and microorganisms in soil. They take soil into their mouths and pass it through their digestive systems, egesting the waste material. Some species of worms leave their waste on the surface, as worm casts. Earthworm waste is richer in bacteria and mineral ions than the rest of the soil, making it an excellent growing medium for plants. The waste material from worms also helps to neutralise acid soils.
- Other types of earthworms feed on leaves and other vegetation that they drag down into the soil (Figure B6e.7). This improves the humus content of the soil, which helps it to hold water. The rotting leaves add to the soil mineral content.
- The burrowing activities of earthworms help to mix up different layers of soil, so plants get access to minerals that might otherwise be too deep down for their roots to reach.

Figure 6e.7 The activities of earthworms help air, water and food for decomposers to get into the soil.

Figure 6e.8 As well as improving the soil, earthworms are an important part of food webs.

Charles Darwin was one of the first people to make a formal study of how earthworms affect soil. In 1881, he published a book called *The Formation of Vegetable Mould* in which he described some of his findings. ('Vegetable mould' is compost or humus.) It was a best-seller. For the first time, people realised how important earthworms were in helping soil to form and in making it good for plant growth. Nevertheless, agricultural practices that did serious damage to earthworm populations still continued for many years into the 20th century – for example, spraying crops with pesticides that killed not only the pests, but also the worms in the soil beneath. Now we understand how important it is to encourage worm activity in soil, because their activities help to keep the soil in good condition and improve crop yields.

Nutrient cycles

In *Gateway Additional Science*, pages 103–106, we saw how carbon and nitrogen are cycled between living organisms and their environment. Many of the organisms that are involved in these cycles live in the soil.

Carbon and nitrogen are not the only elements that are cycled in this way. All the atoms in your body have probably once been part of another organism somewhere, at some time. Although most of these atoms are hydrogen, carbon, oxygen or nitrogen atoms, your body also contains smaller numbers of many others, such as phosphorus – found in all cell membranes – and sulfur, which is found in the keratin that makes up hair and nails. When an organism dies, these elements are returned to the soil or air by the activities of decomposers, many of which are bacteria. Without them, the elements could not be re-used to make the bodies of new organisms.

Figure 6e.9 shows the nitrogen cycle. This diagram is like the one you have already met, in *Gateway Additional Science*, but it contains a little more information about the bacteria that are involved in it.

Some bacteria make a living by feeding on dead animals and parts of plants, as well as on animal faeces. They are **decomposers**. They feed by secreting enzymes that digest the large molecules in these substances, breaking them down into small, soluble ones that they can absorb. This method of nutrition is called **saprotrophism**, and the bacteria are **saprotrophs**. (An alternative name is **saprophytes**.) Fungi feed in the same way, and they, like bacteria, are very important in causing decay.

The saprotrophs, both bacteria and fungi, produce **ammonia** as a waste product from amino acids that they break down. The ammonia is acted on by another group of bacteria, called **nitrifying bacteria**. These bacteria obtain energy by changing ammonia to **nitrites** and then to **nitrates**. Two kinds of bacteria that do this are *Nitrosomonas* and *Nitrobacter*. Ammonia has the formula NH_3, whilst nitrite ions and nitrate ions are NO_2^- and NO_3^- respectively. The process therefore uses oxygen, and can only take place in well-aerated soils. This is the main reason why waterlogged soils tend to be short of nitrate ions, making poor conditions for plant growth.

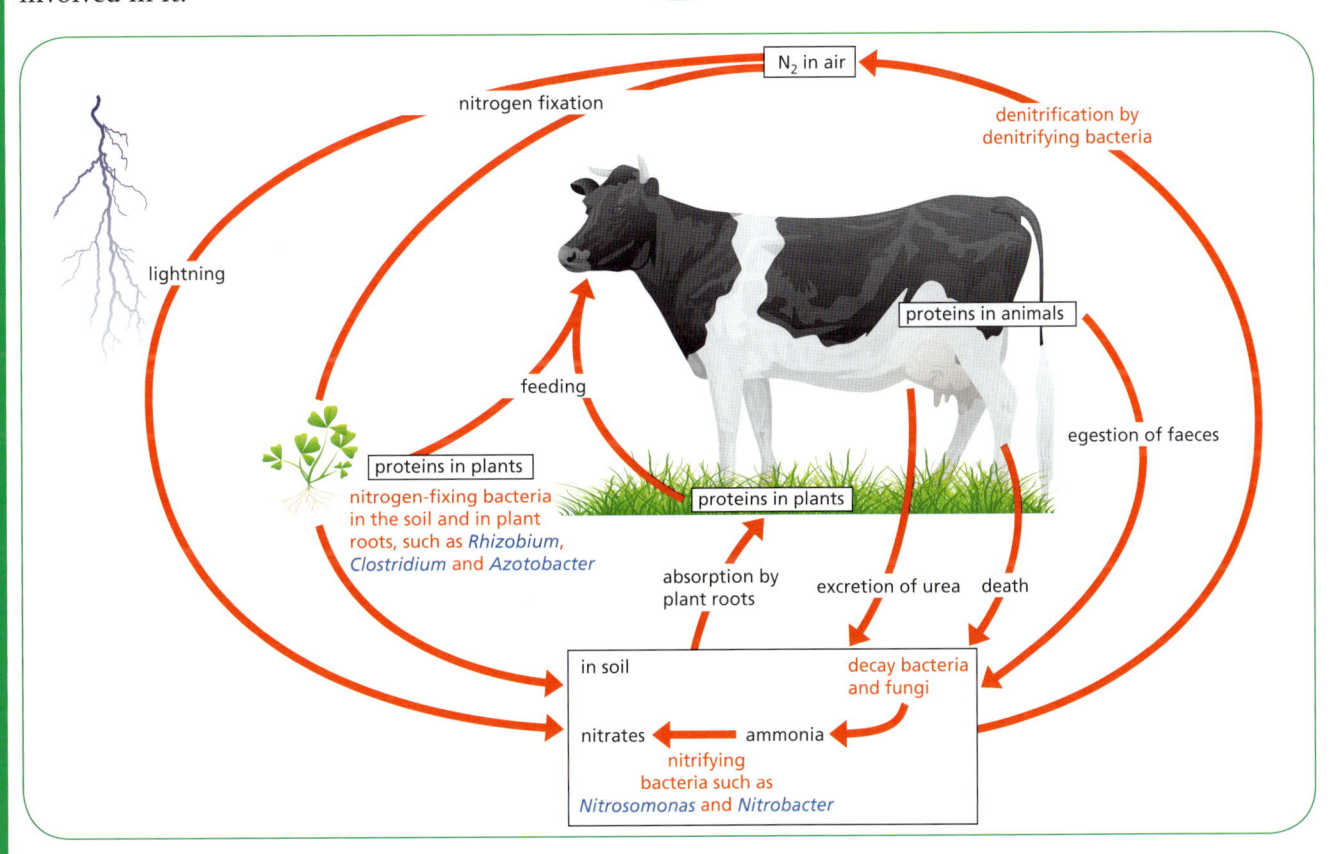

Figure 6e.9 The nitrogen cycle.

Nitrogen-fixing bacteria also help to add nitrogen compounds to the soil. These bacteria are able to convert nitrogen gas from the air into ammonium ions. Some of these bacteria – for example, *Azotobacter* and *Clostridium* – are free-living in the soil. Another important nitrogen-fixer, *Rhizobium*, sometimes lives free in the soil, but often lives in root nodules in particular species of plants. Many of these are **legumes** – plants belonging to the pea and bean family – but there are also many other plants that have *Rhizobium* in their roots. This is a symbiotic relationship, from which both the plants and the bacteria benefit.

Fire warning to gardeners

Every year, the London Fire Brigade is called out to fires in people's gardens. These are not the result of bonfires that have got out of control. The fires just start up all on their own, in the compost heap.

Compost heaps are teeming with organisms, many of them microscopic. All of them are carrying out metabolic reactions, and many of these reactions are exothermic, releasing a lot of heat energy. Given plenty of moisture and poor air circulation, the reactions can get so vigorous that they set fire to the decaying grass, straw or whatever else is in the heap.

The Fire Brigade do their best to educate gardeners on how to avoid this problem – but most gardeners remain completely unaware of it. Leaflets are sent out advising gardeners not to store grass cuttings or other material in bin liners, because this stops air circulating and also traps heat inside the rotting material. The leaflets also advise gardeners to turn their compost heaps regularly – and to site them well away from the garden shed, garage, or back door.

Summary

You should be able to:

- describe the components of soil, and list some of the organisms that live in soil
- state why soil is important for plants
- describe a soil food web, containing herbivores, detritivores and carnivores
- explain why oxygen and water are needed for life in soil
- describe why aeration and drainage can improve soil, and why neutralising acid soils and mixing layers are important
- state that nitrogen, carbon, sulfur and phosphorus are recycled in soil, explain why this is important, and state that different kinds of bacteria are involved
- describe the importance of earthworms to soil fertility
- describe how Darwin highlighted the importance of earthworms in soil
- describe the roles of named types of bacteria in the nitrogen cycle

Questions

1 Explain how each of these activities of earthworms makes soil better for plant growth.

   a making burrows

   b making worm casts

   c dragging leaves into their burrows

2 a List *four* elements that are recycled between living organisms and their environment.

   b For each of these elements, state *one* reason why it is needed by living organisms.

   c Outline the roles that bacteria play in recycling carbon. (You may need to look back at *Gateway Additional Science*, pages 103–104, to help you with this.)

H 3 A gardener noticed that plants grew much better in one part of her garden, area A, compared with the rest. She took a sample from area A, and also from two other areas. She sent them off to be analysed. Table 6e.1 shows the results she received.

Sample area	Total bacterial biomass in µg per gram	Total fungal biomass in µg per gram
A	200	240
B	150	190
C	130	100

Table 6e.1

   a Suggest reasons for the differences in the quantity of bacteria and fungi in the three different soil samples. You may be able to think of four or five different reasons.

   b Explain why plants may grow better in soils where there are plenty of bacteria and fungi.

   c Suggest what the gardener could do to improve the growth of her plants in the areas where samples B and C were taken.

B6f Microscopic life in water

A drop of pond water placed under a microscope reveals a whole world of tiny organisms. Some of them are single-celled – for example, amoebae and bacteria. Some of them are made of several cells linked together, such as the long threads of filamentous algae. Some are made of thousands of cells, and are simply miniature animals or plants. Figures 6f.1 and 6f.2 show some of the organisms you might see in a drop of pond water.

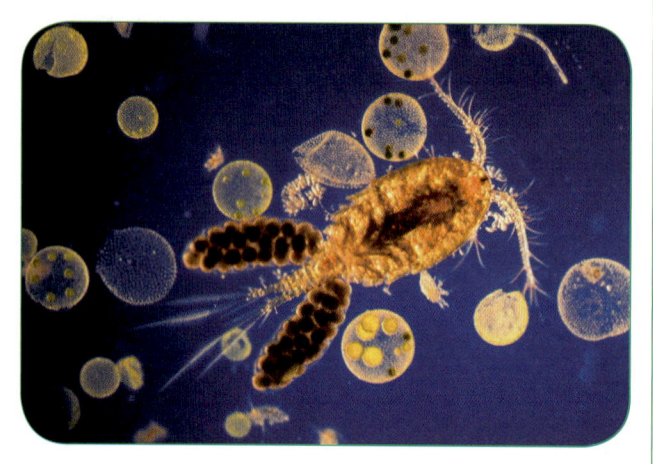

Figure 6f.1 This is a tiny crustacean called *Cyclops*. It is a close relative of crabs and woodlice. The little balls are a kind of alga – very simple plant-like organisms.

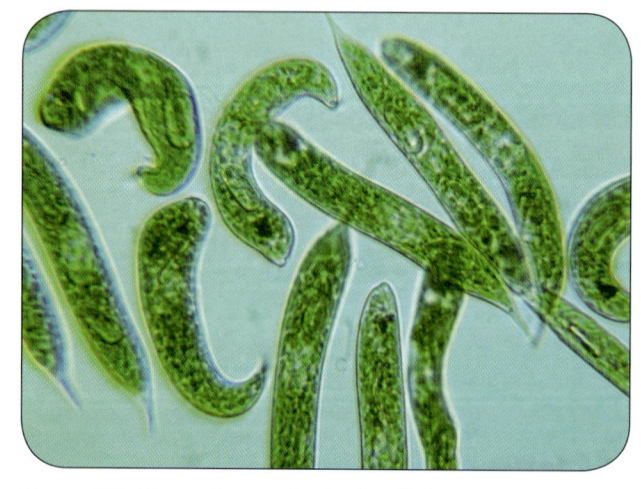

Figure 6f.2 These microscopic, single-celled organisms are called *Euglena*. They can move around like animals, but photosynthesise like plants.

Life in water and on land

We think that life first evolved in water. Water is a much easier environment for organisms to live in. This may not seem so obvious to us, because we belong to a species that has evolved to live on land.

These are some advantages of living in water:
- Aquatic (water-living) organisms are in no danger of running short of water and suffering dehydration. All organisms that live on land have bodies that are adapted to conserve water – for example, by having a waterproof covering and keeping gaseous exchange surfaces deep inside the body.
- Water needs a lot of heat added or removed from it before its temperature changes very much. It has a high specific heat capacity (*Gateway Science*, pages 259–260). This means that its temperature is much more stable than air temperature. Aquatic organisms do not experience the large and sometimes rapid temperature changes that happen on land.
- Water is much denser than air, and it provides support. Aquatic organisms do not need such strong skeletons as land organisms.
- Living in water makes it easy to dispose of your waste products – an aquatic animal can simply release them into the water around it, or let them escape by diffusion.

There are, however, some disadvantages to living in water. For example:
- Water tends to enter the cells of aquatic organisms by osmosis, because the concentration of chemicals in the cytoplasm is often higher than the concentration of chemicals in the water around the cells. This is more of a problem for fresh-water organisms than ones that live in salt water, because salt water has a concentration that is very similar to that of cells.
- Water resists movement through it – it takes much more energy to move forward through water than on land. Many aquatic organisms have evolved streamlined shapes that make this easier.
- Water has much less oxygen dissolved in it than is present in air. It is therefore more difficult to obtain oxygen for respiration in water than in air.

Some living organisms use both environments. Frogs and many other amphibians spend their early lives as tadpoles, living in water. They then change, or **metamorphose**, into land-living adults. It is thought that this life cycle may help to avoid competition for food between the adults and the young. Many insects, such as dragonflies and mosquitoes, also have larvae that live in water before emerging and metamorphosing into flying adults (Figure 6f.3).

Figure 6f.3 Dragonfly nymphs spend up to three years living in water, where they are aggressive predators. The adults are also carnivores, capturing prey on the wing.

SAQ

1 The photographs in Figure 6f.4 show a jellyfish on a beach and in water.

   How does this illustrate one of the advantages of living in water?

Figure 6f.4

Solving osmosis problems

In fresh water, water tends to enter cells by osmosis (*Gateway Additional Science*, pages 63–66). This is not a problem for cells that have a cell wall around them, such as bacteria and plant cells, as the strong wall resists bursting. However, if an animal cell takes in too much water, it can swell and burst.

Amoeba is a single-celled organism which is like an animal cell. (Amoeba is not classified as an animal, because animals are *multicellular* organisms.) Water constantly enters the cell, through its partially permeable cell surface membrane, by osmosis. The amoeba deals with this using a **contractile vacuole**. As the water enters, it is collected into the vacuole, which steadily grows larger. When it reaches a certain size, it bursts and releases the water outside the cell (Figures 6f.5 and 6f.6, overleaf).

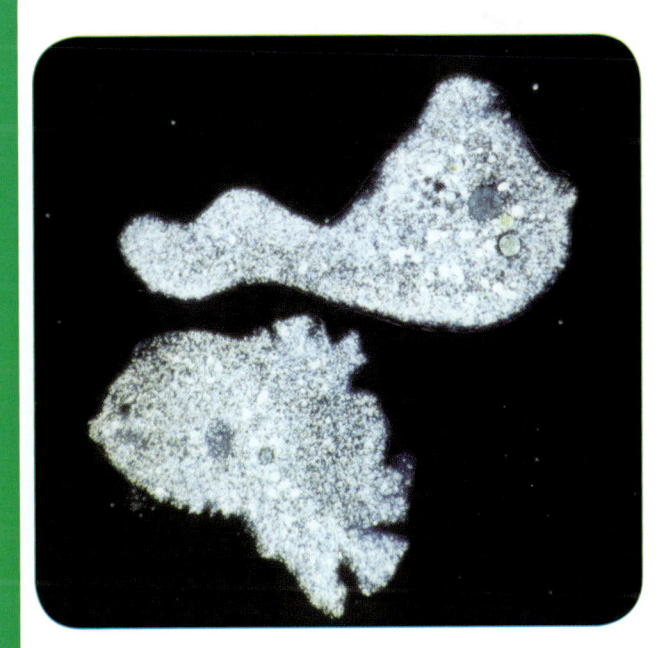

Figure 6f.5 The cell of an amoeba is about 0.2 mm across, so you can just see it with the naked eye.

SAQ

2 An amoeba that usually lives in fresh water is put into sea water. Predict and explain what you would expect to happen to the rate at which its contractile vacuole fills and empties.

There are some organisms, such as salmon and eels, which spend part of their life in the sea and part in fresh water. When in the sea, the sea

96 B6f Microscopic life in water

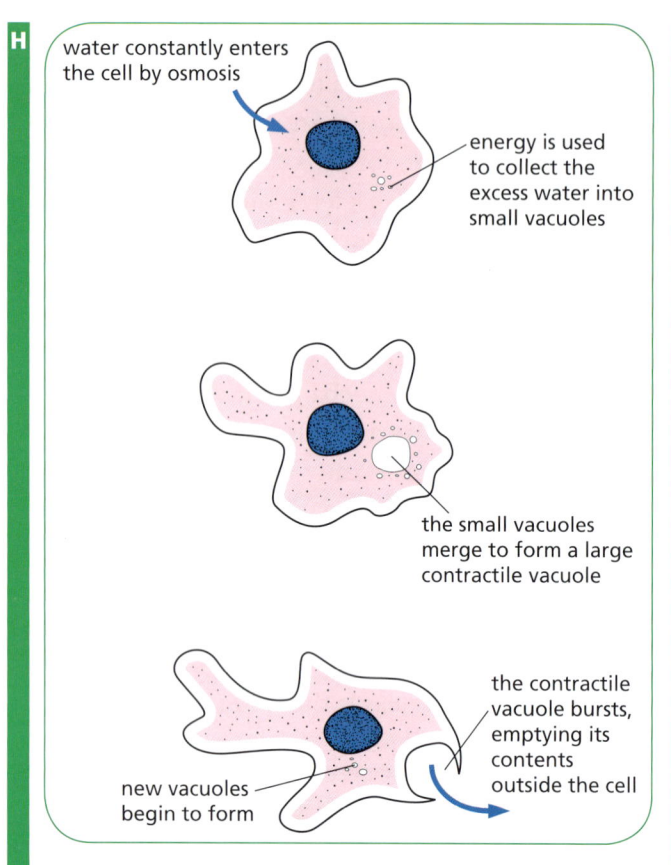

Figure 6f.6 Water regulation in an amoeba.

- water constantly enters the cell by osmosis
- energy is used to collect the excess water into small vacuoles
- the small vacuoles merge to form a large contractile vacuole
- the contractile vacuole bursts, emptying its contents outside the cell
- new vacuoles begin to form

water outside them is more concentrated than the contents of their cells. Ions such as sodium and chloride diffuse into their cells from the sea water, while water moves out by osmosis. The salmon therefore have to excrete excess salt from their body. But when they move into fresh water, they have to remove excess water (Figure 6f.7).

The water and salt regulation is done by the salmon's kidneys and gills. When in the sea, the kidneys produce only a little urine, which is very concentrated. Cells in the gills push sodium and chloride ions out, using active transport. When the fish moves into a freshwater river, the kidneys begin to produce large quantities of dilute urine. The active transport in the gills switches around, so that sodium and chloride ions are moved *into* the cells.

Plankton

Many of the organisms that live in the sea or lakes are so tiny that they cannot swim effectively against any movements of the water. They drift wherever the water currents take them. They are called **plankton**.

Some plankton are plant-like, with cell walls and chloroplasts, in which photosynthesis takes place. They are called **phytoplankton** (Figure 6f.8). Others are animals, for example young larvae of crustaceans (such as crabs, lobsters or barnacles) or coral animals. They are called **zooplankton** (Figure 6f.9).

Figure 6f.8 Phytoplankton – microscopic plant-like organisms that float in the upper layers of the sea.

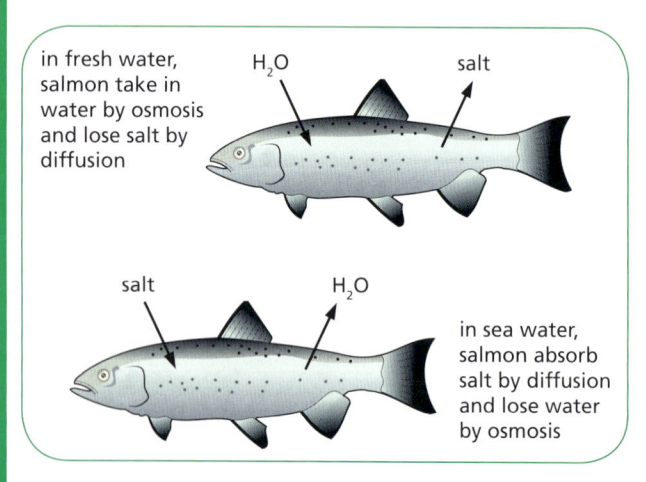

- in fresh water, salmon take in water by osmosis and lose salt by diffusion
- in sea water, salmon absorb salt by diffusion and lose water by osmosis

Figure 6f.7 Problems of water regulation for salmon.

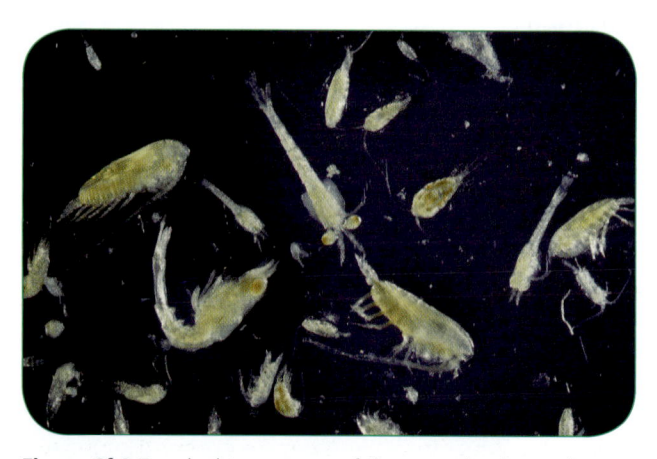

Figure 6f.9 Zooplankton – some of these are tiny larvae that will eventually grow into larger organisms such as crabs or sea anemones.

B6f Microscopic life in water

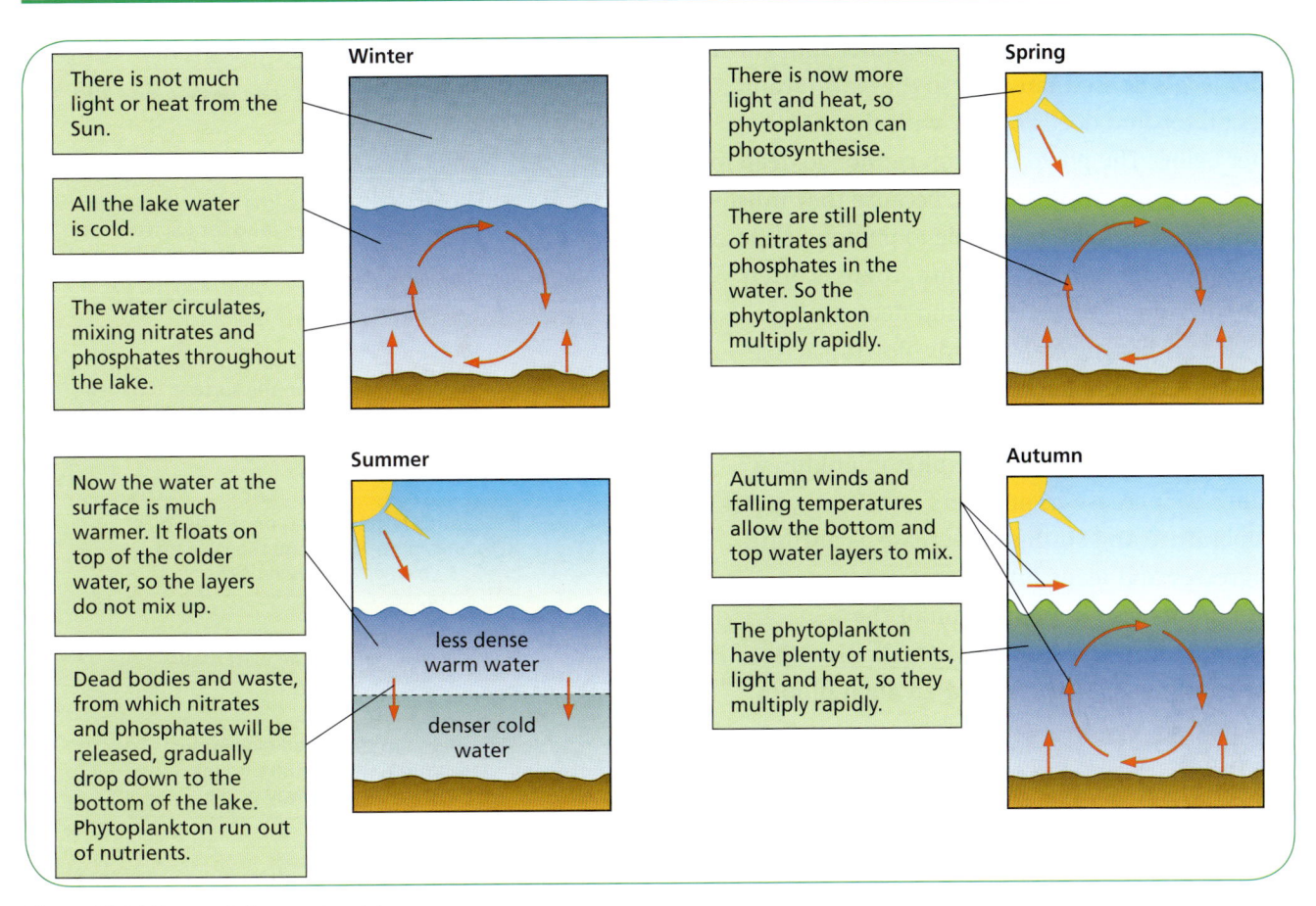

Figure 6f.10 Seasonal changes in a lake.

SAQ

3 Explain why phytoplankton are mostly found near the surface of lakes or the sea.

Although they are very small, the huge numbers of phytoplankton give them a very important role in food chains. They are the main producers in the open ocean and also in many lakes. Like plants on land, their numbers show seasonal variations.

Figure 6f.10 shows the changes in the population of plankton in a lake during one year. The graph in Figure 6f.11 also shows how the temperature and light intensity changed during this time.

Early in the year, days are short, the Sun is low in the sky and temperatures are low. As spring arrives, light levels and day length increase, as does temperature. Photosynthesis can now take place more rapidly, and the populations of phytoplankton increase. There are usually plenty of nitrate ions and phosphate ions in the water at this time of year, which the phytoplankton need to allow them to make proteins and other substances for their new cells.

The rise in phytoplankton numbers does not usually last long. As their numbers increase, so do the numbers of animals that eat them, including zooplankton and also larger animals. This reduces the phytoplankton numbers. They also begin to run out of nitrate and phosphate ions. There may be plenty at the bottom of the lake, but the water at

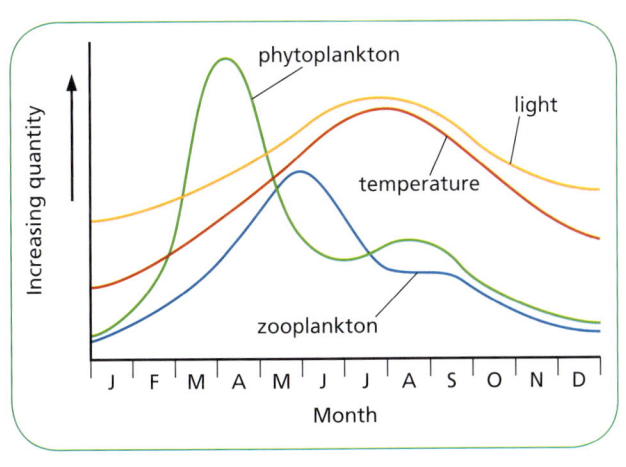

Figure 6f.11 Changes in the populations of plankton, the temperature and the light intensity, in a lake during one year.

the top is now warming in the summer sun. Warm water is less dense than cold, so the warm water stays on top and does not mix with the colder water below. The phytoplankton cannot go deep to reach the minerals they need, because they must stay where it is light, to be able to photosynthesise.

In later summer and autumn, there is often some mixing of the surface and deeper water, which may be caused by winds blowing across the surface of the lake. Nitrate and phosphate ions from the bottom of the lake are brought up to the surface. This gives the phytoplankton a chance to increase in numbers once more, until temperature and sunlight decrease as autumn progresses and leads into winter.

A big increase in population of the phytoplankton can sometimes make the whole lake look green. This is called an **algal bloom**.

Pollution and aquatic life

There are many different substances that can cause water pollution. Untreated sewage, pesticides, detergents, fertilisers and oil (*Gateway Science*, page 167) are all dangerous to micro-organisms and other organisms that live in rivers or the sea. Chemicals known as **PCBs**, which were used in the past for many different applications in industry, caused serious harm before their dangers were recognised and their uses restricted.

In *Gateway Science*, page 125, we saw how untreated sewage that runs into waterways can start off a process that results in shortage of oxygen for the animals that live in the water. This is called **eutrophication**, and it is summarised in Figure 6f.12.

Different species of animals are able to live in water with different concentrations of dissolved oxygen. Some can cope with the almost oxygen-free conditions in badly polluted water, whilst others need high concentrations of oxygen, such as are found in fast-running unpolluted streams and rivers. We can use this fact to check on the pollution status of a river, using certain species of animals as **indicators** (Figures 6f.13 and 6f.14). This is described on page 127 in Gateway Science. A similar approach can be used to check on whether acid rain is causing pollution.

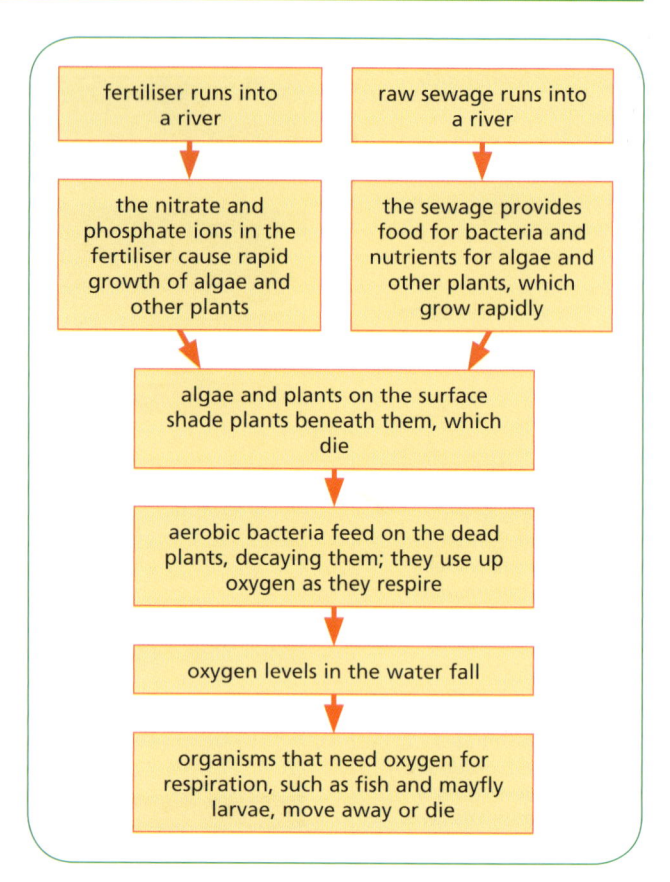

Figure 6f.12 Eutrophication.

Figure 6f.13 This mayfly larva, *Ephemerella ignita*, is only found in water where there is plenty of oxygen and the pH is not too low.

Figure 6f.14 Blood worms, *Chironomus*, are only found in polluted water where there is little oxygen.

Some pollutants are especially dangerous, because they do not break down in the environment or in the bodies of living organisms. This means that they can build up along a food chain, a process known as **bioaccumulation**. The insecticide **DDT**, banned in Europe and the USA since the 1970s, is one such pollutant. Its effects were described in *Gateway Additional Science*, page 88. The DDT could travel huge distances from the places where it was used. Even whales caught hundreds of miles out to sea were sometimes found to have large concentrations of DDT in their blubber. DDT is still used in some parts of the world, where it is an important weapon in the fight against insect-transmitted diseases, especially malaria.

PCBs, like DDT, are organochlorine chemicals. They do not break down naturally, and can undergo bioaccumulation. 'PCBs' stands for polychlorinated biphenyls. In the past, PCBs had a very wide range of uses in industry – for example, as coolants, as hydraulic fluids in machinery and as plasticisers in paint. Aquatic animals at the top of food chains, such as seals and whales, have sometimes been found with large concentrations of PCBs in the fatty tissues in their bodies.

PCBs are known to affect reproduction in fish. There is also some evidence that reproduction might be affected in people who have eaten a lot of fish contaminated with PCBs. For example, the Great Lakes in the USA were badly affected by PCBs in the 1970s, before the harm they were doing was appreciated. Sports fishermen enjoyed fishing there, and often took home the large, predatory fish that they caught, cooking them and eating them. A study of the wives of sports fishermen found that there was a strong correlation between the concentration of PCBs in their blood and their fertility. Women with high PCB concentrations had shorter menstrual cycles, gave birth to babies with smaller birthweights, and were less likely to give birth to boys.

PCBs were banned in the 1970s, and now it is illegal to manufacture them. There are also strict international regulations on disposing of them, from old machinery or other substances in which they are already present.

Very small to very large

Plankton are the smallest organisms in the sea – so small that we cannot see most of them with the naked eye. Yet they form the food of some of the largest animals on Earth, including several species of whales and sharks.

Basking sharks are often seen in the sea around Cornwall and the west coast of Scotland. They are the second largest species of shark, growing up to almost 10 m long. They are completely harmless, feeding by swimming through the sea with their huge mouths open, filtering out plankton from the water. They are called 'basking' sharks because they often swim very near to the surface of the water, sometimes with their dorsal and tail fins breaking the surface. Their noses often stick out of the water as well, making them look like a three-humped animal – perhaps the origin of tales of huge sea serpents.

Figure 6f.15 A basking shark feeding off the coast of Cornwall.

Summary

You should be able to:

- describe the advantages and disadvantages of living in water compared with living on land
- describe how amphibians and some insects exploit both water and land habitats
- **H** explain how osmosis causes problems of water balance, and describe the function of the contractile vacuole in amoebae
- describe the problems for animals that move between salt water and fresh water
- explain what phytoplankton and zooplankton are, and their roles in food webs
- describe and explain why plankton show seasonal variations in numbers
- **H** interpret data on marine food webs
- use data to determine the source of pollution by sewage, oil, PCBs, fertilisers, pesticides and detergents
- explain how untreated sewage and fertilisers can cause eutrophication
- describe how indicator species can be used to estimate pH and oxygen levels in water
- **H** describe the long-term effect of PCBs and DDT on animals such as whales

Questions

1 a List *three* advantages of living in water compared with living on land.

   b List *two* disadvantages of living in water compared with living on land.

   c Name *one* species of organism that lives in water and on land at different stages of its life cycle. Suggest how this benefits the species.

2 a What are plankton?

   b The numbers of phytoplankton usually increase during spring. Explain how each of these factors is involved in this increase:
      i light
      ii temperature
      iii nitrates and phosphates.

continued on next page

Questions - continued

3 Figure 6f.16 shows a food web in the sea.

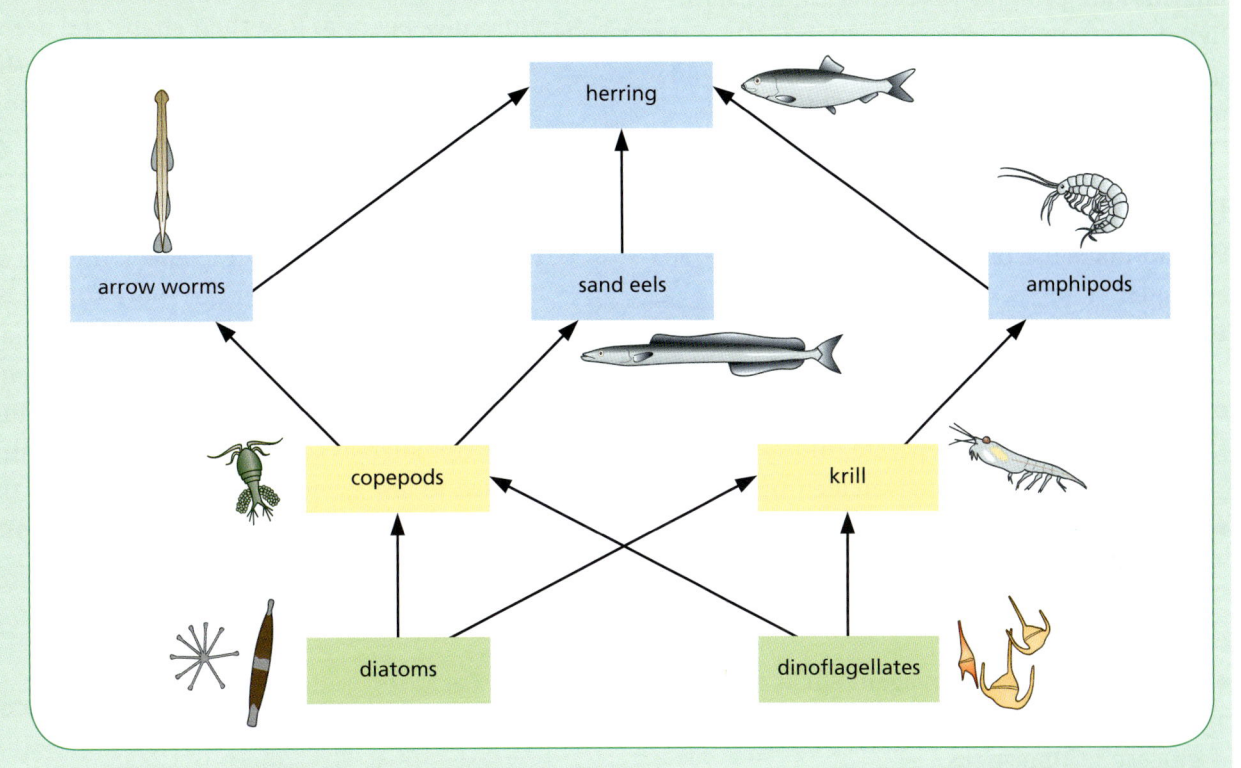

Figure 6f.16

a State the colour that has been used to show each of these trophic levels in the food web:
  i producers
  ii primary consumers (herbivores)
  iii secondary and tertiary consumers (carnivores).

It has been calculated that about 10% of the energy contained within one trophic level is passed on to the next trophic level.

b Explain why energy is lost between trophic levels.

c Calculate how many grams of diatoms and dinoflagellates would be needed to support one gram of herring. Assume that the energy content per gram of each organism in the food web is the same.

B6g Enzymes in action

What are enzymes?
Enzymes are proteins that act as catalysts. All living things make enzymes, which catalyse the metabolic reactions that take place inside and outside their cells.

Using enzymes
We can make use of enzymes in many ways. For example:
- Enzymes are added to washing powders to help to remove stains.
- Enzymes are added to milk to digest proteins. This makes the milk separate into solid curds and liquid whey. Cheese is made from the curds.
- Enzymes that break down cellulose cell walls are added to apples or other fruit. This makes it easier to squeeze juice out of them to make drinks.
- Enzymes are used on reagent sticks to detect substances in urine or blood – for example, glucose testing sticks.
- Enzymes are added to foods to break down substances, altering the taste of the food.

Biological washing powders
Washing powders contain detergents. These help to break fats up into tiny droplets so that they can mix with water. Some washing powders also contain enzymes (*Gateway Additional Science*, pages 192–193.) The enzymes help to remove biological stains such as blood, sweat or grass. These stains may contain proteins and other substances that detergents cannot remove. Different enzymes have different effects:
- **amylase** breaks down starch to maltose
- **lipase** breaks down fats to fatty acids and glycerol
- **protease** breaks down proteins to amino acids.

Maltose, fatty acids, glycerol and amino acids are all soluble in water, so they can be washed away easily.

A little bit more care has to be taken when using biological washing powders rather than ones that contain only detergents. This is because enzymes will only work in certain conditions.

Most of them work best at temperatures below 50 °C and in a neutral pH. Otherwise, the enzyme molecules begin to lose their shape and are said to be **denatured**. Once this has happened, they cannot bind with their substrate and can no longer make a reaction take place.

Glucose test strips
People with diabetes need to check the glucose levels in their blood or urine. There should be no glucose in urine. If there is, this is a sign that the blood glucose level has gone too high, which is dangerous (*Gateway Science*, pages 49–50). Glucose test strips such as Clinistix or Dextrostix can be used to test urine. These are little strips that are dipped into the urine. They change colour according to how much glucose is present.

Figure 6g.1 Colour chart for a glucose test strip. In this strip, light green indicates no or very low glucose, and the more red–brown it becomes, the more glucose is present.

Most people who need to test their blood glucose now use **biosensors**, which give a digital read-out of the actual concentration of glucose in their blood.

Both the strips and the biosensors contain enzymes that catalyse a reaction involving glucose. The reactions produce a coloured product on the test strips, and a small electric current in the biosensor. The electric current provides a reading that indicates the amount of glucose.

SAQ

1 Another way of testing for glucose is to use Benedict's test.
  a Describe how you would use Benedict's test to see if a sample of urine contains glucose.
  b Suggest why test strips or biosensors are better methods than using Benedict's test.

Invertase and sweetness

Most of the sugar that is used to make processed foods comes from sugar cane or sugar beet. This sugar is **sucrose**. Sucrose is a complex sugar or **disaccharide**. Each molecule of sucrose is made of one glucose molecule and one fructose molecule linked together (Figure 6g.2).

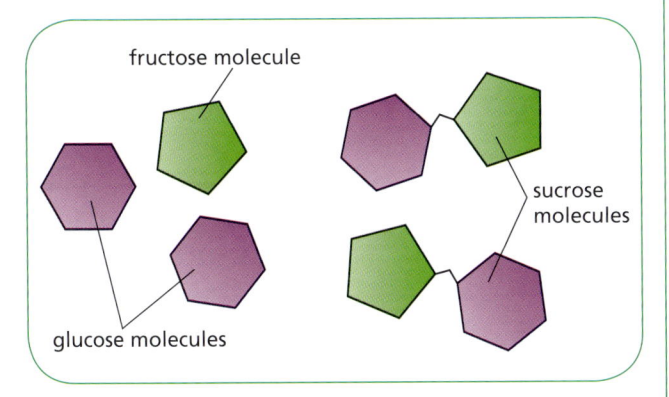

Figure 6g.2 A sucrose molecule is made of a glucose molecule and a fructose molecule linked together.

An enzyme called **invertase** (or **sucrase**) catalyses the breakdown of sucrose to glucose and fructose:

$$\text{sucrose} \xrightarrow{\text{invertase}} \text{glucose} + \text{fructose}$$

Sucrose, like all sugars, tastes sweet. But fructose tastes even sweeter. Breaking down sucrose to glucose and fructose produces more sweetness for the same mass of sugar. This is very useful in the food industry, as it means they can use less sugar in their products.

H Most people like to eat something sweet every now and then. However, too much sugar is not good. In the short term, it can cause tooth decay. In the longer term, a person who eats a lot of sugar runs a higher risk of developing diabetes. Using glucose and fructose, rather than sucrose, can produce sweet foods that are just as attractive to the consumer as foods made with sucrose, but weight for weight contain much less sugar.

Immobilising enzymes

The trouble with using enzymes to change the characteristics of food – for example, using invertase to break down sucrose – is that you can end up with food containing enzymes. This might possibly cause problems for people who eat it. It is also very wasteful; enzymes are expensive to buy, so food manufacturers don't want to use any more than they need.

These two problems can both be solved if the enzymes are **immobilised**. The enzymes are stuck firmly to a surface, or contained within a jelly, so they cannot be washed away. For example, the enzymes on glucose test strips are stuck firmly to the strip, so they cannot wash away when the strip is dipped in the liquid.

Figure 6g.3 shows another method of immobilising enzymes, which you may be able to try out. In this example, the enzyme is **lactase**, which breaks down the lactose sugar in milk to glucose and galactose.

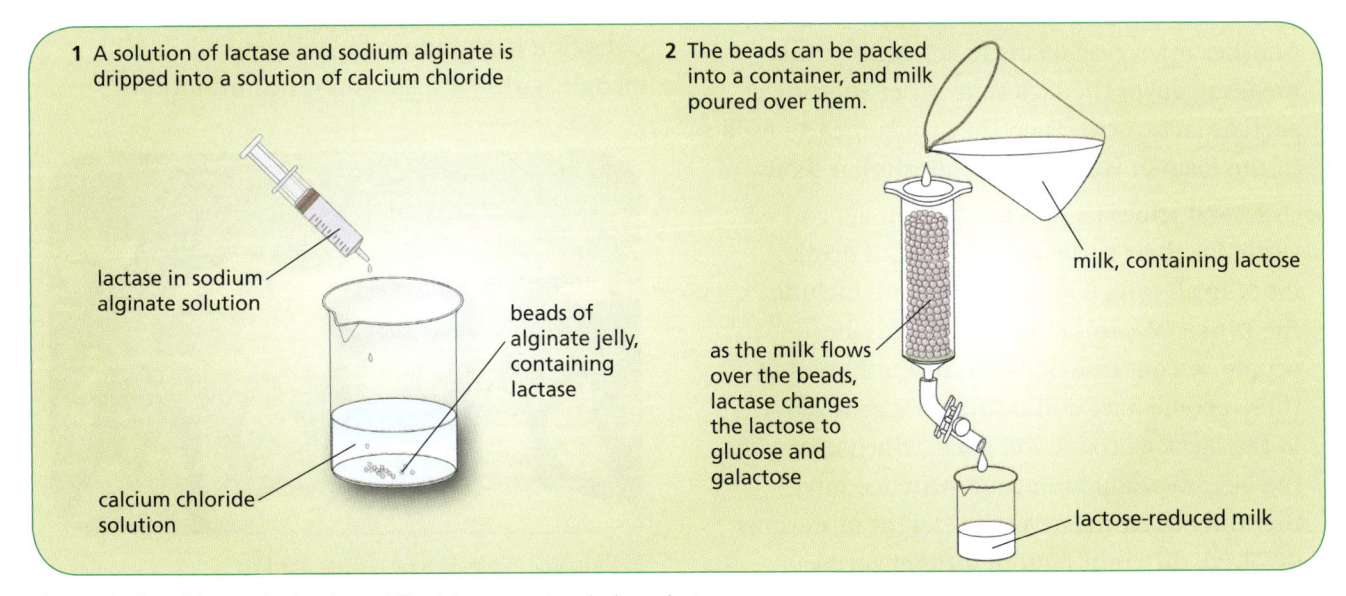

Figure 6g.3 Making and using immobilised lactase to break down lactose.

SAQ

2 Which of these three sugars is a complex sugar (disaccharide) and which are simple sugars (monosaccharides)?

lactose   glucose   galactose

First a solution of the lactase is made, and mixed into a solution of sodium alginate. Then a solution of calcium chloride is made up. Using a pipette, small drops of the lactase–sodium alginate mixture are dropped into the calcium chloride. As the drop hits the calcium chloride, it forms a little jelly bead. The lactase enzyme is trapped inside the bead. It has been immobilised.

Now you can pack the beads into a container, such as a syringe barrel. If milk is poured into the barrel, it runs downwards over the beads.

The enzyme molecules at the surfaces of the beads break down the lactose molecules as they pass over the beads. The milk, now containing glucose and galactose instead of lactose, flows out of the bottom. The enzymes stay in the beads.

The milk is free of enzymes, so the food manufacturer does not have to do anything to remove them from it. Also, because the enzymes stay in place, milk can just flow continuously over them and they will keep on working over and over again. This is called **continuous flow processing**.

SAQ

3 What property of enzymes allows the beads to go on working for a long time?

The evolution of lactose tolerance

Most people do not make lactase when they are adults. We all have a gene that codes for lactase production, but in most people it is switched off when they are weaned. Europeans are an exception to this rule. So are the people in some parts of Africa. These people do keep on producing lactase all their lives.

In parts of the world where people do not produce lactase as adults, they do not normally drink fresh milk. Instead, they allow bacteria or other microorganisms to act on the milk, breaking down the lactose to other substances, such as lactic acid. Then they consume the milk in the form of yoghurt or other similar foods.

Several tribes in Africa, however, depend on cattle for their survival, and they all drink a lot of fresh milk (Figure 6g.4). These include the Dinka, Maasai, Zulu, Swazi and Xhosa people. Recent research has shown that all of these people have a mutation in a gene close to the lactase-production gene, which keeps the lactose-production gene switched on all the time. There are several different mutations involved, different ones in different tribes.

It looks as though people whose ancestors herded cattle have evolved to be able to drink fresh milk. It seems that, when a chance mutation causing long-lasting lactase production occurred, it gave an advantage to anyone who had it, because they were able to drink milk without getting diarrhoea. This would have been especially important in times of drought, when losing water through diarrhoea could have meant rapid death. Natural selection has produced whole populations of people with these advantageous mutations.

Figure 6g.4 A Maasai boy herding cattle.

Why change lactose to glucose and galactose?

When we drink milk, or consume products made from it, the lactose is broken down in the small intestine by lactase. The glucose and galactose are then absorbed into the blood and can be used by the body cells. But many people do not have lactase in their digestive system as adults. We all produce it as babies, when we rely on milk as our only food, but the production of lactase often ceases as we get older. Most people of Asian and African descent do not produce lactase.

This can cause problems when they drink milk. The lactose remains in the digestive system, because its molecules are too big to be absorbed through the wall of the intestines. Here, the lactose provides food for bacteria which can produce gases and diarrhoea. There is therefore a lucrative market for lactose-reduced milk and milk products.

Most cats, too, do not produce lactase, and some people are prepared to pay for lactose-reduced milk to give to their cats.

Summary

You should be able to:

- state some uses of enzymes, and name the enzymes used in biological washing powders
- explain why biological washing powders work best at low temperatures and a neutral pH
- describe how urine can be tested for glucose
- describe how and why invertase is used to break down sucrose to glucose and fructose
- explain how the use of invertase can produce lower calorie foods
- describe how to immobilise enzymes using alginate
- describe some uses of immobilised enzymes, such as reagent sticks
- explain why lactose-free milk is made, and how this is done

Questions

1. a. Name *three* kinds of enzymes that may be present in biological washing powders.
   b. Explain why biological washing powders are able to remove stains that detergents cannot remove.
   c. Explain why most biological washing powders should be used at lower temperatures than non-biological washing powder.

2. a. Explain why a person with diabetes may need to test their urine for glucose.
   b. This test can be done using reagent strips such as Clinistix. How are enzymes involved in the test?

3. a. Describe what is meant by the term *immobilised enzymes*.
   b. Describe *two* advantages of using immobilised enzymes rather than enzymes in solution.
   c. Explain how lactose-reduced milk can be made, and why this is done.

B6h Genetic engineering

Transgenic organisms

Each living organism contains DNA, which carries the organism's genetic information. Long molecules of DNA, called **chromosomes**, contain many different **genes**. A gene can be defined as a length of DNA that codes for the production of a particular protein.

The information carried by the DNA is in the form of a code called the **genetic code**. The sequence of bases on the DNA determines the sequence of amino acids that are joined together to make the protein (*Gateway Additional Science*, pages 3–4).

Different species of organisms have some genes that are the same, but also some that are different. We are now able to take a gene from one organism and insert it into another organism belonging to a different species. This is called **genetic engineering**. The organism with the 'foreign' DNA now has a different genetic code. It is said to be a **transgenic** organism, or a **genetically modified** organism.

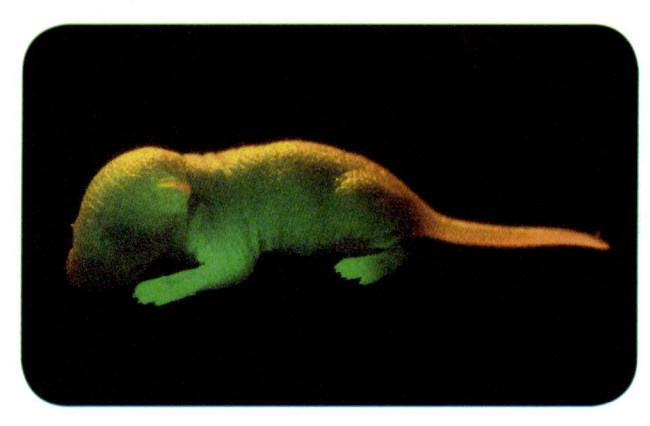

Figure 6h.1 This baby mouse has had genes from a jellyfish added to it, which produce green fluorescence. This gene is often used alongside other genes that researchers want to transfer – if the GM organism has a green glow, then the other gene has probably also been transferred to it.

The reason that a gene from one organism will work in another one is that all organisms use the same genetic code. A new gene can be inserted into a chromosome, and the cell will often be able to follow its instructions just as would have happened in the cell that the gene originally came from.

GM humans?

A genetically modified human would be a person whose cells all contained a gene that had been introduced from another organism. This could be done by inserting the gene into an egg, a sperm cell or an early embryo. That would be the only way of making sure that *all* the cells of the person contained the gene.

Can it be done? If not now, then it will definitely become possible in the near future. The question is whether we *should* do it.

At the moment, we do carry out some very limited 'genetic engineering' on some humans. This is called gene therapy. It has not had very many successes so far, but several children with a genetic disorder that stops them making white blood cells have been treated by giving them cells with a working copy of the missing gene. But this isn't really creating a GM human, because only these white cells have been altered. The gene couldn't be passed on to the next generation, because the white blood cells don't affect the sperm or eggs. We say that they are not present in the 'germ line'.

At some point, society is going to have to decide whether we will accept the idea of putting desirable genes into embryos that are formed as a result of IVF (see page 45). Most people will reject this idea. Some may think that it might be acceptable if it meant that a child had a normal gene, rather than one that caused a serious illness such as cystic fibrosis. But where do we draw the line? If, one day, we can identify the genes that make a person especially intelligent, should it become possible to genetically modify an embryo to have those genes?

Genetic engineering is a rapidly expanding technology. It is being used to improve crops and to produce new medicines. The technology is advancing so rapidly that we are running into problems about deciding what should be allowed to happen and what is best avoided. Many people have strong views about this, but surveys have shown that most people have a very poor understanding of what genetic engineering really is, and how it is done, so their views are not always logical. Some of the advantages and disadvantages of genetic engineering, and some ethical and moral questions relating to it, are described in *Gateway Additional Science*, pages 48–49.

The main stages in genetic engineering

The main steps that need to be followed, in order to transfer a gene from one organism to another, are described in *Gateway Additional Science*, pages 47–48. The process involves these stages:

- Firstly, the desired gene has to be identified. This involves finding the exact length of DNA coding for the characteristic that is wanted – not an easy task amongst the huge quantity of DNA that is present in most cells.
- Next, this gene is cut out from the rest of the DNA.
- Meanwhile, the DNA in the organism that is going to receive the gene is cut open, to make a space for the desired gene to slot into.
- The desired gene is inserted into the DNA of the recipient organism. It is now a transgenic organism.
- A check is made to see if the gene works. Sometimes, other pieces of DNA have to be added, to switch the gene on.
- If the gene works, then the transgenic organism is cloned, producing many organisms each with a copy of the inserted gene.

These processes use several different enzymes. The enzyme that cuts DNA open is called a **restriction enzyme**. There are many different kinds of these enzymes, which are obtained from viruses. Different ones cut DNA at different places, so you need to choose the right one to cut out the particular gene that you want.

When the desired gene is inserted into the recipient's DNA, it needs to link up with the rest of the DNA. This is done by an enzyme called **DNA ligase**.

When this process has been completed, it is probable that only a few of the organisms will have successfully received the new gene. These need to be sorted out from the others. There are several different ways of doing this. One way is to tag a gene that produces a fluorescent protein onto the desired gene that is being transferred. If the organism glows in the dark, then it must have received the fluorescence gene and so it probably has the desired one, too (Figure 6h.1).

It is also important to check that the desired gene is actually working. In the process shown in Figure 6h.2 (overleaf), for example, bacteria that are thought to have successfully received the insulin gene will be cultured, and then the culture will be assayed (tested) to see if insulin is being produced.

GM bacteria for insulin production

Most of the insulin that is used by people with diabetes is now made by bacteria. The bacteria are transgenic organisms. The human gene for insulin has been inserted into them. The bacteria follow its instructions, and produce human insulin.

Figure 6h.2 shows how this is done.

SAQ

1 Identify the stages on Figure 6h.2 where each of the following enzymes is used:
   a restriction enzymes
   b DNA ligase

108 B6h Genetic engineering

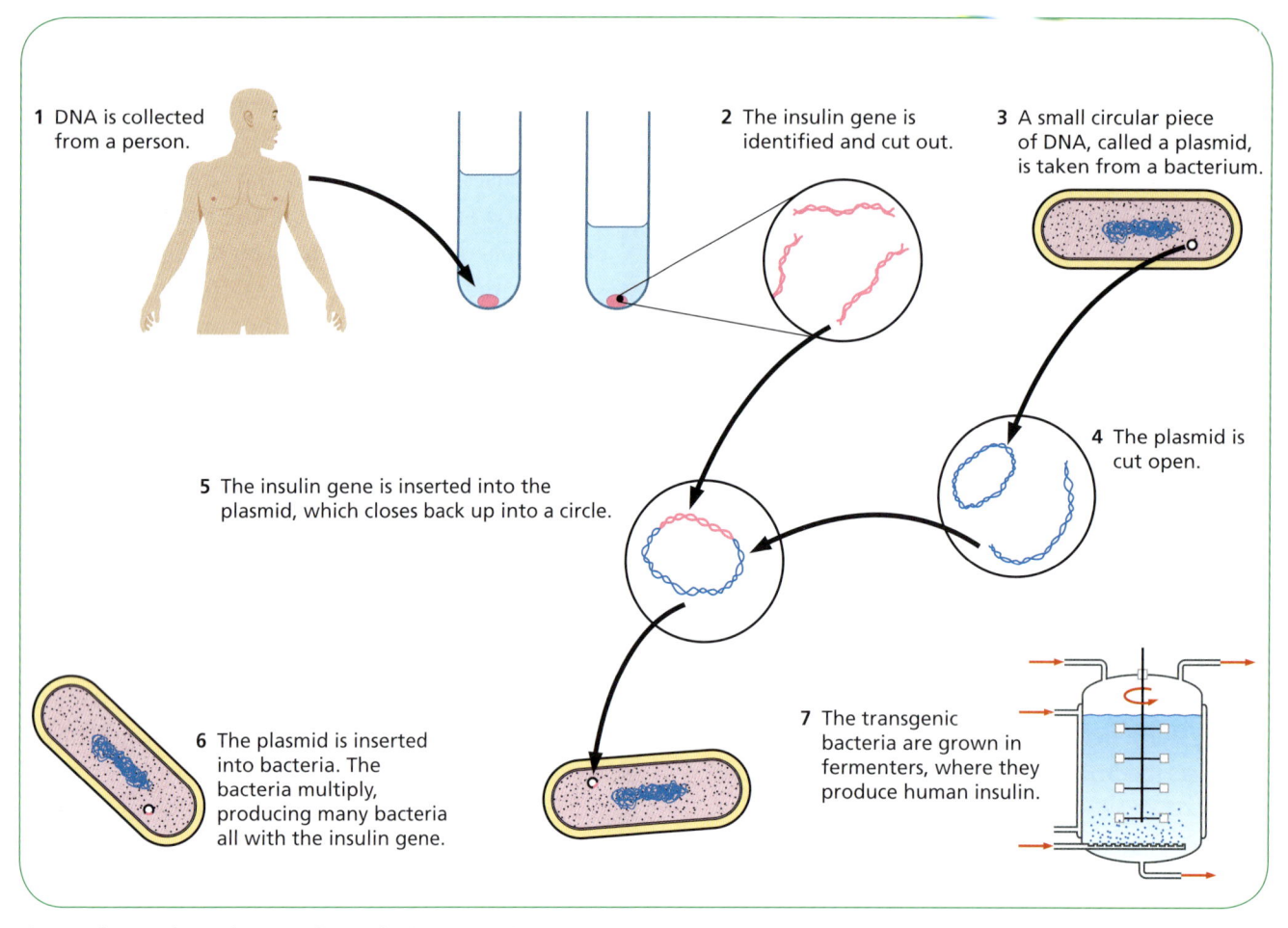

1. DNA is collected from a person.
2. The insulin gene is identified and cut out.
3. A small circular piece of DNA, called a plasmid, is taken from a bacterium.
4. The plasmid is cut open.
5. The insulin gene is inserted into the plasmid, which closes back up into a circle.
6. The plasmid is inserted into bacteria. The bacteria multiply, producing many bacteria all with the insulin gene.
7. The transgenic bacteria are grown in fermenters, where they produce human insulin.

Figure 6h.2 Producing bacteria that make human insulin.

GM crop plants

There are already many examples of genetically modified crop plants that are grown in different parts of the world. **Soya beans** and **maize** have been given a gene that makes them resistant to a herbicide (weedkiller), which makes it much easier and cheaper for the farmer to control weeds growing amongst the crop (Figure 6h.3). **Cotton** has been engineered to make a toxin that kills insects that try to eat it (Figure 6h.4).

Figure 6h.3 This maize has had a gene inserted into it that makes it resistant to a weedkiller. The farmer has sprayed the weedkiller on the field, killing the weeds and leaving the maize growing strongly.

Figure 6h.4 These cotton plants contain a gene that makes them poisonous to an insect pest called the cotton boll weevil.

These examples help a farmer to get an increased yield from his crop. Getting rid of weeds reduces competition for light, water and mineral ions from the soil. Killing pests helps the cotton plants to grow vigorously and produce more cotton. There is another potential benefit from this – the farmer does not need to spray pesticides on the crop, which could have harmful effects on other living organisms that live in or near the field.

Some crop plants have been genetically engineered to produce different or more nutrients. For example, rice has been modified to produce vitamin A (*Gateway Additional Science*, pages 46–47). Research is being carried out to try to produce genetically modified plants that can survive in difficult conditions, such as very dry or salty soil. Salt can be very damaging to plants, but they may be able to survive it better if they contain a gene for making a substance called glycine betaine. Tomatoes have been genetically engineered to make glycine betaine, and they are much better at growing in salty soil than other tomatoes.

SAQ

2 When a new GM crop, such as the tomato that makes glycine betaine, has been produced, it has to undergo stringent tests before it can be grown in open fields or sold as food. Suggest what these tests might include.

Summary

You should be able to:

- describe how genetic engineering changes the genetic code of an organism, producing a transgenic organism
- describe the main stages in genetic engineering
- (H) describe the functions of restriction enzymes and DNA ligase
- describe how bacteria have been genetically modified to produce human insulin
- describe the advantages of growing genetically modified soya beans, maize and cotton
- describe how genetic engineering can improve plants by increasing yield, producing resistance to weedkillers or insect pests, allowing them to grow in difficult conditions, or making them produce useful chemicals
- (H) discuss the advantages and disadvantages of genetic engineering

Questions

1. **a** What is meant by a *transgenic organism*?

   b Explain why genes from one organism can work in another organism.

   c Describe *three* examples of genetically modified plants. For each example, explain why farmers might want to grow them.

2. Table 6h.1 shows how many millions of hectares of land were used to grow genetically modified crops between 1996 and 2005.

Genetic modification	Area of land used to grow each type of crop worldwide, in millions of hectares									
	1996	1997	1998	1999	2000	2001	2002	2003	2004	2005
herbicide tolerance	0.6	6.9	19.8	28.1	32.7	40.6	44.2	49.7	58.6	63.7
insect resistance	1.1	4.0	7.7	8.9	8.3	7.8	10.1	12.2	15.6	16.2

Table 6h.1

   a Display these figures as a graph. Use the same axes for both sets of figures.

   b Describe the trend in the growth of crops that have been genetically engineered to be resistant to herbicides.

   c Compare the trend you have described in **b** with the trend in the growing of insect-resistant crops.

   d Explain why growing crops that are resistant to herbicides could make the crops cheaper to buy.

3. Researchers are producing genetically engineered bananas that manufacture vaccines against diseases such as cholera. A person could become vaccinated against the disease just by eating a banana.

   Discuss the advantages and disadvantages of this idea. Do you think people would accept it?

5a Moles and empirical formulae

The amount of a substance

There are many different ways to measure an object. We can measure its height or its mass, for example.

When a chemist is given a sample of a pure substance, he or she will often need to measure something different. A chemist may need to know how many particles there are in the sample. The number of particles in a sample of a substance is known as the **amount** of the substance. Chemists use the word *amount* very precisely. When chemists talk about the *amount of copper sulfate* in a beaker of solution, they mean the *number of copper sulfate formula units* dissolved in the solution in the beaker (Figure 5a.1).

Figure 5a.1 The solution in the right-hand beaker has twice as much copper sulfate dissolved in it as the solution in the left-hand beaker. The right-hand beaker contains twice the *amount* of copper sulfate – twice the number of formula units.

Moles

Chemists measure the amount of a substance in **moles**. One mole of a substance contains just over six hundred thousand million million million particles. This number can also be written 6×10^{23}. One mole of any substance contains this number of particles.

Each element has a relative atomic mass. If the relative atomic mass of an element is measured out in grams then the sample we get consists of one mole of atoms. The mass of this sample is called the **molar mass** of that element.

Worked example 1

What is the mass of one mole of magnesium atoms?

Step 1: the relative atomic mass of magnesium is 24. (You can find this on the Periodic Table on page 359. You will need to use this table as a source of data for many exercises in this item.)

Step 2: therefore, 24 g is the molar mass of magnesium.

Answer: the mass of one mole of magnesium atoms is 24 g (Figure 5a.2).

Worked example 2

What is the mass of one mole of iron atoms?

Step 1: the relative atomic mass of iron is 56 (from the Periodic Table).

Step 2: therefore, 56 g is the molar mass of iron.

Answer: the mass of one mole of iron atoms is 56 g (Figure 5a.2).

Figure 5a.2 The sample of magnesium on the left contains one mole of magnesium atoms. The sample of iron on the right contains one mole of iron atoms. Although the samples have different masses, they both have the same number of atoms.

SAQ

1. Use the Periodic Table on page 359 to find the relative atomic mass of each of the following elements:

   hydrogen, carbon, sulfur, zinc, lead

2. What is the mass of one mole of:
   a. hydrogen atoms?
   b. carbon atoms?
   c. sulfur atoms?
   d. zinc atoms?
   e. lead atoms?

3. How many atoms are there in 24 g of magnesium? Answer in three different ways.

Relative formula mass

The formula of a compound tells us the atoms in one *formula unit*. These atoms make up the smallest possible unit of that compound. We can find the relative mass of one formula unit of a compound by adding together the relative atomic masses of the atoms in one formula unit. This mass is called the **relative formula mass** of the compound.

If the relative formula mass of a compound is measured out in grams then the sample you get contains one mole of formula units.

Worked example 3

What is the mass of one mole of silicon dioxide formula units?

The formula of silicon dioxide is SiO_2.

Step 1: the relative atomic mass of silicon is 28, and the relative atomic mass of oxygen is 16.

Step 2: the formula of silicon dioxide is SiO_2. Therefore, 28 + 16 + 16 is the relative formula mass of silicon dioxide.

28 + 16 + 16 = 60

Therefore, 60 g is the molar mass of silicon dioxide.

Answer: the mass of one mole of silicon dioxide formula units is 60 g (Figure 5a.3).

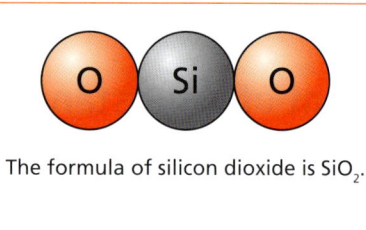

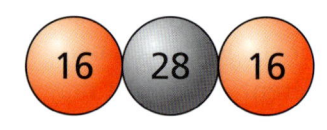

The formula of silicon dioxide is SiO_2.

The atoms have these masses, so the formula mass of SiO_2 is 60.

One mole of SiO_2 formula units has a mass of 60 g.

Figure 5a.3

For a covalent compound, the formula unit is also called a **molecule**. (Look at Item C3c *Covalent bonding and the structure of the Periodic Table* in *Gateway Additional Science* to remind yourself about covalent bonding.) For these compounds, the relative formula mass is also called the **molecular mass**.

Worked example 4

What is the mass of one mole of water molecules?

The formula of water is H_2O.

Step 1: the relative atomic mass of hydrogen is 1, and the relative atomic mass of oxygen is 16.

continued on next page

Worked example 4 - continued

Step 2: the formula of water is H₂O. Therefore, 1 + 1 + 16 is the relative formula mass of water.

1 + 1 + 16 = 18

Therefore, 18 g is the molar mass of water.

Answer: the mass of one mole of water molecules is 18 g (Figure 5a.4).

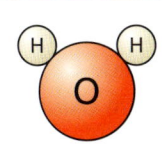

The formula of water is H₂O. This is one water molecule.

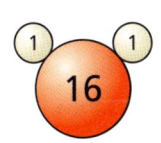

The atoms have these masses, so the molecular mass of water is 18.

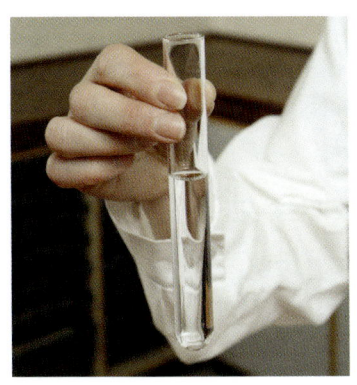

One mole of water molecules has a mass of 18 g.

Figure 5a.4

SAQ

4 What are the relative formula mass and the molar mass of each of the following ionic compounds?
   a sodium hydroxide, NaOH
   b iron(III) oxide, Fe₂O₃
   c calcium carbonate, CaCO₃
   d iron(III) hydroxide, Fe(OH)₃
   e calcium nitrate, Ca(NO₃)₂

5 What are the relative formula mass and the molar mass of each of the following covalent compounds?
   a carbon dioxide, CO₂
   b ammonia, NH₃
   c methane, CH₄
   d sulfur dioxide, SO₂
   e glucose, C₆H₁₂O₆

Calculating a number of moles

If we know the mass of a sample of a pure substance, we can calculate the number of moles of the substance. We use the equation:

$$\text{number of moles} = \frac{\text{mass of sample}}{\text{molar mass}}$$

Worked example 5

How many moles of calcium carbonate, CaCO₃, are there in 400 g?

Step 1: the relative formula mass of calcium carbonate is 40 + 12 + 16 + 16 + 16 = 100.

So the molar mass of calcium carbonate is 100 g.

Step 2: the number of moles of calcium carbonate in the sample is found by using:

$$\text{number of moles} = \frac{\text{mass of sample}}{\text{molar mass}}$$

$$= \frac{400\,g}{100\,g}$$

$$= 4$$

Answer: 400 g is 4 moles of calcium carbonate.

SAQ

6 How many moles of magnesium sulfate, MgSO₄, are there in each of the following samples?
   a 360 g
   b 15 g
   c 4.92 g

114 C5a Moles and empirical formulae

H If we know how many moles of a compound we have, we can calculate the mass of each element that is present.

Worked example 6

What mass of iron is present in 10 moles of iron(III) oxide, Fe_2O_3?

Step 1: in 1 mole of iron(III) oxide there is 56 g + 56 g = 112 g of iron.

Step 2: in 10 moles of iron(III) oxide there is 10 × 112 g = 1120 g of iron.

Answer: there is 1120 g of iron present in 10 moles of iron(III) oxide.

SAQ

7 What mass of bromine is present in each of the following amounts of sodium bromide, NaBr?
   a 6 moles
   b 25 moles
   c 0.72 moles

Relative atomic mass

We can find the **relative atomic mass** of an element by looking at the Periodic Table. For each element, this number compares the average mass of an atom of the element with the mass of a carbon-12 atom. More specifically, it compares the average mass of an atom of the element with one-twelfth of the mass of a carbon-12 atom. A carbon-12 atom has six protons, six neutrons and six electrons. The six protons and six neutrons give this atom a mass number of 12, so it is called carbon-12. The relative atomic mass of magnesium is 24. This means the average mass of a magnesium atom is 24 times one-twelfth the mass of a carbon-12 atom.

Mass is conserved in a chemical reaction

In every chemical reaction, the total mass of the products is equal to the total mass of the reactants used to form them. The total mass stays the same – we say that mass is **conserved** in a chemical reaction. Mass is conserved because the atoms of the substances that react are not destroyed during the reaction. New atoms are not created during the reaction. The products are made of exactly the same number of atoms, of the same types, as the reactants were.

This is called the **principle of conservation of mass**. We can use this principle to work out some masses that are difficult to measure directly.

Worked example 7

When calcium hydroxide is heated, it decomposes to calcium oxide and water. The word equation for this reaction is:

calcium hydroxide → calcium oxide + water

If 74 g of calcium hydroxide is heated in a crucible, 56 g of calcium oxide is left in the crucible when the reaction is finished. What mass of water escaped into the air?

Step 1: mass is conserved in this chemical reaction. The total mass of products must be the same as the total mass of reactants.

The mass of the calcium hydroxide was 74 g, so the total mass of the calcium oxide and water must be 74 g.

Step 2: the mass of calcium oxide was 56 g.

74 g − 56 g = 18 g

Answer: the mass of water that escaped into the air was 18 g.

This would be very difficult to measure directly.

74 g of calcium hydroxide is put into a crucible and heated strongly. → Afterwards 56 g of calcium oxide is left in the crucible. 18 g of water escaped into the air.

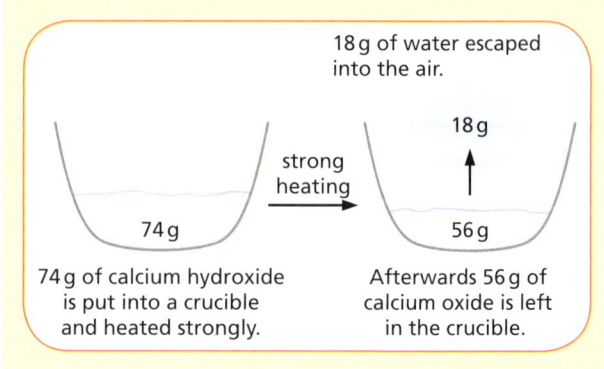

Figure 5a.5

SAQ

8 What mass of water is formed in each case?

   a When 37 g of calcium hydroxide is heated, 28 g of calcium oxide is formed.

   b When 222 g of calcium hydroxide is heated, 168 g of calcium oxide is formed.

   c When 7.4 g of calcium hydroxide is heated, 5.6 g of calcium oxide is formed.

Some metals undergo *combustion* reactions – they burn in air to form an oxide. The mass of the metal oxide product is more than the mass of the metal reactant at the start. In these reactions, the metal combines with oxygen from the air. By knowing the mass of the metal reactant at the start and the mass of product, we can work out the mass of oxygen that reacted.

Worked example 8

When lithium is heated in air, it combines with oxygen to form lithium oxide. The word equation for this reaction is:

lithium + oxygen → lithium oxide

If 14 g of lithium is heated in a crucible, 30 g of lithium oxide is left in the crucible when the reaction is finished. What mass of oxygen has been gained?

Step 1: mass is conserved in this chemical reaction, so the total mass of products must be the same as the total mass of reactants.

The mass of the lithium oxide was 30 g, so the total mass of the lithium and oxygen must be 30 g.

Step 2: the mass of lithium was 14 g.

30 g − 14 g = 16 g

Answer: the mass of oxygen that reacted with the lithium was 16 g.

This would be very difficult to measure directly.

---

16 g of oxygen combine with the lithium.

16 g

14 g of lithium is put into a crucible and heated.

combustion →

Afterwards 30 g of lithium oxide is left in the crucible.

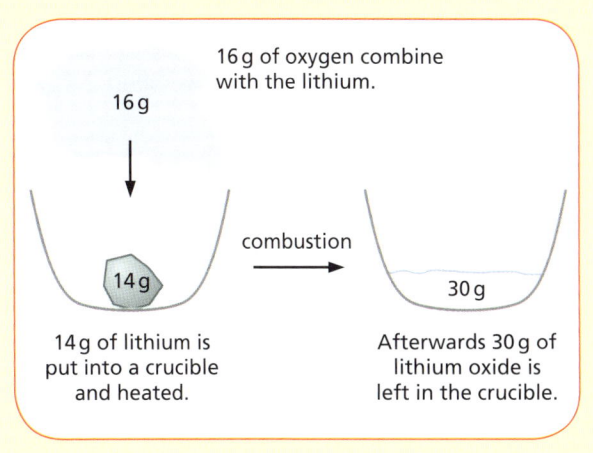

Figure 5a.6

If twice the mass of lithium had been involved (28 g), then twice the mass of lithium oxide (60 g) would have formed and twice the mass of oxygen (32 g) would have reacted.

SAQ

9 If the following masses of lithium are burned, what mass of lithium oxide forms in each case? What mass of oxygen reacts in each case? (Hint: compare the masses below with the mass of lithium burned in Worked example 8. The masses of lithium oxide and oxygen involved will be in the same ratio.)

   a 1.4 g of lithium

   b 70 g of lithium

   c 2.8 g of lithium

SAQ 9 involved quite simple ratios. We can solve more complicated problems by calculations using moles. Worked example 9 (on the next page) shows how this is done.

Worked example 9

What mass of calcium oxide forms when 2.4 g of calcium burns in oxygen?

Step 1: first a balanced symbol equation is needed.

$2Ca + O_2 \rightarrow 2CaO$

This equation says that the number of CaO formula units formed is the same as the number of Ca atoms that react.

This means that the number of *moles* of CaO formula units formed is the same as the number of *moles* of Ca atoms that react.

Step 2: the number of moles of calcium atoms that react is found by using.

$$\text{number of moles} = \frac{\text{mass of sample}}{\text{molar mass}}$$

The molar mass of calcium is 40 g so:

$$\text{number of moles} = \frac{2.4\,g}{40\,g} = 0.06 \text{ moles}$$

This means that 0.06 moles of calcium react, so 0.06 moles of calcium oxide form.

Step 3: the molar mass of calcium oxide is 56 g (40 + 16 = 56).

0.06 moles of calcium oxide have a mass of 0.06 × 56 g = 3.36 g

Answer: when 2.4 g of calcium is burned in oxygen, 3.36 g of calcium oxide forms.

SAQ

10 What mass of calcium oxide forms when the following masses of calcium are burned in oxygen?
   a 3.0 g
   b 7.6 g
   c 6.4 g

Reduction

In the Middle Ages, making metals was a very important job. Better metals meant better weapons, which meant a better army. Unfortunately the metal workers did not understand the chemistry of what they were doing. They knew that they had to take a special mineral called an ore and heat it with coal or charcoal in a very hot furnace. They knew that this produced the metal they wanted. But they didn't know what was going on.

Many metal workers thought that what they were doing was magic, and superstitions built up around the art of smelting steel in particular. If the smelters made good steel one day when the chief smelter was wearing a particular hat, for example, then the hat had to be worn every time they smelted steel in future. There is even a story from Turkey of a furnace producing superb steel during an afternoon when a little girl cried constantly nearby. Next time the smelters fired up the furnace they made sure to use exactly the same conditions including the crying child! We hope they didn't make her cry on purpose...

One of the things that puzzled the smelters most was that they could never get one tonne of metal from one tonne of ore. It didn't matter what they wore or who was crying, they could never do it. Because of this 'disappearing mass' they called the smelting process 'reduction' because the mass of ore was 'reduced' into a smaller mass of metal. The term 'reduction' is still used today.

continued on next page

Reduction – *continued*

Now that we understand the process, we can explain the 'disappearing mass'. The ore contains a metal compound – for example, iron oxide. In order to get the iron, you have to take the oxygen away. Therefore, the iron you get out of the furnace has less mass than the iron oxide that you put in. All of the mass of the iron oxide is conserved during smelting. The early smelters did not realise that part of their ore (oxygen) was combining with carbon and leaving their furnaces as invisible carbon dioxide gas.

Molecular formulae and empirical formulae

The formula of a compound tells us which atoms are present in one formula unit of the compound. It also tells us how many of each type of atom are present in one formula unit. We often refer to the formula as the **molecular formula** if the compound is covalently bonded. For example, lactic acid is the acid present in sour milk. The formula of lactic acid is $C_3H_6O_3$. This means that one formula unit, or molecule, of lactic acid consists of three carbon atoms, six hydrogen atoms and three oxygen atoms (Figure 5a.7). $C_3H_6O_3$ is the molecular formula of lactic acid.

one molecule of lactic acid

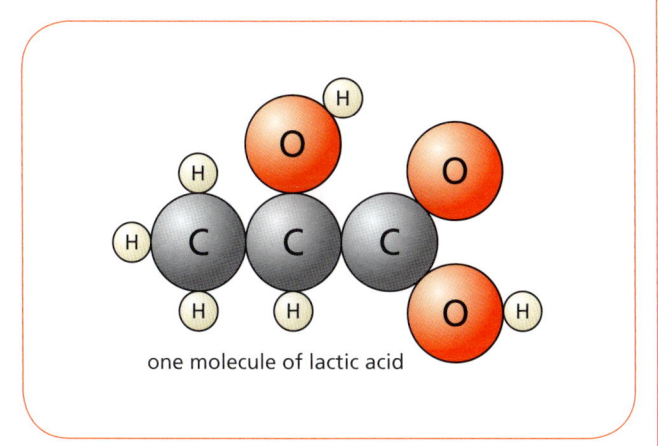

Figure 5a.7 This is the displayed formula of lactic acid. Its molecular formula is $C_3H_6O_3$.

The ratio of carbon : hydrogen : oxygen in lactic acid is 3 : 6 : 3. These numbers are all multiples of 3. To get them in their simplest ratio we divide them all by 3. The simplest ratio of carbon : hydrogen : oxygen in lactic acid is 1 : 2 : 1. Writing the formula using this simplest ratio gives us CH_2O. CH_2O is called the **empirical formula** of lactic acid.

The empirical formula of a compound tells us the atoms in one formula unit in their simplest whole-number ratio. For some compounds, the empirical formula and the molecular formula are the same. The molecular formula of sulfuric acid is H_2SO_4. This is also the empirical formula, because the molecular formula already gives the simplest whole-number ratio.

SAQ

11 Figure 5a.8 shows the displayed formula of ethanoic acid.
   a What is the molecular formula of ethanoic acid?
   b What is the empirical formula of ethanoic acid?

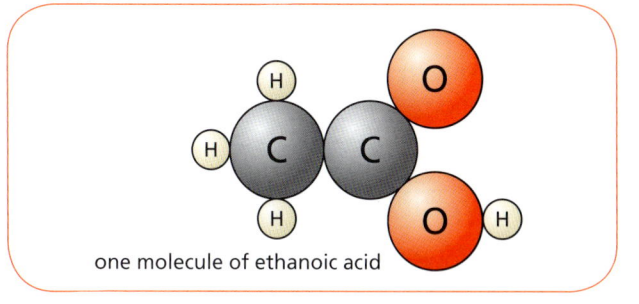

one molecule of ethanoic acid

Figure 5a.8 This is the displayed formula of ethanoic acid.

12 Parts **a**–**d** below each give the name and molecular formula of a covalent compound. What is the empirical formula of each compound?
   a hydrogen peroxide, H_2O_2
   b carbon dioxide, CO_2
   c ethanol, C_2H_6O
   d glucose, $C_6H_{12}O_6$

13 Which two compounds from SAQs **11** and **12** have the same empirical formula as lactic acid?

Calculating an empirical formula

We can calculate the empirical formula of a compound if we know what percentage each of its elements contributes to the mass of one of its formula units. The data used here is called *percentage composition by mass* data.

Worked example 10

An oxide of nitrogen consists of 30.4% nitrogen and 69.6% oxygen. What is the empirical formula?

Step 1: write down the percentage of each element present.

Nitrogen: 30.4%

Oxygen: 69.6%

Step 2: divide each percentage by the relative atomic mass of that element.

Nitrogen: $\dfrac{30.4}{14} = 2.17$

(14 is the relative atomic mass of nitrogen.)

Oxygen: $\dfrac{69.6}{16} = 4.35$

(16 is the relative atomic mass of oxygen.)

Step 3: divide by the lowest number to give the simplest ratio.

Nitrogen: $\dfrac{2.17}{2.17} = 1$

(2.17 is the lowest of 2.17 and 4.35)

Oxygen: $\dfrac{4.35}{2.17} = 2$

So the ratio of atoms nitrogen:oxygen is 1:2.

Answer: therefore, the empirical formula of the oxide of nitrogen is NO_2.

SAQ

14 Find the empirical formula of each oxide of sulfur from the following percentage composition data.
   a 40% sulfur and 60% oxygen
   b 50% sulfur and 50% oxygen

The empirical formula of a compound can also be calculated if we know what mass each of its elements contributes to the mass of a sample of the compound.

Worked example 11

8.80 g of an oxide of nitrogen consists of 5.60 g nitrogen and 3.20 g oxygen. What is the empirical formula?

Step 1: divide the mass of each element by the relative atomic mass of that element.

Nitrogen: $\dfrac{5.60}{14} = 0.40$

(14 is the relative atomic mass of nitrogen.)

Oxygen: $\dfrac{3.20}{16} = 0.20$

(16 is the relative atomic mass of oxygen.)

Step 2: look for the simplest ratio. The simplest ratio of nitrogen:oxygen is 2:1.

Answer: the formula of the oxide of nitrogen is N_2O.

SAQ

15 Find the empirical formula of each hydrocarbon from the following data.
   a 120 g of ethane consists of 96 g of carbon combined with 24 g of hydrogen.
   b 390 g of benzene consists of 360 g of carbon combined with 30 g of hydrogen.

Summary

You should be able to:

- recall that the unit for the amount of substance is the mole
- calculate the relative formula mass and the molar mass of a substance
- **H** recall and use the relationship between molar mass, number of moles and mass of sample
- know that the relative atomic mass of an element is given as a comparison with the mass of a carbon-12 atom
- recall that mass is conserved in a chemical reaction and use this to calculate the mass of gas gained or lost in a chemical reaction
- use ratios to help calculate reacting masses
- **H** use the concept of the mole to help calculate reacting masses
- understand the term *empirical formula* and work out the empirical formula of a compound from its molecular formula
- **H** calculate the empirical formula of a compound from suitable data

Questions

1. Calcium hydrogencarbonate is commonly found dissolved in the tap-water in southern England. When it is heated, it decomposes to form calcium carbonate, carbon dioxide and water. The balanced symbol equation for this is:

   $Ca(HCO_3)_2 \rightarrow CaCO_3 + CO_2 + H_2O$

   a How many elements are there in calcium hydrogencarbonate?

   b How many atoms are there in one formula unit of calcium hydrogencarbonate?

   c What are the relative formula masses of calcium hydrogencarbonate, calcium carbonate, carbon dioxide and water?

   d What are the molar masses of calcium hydrogencarbonate, calcium carbonate, carbon dioxide and water?

2. When 162 g of calcium hydrogencarbonate decomposes, 100 g of calcium carbonate forms and 18 g of water forms.

   a What mass of carbon dioxide will form when 162 g of calcium hydrogencarbonate decomposes?

   b What masses of calcium carbonate, carbon dioxide and water will form when 810 g of calcium hydrogencarbonate decomposes?

continued on next page

Questions - continued

3. What mass of chromium is there in each of the following samples?

    a   1 kg of chromium(IV) oxide, which contains 381 g of oxygen

    b   1 kg of chromium(III) oxide, which contains 316 g of oxygen

    c   1 kg of chromium(III) chloride, which contains 672 g of chlorine

    d   1 kg of chromium(III) sulfide, which contains 480 g of sulfur

4. Calculate the empirical formula of each chromium compound in question 3.

5. The chemical barium sulfate can be made by mixing zinc sulfate solution and barium chloride solution. The balanced symbol equation for this reaction is:

   $ZnSO_4 + BaCl_2 \rightarrow ZnCl_2 + BaSO_4$

    a   Calculate the formula mass of each of the four substances in this equation.

    b   What is the molar mass of each substance?

   A chemist wants to make barium sulfate using this reaction. She starts by dissolving 20.8 g of barium chloride in water.

    c   How many moles of barium chloride are there in 20.8 g of barium chloride?

    d   How many moles of barium sulfate could be made starting with this mass of barium chloride?

    e   What mass of barium sulfate could be made starting with this mass of barium chloride?

5b Electrolysis

Decomposition by electrolysis

If a direct current (**DC**) passes through a molten ionic compound, a chemical change takes place. The substance decomposes, or splits up – often into the elements it is made of. This is a very useful way to make new substances. We call the process **electrolysis**. Electrolysis is the decomposition of a liquid caused by passing an electric current through it.

One substance that undergoes electrolysis is molten potassium chloride. Figure 5b.1 shows the apparatus for performing this electrolysis. Have a look at page 144 of *Gateway Additional Science* to remind yourself what terms like *cathode* and *anode* mean.

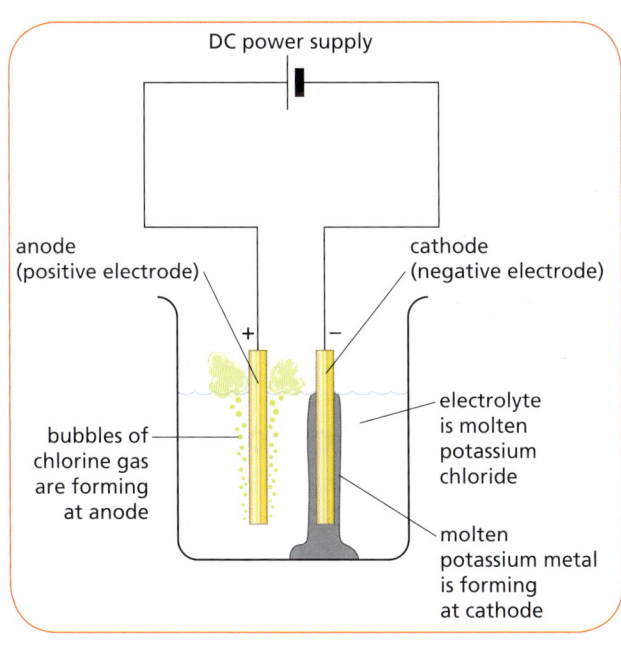

Figure 5b.1 Electrolysing molten potassium chloride. The potassium chloride decomposes into potassium metal at the cathode and chlorine gas at the anode.

When a direct current passes through an electrolyte, we say there is a flow of electric charge through the electrolyte. *Flow of electric charge* is another way to describe an electric current.

The discharge of ions

The flow of charge through the electrolyte happens because negative ions in the electrolyte move to the anode and positive ions move to the cathode. (They are each attracted to the electrode with the opposite charge, as Figure 5b.2 shows.) This movement is not possible in a *solid* ionic compound. In a solid, the ions are bound strongly in position. However, the ions *can* move if the substance is melted. This is why a *molten* ionic compound can conduct an electric current.

When the ions reach the electrodes, they can change into uncharged atoms. When they do this, they leave the electrolyte. They leave either by forming a solid coating on one of the electrodes, or by forming a puddle of liquid at the electrode, or by bubbling out as a gas. This change in the ions, which results in the ions leaving the electrolyte, is known as the **discharge of ions**.

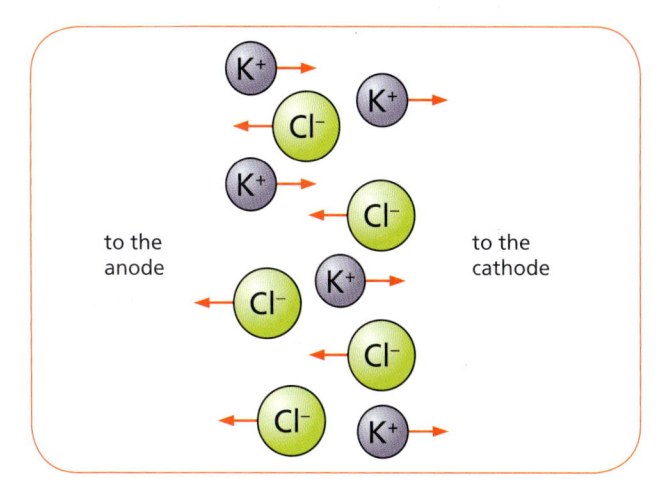

Figure 5b.2 The movement of the ions in the electrolyte.

Three other examples of electrolytes that can be used in the apparatus in Figure 5b.1 are molten aluminium oxide, molten lead bromide and molten lead iodide. Like potassium chloride, each of these compounds decomposes into its elements during electrolysis. Table 5b.1 summarises this.

Electrolyte	Product at anode	Product at cathode
molten potassium chloride	chlorine	potassium
molten aluminium oxide	oxygen	aluminium
molten lead bromide	bromine	lead
molten lead iodide	iodine	lead

Table 5b.1 The substances formed at each electrode during the electrolysis of four molten electrolytes.

122 C5b Electrolysis

You can read more about the electrolysis of molten aluminium oxide on pages 144 to 146 of *Gateway Additional Science*.

SAQ

1. What do we mean by the terms *flow of charge* and *discharge of ions*?
2. a Why doesn't solid potassium chloride conduct electricity?
   b When molten potassium chloride is electrolysed, why do the potassium ions go to the cathode?
   c Why do the chloride ions go to the anode?
3. a Draw a fully labelled diagram to show the electrolysis of molten lead bromide.
   b What forms at each electrode when molten lead bromide is electrolysed?

How ions turn into atoms

When we electrolyse potassium chloride, potassium ions move to the cathode. At the cathode they gain electrons and become potassium atoms (Figure 5b.3). We can describe this process using a half-equation:

$K^+ + e^- \rightarrow K$

Chloride ions move to the anode. At the anode, they lose electrons and become chlorine atoms. The chlorine atoms always bond covalently in pairs, forming chlorine molecules. We can describe this process using a half-equation:

$2Cl^- - 2e^- \rightarrow Cl_2$

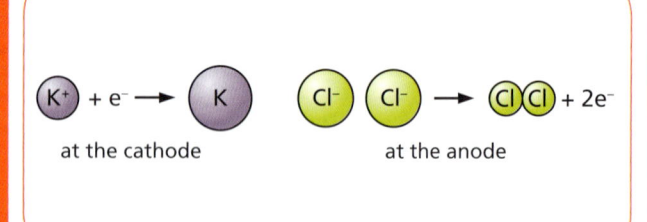

Figure 5b.3 How ions change into atoms at the electrodes.

The changes at each electrode are similar when molten aluminium oxide, molten lead bromide, and molten lead iodide are electrolysed. The metal ions gain electrons at the cathode and the non-metal ions lose electrons at the anode. Table 5b.2 summarises this.

Electrolyte	Anode half-equation	Cathode half-equation
molten potassium chloride	$2Cl^- - 2e^- \rightarrow Cl_2$	$K^+ + e^- \rightarrow K$
molten aluminium oxide	$2O^{2-} - 4e^- \rightarrow O_2$	$Al^{3+} + 3e^- \rightarrow Al$
molten lead bromide	$2Br^- - 2e^- \rightarrow Br_2$	$Pb^{2+} + 2e^- \rightarrow Pb$
molten lead iodide	$2I^- - 2e^- \rightarrow I_2$	$Pb^{2+} + 2e^- \rightarrow Pb$

Table 5b.2 The half-equations for the process occurring at each electrode during the electrolysis of four molten electrolytes.

SAQ

4. Use Table 5b.2 to produce diagrams, similar to those in Figure 5b.3, to show what happens at each electrode during the electrolysis of:
   a molten aluminium oxide
   b molten lead bromide
   c molten lead iodide.

Electrolysis of aqueous solutions

Electrolysis also takes place if a direct current is passed through an aqueous solution of an ionic compound. The aqueous solution acts as the electrolyte. In this case, *either* the substance dissolved in the solution (the solute) or the water decomposes.

Discharge of ions from aqueous electrolytes

If the electrolyte is an aqueous solution, it contains ions of the solute, and hydrogen ions and hydroxide ions because of the presence of water. In some cases, the ions discharged during electrolysis are the hydrogen ions and the hydroxide ions. Here the products of the electrolysis are hydrogen gas and oxygen gas. This happens when potassium nitrate solution is electrolysed, and is shown in Figure 5b.4.

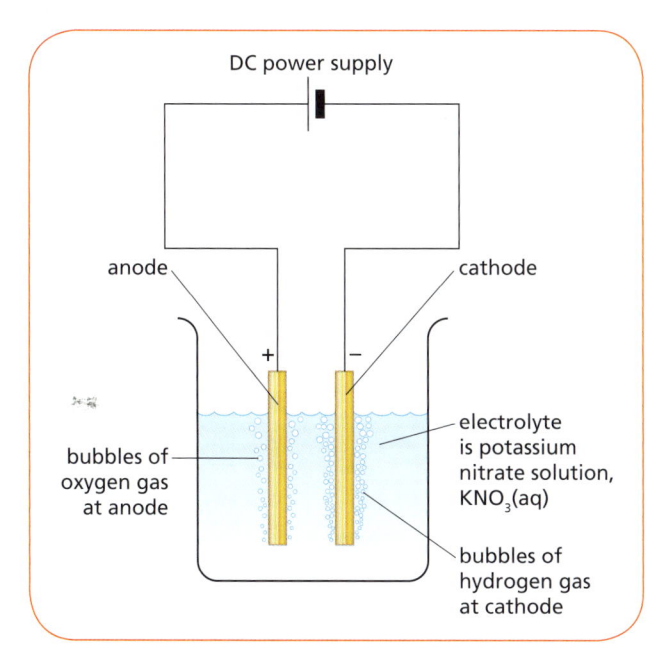

Figure 5b.4 The electrolysis of potassium nitrate solution.

SAQ

5 Figure 5b.4 shows the electrolysis of potassium nitrate solution.

  a What is the formula of potassium nitrate?
  b Which substance forms at the cathode?
  c Which substance forms at the anode?
  d What term is used to describe the changing of ions into atoms?

6 When a direct current passes through an aqueous solution of potassium sulfate (K_2SO_4), electrolysis takes place. The product at each electrode is the same as it is in the electrolysis of aqueous potassium nitrate. Draw a fully labelled diagram to show this process taking place.

Ions present in the electrolyte

When potassium nitrate solution is electrolysed, oxygen and hydrogen are discharged at the electrodes. The electrolyte contains two types of positive ion: potassium ions and hydrogen ions. The hydrogen ions are present because of the water. The electrolyte contains two types of negative ion: nitrate ions and hydroxide ions. The hydroxide ions are present because of the water.

To be discharged at the cathode, a positive ion must gain electrons. Hydrogen ions gain electrons more easily than potassium ions, so hydrogen bubbles out at the cathode (Figure 5b.5). We can describe hydrogen ions gaining electrons with a half-equation:

$$2H^+ + 2e^- \rightarrow H_2$$

To be discharged at the anode, a negative ion must lose electrons. Hydroxide ions lose electrons more easily than nitrate ions. When hydroxide ions lose electrons, oxygen gas and water are formed, so oxygen bubbles out at the cathode (Figure 5b.6). We can describe hydroxide ions losing electrons with a half-equation:

$$4OH^- - 4e^- \rightarrow O_2 + 2H_2O$$

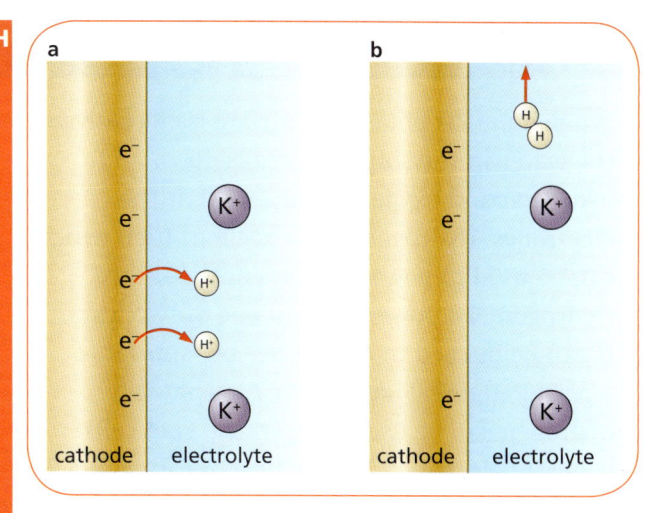

Figure 5b.5 a Hydrogen ions and potassium ions are both attracted to the cathode. The hydrogen ions gain electrons more easily. b Hydrogen gas (H_2) forms and bubbles out.

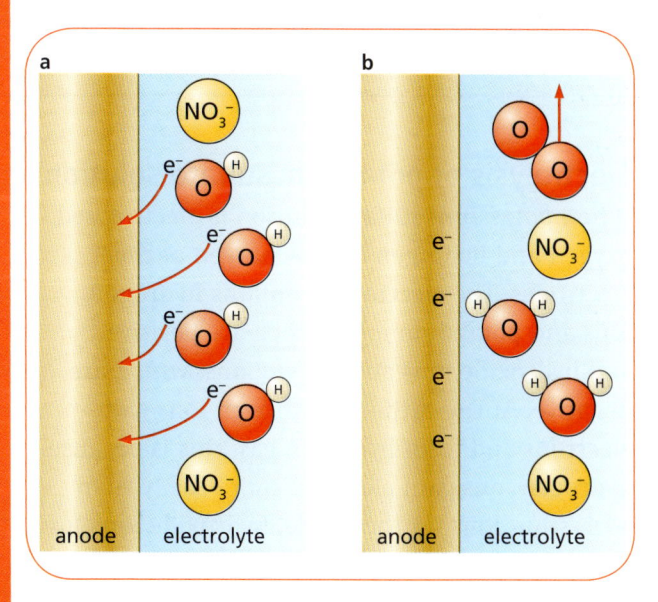

Figure 5b.6 a Hydroxide ions and nitrate ions are both attracted to the anode. The hydroxide ions lose electrons more easily. b Oxygen gas (O_2) forms and bubbles out.

Tomorrow's fuel?

When aqueous potassium nitrate is electrolysed, the water in the solution breaks down and hydrogen and oxygen are given off at the electrodes. Hydrogen is an excellent fuel, which can be used to power the cars of the future. Hydrogen can power a car in one of two ways. It can be burned in an engine very similar to a petrol engine (Figure 5b.7). Alternatively, hydrogen can be used in a *fuel cell*. Fuel cells generate electricity, so this type of car would have an electric motor.

In both types of car, the hydrogen will combine with oxygen, forming water! Compared with petrol engines, which produce carbon dioxide – a major 'greenhouse' gas contributing to global warming – this sounds marvellous. Sadly there is one catch. The hydrogen is produced by electrolysis, the electrolysis needs electric power, and generating electric power is currently a major cause of pollution.

Figure 5b.7 This car's six-litre engine is powered by burning hydrogen.

SAQ

7   This question is about the electrolysis of potassium nitrate solution.
   a   Which ions go to the anode?
   b   Which ions are discharged at the anode?
   c   Write a half-equation for this process.
   d   Which ions go to the cathode?
   e   Which ions are discharged at the cathode?
   f   Write a half-equation for this process.

SAQ

8   This question is about the electrolysis of potassium sulfate solution.
   a   Why are hydrogen ions discharged at the cathode, but potassium ions are not?
   b   Why are hydroxide ions discharged at the anode, but sulfate ions are not?

Electrolysis of potassium sulfate solution

When potassium sulfate solution is electrolysed, oxygen and hydrogen are discharged at the electrodes. The electrolyte contains hydrogen ions and potassium ions, which are positive, and hydroxide ions and sulfate ions, which are negative. At the cathode, hydrogen ions gain electrons and form hydrogen gas; this is because hydrogen ions gain electrons more easily than potassium ions. At the anode, hydroxide ions lose electrons and form oxygen gas and water; this is because oxygen ions gain electrons more easily than sulfate ions.

Electrolysing copper(II) sulfate solution

Figure 5b.8 shows how copper(II) sulfate solution can be electrolysed with copper electrodes. When this is done, the copper anode – the positive electrode – gradually dissolves. At this electrode, copper atoms from the electrode turn into copper ions and go into the solution.

At the same time, the copper cathode – the negative electrode – gradually increases in mass. At this electrode, copper ions from the solution turn into copper atoms and plate the electrode.

When this happens, the cathode always gains exactly the same mass as the anode loses. The actual mass changes depend on two factors: the size of the electric current and the time that the

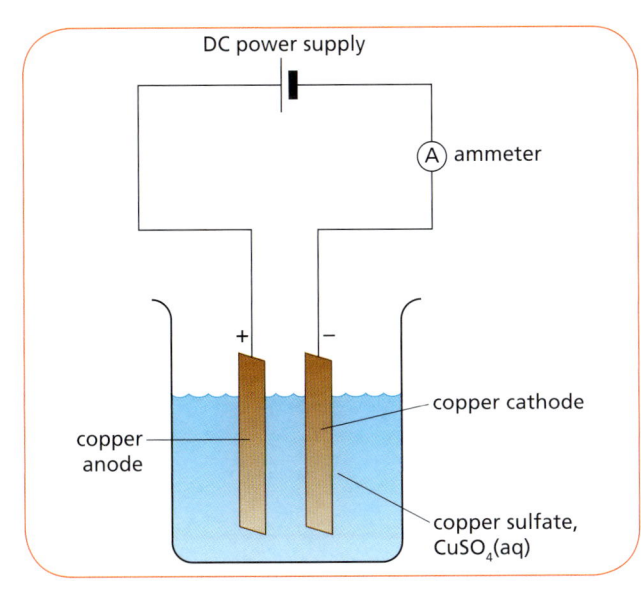

Figure 5b.8 The electrolysis of copper sulfate solution using copper electrodes.

current flows for. A larger current, or a longer time, will each result in a greater mass change.

SAQ

9 If the apparatus in Figure 5b.8 is switched on and allowed to run for 5 minutes, with a current of 2 amperes (A) flowing, the anode loses 0.2 g of copper.
   a How would the current be measured, according to Figure 5b.8?
   b What will happen to the cathode?
   c Give two ways of increasing the loss of mass at the anode.

The electrode processes

When copper sulfate solution is electrolysed with copper electrodes, the anode loses copper atoms. The copper atoms themselves lose two electrons each and become Cu^{2+} ions, which go into the solution (Figure 5b.9). We can describe this with a half-equation:

$$Cu - 2e^- \rightarrow Cu^{2+}$$

The cathode gains copper atoms. This is because Cu^{2+} ions in the solution gain electrons and become copper atoms. We can describe this with a half-equation:

$$Cu^{2+} + 2e^- \rightarrow Cu$$

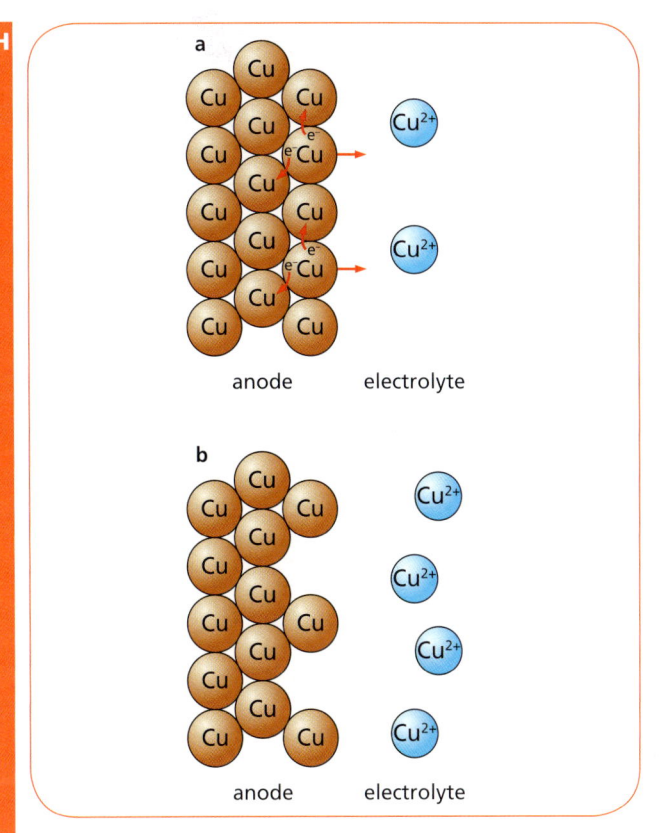

Figure 5b.9 a At the anode, copper atoms lose electrons. **b** They become copper ions in the electrolyte.

SAQ

10 Figure 5b.9a shows two copper atoms each losing two electrons. What do you think happens to these electrons?

Pure copper

Copper has been important to humans ever since the Bronze Age. One of its main uses today is for electric wiring. The copper used for this purpose has to be pure. It is purified by the electrolysis of copper sulfate solution using copper electrodes.

At the start of the electrolysis, the anode is made of a large piece of impure copper and the cathode is made of a small piece of pure copper. When the electrolysis is finished, the anode is now very small, and the cathode consists of a large piece of highly pure copper. All the impurities are left dissolved in the electrolyte or in a sludge at the bottom of the electrolyte tank.

C5b Electrolysis

Calculating electric charge

The actual mass of copper that is gained by the cathode and lost by the anode depends on the amount of electric charge that has flowed. We use the symbol Q to represent electric charge, and it is measured in units called coulombs (C). The amount of electric charge that has flowed can be calculated by multiplying the current (symbol I, measured in amperes, A) by the time (symbol t, measured in seconds, s). The equation for this is:

$$Q = It$$

When electrolysis is performed, the amount of a substance that appears (or disappears) at each electrode is proportional to the number of coulombs of electric charge that have flowed.

Worked example 1

Potassium nitrate solution is electrolysed. A current of 3 A is passed for 20 minutes. 450 cm³ of hydrogen is collected at the cathode and 225 cm³ of oxygen is collected at the anode. How much of each gas would be collected if the current had been 2 A and the time had been 1 hour?

Step 1: 3 A flowed for 20 minutes (which is 20 × 60 seconds).

The charge that flowed is calculated using $Q = It$:

$Q = 3 \times 20 \times 60 = 3600\,C$

Step 2: if the current had been 2 A and the time had been 1 hour:

$Q = 2 \times 60 \times 60 = 7200\,C$

7200 C is twice as many coulombs as 3600 C so twice as much hydrogen and oxygen would have formed.

Answer: 900 cm³ of hydrogen and 450 cm³ of oxygen would be collected.

SAQ

11 What volumes of hydrogen and oxygen will form if potassium nitrate solution is electrolysed with the following currents for the following times?

   a 0.3 A for 20 minutes
   b 6 A for 80 minutes
   c 1 A for 1 hour

Summary

You should be able to:

- describe electrolysis in terms of flow of charge and the discharge of ions
- label the apparatus needed to electrolyse a molten substance or an aqueous solution
- predict the products of the electrolysis of four molten electrolytes and explain why they don't conduct electricity when solid
- state the products of electrolysing aqueous potassium nitrate and aqueous potassium sulfate

continued on next page

Summary - *continued*

- **[H]** explain that in some cases it is easier to discharge ions from the water in the solution than from the solute itself, with reference to the electrolysis of aqueous potassium nitrate and aqueous potassium sulfate

- describe the electrolysis of copper(II) sulfate with copper electrodes, including the change in mass at the electrodes

- describe the factors that affect the change in the amount of a substance at the anode or cathode during electrolysis

- **[H]** use the relationship $Q = It$ to perform calculations based on charge, current and time to predict the change in the amount of a substance at the anode or cathode

- write half-equations for the electrode processes occurring during all the electrolysis reactions studied in this item

Questions

1. Lead and iodine can be made by electrolysis.

    a. Which substance should be electrolysed to do this?

    b. Draw a labelled diagram of this process.

    c. What two factors affect the amounts of lead and iodine that are obtained?

2. Name the substances that are formed at each electrode during the electrolysis of the following substances.

    a. aqueous potassium sulfate

    b. molten aluminium oxide

    c. aqueous potassium nitrate

3. **[H]** A current of 0.7 A is passed through molten potassium chloride for 10 minutes. This produces 0.17 g of potassium and 0.16 g of chlorine.

    a. How many coulombs of charge flowed during the 10 minutes?

    b. How many coulombs of charge would have flowed if a current of 2.0 A had been used for 14 minutes?

    c. What masses of potassium and chlorine would have been produced in this case?

4. a. Describe the process happening at each electrode when copper(II) sulfate solution is electrolysed with copper electrodes.

    b. At which electrode is copper oxidised? Explain your answer. (If you cannot remember learning this, then *Gateway Additional Science* page 135 will help.)

    c. At which electrode is copper reduced? Explain your answer. (See *Gateway Additional Science* page 141.)

C5c Quantitative analysis

Chemists often need to know exact details about the amount (or *quantity*) of a particular substance in a solid, in a solution, or sometimes even in a sample of food or drink. Measuring these quantities is called **quantitative analysis**. This item is an introduction to some of the ideas and methods involved in quantitative analysis, particularly those involving the concentration of a solution.

Recommended daily amounts

The labels on packets of food tell us a lot about what is in the packet. Figure 5c.1 shows the ingredients list and nutrition information from a packet of breakfast cereal.

Nutrition Information

	Typical value per 100 g		30 g serving with 125 ml of semi-skimmed milk	
ENERGY	1681 kJ	397 kcal	755 kJ*	178 kcal
PROTEIN	6 g		6 g	
CARBOHYDRATE	82 g		31 g	
of which sugars	35 g		17 g	
starch	47 g		14 g	
FAT	5 g		3.5 g*	
of which saturates	0.9 g		1.5 g	
FIBRE	2.5 g		0.8 g	
SODIUM	0.45 g		0.2 g	
SALT	1.15 g		0.5 g	
VITAMINS:		(% RDA)		(% RDA)
THIAMIN (B₁)	1.2 mg	(83)	0.4 mg	(29)
RIBOFLAVIN (B₂)	1.3 mg	(83)	0.7 mg	(44)
NIACIN	14.9 mg	(83)	4.7 mg	(26)
VITAMIN B₆	1.7 mg	(83)	0.6 mg	(29)
FOLIC ACID	334 µg	(167)	108 µg	(54)
VITAMIN B₁₂	0.83 µg	(83)	0.77 µg	(77)
MINERALS:				
IRON	8.0 mg	(57)	2.4 mg	(17)

Figure 5c.1 Many food packages give detailed information about the substances in the food. This can help you plan a healthy balanced diet.

A typical bowl of this cereal is 30 g of cereal with 125 ml (125 cm³) of semi-skimmed milk poured onto it. The blue panel on the packet tells you what you get from this bowl of cereal and milk. For example, you would get 6 g of protein and 0.8 g of fibre. The panel also says you would get 0.4 mg of a vitamin called thiamin, and it says that this is *29% of your RDA*.

This means that the bowl of cereal would give you 29% of your **recommended daily amount** (RDA) of this vitamin. The term *recommended daily allowance* is sometimes used instead. Your recommended daily amount of thiamin is 1.4 mg. The RDA is the amount of this vitamin that everyone needs each day in order to stay healthy. One bowl of the cereal gives you 29% of this amount, so four bowls would give you 116% of your RDA of thiamin – slightly more than you need each day.

SAQ

1. **a** What percentage of your RDA of folic acid do you get if you eat 30 g of the cereal from the packet in Figure 5c.1, with 125 cm³ of semi-skimmed milk?
   b How many 30 g bowls of this cereal do you need to get 100% of your RDA of folic acid?

Salt

Too much salt in your diet can cause you to have heart problems. There is an RDA for salt, but unlike the RDA for a vitamin you should try to keep *below* the RDA for salt. The RDA for a vitamin is a minimum value, but for salt it is a maximum value. Your RDA for salt is 6 g.

SAQ

2. Use Figure 5c.1 to find out how many 30 g bowls of cornflakes you need to eat to go over your RDA for salt.

Sodium in salt

Salt is sodium chloride; its formula is NaCl. The relative atomic mass of sodium is 23. The relative atomic mass of chlorine is 35.5. The relative formula mass of sodium chloride is 23 + 35.5 = 58.5. Therefore, one mole of salt has a mass of 58.5 g, and contains 23 g of sodium. 58.5 g is approximately 2.5 times 23 g (58.5 ÷ 23 = 2.54). So, if you know the mass of sodium in a portion of food, you can calculate the approximate mass of salt it contains by multiplying the mass of sodium by 2.5.

(The relative atomic mass values needed in this item can be found in the Periodic Table on page 359.)

SAQ

3 a What mass of sodium is there in one 30 g bowl of the cereal in the packet in Figure 5c.1?

b What do you get if you multiply this mass by 2.5?

c Is this the same as the mass of salt in the bowl of cereal?

Multiplying by 2.5 is a good *rule-of thumb* for converting a mass of sodium into the mass of salt that contains it. It is not completely reliable though. Sodium chloride is not the only sodium compound found in foods. It is possible that a portion of food that contains 0.2 g of sodium does not in fact contain 0.5 g of salt.

The concentration of a solution

When a **solute** is dissolved in a **solvent**, such as water, a solution is formed. The amount of solute dissolved in a certain volume of solution is called the **concentration** of the solution. The volume of a solution is usually measured in dm^3, which is pronounced 'D M cubed' or 'decimetre cubed'. $1 dm^3$ is also known as 1 litre. There are $1000 cm^3$ in $1 dm^3$.

- A solution that is more concentrated has more solute particles (molecules or formula units) dissolved in $1 dm^3$ of solution.
- A solution that is less concentrated has fewer solute particles (molecules or formula units) dissolved in $1 dm^3$ of solution.

The solute particles are closer together in a more concentrated solution than they are in a less concentrated solution (Figure 5c.3, over the page).

Another way of describing a solution that is less concentrated is to say that it is *more dilute*.

Vital chemicals

Cereal packets tell us how much of our recommended daily amount of each vitamin we will get if we eat a bowl of the cereal in the packet. Do we really need these vitamins? Is the cereal manufacturer just trying to con us into buying their product because it is 'healthy'?

Thiamin is also called Vitamin B_1. Lack of thiamin can cause you to become easily confused, to have low powers of concentration, to become exhausted easily, and to have poor balance (Figure 5c.2). Thiamin sounds very important for a successful day at school!

We need folic acid in order to make red blood cells. Lack of red blood cells leads to the condition known as *anaemia*. Anaemia results in tiredness and lethargy. It is particularly important for a woman to have a good intake of folic acid if she is pregnant, as it is very good for the health of the baby. The RDA for folic acid is 200 μg, but for women in early pregnancy it is 400 μg.

The start of the word *vitamin* comes from the word *vital*, which means *essential*. Vitamins really are essential for a healthy life.

Figure 5c.2 Not enough thiamin?

130 **C5c** Quantitative analysis

a More concentrated solution   b Less concentrated solution

solvent particle

solute particle

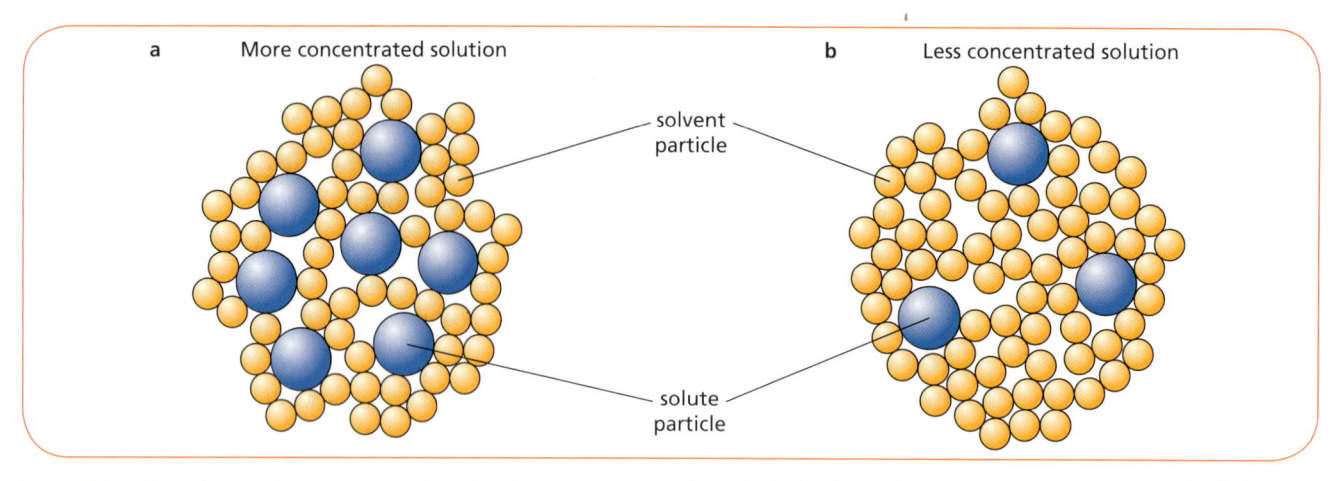

Figure 5c.3 The solute particles are closer together in the more concentrated solution than they are in the less concentrated solution.

SAQ

4 a Which of the solutions in Figure 5c.3 is more concentrated?

b Which of the solutions in Figure 5c.3 is more dilute?

The units we use to measure concentration are grams per dm^3 or moles per dm^3.

Grams per dm^3 is usually written g per dm^3 or g/dm^3. If the concentration of a solution is given as $24 g/dm^3$, for example, this means there are 24 g of solute dissolved in each dm^3 of solution.

Moles per dm^3 is usually written mol per dm^3 or mol/dm^3. If the concentration of a solution is given as $0.5 mol/dm^3$ this means there are 0.5 moles of solute dissolved in each dm^3 of solution.

SAQ

5 Solutions A and B are made by dissolving sodium nitrate in water. Solution A has a concentration of $10 g/dm^3$ and solution B has a concentration of $40 g/dm^3$.

a Which substance is the solute in these solutions?

b Which substance is the solvent in these solutions?

c What do we mean if we say a solution has a concentration of $10 g/dm^3$?

d Which solution is more concentrated, A or B?

e Which solution is more dilute?

f In which solution are the sodium nitrate particles more crowded?

Converting between dm^3 and cm^3

There are $1000 cm^3$ in $1 dm^3$ (Figure 5c.4).

- If you have a volume in dm^3 and need to know how many cm^3 this is, multiply the volume in dm^3 by 1000.
- If you have a volume in cm^3 and need to know how many dm^3 this is, divide the volume in dm^3 by 1000.

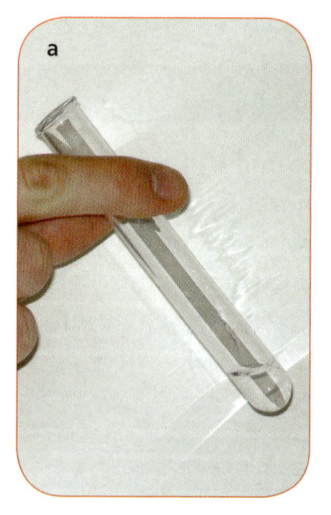

Figure 5c.4 a This test tube contains $1 cm^3$ of water. **b** This beaker contains $1 dm^3$ of water. There are $1000 cm^3$ of water in the beaker.

SAQ

6 How many cm^3 are there in:

a $2 dm^3$?   **b** $12 dm^3$?

c $0.5 dm^3$?   **d** $0.009 dm^3$?

7 How many dm^3 are there in:

a $8000 cm^3$?   **b** $1700 cm^3$?

c $400 cm^3$?   **d** $23 cm^3$?

Dilution

Sometimes we need to change the concentration of a solution. Some substances are made as highly concentrated solutions, then distributed around the world. When they arrive at their destination, water is added to them to give the lower concentration that is right for their use (Figure 5c.5). This avoids unnecessarily transporting water. Adding a solvent (e.g. water) to a solution to make it less concentrated is called **dilution**.

Liquid medicines and baby milk may be transported in a concentrated state then diluted before using. There are risks involved here. If a medicine such as an antibiotic is over-diluted its concentration may become too low for it to be effective. In this case, the bacteria it is designed to kill may survive, and continue to make the patient ill. If baby milk is over-diluted it will not be nourishing enough, so it will leave the baby feeling hungry. Alternatively, if baby milk is not diluted enough then the baby will be fed a solution which is too concentrated. This can damage the baby's kidneys.

SAQ

8 a Describe three things that can be transported as a highly concentrated solution and then diluted to a lower concentration before using.

b Give two situations in which diluting a solution too much can be dangerous.

Performing an accurate dilution

Solutions can be diluted accurately by using appropriate apparatus. Figure 5c.6 shows 1.0 mol/dm^3 nitric acid solution being diluted carefully to give 0.1 mol/dm^3 nitric acid solution.

Because 50 cm^3 of solution was diluted to give ten times that volume, the final concentration is one-tenth of the starting concentration. The 1.0 mol/dm^3 solution was diluted to give a 0.1 mol/dm^3 solution.

Figure 5c.5 This orange juice was transported in a very concentrated state, then diluted and packaged in the country where it was sold to consumers.

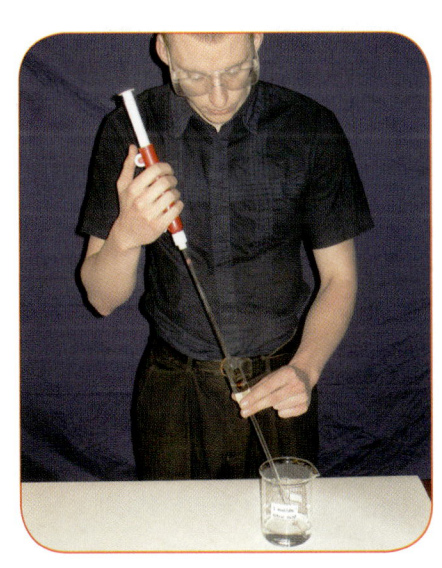

Figure 5c.6a A graduated pipette is used to measure 50 cm^3 of 1.0 mol/dm^3 nitric acid solution.

Figure 5c.6b The 50 cm^3 of 1.0 mol/dm^3 nitric acid solution is put into a 500 cm^3 volumetric flask.

Figure 5c.6c Water is added to the volumetric flask until it contains exactly 500 cm^3 of solution. The concentration of the nitric acid solution is now 0.1 mol/dm^3.

132 C5c Quantitative analysis

Calculating a concentration in mol/dm³

To calculate the concentration of a solution, we divide the amount of solute by the volume of solvent. The amount of solute will be in moles and the volume of solvent will be in dm³, so the answer will be in mol/dm³.

Worked example 1

If 4 moles of HCl are dissolved in 2 dm³ of water, what is the concentration of the solution in mol/dm³?

$$\frac{4 \text{ moles}}{2 \text{ dm}^3} = 2 \text{ mol/dm}^3$$

Answer: the concentration is 2 mol/dm³.

SAQ

9 What are the concentrations, in mol/dm³, of the following hydrochloric acid solutions?
   a 4 moles of hydrochloric acid are dissolved in 1 dm³ of water.
   b 0.8 moles of hydrochloric acid are dissolved in 2 dm³ of water.
   c 0.04 moles of hydrochloric acid are dissolved in 500 cm³ of water. (Hint: convert 500 cm³ into dm³ first.)

Calculating a concentration in g/dm³

The concentration of a solution can also be calculated in g/dm³. To do this, we divide the mass of solute by the volume of solvent. If the mass of solute is in grams and the volume of solvent is in dm³ the answer will be in g/dm³.

Worked example 2

If 60 g of NaOH are dissolved in 3 dm³ of water, what is the concentration of the solution in g/dm³?

$$\frac{60 \text{ g}}{3 \text{ dm}^3} = 20 \text{ g/dm}^3$$

Answer: the concentration is 20 g/dm³.

SAQ

10 What are the concentrations, in g/dm³, of the following sodium hydroxide solutions?
    a 75 g of sodium hydroxide are dissolved in 5 dm³ of water.
    b 16 g of sodium hydroxide are dissolved in 8 dm³ of water.
    c 20 g of sodium hydroxide are dissolved in 500 cm³ of water. (Hint: convert 500 cm³ into dm³ first.)

Converting a concentration in g/dm³ into mol/dm³

To convert the mass of a sample in grams into a number of moles, we divide the mass in grams by the mass of one mole of the substance:

$$\text{number of moles} = \frac{\text{mass of sample}}{\text{molar mass}}$$

(See Item C5a, *Moles and empirical formulae*, page 113.)

So, to convert the concentration of a solution in g/dm³ into a concentration in mol/dm³ we divide by the molar mass of the solute. The equation is:

$$\text{concentration in mol/dm}^3 = \frac{\text{concentration in g/dm}^3}{\text{molar mass of solute}}$$

Worked example 3

If 120 g of NaOH are dissolved in 3 dm³ of water, what is the concentration of the solution in mol/dm³?

Step 1: first find the concentration in g/dm³.

$$\frac{120 \text{ g}}{3 \text{ dm}^3} = 40 \text{ g/dm}^3$$

Step 2: then divide the answer by the molar mass of NaOH. The relative formula mass of NaOH is 23 + 16 + 1 = 40, so the molar mass of NaOH is 40 g.

$$\frac{40 \text{ g/dm}^3}{40 \text{ g}} = 1 \text{ mol/dm}^3$$

Answer: the concentration is 1 mol/dm³.

SAQ

11 What are the concentrations, in mol/dm³, of the following sodium hydroxide solutions?

   a 80 g of sodium hydroxide are dissolved in 4 dm³ of water.

   b 20 g of sodium hydroxide are dissolved in 2 dm³ of water.

   c 8 g of sodium hydroxide are dissolved in 100 cm³ of water. (Hint: convert 100 cm³ into dm³ first.)

Calculating the number of moles of solute in a sample of a solution

To calculate the number of moles of solute dissolved in a sample, multiply the volume of the sample by its concentration.

number of moles = volume × concentration

$$N = V \times C$$

For this calculation to work:
- the volume must be in dm³ (not cm³)
- the concentration must be in mol/dm³ (not g/dm³).

SAQ

12 How many moles of salt (NaCl) are there in:

   a 2 dm³ of 3 mol/dm³ salt solution?

   b 65 cm³ of 4 mol/dm³ salt solution?

   c 200 cm³ of 117 g/dm³ salt solution?

Summary

You should be able to:

- interpret information on food packaging, understanding what is meant by an RDA
- convert amounts of sodium into amounts of salt, appreciating one limitation of this approach
- describe high and low concentration in terms of the closeness of solute particles
- recall the units used to measure concentration – g/dm³ and mol/dm³
- use dm³ and cm³ as units to measure volume, and convert between them
- convert between concentrations given in g/dm³ and mol/dm³
- use the expressions $N = V \times C$ and $C = N \div V$ in calculations
- describe how a solution can be diluted
- recall three situations where dilution can be necessary, describing some possible dangers if dilution is done wrongly
- perform calculations relating to dilution

Questions

1. Figure 5c.7 shows the nutrition information from a wholewheat biscuit breakfast cereal.

    a What percentage of your RDA of thiamin and folic acid do you get from a 36 g bowl of this cereal with 150 cm³ of milk?

    b How does this compare with the amount of thiamin and folic acid that you get from a 30 g bowl of cereal from the packet in Figure 5c.1, with 125 cm³ of milk?

    c Why is this not a fair comparison?

    d Explain what is meant by *RDA*.

Nutrition		
Typical values	Per 36g with 150ml semi-skimmed milk	Per 100g
Energy	811 kJ	1432 kJ
	192 kcal	338 kcal
Protein	9.1g	11.5g
Carbohydrate	31.8g	68.1g
of which sugars	8.8g	4.4g
of which starch	23.0g	64.0g
Fat	3.1g	2.0g
of which saturates	1.7g	0.6g
of which mono-unsaturates	0.8g	0.2g
of which polyunsaturates	0.4g	1.1g
Fibre	3.6g	10.0g
Salt	0.4g	0.7g
of which sodium	0.2g	0.3g
Thiamin (B1)	0.5mg (35% RDA)	1.2mg (85% RDA)
Riboflavin (B2)	0.8mg (47% RDA)	1.4mg (85% RDA)
Niacin	5.7mg (31% RDA)	15.3mg (85% RDA)
Folic Acid	70.2μg (35% RDA)	170μg (85% RDA)
Iron	4.4mg (31% RDA)	11.9mg (85% RDA)

Figure 5c.7

2. A chemist has 50 cm³ of vitamin C solution. Its concentration is 5 g/dm³. The chemist takes 10 cm³ of the solution and dilutes it with water so that its total volume is 1000 cm³.

    a What is another name for *1000 cm³*?

    b Describe how the chemist would perform the dilution, naming any equipment that would be used.

    c What is the concentration of the new solution?

    d The RDA for vitamin C is 0.060 g. Would a person get this much vitamin C by drinking all of the new solution?

3. 45 g of glucose are dissolved in 500 cm³ of solution. The formula of glucose is $C_6H_{12}O_6$.

    a Convert 500 cm³ into dm³.

    b What is the concentration of the glucose solution in g/dm³?

    c What is the molar mass of glucose?

    d What is the concentration of the glucose solution in mol/dm³?

4. How many moles of magnesium sulfate are there in 50 cm³ of a solution whose concentration is 30 g/dm³? The formula of magnesium sulfate is $MgSO_4$.

5d Titrations

What is a titration?

We can measure the exact volume of acid needed to neutralise a measured volume of alkali, by titration. Titration is one of the commonest techniques that chemists use in quantitative analysis. If we know the concentration of either the acid or the alkali, we can calculate the other concentration from the titration results.

Performing a titration

During titration, it is important to measure the volumes accurately. For this reason, chemists use a **burette** and a **graduated pipette**. These are the most accurate volume-measuring devices in most labs. In an acid–alkali titration, an indicator must be used. The indicator changes colour at the **end-point** of the titration.

Figures 5d.1 to 5d.4 show a titration between hydrochloric acid and sodium hydroxide. The indicator used in the titration is phenolphthalein. This is an excellent indicator to use in a titration; it is colourless in acids and purpley-pink in alkalis. Its colour change at the end-point of the titration is unmistakable.

- The acid is put into the burette until it reaches the $0.0\,cm^3$ line (Figure 5d.1).
- $25\,cm^3$ of sodium hydroxide are measured out using a $25\,cm^3$ graduated pipette. A pipette filler is used to suck the alkali into the pipette. The alkali is then put into a conical flask (Figure 5d.2).
- Phenolphthalein indicator is added to the alkali in the conical flask (Figure 5d.3).

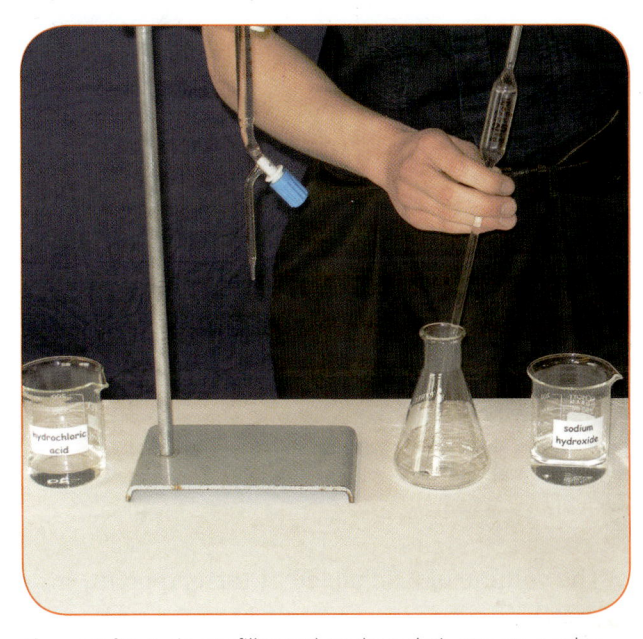

Figure 5d.2 A pipette filler and graduated pipette are used to measure $25\,cm^3$ of sodium hydroxide into a conical flask.

Figure 5d.1 A funnel is used to fill the burette with dilute hydrochloric acid.

Figure 5d.3 Phenolphthalein indicator is added to the sodium hydroxide solution, which turns purple.

136 C5d Titrations

- Acid from the burette is added slowly to the alkali in the conical flask. When just enough acid has been added to neutralise the alkali, the phenolphthalein goes colourless (Figure 5d.4). No more acid is added. The volume reading on the burette is noted down.

	Titration 1	Titration 2	Titration 3
Burette reading at end-point	14.6 cm^3	29.0 cm^3	43.5 cm^3
Initial burette reading	0.0 cm^3	14.6 cm^3	29.0 cm^3
Titre	14.6 cm^3		

Table 5d.1

a Calculate the titre value for Titration 2 and Titration 3.

b Why did the experimenter repeat the titration three times?

c Were the repeats successful?

d What was the average titre value?

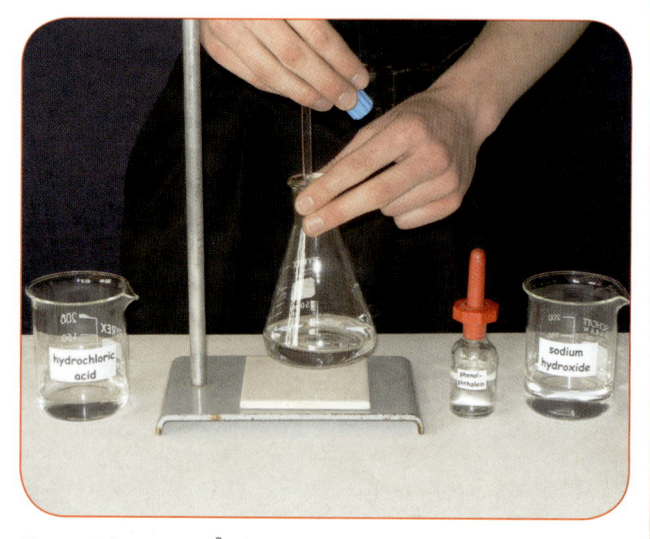

Figure 5d.4 14.6 cm^3 of hydrochloric acid from the burette have been added to the 25 cm^3 of sodium hydroxide in the conical flask. The indicator has just gone colourless. This is the end-point of the titration.

The volume of acid added at the end-point, known as the **titre**, is 14.6 cm^3. The titration should be repeated until two titre values in good agreement with each other are obtained. To be *in good agreement* the two titre values should be within 0.1 cm^3 of each other.

SAQ

1 Read this item again up to this point, close the book, and then describe in your own words how you would perform a titration.

2 In Table 5d.1, the results of the titration shown in Figures 5d.1 to 5d.4 are recorded in the Titration 1 column. The experimenter performed Titration 2 and Titration 3 using fresh 25 cm^3 samples of alkali, but did not refill the burette with acid each time.

Different indicators

Phenolphthalein is not the only indicator that can be used for a titration. **Litmus** and **screened methyl orange** can also be used (Figures 5d.5 and 5d.6).

Figure 5d.5 Litmus indicator has been added to both flasks.

Figure 5d.6 Screened methyl orange indicator has been added to both flasks.

Phenolphthalein, litmus and screened methyl orange are *single indicators*. They give a sharp colour change at the titration end-point. Universal indicator is not a single indicator – it is a mixture of many indicators. This means it gives an excellent range of colours, so we can use it to estimate the pH number of a solution. Universal indicator is not suitable for titrations though, because it does not give a sharp colour change at the end-point of the titration.

SAQ

3 What colour is each of the following indicators in acid and in alkali?

   a phenolphthalein

   b screened methyl orange

   c litmus

4 Why is universal indicator unsuitable for use in a titration?

How pH changes during a titration

As the acid is added to the alkali during a titration, the pH in the conical flask changes. The pH starts off at around 14 – sodium hydroxide is a strong alkali. As we add acid from the burette, the alkali is gradually neutralised. The acid and alkali react together to give a salt and water, which are both neutral. At the end-point, the pH of the salt solution is 7. If we carry on adding acid from the burette, the acid becomes in excess. This causes the pH to continue to fall until it reaches 1 or 2 – hydrochloric acid is a strong acid.

These pH changes do not happen in a steady fashion. To begin with, the pH of the solution in the conical flask changes very little as acid is added. It remains at around 14. When the end-point is near, the pH begins to fall rapidly. At the end-point, the pH changes by several units very suddenly. This is shown in Figure 5d.7.

The volume of acid needed to neutralise the alkali – that is, the volume that has been added at the end-point – can be easily identified from the graph. It is the volume that causes the very large change in pH. The graph in Figure 5d.7 shows that

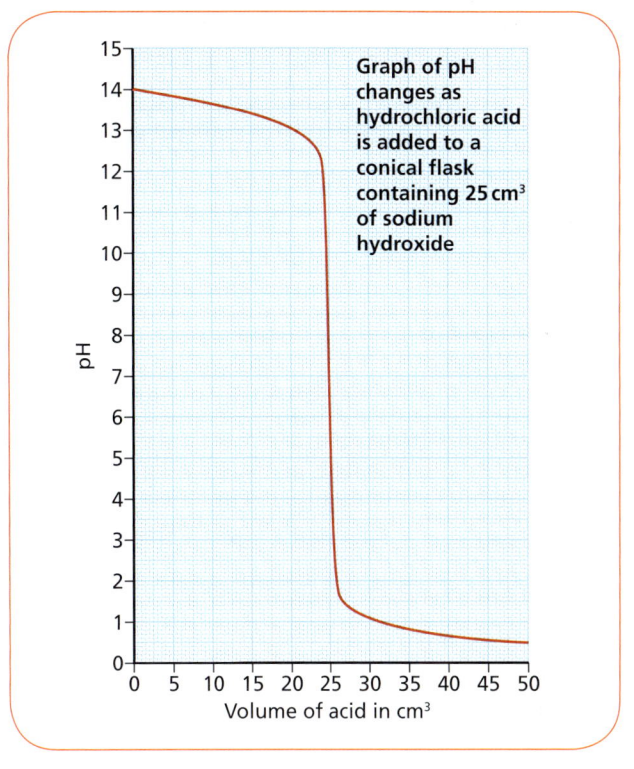

Figure 5d.7 This graph shows how the pH in the flask changes during the titration described on this page.

the end-point of this titration was reached when 25.0 cm³ of acid had been added.

SAQ

5 Why does the pH of the alkali in the flask decrease as the acid is added to it from the burette?

6 Complete this word equation for the reaction happening in the conical flask:

   acid + alkali → _____ + _____

7 Use the graph in Figure 5d.7 to answer the questions below.

   a What volume of acid had been added when the pH was exactly 13?

   b What was the pH when 30 cm³ of acid had been added?

   c What volume of acid had been added at the end-point?

Trading standards

Figure 5d.8 The nutrition information panel from a carton of orange juice.

The supplier of the orange juice in Figure 5d.8 has to be certain that the juice contains 50 mg of vitamin C per 200 ml serving, as claimed on the packaging. This is equal to a concentration of 0.25 g/dm³. To check that this is true, the supplier titrates a sample of the juice. A small volume of the juice is measured using a graduated pipette and put into a conical flask. Then a solution of a dark blue chemical known as 'DCPIP' is put into a burette. DCPIP reacts with vitamin C.

The DCPIP solution is slowly added to the juice in the flask. As long as there is vitamin C in the juice, the blue colour disappears, but as soon as all the vitamin C has reacted, the orange juice turns blue. The volume of DCPIP solution that has been added when a permanent blue colour appears in the juice is noted down. This volume is used to calculate the concentration of vitamin C in the original fruit juice.

As long as the vitamin C concentration is at least as high as 0.25 g/dm³, the juice can be sold legally. This is a good example of quantitative analysis. Most manufacturers of food and drink have a Quality Control Laboratory where the analysis takes place. It is the job of the QCL to ensure that the information on the packaging is true. If it is not true, the manufacturer can find itself taken to court by the Trading Standards Authority – with potentially costly results.

Figure 5d.9 Does this juice contain all the vitamin C it should? A titration test could be used to find out.

Using titration to calculate a concentration

We can use the results of a titration to find an unknown concentration. The titration shown in Figures 5d.1 to 5d.4 involves two solutions: an acid and an alkali. If we know the concentration of one of these, we can calculate the concentration of the other solution from the titration results.

The titration shown is between hydrochloric acid and sodium hydroxide. The equation for the neutralisation reaction is:

$HCl + NaOH \rightarrow NaCl + H_2O$

One HCl formula unit reacts with one NaOH formula unit. *At the titration end-point* the number of moles of HCl added is the same as the number of moles of NaOH that were originally pipetted into the flask.

Worked example

A hydrochloric acid solution of unknown concentration was titrated against 0.1 mol/dm³ sodium hydroxide solution. 25.0 cm³ of the sodium hydroxide was pipetted into the flask at the start and an indicator was added. 16.7 cm³ of hydrochloric acid had been added from the burette when the end-point was reached. What is the concentration of the acid?

Step 1: first calculate the number of moles of alkali that were used, remembering to convert the volume into dm³ (25.0 cm³ = 0.0250 dm³). To do this we use:

number of moles = volume × concentration

or $N = V \times C$

number of moles of alkali
$= 0.0250 \, dm^3 \times 0.1 \, mol/dm^3$
$= 0.00250$ moles

Step 2: the number of moles of alkali is the same as the number of moles of acid at the end-point, so we know the number of moles of acid was also 0.00250 moles. To calculate the concentration of the acid we use $C = N \div V$.

concentration of acid $= \dfrac{0.00250 \text{ moles}}{0.0167 \, dm^3}$

$= 0.150 \, mol/dm^3$

Answer: concentration of acid = 0.150 mol/dm³.

SAQ

8 At the end-point of a titration, 32.2 cm³ of hydrochloric acid solution of unknown concentration have neutralised 25.0 cm³ of 0.40 mol/dm³ sodium hydroxide solution.

   a How many moles of sodium hydroxide are there in 25.0 cm³ of 0.40 mol/dm³ solution?

   b How many moles of hydrochloric acid have reacted with this number of moles of sodium hydroxide at the end-point?

   c Calculate the concentration of the hydrochloric acid solution. Give your answer to two significant figures (2 s.f.).

Summary

You should be able to:

♦ describe how to perform a titration, labelling and identifying the apparatus used

♦ name three indicators suitable for a titration between hydrochloric acid and sodium hydroxide, giving the colour of each in acid and in alkali

♦ explain why universal indicator is unsuitable for an acid–alkali titration

♦ describe and explain the pH changes during the reaction of an acid with an alkali

♦ interpret a simple pH curve

♦ sketch a pH titration curve for the titration of an acid with an alkali

♦ calculate the concentration of an acid or an alkali from titration results

♦ state and use the relationships $N = V \times C$, $C = N \div V$ and $V = N \div C$ (Figure 5d.10)

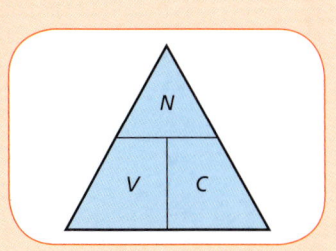

Figure 5d.10

Questions

1 Stephen performed a titration between hydrochloric acid and sodium hydroxide. He measured 25 cm³ of sodium hydroxide using a measuring cylinder and poured it into a beaker. He added two drops of universal indicator to his bottle of acid, which went red, and then poured the acid from its bottle straight into a burette. He added the acid to the alkali in the beaker until the indicator went green, and then packed his equipment away.

   a What did Stephen get wrong? (There are at least six errors in his technique.)

   b How should he have performed the titration?

2 a If you add universal indicator to a solution and get a yellow colour, what is the pH number of the solution? (Look at *Gateway Additional Science* pages 161–162 to remind yourself about universal indicator solution, if you need to.)

   b What colour do you see if universal indicator is added to a solution of pH 10?

3 25 cm³ of sodium hydroxide was put into a conical flask. Phenolphthalein was added. Hydrochloric acid was then added from a burette. The change in pH of the solution in the flask is shown by the graph in Figure 5d.11.

   a What volume of acid had been added when the pH was exactly 13.5?

   b What was the pH when 38 cm³ of acid had been added?

   c What volume of acid had been added when the phenolphthalein changed colour?

   d What do we call this event during a titration?

Graph of pH changes as hydrochloric acid is added to a conical flask containing 25 cm³ of sodium hydroxide

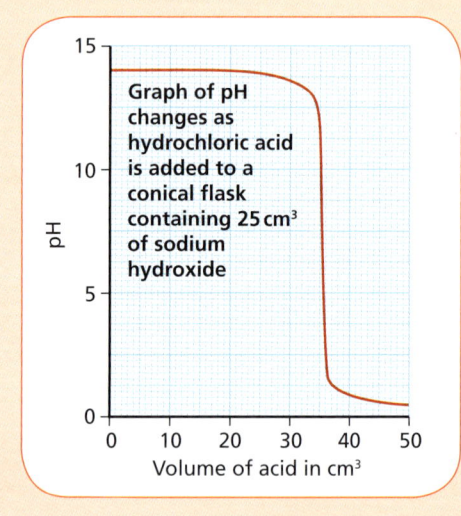

Figure 5d.11

4 20.0 cm³ of hydrochloric acid are put into a conical flask. It has a pH of 1.0. Indicator is added. Sodium hydroxide is added from a burette. The indicator changes colour when 16.2 cm³ of sodium hydroxide has been added. The pH of the solution is 12.4 when 25.0 cm³ of sodium hydroxide has been added.

Use this information to sketch a graph, showing how the pH of the solution in the flask changes as 25.0 cm³ of alkali is added slowly to the flask.

5 You have a solution of known concentration and wish to measure out a particular number of moles of solute. The volume of solution needed can be calculated by using the relationship $V = N \div C$, or:

$$\text{volume in dm}^3 = \frac{\text{number of moles}}{\text{concentration in mol/dm}^3}$$

Use this relationship to calculate what volume of 0.080 mol/dm³ sulfuric acid solution contains exactly 0.0016 moles of sulfuric acid. Convert your answer into cm³. You should give your answer to three significant figures (3 s.f.).

Gas volumes

Measuring gases

Many chemical reactions produce a gas as one of the products. It is useful to be able to measure the volume, or mass, of gas produced. Figures 5e.1 to 5e.4 show an example of a reaction in which a gas is produced – the catalytic decomposition of hydrogen peroxide. Hydrogen peroxide solution will decompose if a catalyst called manganese dioxide is added to it. The products of the reaction are water and oxygen gas:

hydrogen peroxide → water + oxygen

Figures 5e.1 to 5e.4 show four ways of measuring the amount of oxygen gas given off in the reaction.

Measuring cylinder

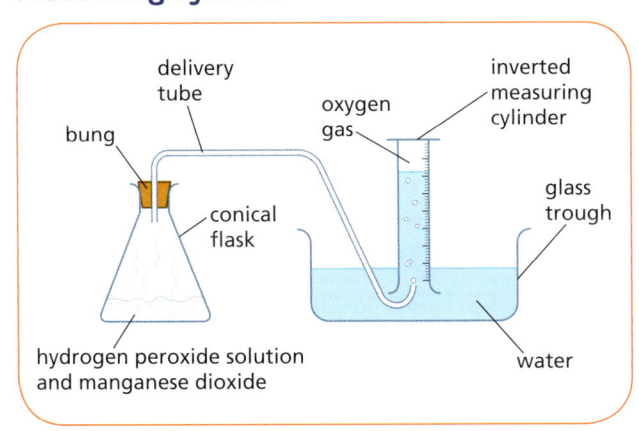

Figure 5e.1 Collecting a gas through water into an inverted (upside down) measuring cylinder.

This technique is simple to use, as measuring cylinders are quite small in size and often made of plastic. It is not particularly accurate, however, because the volumes measured by a measuring cylinder are only approximate.

Burette

This technique is more difficult to use because of the way a burette is **graduated**. *Graduated* here means that lines and numbers are marked on so that volumes can be read. The **graduations** on a burette are designed for the burette to be used upright. By contrast the graduations on a measuring cylinder work well upright *or* upside-down. If used correctly, however, a burette is much more accurate than a measuring cylinder.

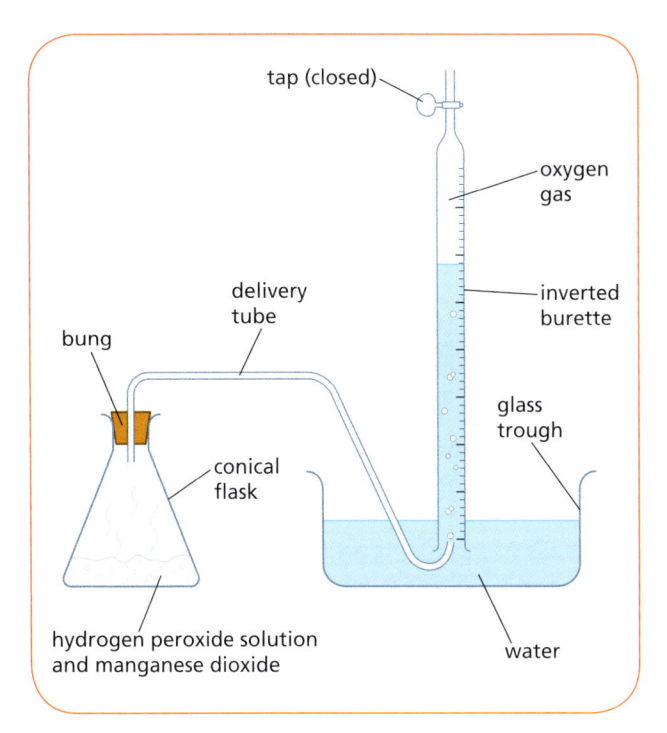

Figure 5e.2 Collecting a gas through water into an inverted burette.

Gas syringe

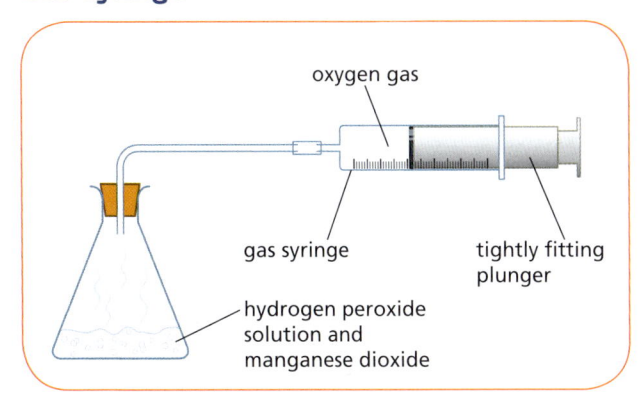

Figure 5e.3 Collecting a gas in a gas syringe.

Gas syringes are straightforward to use and are also very accurately graduated. The plunger of the syringe can stick sometimes, but this can be prevented by rotating it occasionally with your fingers.

SAQ

1 a Describe three methods of measuring the volume of gas produced in a reaction.

b Give one advantage and one disadvantage of each method.

142 C5e Gas volumes

Top-pan balance

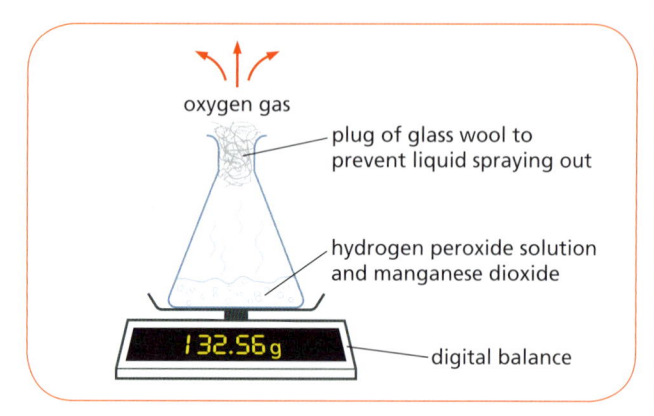

Figure 5e.4 Measuring the mass of a gas given off using a digital top-pan balance.

The mass of gas given off can be measured by placing the conical flask on a top-pan balance. As the oxygen gas is released from the conical flask, the mass of the conical flask and its contents goes down.

Sometimes the reaction is vigorous enough to cause some of the solution to spray out. A plug of glass wool is often placed in the neck of the flask to prevent this.

SAQ

2 Why is it necessary to stop solution spraying out?

Time (seconds)	Volume of oxygen collected (cm^3)
0	0
20	34
40	54
60	68
80	78
100	87
120	94
140	99
160	102
180	104
200	105
220	105
240	105

Table 5e.1 The volume of oxygen gas collected over time from the catalytic decomposition of hydrogen peroxide (Figure 5e.1).

Presenting data

If the volume of gas produced in a reaction is measured at regular time intervals, the data obtained can be presented in a table.

The reaction and apparatus shown in Figure 5e.1 were used to produce the data in Table 5e.1. This data was used to plot the graph shown in Figure 5e.5. Table 5e.1 and the graph in Figure 5e.5 tell us several things.

- **When was the reaction fastest?** From Table 5e.1, we can see that 34 cm^3 of gas were collected in the first 20 seconds. In every other 20-second period, there was less gas collected. On the graph, the line is steepest at the beginning. Both of these observations mean the reaction was fastest at the start.

- **What was the final volume of gas produced?** From Table 5e.1, it is clear that 105 cm^3 of gas was collected after 200 seconds. No more gas was collected after this. On the graph, we can see that the line stops rising at 105 cm^3. Both of these observations mean the reaction produced a final volume of gas of 105 cm^3.

- **How long did the reaction take to stop?** From Table 5e.1, it is clear that the final volume of gas collected was 105 cm^3. This volume was first recorded after 200 seconds. On the graph, the line stops rising and becomes flat (parallel to the bottom axis) after 200 seconds. Both

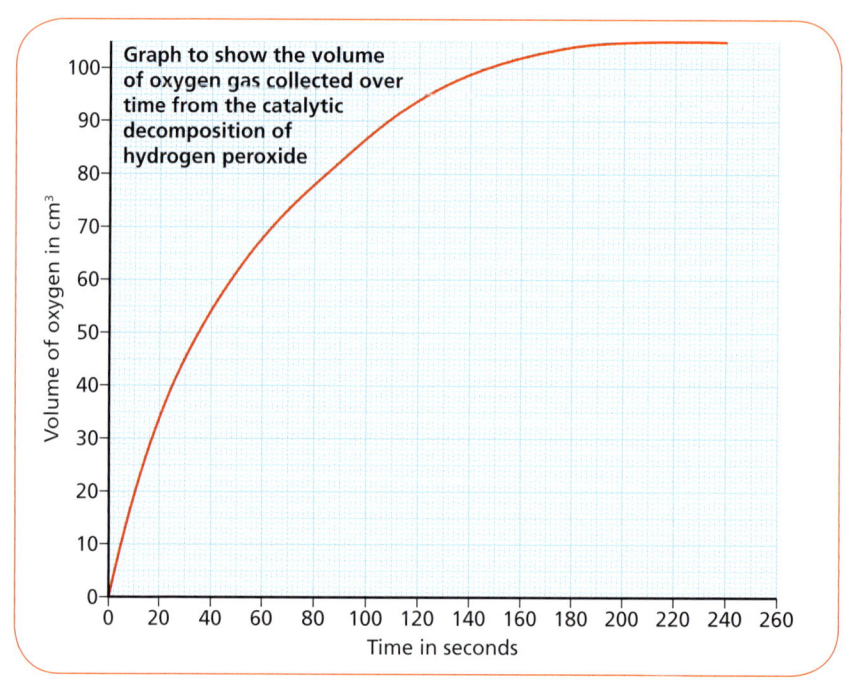

Figure 5e.5 This graph was plotted using the data in Table 5e.1.

of these observations mean the reaction had stopped after 200 seconds.

SAQ

3 This question refers to the data presented in Table 5e.1 and Figure 5e.5.
   a What was the final volume of gas collected in the experiment? How do Table 5e.1 and Figure 5e.5 show this?
   b How long did the reaction take to stop?
   c How long did it take for 75 cm³ of gas to be collected?
   d How much gas had been collected after 150 seconds?
   e Did you use Table 5e.1 or Figure 5e.5 to answer parts c and d?

4 Some manganese dioxide is added to a small amount of hydrogen peroxide solution. The oxygen gas given off is collected in an inverted burette. The volume of the gas is noted every 30 seconds.

There is no oxygen in the burette at the start.

There is 20cm³ of oxygen in the burette after the first 30 seconds.

The reaction stops 120 seconds after it started.

A total of 48cm³ of oxygen is collected.

Use these facts to produce a graph of volume of oxygen (y-axis) against time (x-axis) for this reaction.

Limiting reactant

When a chemical reaction finishes, a certain amount of product has been formed. This amount depends on the amounts of the reactants that were put together at the start. If more of both reactants are put together, more product will form. In reactions that produce a gas, more gas forms if more of the reactants are allowed to react.

If a reaction is started and one of the reactants is in short supply compared with the others, the reaction will stop when this reactant has been used up. We call this the **limiting reactant**.

Carbide lamps

Reactions that make gases have had some interesting uses over the years. People have always wanted bright lights, but electric lights are a comparatively recent invention. Before then, when very bright light was needed, people used carbide lamps (Figure 5e.6).

A carbide lamp worked by dripping water onto a small container of calcium carbide. The reaction between the water and the calcium carbide produced ethyne gas (acetylene), which was burned to provide the light. The carbide lamp was a good solution to the problem. Just 10g of calcium carbide could produce almost 4dm³ of ethyne. This much ethyne could light the equivalent of a 100W light bulb for half an hour.

Carbide lamps are now a thing of the distant past. This is just as well. The people who used them seem to remember, above all else, that they smelt awful.

Figure 5e.6 A carbide lamp.

144 C5e Gas volumes

Figure 5e.7 The hydrochloric acid 0.1 mol/dm^3 and the marble will react together to give 120 cm^3 of carbon dioxide gas. The acid is the limiting reactant.

Figure 5e.8 This time there is twice as much acid. The acid and the marble will react together to give 240 cm^3 of carbon dioxide gas. Doubling the amount of the limiting reactant doubles the amount of product.

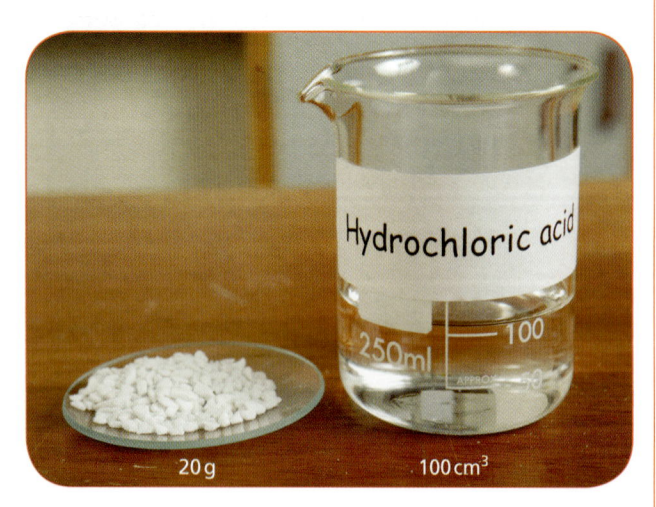

Figure 5e.9 The acid and the marble will react together to give 120 cm^3 of carbon dioxide gas. The acid is the limiting reactant so using more marble chips has no effect on the amount of product.

If more of the limiting reactant is used, more product will form. If less of the limiting reactant is used, less product will form (Figures 5e.7 to 5e.9). Changing the amount of any other reactant will not affect the amount of product formed (unless it is reduced so much that it becomes the limiting reactant).

The amount of product formed is directly proportional to the amount of the limiting reactant. If the amount of the limiting reactant is halved, for example, the amount of product formed is also halved.

SAQ

5 What do we mean by a limiting reactant?
6 How much carbon dioxide gas would form from the reaction between the following amounts of hydrochloric acid and marble chips? (Hint: use the information in Figures 5e.7 to 5e.9.)

   a 300 cm^3 of 0.1 mol/dm^3 hydrochloric acid and 10 g of marble chips
   b 50 cm^3 of 0.1 mol/dm^3 hydrochloric acid and 10 g of marble chips
   c 100 cm^3 of 0.1 mol/dm^3 hydrochloric acid and 50 g of marble chips
   d 200 cm^3 of 0.1 mol/dm^3 hydrochloric acid and 50 g of marble chips
   e Explain the reasoning behind each of your answers.

H One mole of any gas occupies 24 dm^3

One mole of any gas occupies the same volume at room temperature and pressure. This volume is 24 dm^3. Because of this, it is possible to work out the volume of a sample of gas.

Worked example 1

The molar mass of oxygen (O_2) is 32. What is the volume of 4.0 g of oxygen, at room temperature and pressure?

continued on next page

Worked example 1 - continued

Step 1: calculate the number of moles of oxygen in 4.0 g.

$$\text{number of moles} = \frac{\text{mass of sample}}{\text{molar mass}}$$

$$= \frac{4.0}{32}$$

$$= 0.125 \text{ moles}$$

Step 2: multiply the number of moles of gas by 24 dm^3 to find the volume.

volume = 0.125 × 24 = 3.0 dm^3

Answer: 4.0 g of oxygen has a volume of 3.0 dm^3.

SAQ

7 What is the volume, at room temperature and pressure, of the following masses of oxygen?

   a 320 g
   b 0.40 g
   c 1.0 g
   d 0.16 g

It is possible to calculate the volume of gas produced in a chemical reaction. We need to know the amount of the limiting reactant and we need to have a balanced symbol equation for the reaction. This calculation is possible because one mole of gas occupies a volume of 24 dm^3 at room temperature and pressure.

Worked example 2

5.6 g of iron is put into an excess of dilute sulfuric acid. What volume of hydrogen gas is formed, at room temperature and pressure?

The balanced symbol equation for the reaction is:

$Fe + H_2SO_4 \rightarrow FeSO_4 + H_2$

Step 1: the acid is in excess. This tells us that the limiting reactant is the iron. The equation says that each iron atom that reacts produces one hydrogen molecule. Therefore, the number of moles of iron that react is the same as the number of moles of hydrogen that are produced.

Since the relative atomic mass of iron is 56, the molar mass of iron is 56 g.

$$\text{number of moles} = \frac{\text{mass of sample}}{\text{molar mass}}$$

$$= \frac{5.6}{56}$$

$$= 0.10 \text{ moles}$$

So 0.10 moles of hydrogen will be produced.

Step 2: since one mole of gas occupies a volume of 24 dm^3, 0.10 moles of hydrogen has a volume of 2.4 dm^3.

Answer: 2.4 dm^3 or 2400 cm^3 of hydrogen is formed.

SAQ

8 What volume of hydrogen forms, at room temperature and pressure, when the following masses of iron are put into excess dilute sulfuric acid? Give your answers to two significant figures (2 s.f.). (Hint: use the balanced symbol equation shown in Worked example 2.)

   a 56 g
   b 2.8 g
   c 1.0 g
   d What is the limiting reactant in these reactions?

Summary

You should be able to:

- describe how you would investigate a gas reaction, measuring the volume, or mass, of gas given off
- explain why a reaction eventually stops
- understand the term *limiting reactant* and describe how the amount of limiting reactant affects the amount of product
- interpret graphs and results tables that show the volume of gas produced during the course of a reaction
- **H** produce a sketch graph to show the volume of gas produced during the course of a reaction, given appropriate details
- use the fact that one mole of any gas occupies 24 dm^3 at room temperature and pressure to perform calculations involving gas volumes and number of moles

Questions

1. a Write a word equation for the decomposition of hydrogen peroxide.

   b What role does manganese dioxide play in this reaction?

   c What is the limiting reactant in this reaction?

2. You add manganese dioxide to some hydrogen peroxide. When the reaction has stopped, 0.84 g of oxygen gas has been given off.

   a How would you investigate this reaction and obtain this result?

   b What mass of oxygen would be given off if you performed a second experiment, using twice the amount of hydrogen peroxide, but keeping everything else the same as in part **a**?

   c What mass of oxygen would be given off if you performed a third experiment, using twice the amount of manganese dioxide, but keeping everything else the same as in part **a**?

   continued on next page

Questions – *continued*

3 Some pieces of iron(II) carbonate are put into hydrochloric acid. The iron(II) carbonate and the acid react to give carbon dioxide gas. The acid is in excess. The volume of gas given off was recorded every minute for 10 minutes. The results are shown in Table 5e.2.

Time in minutes	Volume of gas in cm³
0	0
1	48
2	52
3	54
4	55
5	56
6	56
7	56
8	56
9	56

Table 5e.2

a Plot a graph of these results, with time on the *x*-axis and volume on the *y*-axis.

b How might the volume of carbon dioxide have been measured?

c Estimate how long it took to collect 50 cm³ of carbon dioxide.

d Estimate the volume of carbon dioxide collected after 30 seconds.

e How long did it take for the reaction to stop?

f Why did the reaction stop?

H 4 How many moles of gas are there in the following samples, which are at room temperature and pressure?

a 120 dm³ of nitrogen

b 120 cm³ of nitrogen

c 6000 cm³ of carbon monoxide

d 24 cm³ of neon

5 The molar mass of argon is 40 g. For each of the following samples, work out how many moles of argon there are, and then calculate the volume of the sample at room temperature and pressure.

a 160 g

b 20 g

c 1.0 g

d 0.075 g

C5f Equilibria

Reversible reactions

In many chemical reactions, the reaction is simple to perform, and quite simple to understand. Reactants are mixed, they react and products are formed. Not all reactions do this. In some reactions the product, or products, can react to re-form the reactants. A reaction like this is called a **reversible reaction**.

The reaction of hydrogen with nitrogen to make ammonia is a reversible reaction. Once formed, the ammonia reacts to re-form hydrogen and nitrogen. This is the reaction in the Haber process (see Item C4d, *Making ammonia – Haber process and costs*, in *Gateway Additional Science*). Because the reaction is reversible, the word equation uses the ⇌ arrow:

hydrogen + nitrogen ⇌ ammonia

Dynamic equilibrium

When the Haber process is started, hydrogen and nitrogen react and some ammonia is formed. This is called the forward reaction (Figure 5f.1).

hydrogen + nitrogen → ammonia

Figure 5f.1 The forward reaction.

Some of the ammonia that has formed by the forward reaction then reacts and re-forms some hydrogen and nitrogen. This is called the **backward reaction** (Figure 5f.2).

ammonia → hydrogen + nitrogen

Figure 5f.2 The backward reaction.

The forward reaction and the backward reaction both take place, because the reaction is reversible. After a little time, a situation is reached where hydrogen and nitrogen are reacting to make ammonia, and ammonia is reacting to make hydrogen and nitrogen, at the *same rate*. The ammonia is being made by the forward reaction, and is being consumed by the backward reaction, at the *same rate*. This situation is called **dynamic equilibrium** (Figure 5f.3):

- *dynamic* – both the forward reaction and the backward reaction are happening, at the same rate
- *equilibrium* – the reaction has reached a point where the amounts of reactants and products do not change overall.

hydrogen + nitrogen ⇌ ammonia

Figure 5f.3 When dynamic equilibrium is reached, the forward reaction and the backward reaction happen at the same rate. The concentrations of the reactants and products do not change once dynamic equilibrium has been reached.

SAQ

1. What do we mean by the following terms?
   a. reversible reaction
   b. dynamic equilibrium
2. Use the reaction that occurs in the Haber process to explain the meaning of the terms *forward reaction* and *backward reaction*.

Position of equilibrium

Dynamic equilibrium can be reached with large amounts of reactants remaining and very small amounts of products present (Figure 5f.4). In this case, we say *the position of equilibrium is on the left*. Here *left* means on the left of the equation describing the reaction – the left is the 'reactants' side.

Dynamic equilibrium can also be reached when large amounts of products have formed and very little of the reactants remain (Figure 5f.5). In this case, we say *the position of equilibrium is on the right*. Here *right* means on the right of the equation describing the reaction – the right is the 'products' side.

C5f Equilibria   149

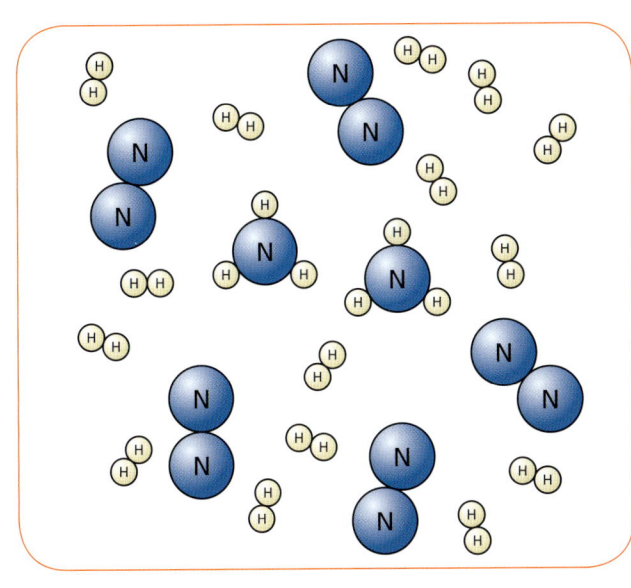

Figure 5f.4 A possible equilibrium position in the Haber process. The position of equilibrium is on the left.

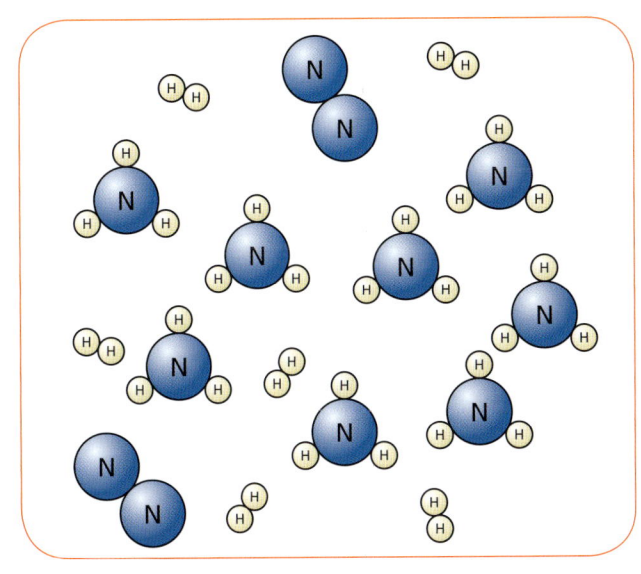

Figure 5f.5 Another possible equilibrium position in the Haber process. The position of equilibrium is on the right. This position is different from that in Figure 5f.4, and different conditions are needed to reach it.

Why equilibrium is reached

When a reversible reaction begins, the concentration of each reactant is high, and the concentration of each product is zero. Because of this, the forward reaction has a high rate, and the backward reaction has no rate. As the reactants are used up, their concentrations decrease and the rate of the forward reaction decreases. As the products are formed, their concentrations increase and the rate of the backward reaction increases.

There comes a time when the forward and backward reactions have equal rates. Reactants are becoming products, and products are becoming reactants, at the same rate. Dynamic equilibrium has now been reached.

SAQ

3 Rewrite the two paragraphs above, using the reaction that makes ammonia as an example. Replace all the general references to *reactants* and *products* with the names of the specific substances in the reaction that makes ammonia. Begin with *When the Haber process begins …*

Equilibrium positions are not fixed

The amounts of reactants and products that are present in a mixture at equilibrium are affected by temperature and pressure. For example, in the reaction:

hydrogen + nitrogen ⇌ ammonia

changing the temperature or pressure will change the percentage of ammonia in the equilibrium mixture. We say the **yield** of ammonia is affected.

At a pressure of 1 atmosphere and a temperature of 300 °C, the percentage of ammonia at equilibrium is 2.2%. At a pressure of 100 atmospheres and a temperature of 300 °C, it is 51.0%. At a pressure of 1 atmosphere and a temperature of 400 °C, it is 0.5% (Table 5f.1).

Pressure in atmospheres	Temperature in °C	Percentage of ammonia at equilibrium (yield)
1	300	2.2
100	300	51.0
1	400	0.5

Table 5f.1 In the Haber process, the percentage of ammonia in the equilibrium mixture depends on the temperature and the pressure.

SAQ

4 In the Haber process, what happens to the *yield* of ammonia (the percentage of ammonia in the equilibrium mixture) as:

   a the pressure is increased?
   b the temperature is increased?

150 C5f Equilibria

5 Use the graph in Figure 5f.6 to answer the following questions.
   a What is the percentage yield of ammonia at 300 °C and 350 atmospheres?
   b What pressure gives a yield of ammonia of 80% at 300 °C?
   c Does the extra data in this graph confirm your answer to SAQ 4a?

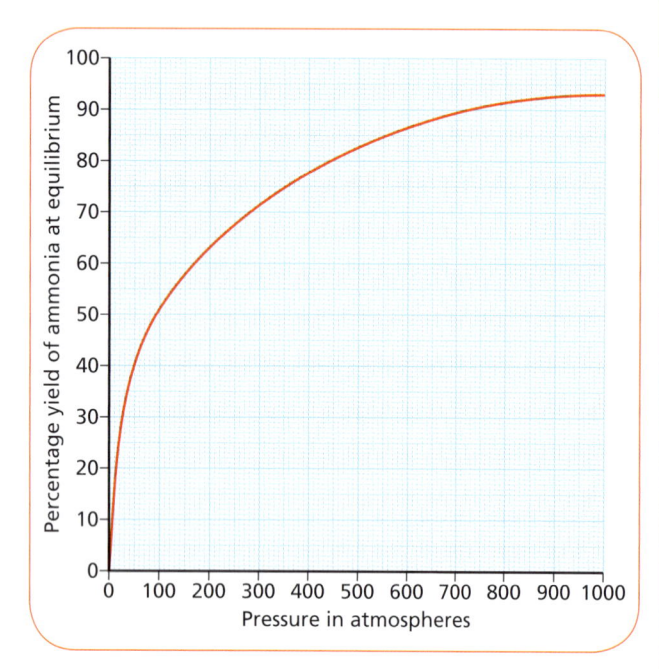

Figure 5f.6 This graph shows the percentage yield of ammonia at equilibrium at different pressures. The temperature in every case is 300 °C.

Disturbing an equilibrium

The amounts of reactants and products in a reaction that has reached equilibrium will stay constant, unless the reaction is disturbed. A reaction may be disturbed in a number of ways.
- Adding more products causes the equilibrium position to shift to the left (more reactants are formed).
- Removing products causes the equilibrium position to shift to the right (more products are formed).
- Adding more reactants causes the equilibrium position to shift to the right (more products are formed).
- Removing reactants causes the equilibrium position to shift to the left (more reactants are formed).

If a reaction is at equilibrium in a situation where reactants and products cannot be added or removed, we say it is in a **closed system**. Without a closed system, the reaction cannot reach a stable equilibrium. If the system is not closed, then the position of equilibrium will constantly shift as substances enter or leave the system.

SAQ

6 What do we mean by a *closed system*?
7 Once the Haber process has reached equilibrium, what will happen to the position of equilibrium if the following changes are made?
   a Hydrogen and nitrogen are added to the system.
   b Ammonia is added to the system.
   c Hydrogen is removed from the system.
   d Ammonia is removed from the system.

The equilibrium in the Haber process is disturbed if the pressure is changed. The balanced symbol equation for the reaction is:

$$3H_2 + N_2 \rightleftharpoons 2NH_3$$

Increased pressure favours the side with the *fewest* gas molecules. There are four gas molecules on the left of this equation and two on the right. So increased pressure shifts the position of equilibrium to the right – so *more* ammonia is formed.

Decreased pressure favours the side with the *most* gas molecules. There are four gas molecules on the left of the equation and two on the right. So decreased pressure shifts the position of equilibrium to the left – so *less* ammonia is formed.

SAQ

8 a Why does increased pressure increase the yield of ammonia in the Haber process?
   b Why does decreased pressure decrease the yield of ammonia in the Haber process?

The Haber process – making it work

The Haber process is a large-scale, continuous process, making millions of tonnes of ammonia every year. It is an extremely important process, as ammonia is used as a raw material for many other products and processes (Figures 5f.7 and 5f.8).

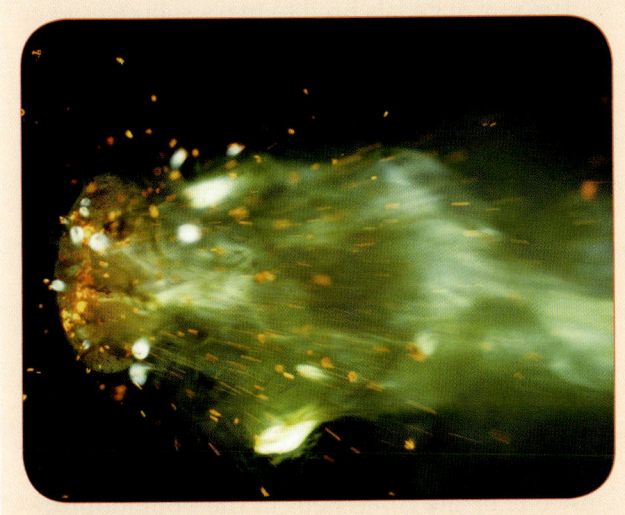

Figure 5f.7 Ammonia is used to make nitric acid, which is one of the raw materials used to make this explosive.

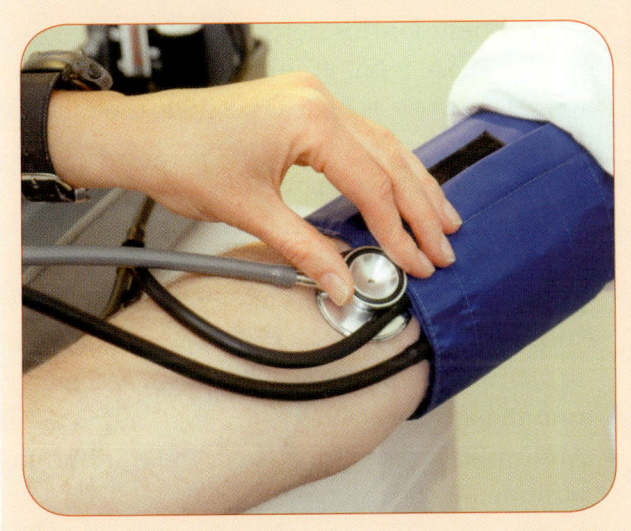

Figure 5f.8 This cuff is used when measuring blood pressure. It is made of nylon. Ammonia was used to make the nylon.

The reaction is naturally very slow. It is speeded up by using an iron catalyst, high pressure, and a raised temperature. This is explained in more detail in Item C4d, *Making ammonia – Haber process and costs*, in *Gateway Additional Science* pages 186 and 187.

However, since the reaction:

$$3H_2 + N_2 \rightleftharpoons 2NH_3$$

is a reversible reaction that reaches dynamic equilibrium, there are two more problems.

1. **How can we ensure that all the hydrogen and nitrogen eventually react to produce ammonia?** After the hydrogen and nitrogen reactants have produced some ammonia, the ammonia is removed. Removing the product makes the equilibrium position shift to the right, producing more ammonia. At the same time, more hydrogen and nitrogen reactants are added. Adding more reactants also makes the equilibrium position shift to the right, producing more ammonia.

2. **How can we stop the ammonia turning back into nitrogen and hydrogen once it has been produced?** The ammonia is removed by cooling it down so that it liquefies (Figure 5f.9). Its temperature is lowered and it is taken away from the iron catalyst. Each of these changes greatly slows down the reaction by which ammonia decomposes back to nitrogen and hydrogen. It can be stored without decomposition.

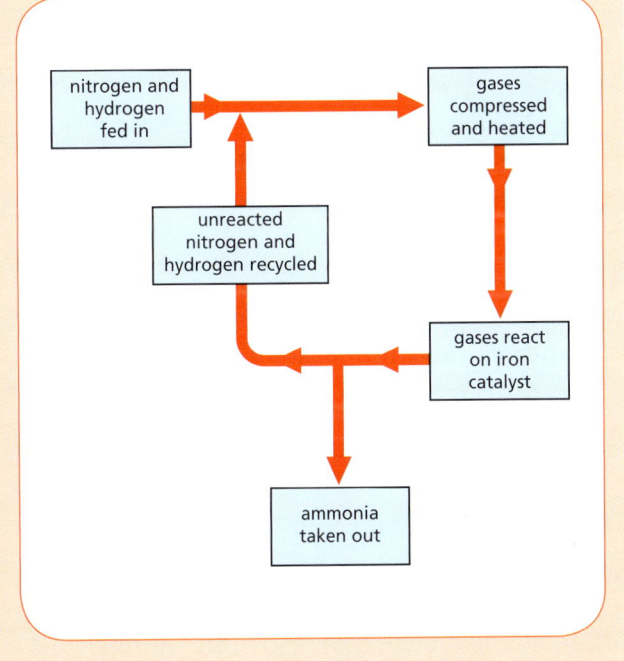

Figure 5f.9 Flow diagram of the Haber process.

C5f Equilibria

The Contact process

One of today's most important industrial chemicals is sulfuric acid. Sulfuric acid is made in a multi-step process called the **Contact process** (Figure 5f.10).

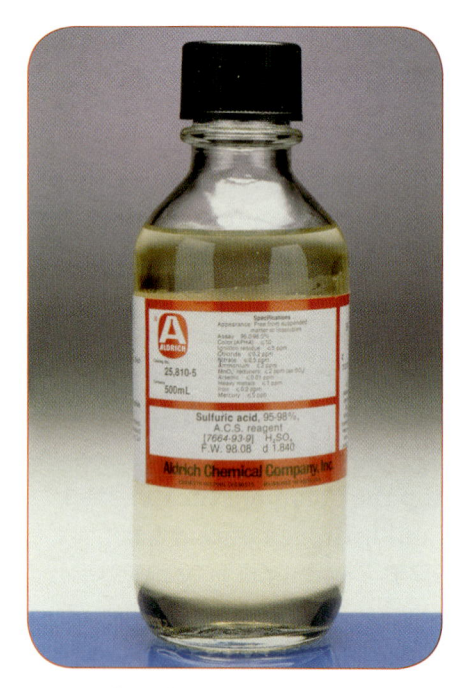

Figure 5f.10 The sulfuric acid in this bottle was made by the Contact process.

Step 1

In Step 1, sulfur reacts with oxygen to make sulfur dioxide:

sulfur + oxygen → sulfur dioxide

$S + O_2 \rightarrow SO_2$

The sulfur is obtained from under the ground by mining. The oxygen is obtained from the air.

Step 2

In Step 2, the sulfur dioxide reacts with more oxygen from the air to make sulfur trioxide:

sulfur dioxide + oxygen ⇌ sulfur trioxide

$2SO_2 + O_2 \rightleftharpoons 2SO_3$

The ⇌ symbol is used because this is a reversible reaction. The conditions used for this reaction are a temperature of 450 °C, a catalyst made of vanadium(V) pentoxide (V_2O_5), and normal atmospheric pressure.

Step 3

In Step 3, the sulfur trioxide reacts with water to give sulfuric acid:

sulfur trioxide + water → sulfuric acid

$SO_3 + H_2O \rightarrow H_2SO_4$

SAQ

9 The raw materials used to make sulfuric acid are sulfur, oxygen and water.
   a How is each of these substances obtained?
   b Which step, or steps, in the Contact process involves each raw material?
   c Write a word equation for each step in the Contact process
   d Write a balanced symbol equation for each step in the Contact process

Choosing the right conditions

Only one of the three steps in the Contact process is reversible – the second step:

$2SO_2 + O_2 \rightleftharpoons 2SO_3$

The conditions for this step have to be chosen carefully. The amount of sulfur trioxide produced (the *yield*) must be as high as possible, the rate of the reaction must be as high as possible, and the costs must be as low as possible.

- **A temperature of 450 °C is used**. A raised temperature increases the rate of the reaction but unfortunately it also lowers the yield of sulfur trioxide. 450 °C is known as a *compromise* or *optimum* temperature – it gives an acceptable rate and an acceptable yield.
- **Atmospheric pressure is used**. The equilibrium position for the reaction:

  $2SO_2 + O_2 \rightleftharpoons 2SO_3$

  lies well over to the right at atmospheric pressure. The yield is high. A higher pressure would give a higher yield, but using a high pressure is expensive.
- **A vanadium(V) pentoxide catalyst is used**. The catalyst has no effect on yield, but it does increase the rate of the reaction.

Summary

You should be able to:

- describe a reversible reaction, using and recognising the ⇌ symbol
- explain *dynamic equilibrium* in terms of the rates of the forward and backward reactions
- use the term *position of equilibrium* accurately
- **[H]** use reaction rates and the concept of a closed system to explain why a reaction may reach equilibrium
- interpret data in graphical or table form about equilibrium composition
- **[H]** explain how some factors affect the position of equilibrium
- describe the Contact process for making sulfuric acid, naming the three raw materials, and using word equations
- state the conditions used in the reaction that produces sulfur trioxide
- **[H]** describe the Contact process using balanced symbol equations
- explain the conditions used in the reaction that produces sulfur trioxide

Questions

1. In the Contact process, the reaction:

    sulfur dioxide + oxygen ⇌ sulfur trioxide

    is reversible, and reaches dynamic equilibrium. When equilibrium is reached there is much more sulfur trioxide present than sulfur dioxide and oxygen.

    a How can you tell from the word equation that the reaction is reversible?

    b What is the forward reaction here?

    c What is the backward reaction here?

    d What is meant by *dynamic equilibrium*?

    e Does the position of equilibrium lie to the left or to the right in this reaction?

    continued on next page

Questions – *continued*

2 Table 5f.2 shows the percentage yield of ammonia at equilibrium over a range of temperatures. The pressure in every case is 200 atmospheres.

Temperature in °C	Percentage yield of ammonia
300	62.8
400	36.3
500	17.5
600	8.2
700	4.1

Table 5f.2

Use the data in the table to answer the following questions.

a What is the percentage yield of ammonia at 350 °C and 200 atmospheres?

b What temperature gives a yield of ammonia of 20% at 200 atmospheres pressure?

c Does the extra data in this table confirm your answer to SAQ 4b?

Use the data in the table to plot a graph of temperature (*x*-axis) against percentage of ammonia at equilibrium (*y*-axis). Use your graph to answer parts **a** and **b** again. Are you more confident of your answers from the table, or from the graph? Explain your decision.

3 a What temperature, pressure and catalyst are used in the step of the Contact process that produces sulfur trioxide?

H b What would be the effect of using a very high pressure on the yield of sulfur trioxide from this step?

c Explain why a higher temperature and a higher pressure are not used in this step.

d Explain why a catalyst is used in this step.

5g Strong and weak acids

Ethanoic acid and hydrochloric acid

Ethanoic acid is a **weak acid**. Hydrochloric acid is a **strong acid**. If we make up a solution of each acid of identical concentrations, their pHs are not the same. The hydrochloric acid has a lower pH, probably around 1 or 2. The ethanoic acid has a higher pH, probably around 3 to 5 (Figure 5g.1).

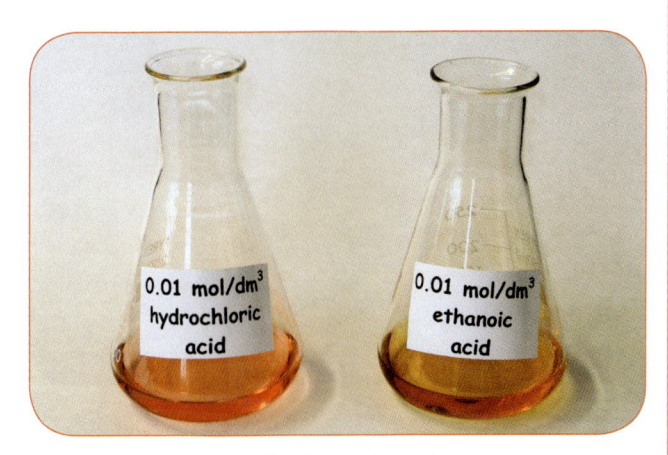

Figure 5g.1 The two acid solutions have the same concentration, but adding universal indicator solution shows that the hydrochloric acid has a lower pH.

Acids produce H⁺ ions in water

Acids are covalent compounds. Pure HCl, for example, is a gas called hydrogen chloride, with the hydrogen atom and the chlorine atom covalently bonded together. When put into water, the HCl molecules ionise, producing H^+ ions and Cl^- ions. Every HCl molecule does this when HCl gas dissolves in water (Figure 5g.2a). HCl *completely* ionises. This is why HCl is a strong acid. Nitric acid and sulfuric acid are also strong acids.

Pure ethanoic acid is a covalent compound with the formula CH_3CO_2H. When put into water, some of the CH_3CO_2H molecules ionise, producing H^+ ions and $CH_3CO_2^-$ ions. However, only a small fraction of the CH_3CO_2H molecules do this when ethanoic acid dissolves in water (Figure 5g.2b). CH_3CO_2H *partially* ionises. This is why ethanoic acid is a weak acid.

Every acid produces H^+ ions when it is dissolved in water. This is the property that makes it an acid. However, if we compare a solution of a strong acid and a solution of a weak acid of the same concentration, the strong acid solution has a higher concentration of H^+ ions than the weak acid.

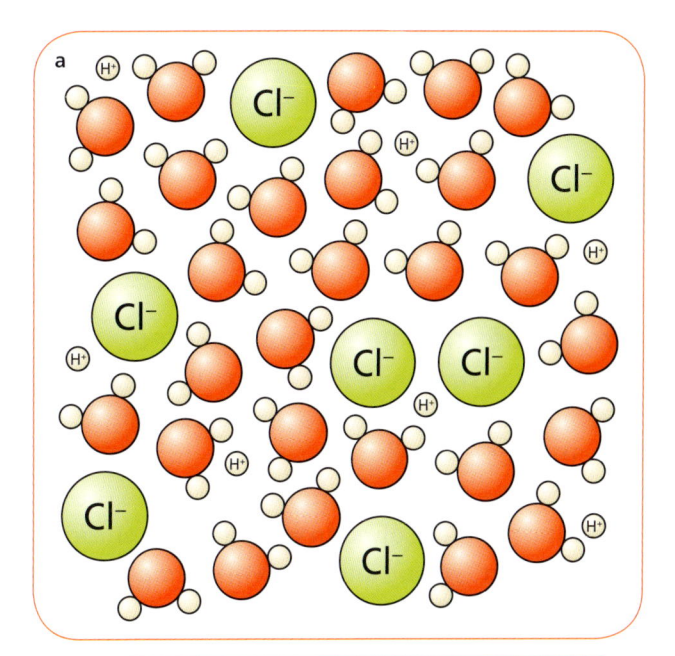

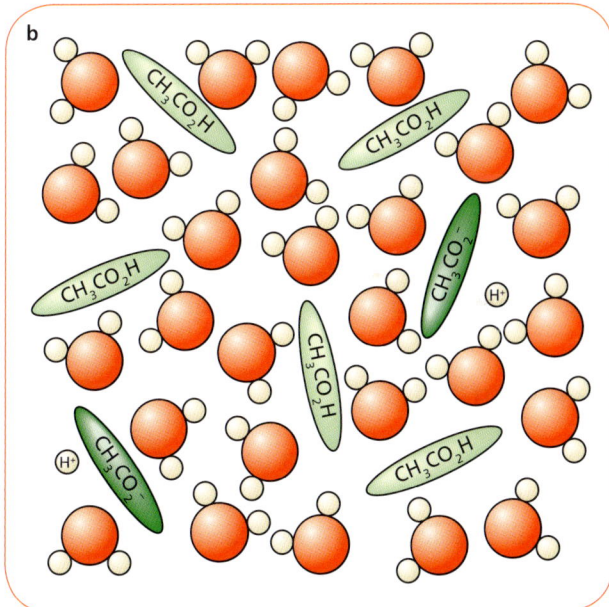

Figure 5g.2 a In HCl solution, every HCl molecule has ionised. **b** In CH_3CO_2H solution, only a small proportion of the CH_3CO_2H molecules have ionised.

This is because the ionisation of a weak acid is a reversible reaction. CH_3CO_2H molecules ionise in water to give H^+ ions and $CH_3CO_2^-$ ions, but H^+ ions and $CH_3CO_2^-$ ions re-join readily to give CH_3CO_2H molecules. When ethanoic acid is put into water, an equilibrium mixture is produced, consisting of a lot of CH_3CO_2H molecules and a smaller number of H^+ ions and $CH_3CO_2^-$ ions.

156 C5g Strong and weak acids

SAQ

1. What property of HCl makes it acidic?
2. **a** Explain why hydrochloric acid is a strong acid and ethanoic acid is a weak acid.
   b What experiment could you do to show these different strengths?
   c What results would you get?

Strong does not mean concentrated

We use the terms *strong* and *weak* to describe different acids. They refer to the degree of ionisation of the acids.

We use the term *concentration* to describe the amount of acid dissolved in each dm^3 of solution.

Remember: *strong* and *concentrated* do not mean the same thing.

- A dilute solution of hydrochloric acid has a *low concentration*, but it is still a *strong acid*, because the HCl molecules are completely ionised.
- A concentrated solution of ethanoic acid has a *high concentration*, but it is still a *weak acid*, because the CH_3CO_2H molecules are only partially ionised.

Since the ionisation of CH_3CO_2H molecules is a reversible reaction, we write it as an equation like this:

$$CH_3CO_2H \rightleftharpoons H^+ + CH_3CO_2^-$$

The position of equilibrium lies to the left. A small proportion of the CH_3CO_2H molecules ionise. Ethanoic acid is a weak acid.

The ionisation of HCl molecules is not a reversible reaction, so we write it as an equation like this:

$$HCl \rightarrow H^+ + Cl^-$$

SAQ

3. Explain the following statements as fully as you can, concentrating on the italic terms.
   a Acid rain is a *very dilute* solution of the *strong acids* sulfuric acid and nitric acid.
   b A red ant squirts out a *high concentration* of the *weak acid* methanoic acid.

Same reaction, different rates

Hydrochloric acid and ethanoic acid both react with calcium carbonate powder. Both reactions make carbon dioxide gas. However, if we use hydrochloric acid and ethanoic acid solutions of the same concentration, and put the same mass of calcium carbonate powder into each, the calcium carbonate reacts faster with the hydrochloric acid (Figure 5g.3).

Figure 5g.3 The calcium carbonate is reacting more quickly with the hydrochloric acid than it is with the ethanoic acid.

The hydrochloric acid reacts more quickly with the calcium carbonate than the ethanoic acid does. However, when the two reactions are finished, they will have produced identical volumes of carbon dioxide gas. This is because the same amounts of acid and calcium carbonate were used in each case.

SAQ

4. How can you tell from Figure 5g.3 that the calcium carbonate is reacting more quickly with the hydrochloric acid than it is with the ethanoic acid?

If we put magnesium powder into hydrochloric acid solution, and into ethanoic acid solution of the same concentration, a gas is produced. The gas given off in these reactions is hydrogen. The magnesium reacts more quickly with the hydrochloric acid than it does with the ethanoic acid (Figure 5g.4).

If the same amounts of acid and magnesium are used in each case, both reactions will have given off the same volume of hydrogen when they are finished.

Figure 5g.4 The magnesium is reacting more quickly with the hydrochloric acid than it is with the ethanoic acid.

SAQ

5 Look at Figure 5g.4.
   a What factor affects the rate of these reactions?
   b What factors affect the volume of hydrogen that will have been given off when the reactions are finished?

Acids conduct electricity

Aqueous solutions of hydrochloric acid and ethanoic acid both conduct an electric current. If the electric current is DC, the acids are chemically changed by the electric current. We say the acids act as *electrolytes* and undergo *electrolysis* (see Item C5b, *Electrolysis*). During electrolysis, both acids give off hydrogen gas at the cathode (Figure 5g.5). The hydrogen ions in the acids are positive, so they move to the cathode.

Solutions of hydrochloric acid and ethanoic acid do not have the same electrical conductivity. If the two solutions have the same concentration, and the voltage used is the same, the electric current that flows through the hydrochloric acid is greater. The hydrochloric acid solution is said to have a greater electrical conductivity.

SAQ

6 When hydrochloric acid solution is electrolysed, why does the hydrogen gas come off at the cathode, not at the anode?

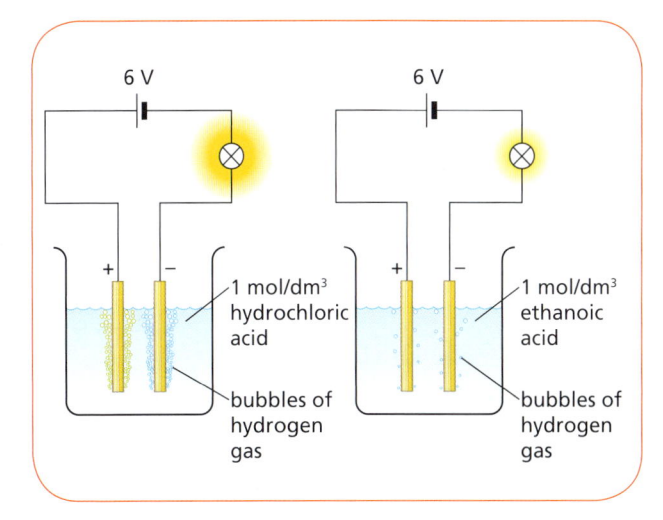

Figure 5g.5 The hydrochloric acid and the ethanoic acid both undergo electrolysis. Both give off hydrogen gas at the cathode.

SAQ

7 Look at Figure 5g.5. Which has the greater electrical resistance, the hydrochloric acid or the ethanoic acid?

8 In what two ways does Figure 5g.5 show that the hydrochloric acid solution has a greater electrical conductivity than the ethanoic acid solution?

Degree of ionisation can explain behaviour

If we have solutions of hydrochloric acid and ethanoic acid of the same concentration, the concentration of H^+ ions in the two solutions will not be the same. Hydrochloric acid is a strong acid while ethanoic acid is a weak acid, so the hydrochloric acid will have a higher concentration of H^+ ions because it is more fully ionised. This fact helps us to understand why hydrochloric acid and ethanoic acid behave differently.

Hydrochloric acid reacts more quickly with calcium carbonate than ethanoic acid does. This is because there are more H^+ ions in each cm^3 of solution, so they will collide more frequently with calcium carbonate formula units. The more frequent the collisions are, the faster the reaction goes.

Hydrochloric acid reacts more quickly with magnesium than ethanoic acid does. Again, this is because there are more H^+ ions in each cm^3 of solution, so they will collide more frequently with

158 C5g Strong and weak acids

magnesium atoms – and the more frequent the collisions are, the faster the reaction goes.

A higher concentration of ions means that there are more charge carriers present, which means the solution conducts electric current better. Hydrochloric acid is therefore a better conductor of electricity than ethanoic acid is.

Using weak acids

Using a kettle to boil water sometimes leads to a build-up of solid calcium carbonate in the kettle (Figure 5g.6). This build-up is known as *scale*. You can read more about this in Item C6f, *Hardness of water*.

Scale can be removed from the inside of a kettle by using a *descaling agent* consisting of a weak acid. The weak acid reacts with the calcium carbonate to produce a soluble salt, carbon dioxide and water. A strong acid is less suitable for this job, as the higher concentration of H^+ ions is much more likely to attack metal parts and damage the kettle.

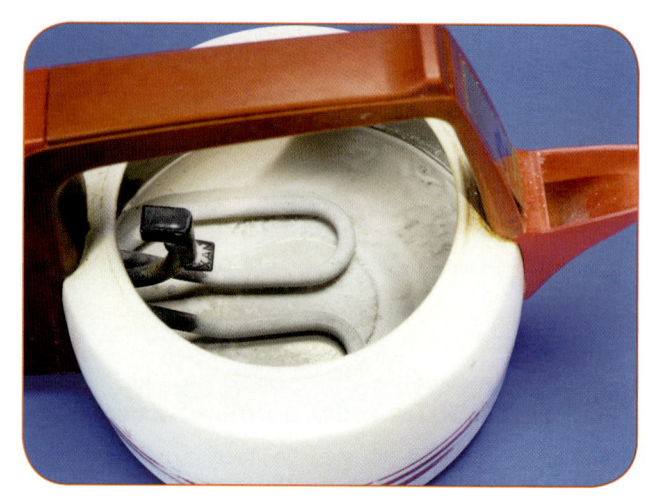

Figure 5g.6 Kettle scale is a build-up of calcium carbonate.

Weak acids

Many natural substances, including many of the things that we eat and drink, are weak acids.

Figure 5g.7 This fizzy drink contains citric acid …

Figure 5g.8 … and so do the pineapples on this stall.

Figure 5g.9 This chutney contains vinegar – ethanoic acid.

Figure 5g.10 The tea made from the leaves being harvested here will contain tannic acid.

continued on next page

Weak acids - *continued*

Figure 5g.11 The stinging hairs of this nettle contain methanoic acid.

The five acids illustrated by Figures 5g.7 to 5g.12 are all weak acids. They produce low pH values, often as low as 3 or 4, but not much lower. If any of these acids were strong they would be potentially dangerous.

Figure 5g.12 Lactic acid has been formed by anaerobic respiration in this athlete's muscles.

Summary

You should be able to:

- name some strong acids and some weak acids
- state that a strong acid ionises completely in water to produce H⁺ ions
- state that a weak acid ionises reversibly, producing an equilibrium mixture
- state that the pH of a weak acid is higher than the pH of a strong acid of the same concentration
- **H** understand and use the terms *acid strength* and *acid concentration*
- write equations for the ionisation of strong and weak acids
- describe how hydrochloric acid and ethanoic acid react with magnesium and with calcium carbonate, explaining the difference in rate of reaction and understanding the reasons why a particular volume of gas forms
- describe the electrolysis of hydrochloric acid and of ethanoic acid, explaining why hydrogen forms at the cathode
- explain why hydrochloric acid is more electrically conductive than ethanoic acid
- explain why weak acids are more useful as descaling agents than strong acids

Questions

1. Name three strong acids.

2. Figure 5g.3 allows you to judge by eye that calcium carbonate reacts more quickly with hydrochloric acid than it does with ethanoic acid of the same concentration. Describe in as much detail as possible how you could conduct a better experiment, taking measurements and ensuring fair testing, to prove this observation true.

3. Give one use of acids for which a weak acid is preferred to a strong acid, and explain why the weak acid is the better choice.

4. You are given two beakers of acid. Both have a concentration of 0.1 mol/dm^3. One is a weak acid and one is a strong acid, but they are not labelled.

   a. List four experiments that would enable you to tell which is which.

   b. What observations would you make, with each acid, in each experiment?

5. 2.4 g of magnesium powder is added to 500 cm^3 of 1.0 mol/dm^3 hydrochloric acid. The magnesium is the limiting reactant. The molar mass of magnesium is 24 g. One mole of any gas occupies 24 dm^3 at room temperature and pressure. The balanced symbol equation is:

   $2HCl + Mg \rightarrow MgCl_2 + H_2$

   a. What does *limiting reactant* mean?

   b. How many moles of magnesium are there in 2.4 g?

   c. How many moles of hydrogen are produced?

   d. What volume of hydrogen is produced?

   e. What mass of hydrogen is produced?

   f. The experiment is repeated with 2.4 g of magnesium powder and 1000 cm^3 of 1.0 mol/dm^3 ethanoic acid. How will the results compare to the original results?

6. a. Write an ionic equation for the reaction at the cathode during the electrolysis of dilute hydrochloric acid.

   b. Why is the ionic equation for the reaction at the cathode during the electrolysis of dilute ethanoic acid the same as this?

5h Ionic equations

Identifying ions in a solution

A solution of a substance dissolved in water is called an **aqueous solution**. Some aqueous solutions are coloured, but most look like water – clear and colourless. We can find out whether or not certain ions are present in a solution by adding other solutions to it.

Testing for halide ions

The Group 7 elements include chlorine, bromine and iodine. They are called the **halogens**. The halogens form many compounds in which they are present as singly charged negative ions, called **halide** ions. The halide ions formed by chlorine, bromine and iodine are called chloride, bromide and iodide. Their symbols are Cl^-, Br^- and I^-. To test a solution to see if it contains halide ions, we add silver nitrate solution.

If silver nitrate solution is added to a solution that contains chloride ions, a white solid forms immediately (Figure 5h.1). The white solid that forms is silver chloride. If silver chloride does not form, then the original solution did not contain chloride ions. A solid that forms in this way when two solutions are mixed is called a **precipitate** and the reaction is called a **precipitation** reaction. In a precipitation reaction, ions from one solution react with ions from another solution to form a precipitate.

If silver nitrate solution is added to another solution, which contains bromide ions, a cream precipitate forms immediately (Figure 5h.2). The cream precipitate that forms is silver bromide.

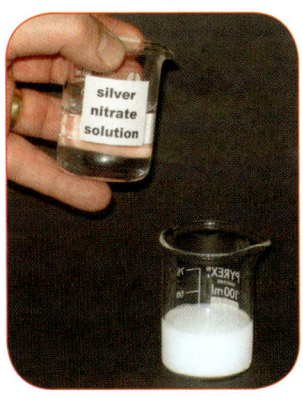

Figure 5h.1 The white precipitate shows that this solution contained dissolved chloride ions.

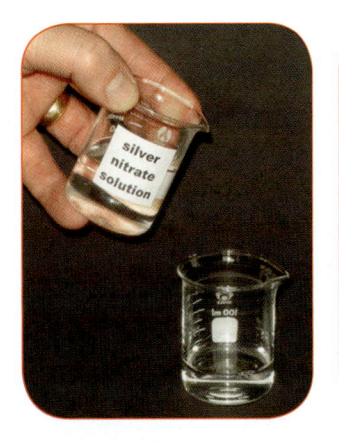

Figure 5h.2 The cream precipitate shows that this solution contained dissolved bromide ions.

If silver bromide does not form, then the original solution did not contain bromide ions.

If silver nitrate solution is added to another solution, which contains iodide ions, a yellow precipitate forms immediately (Figure 5h.3). The yellow precipitate that forms is silver iodide. If silver iodide does not form, then the original solution did not contain iodide ions.

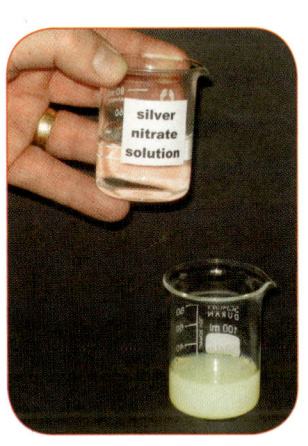

Figure 5h.3 The yellow precipitate shows that this solution contained dissolved iodide ions.

SAQ

1 What does it tell us if, when we add silver nitrate solution to an unidentified solution, we get the following results?
   a A yellow precipitate forms.
   b A white precipitate forms.
   c A cream precipitate forms.
   d No precipitate forms.

Using barium chloride

An aqueous solution may contain sulfate ions. To test a solution to see if it contains sulfate ions, we add barium chloride solution. If we add barium chloride solution to another solution that contains sulfate ions, a white solid forms immediately (Figure 5h.4). The white precipitate that forms is barium sulfate. If barium sulfate does not form, then the original solution did not contain sulfate ions.

Figure 5h.4 The white precipitate shows that this solution contained dissolved sulfate ions.

In all of these precipitation reactions, the substances that form as precipitates are insoluble in water. A soluble substance will not form as a precipitate. An insoluble substance will form as a precipitate.

Word equations

If we mix sodium chloride solution with silver nitrate solution, a white precipitate of silver chloride is formed. The word equation for this reaction is:

sodium chloride (aq) + silver nitrate (aq)
→ sodium nitrate (aq) + silver chloride (s)

The letters in brackets – (aq) and (s) – are **state symbols**. They tell us what state each reactant or product is in:
- (aq) means *aqueous*, which means *in aqueous solution*, or *dissolved in water*
- (s) means *solid*.

There are two more state symbols, which are not used in this equation:
- (l) means *liquid*
- (g) means *gas*.

Word equations, and balanced symbol equations, can be written without state symbols, but including state symbols gives more information.

SAQ

2 Complete the word equations, including state symbols, for the following precipitation reactions:
   a sodium iodide (aq) + silver nitrate (aq) →
   b sodium bromide (aq) + silver nitrate (aq) →
   c sodium sulfate (aq) + barium chloride (aq) →

How precipitation works

Solid silver nitrate will not react with solid sodium chloride. In these ionic solids, the ions are bound in fixed positions by strong ionic bonds. The silver ions and chloride ions cannot move, so they cannot collide, so they will not react to form silver chloride.

If the ionic solids are melted, or aqueous, the ions are able to move. If they can move they can collide, and if they can collide they can react. The reaction is most conveniently performed in aqueous solution.

A precipitation reaction, like the one illustrated in Figure 5h.5, is very fast. In fact most precipitation reactions appear to occur instantaneously.

SAQ

3 **a** Why doesn't solid silver nitrate react with solid sodium chloride?
  b Why does dissolving the two solids in water make a difference?
  c Write a word equation for the reaction that happens when solutions of silver nitrate and sodium chloride are mixed.

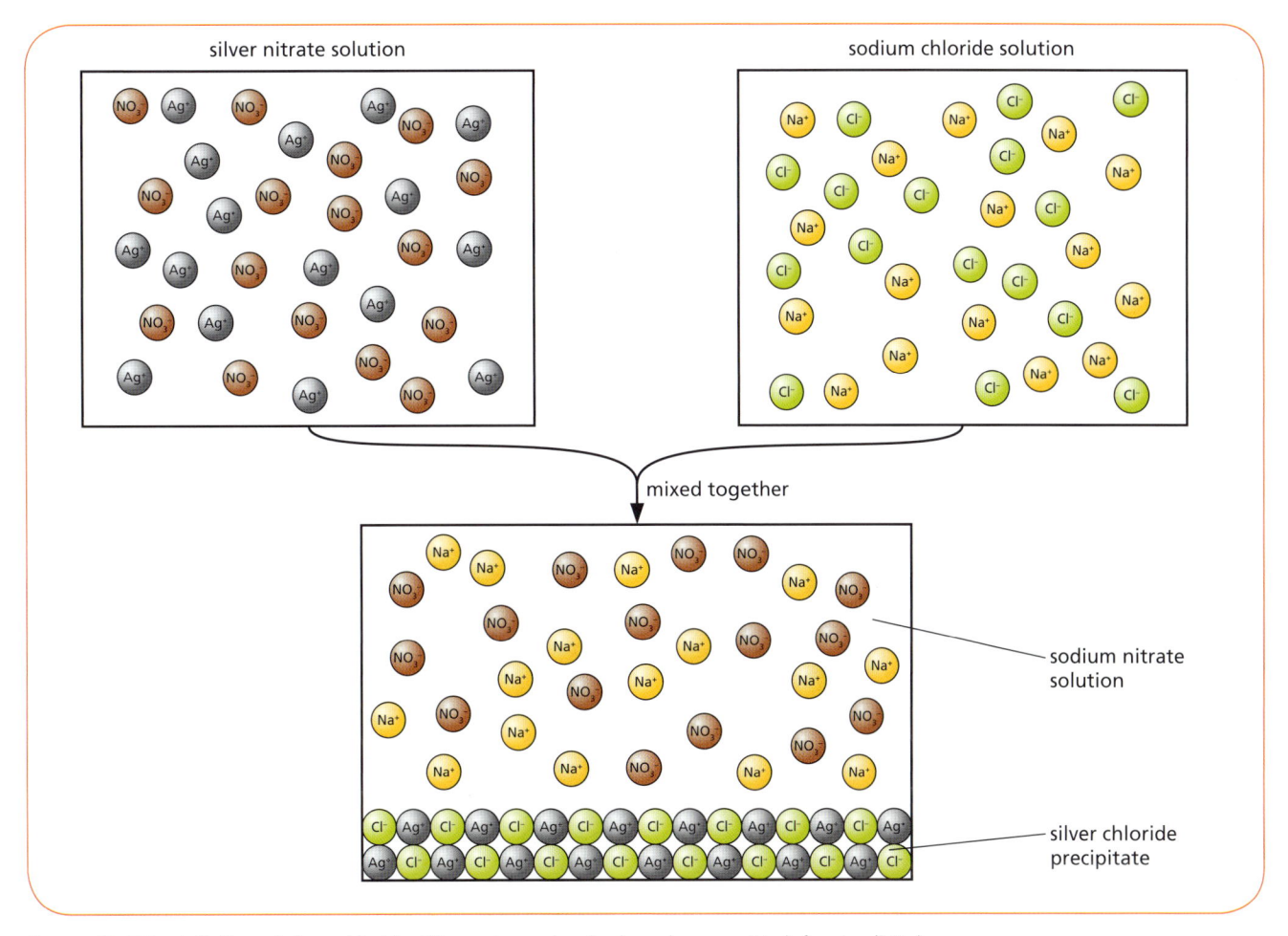

Figure 5h.5 Precipitation of silver chloride. (The water molecules have been omitted, for simplicity.)

Ionic equations

Balanced symbol equations can be written for the precipitation reactions in this item. For example, for the reaction illustrated in Figure 5h.5:

$$AgNO_3(aq) + NaCl(aq) \rightarrow AgCl(s) + NaNO_3(aq)$$

For a substance in solution, the ions are separate, so the equation can be rewritten like this:

$$Ag^+(aq) + NO_3^-(aq) + Na^+(aq) + Cl^-(aq) \rightarrow AgCl(s) + Na^+(aq) + NO_3^-(aq)$$

It is clear from this that the sodium ions and nitrate ions do not change. If these ions are omitted from both sides of the equation we get an **ionic equation**, which looks like this:

$$Ag^+(aq) + Cl^-(aq) \rightarrow AgCl(s)$$

The ionic equation differs from the balanced symbol equation in two ways:

1 It omits the sodium ions, the nitrate ions and the sodium nitrate. Figure 5h.5 shows that these ions do not undergo any change. They are independent, aqueous ions before and after mixing. Ionic equations omit ions that do not change.

2 It includes the ionic charges.

The ions that do not undergo change – in this case the sodium ions and the nitrate ions – are known as **spectator ions**.

SAQ

4 Use the following balanced symbol equations to produce ionic equations for the formation of silver bromide and silver iodide:

  a $AgNO_3(aq) + NaBr(aq) \rightarrow AgBr(s) + NaNO_3(aq)$

  b $AgNO_3(aq) + NaI(aq) \rightarrow AgI(s) + NaNO_3(aq)$

5 Barium ions are Ba^{2+}, and sulfate ions are SO_4^{2-}. Write an ionic equation, with state symbols, for the formation of barium sulfate.

Making a salt by precipitation

Mixing two solutions is a good way to make an insoluble *salt*. (A salt is a neutral ionic compound – see Item C4a, *Acids and bases* in *Gateway Additional Science*.) The insoluble product can be removed from excess reactants, and from the soluble product, by filtration. After filtration, the precipitate must be washed and dried.

Lead iodide is a bright yellow salt that is very insoluble in water. It can be made by mixing a solution of a soluble lead compound, like lead nitrate, with a solution of a soluble iodide compound, like sodium iodide (Figure 5h.6).

Figure 5h.6a Yellow lead iodide forms as the lead nitrate and sodium iodide react.

Figure 5h.6b The insoluble lead iodide is filtered out using a filter funnel and filter paper.

Figure 5h.6c The insoluble lead iodide is washed with distilled water.

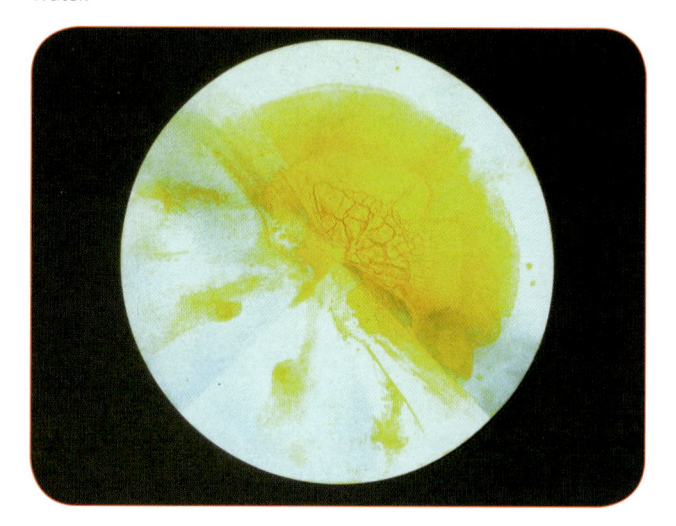

Figure 5h.6d The filter paper and lead iodide are dried.

SAQ

6 Use the series of photos in Figure 5g.6 to describe in your own words how you could make a sample of lead iodide.

Barium meal

If a patient has something wrong with their intestines, a non-invasive procedure like an X-ray is a much safer way to start a diagnosis than an operation. Unfortunately, doctors have a problem if they wish to X-ray the patient's intestines. Unlike bones, these parts of the body do not block X-rays, so they don't leave a shadow on the X-ray film.

The answer to this problem is to give the patient a *barium meal* before the X-ray. The patient has to eat a sort of grey, tasteless, porridge containing a lot of barium sulfate. Barium sulfate blocks X-rays, so when the patient's abdomen is X-rayed the barium sulfate leaves a shadow on the film. This shadow is the same shape as the patient's intestines, complete with any damage, pouches or blockages that the doctor may need to know about (Figure 5h.7).

But barium compounds are poisonous! How can the patient eat a barium meal without being poisoned? Fortunately, barium sulfate is *extremely* insoluble. Only 2.5 *thousandths* of a gram of barium sulfate will dissolve in 1 dm^3 of water. This low solubility is the reason why barium sulfate can be made by precipitation, and why barium chloride can be used as a test for sulfate ions. It is also the reason why a barium meal isn't poisonous – the barium sulfate won't dissolve and get into the patient's blood.

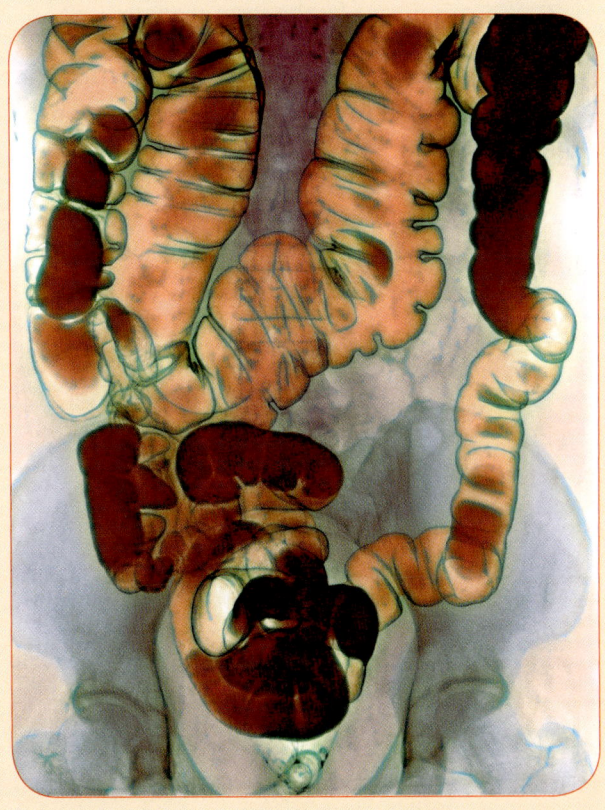

Figure 5h.7 Barium sulfate porridge has made the large intestine show up clearly on this X-ray photograph.

Summary

You should be able to:

- describe how an aqueous solution can be tested for the presence of chloride, bromide, iodide and sulfate ions

- construct word equations for precipitation reactions, including state symbols

- describe a precipitation reaction, explaining why it will work with two solutions but not with two solids

- **H** construct balanced symbol equations and ionic equations for precipitation reactions, including state symbols

- describe how a dry sample of an insoluble compound can be made using a precipitation reaction

Questions

1. Explain the meaning of each state symbol in the following word equation.

    hydrochloric acid (aq) + calcium carbonate (s) → calcium chloride (aq) + water (l) + carbon dioxide (g)

2. What is meant by the term *spectator ion*?

3. Draw a diagram to show how lead iodide can be filtered out of a mixture. (This is shown in Figure 5h.6b.) Label your diagram as fully as you can.

4. 200 cm^3 of 0.10 mol/dm^3 sodium sulfate solution was put into a beaker. An excess of barium chloride solution was added. The precipitate that formed was filtered out, washed and weighed. The final mass of filtered and washed precipitate was 3.78 g. The balanced chemical equation for the precipitation reaction is:

    $Na_2SO_4(aq) + BaCl_2(aq) \rightarrow 2NaCl(aq) + BaSO_4(s)$

    a. Name the precipitate.

    b. Which reactant is the limiting reactant?

    c. How many moles of limiting reactant were used?

    d. What is the molar mass of the precipitate?

    e. How many moles of precipitate were formed?

    f. How many moles of precipitate should have formed?

    g. What was the percentage yield of precipitate? Give your answer to two significant figures (2 s.f.).

    h. Give possible reasons why the percentage yield was less than 100%.

5. The formula of iron(III) hydroxide is $Fe(OH)_3$. It is insoluble in water and can be made by a precipitation reaction. It is formed if sodium hydroxide (NaOH) solution and iron(III) chloride ($FeCl_3$) solution are mixed together.

    a. Write a word equation for this reaction.

    b. Write a balanced symbol equation for this reaction.

    c. Write an ionic equation for this reaction.

Energy transfers – fuel cells

Alternative energy

Ever since the industrial revolution, the world has relied heavily on fossil fuels. Coal, oil and gas have all been burned in huge amounts, by industry, in our homes, and in our vehicles. The fossil fuel era is now coming to an end. The fossil fuels are running out – and when they are gone, they will be gone for ever. There are major concerns over the pollution they cause. When fossil fuels burn, carbon dioxide is released. This gas is causing the world's climate to change.

We need to rely on different energy sources. We need alternatives that won't run out and that cause less pollution. One of the most promising alternative energy sources is hydrogen. There are three main reasons for this:

- As long as we can generate electricity, we can produce hydrogen. It won't run out. (See *Gateway Additional Science*, page 146.)
- When hydrogen reacts with oxygen, the reaction is exothermic. This means energy is released.
- When hydrogen reacts with oxygen, the only product is water – and water is *not* a pollutant.

SAQ

1 a Is the reaction between hydrogen and oxygen endothermic or exothermic?
   b Write a word equation for the reaction between hydrogen and oxygen.

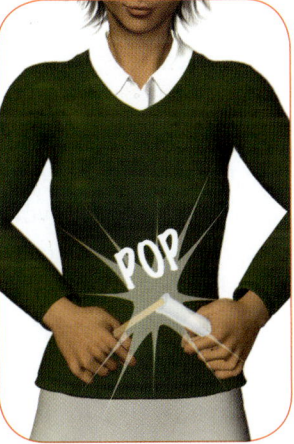

Figure 6a.1 A successful pop test. The gas in the tube was hydrogen.

Identifying hydrogen

If you place a lighted splint into a test tube filled with hydrogen, you will hear a squeaky pop (Figure 6a.1). This test is called the 'pop test'. If the gas in the test tube does not make a squeaky pop when a lighted splint is put in it, then the gas is not hydrogen.

Identifying oxygen

If you place a glowing splint into a test tube filled with oxygen, the splint re-lights (Figure 6a.2). This test is called the 'glowing splint test'. If the gas in the test tube does not re-light the glowing splint, then the gas is not oxygen.

Figure 6a.2 A successful glowing splint test. The gas in the tube was oxygen.

SAQ

2 What would you see or hear if you placed:
   a a glowing splint into a test tube full of oxygen?
   b a glowing splint into a test tube full of nitrogen?
   c a glowing splint into a test tube full of hydrogen?
   d a lighted splint into a test tube full of oxygen?
   e a lighted splint into a test tube full of nitrogen?
   f a lighted splint into a test tube full of hydrogen?

Energy level diagrams

An energy level diagram gives information about a chemical reaction.

The energy level diagram includes an equation for the reaction it describes. As usual, this is done with reactants on the left, and products on the right.

The energy level diagram in Figure 6a.3 shows an exothermic reaction. It does so by showing the products at a lower energy level than the reactants. When the reactants form the products, they go to a lower energy state – therefore, energy is being given out. This means the reaction is exothermic.

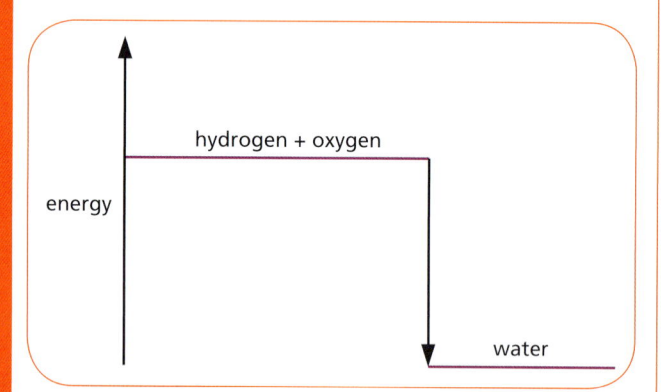

Figure 6a.3 An energy level diagram for the reaction between hydrogen and oxygen.

SAQ

3 Look at Figure 6a.3.
   a What are the reactants in this reaction?
   b What is the product in this reaction?
   c How does the energy level diagram show that this reaction is exothermic?

Fuel cells

Hydrogen is an excellent fuel. When it is burned in air or oxygen, energy is given out and water is produced. Burning is not the only way of using hydrogen as an energy source. Hydrogen can be reacted with oxygen in a special device called a **fuel cell**. In a fuel cell, the energy is not released as heat, or as a popping noise – it is released as electrical energy. This electrical energy produces an electric current (Figure 6a.4).

The fuel cell has to be supplied with hydrogen and oxygen. Sometimes air is used as the oxygen source. The hydrogen and oxygen react inside the fuel cell. The energy released when they react creates a potential difference. A potential difference can also be called a voltage. The voltage can cause an electric current to flow. This current can do useful work if the fuel cell is connected to a bulb, or an electric motor, for example. A fuel cell is a very efficient way of producing electrical energy.

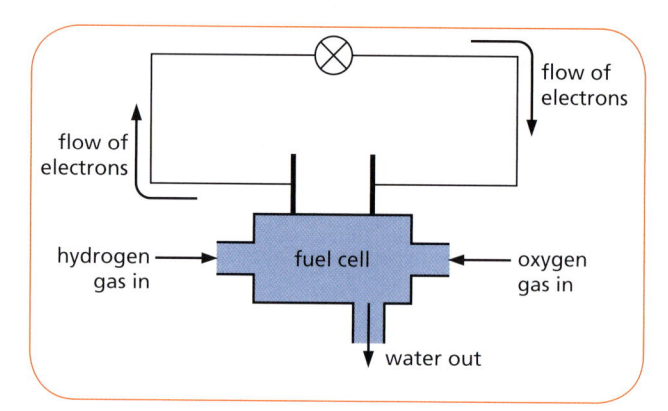

Figure 6a.4 A simple representation of a fuel cell.

SAQ

4 a What two energy forms are released when oxygen and hydrogen react together by burning?
   b What energy form is released when oxygen and hydrogen react together in a fuel cell?
   c Write a word equation for the chemical reaction that takes place in a fuel cell.

The electrode reactions in a fuel cell

The overall reaction in a fuel cell is hydrogen and oxygen reacting together to make water:

$$2H_2 + O_2 \rightarrow 2H_2O$$

In a fuel cell, the hydrogen and oxygen do not react directly. The hydrogen gas reacts at one electrode, which is called the anode. The oxygen gas reacts at the other electrode, the cathode.

Every hydrogen atom loses its outer electron at the anode. Since they lose electrons, the hydrogen atoms are being **oxidised** at the anode. The ionic half-equation for the anode reaction is:

$$H_2 \rightarrow 2H^+ + 2e^-$$

H The H⁺ ions that are produced diffuse through a membrane to the cathode (Figure 6a.5). The electrons that are produced travel round the external circuit to the cathode.

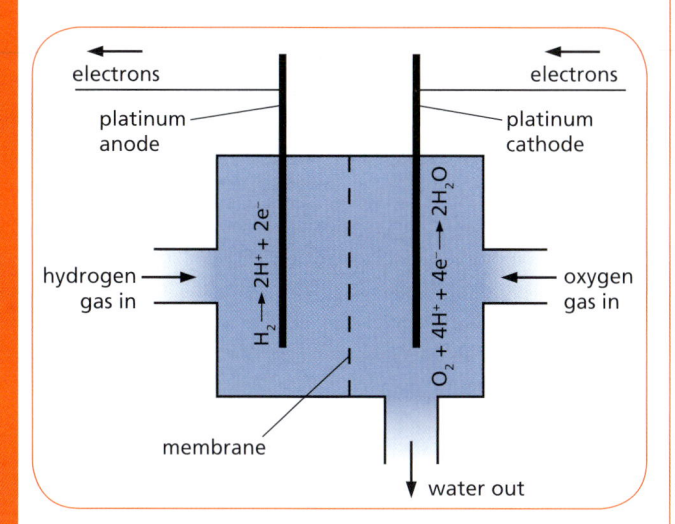

Figure 6a.5 The electrode reactions in a fuel cell.

At the cathode, the oxygen gas reacts with the H⁺ ions that have diffused from the anode, at the same time gaining the electrons that have travelled through the external circuit. Since they gain electrons, the oxygen atoms are being **reduced** at the cathode. The ionic half-equation for the cathode reaction is:

$$O_2 + 4H^+ + 4e^- \rightarrow 2H_2O$$

In a fuel cell, the anode is negative and the cathode is positive. This is the opposite way round to an electrolysis cell, because a fuel cell works in the opposite direction. A fuel cell *supplies* electrical energy, while an electrolysis cell *receives* electrical energy.

SAQ

5 **a** Write an ionic half-equation for the anode reaction in a fuel cell.

   b Why is it correct to say that oxidation takes place at the anode?

   c Write an ionic half-equation for the cathode reaction in a fuel cell.

   d Why is it correct to say that reduction takes place at the cathode?

   e Write a balanced symbol equation for the overall reaction in a fuel cell.

Uses of fuel cells

Fuel cells are likely to be the power sources in the next generation of cars (Figure 6a.6). The cars will have electric motors, and the fuel cells will provide the electrical energy. Unlike petrol, the supply of hydrogen will not run out. The fuel cells will produce much less pollution than petrol engines, and the electric motors will make the cars much quieter.

Figure 6a.6 The Toyota FINE-S prototype is powered by hydrogen–oxygen fuel cell technology.

Fuel cells are already used to provide electrical power for spacecraft (Figure 6a.7).
- They are much lighter than conventional batteries.
- Solar panels on the outside of a spacecraft can be damaged by rocks and other space-debris. A fuel cell can be protected inside the spacecraft.
- The water produced by a fuel cell is used as a source of drinking water by the astronauts.

Figure 6a.7 The space shuttle gets its electrical energy from fuel cells. The astronauts drink the water the fuel cells produce.

Using fuel cells as a source of electricity

A conventional power station generates electricity by a multi-step process:
- heat is generated, either by a nuclear reaction or by burning a fossil fuel
- the heat is used to boil water, creating steam
- the steam is used to drive turbines
- the turbines turn generators, which generate electricity.

By contrast, a fuel cell works like this:
- hydrogen and oxygen react, which generates electricity.

The fuel cell clearly involves far fewer stages. Because it transfers the chemical energy in the fuel directly to electrical energy, the fuel cell is much more efficient. Less energy is lost – in particular, less heat energy is lost. As we've already seen, the fuel cell also causes much less pollution than a conventional power station.

Unfortunately, the output voltage of a single fuel cell is low. Although it is possible to use a large number of fuel cells to generate sufficient voltage to power a small car, they would be unlikely to replace conventional power stations.

Table 6a.1 summarises the advantages of each method of generating electricity.

Advantages of fuel cells	Advantages of conventional power stations
<ul><li>fewer stages</li><li>direct transfer from chemical energy to electrical energy – less energy is lost</li><li>greater efficiency</li><li>less pollution</li><li>fuel will not run out</li></ul>	<ul><li>greater power output</li></ul>

Table 6a.1

Fuel cells everywhere?

It looks very likely that most of us will encounter fuel cells in the next generation of cars, but will that be the only place? Given that very few of us will get to fly in the space shuttle, where else might we use fuel cells?

Manufacturers are developing fuel-cell-powered laptop computers (Figure 6a.8). Each refuelling is hoped to give it 20 hours of use before the next refuelling – a big advance on rechargeable batteries, which last only two or three hours. If this sort of research is successful, there are many other similar-sized devices that could also use fuel cells. There is even a suggestion that they will become small enough to replace the batteries in mobile phones.

Larger fuel cell packs are being developed that could provide enough output power for an entire household. We will probably have to wait a while for even more powerful fuel cell packs that can power factories or schools, but this is expected to happen. At the moment, we make very little use of fuel cells. As fossil fuels run out and environmental issues become more important, this is bound to change.

Figure 6a.8 This laptop is powered by a fuel cell instead of a rechargeable battery.

Summary

You should be able to:

- describe the tests for oxygen and hydrogen
- state that the reaction of oxygen with hydrogen is exothermic
- write a word equation for the reaction of oxygen with hydrogen
- describe how a hydrogen–oxygen fuel cell is used to generate electricity
- **H** draw and interpret an energy level diagram for the reaction of oxygen with hydrogen
- write a balanced symbol equation for the reaction of oxygen with hydrogen
- explain the reaction at each electrode in a hydrogen–oxygen fuel cell
- identify where oxidation and reduction take place in a hydrogen–oxygen fuel cell
- explain why fuel cells are being developed for cars
- state some advantages of using fuel cells as a source of electrical power in a spacecraft
- **H** explain some advantages of fuel cells over conventional methods of generating electricity

Questions

1. a Name the two reactants in a fuel cell.
   b Write a word equation for the chemical reaction in the fuel cell.
   c What type of energy is produced by the fuel cell?

2. Four different gases – A, B, C and D – are tested with a lighted splint, a glowing splint, and limewater. The results are shown in Table 6a.2.

	Lighted splint test result	Glowing splint test result	Limewater test result
A	splint goes out	splint goes out	limewater goes cloudy
B	splint burns brightly	splint re-lights	no change
C	splint goes out	splint goes out	no change
D	squeaky pop heard	splint goes out	no change

Table 6a.2

What can you say about the identity of the four gases?

3. a What will be the advantages of using fuel cells rather than a petrol engine to power a car?
   b Give two reasons why fuel cells are a good source of electrical power in the space shuttle.

H 4. Electrical power generated by a coal-fired power station might one day be replaced by electrical power generated by fuel cells. Give three advantages and one disadvantage of using fuel cells for this purpose.

C6b Redox reactions

Rusting is a redox reaction

When a metallic element reacts with another substance, a metal compound is formed. Reactions of this sort are called **redox reactions**. The element or compound the metal reacts with is reduced, and the metal element is oxidised. The rusting of iron is a well-known example of this sort of reaction. When iron rusts, it reacts with oxygen and water, forming hydrated iron(III) oxide:

iron + oxygen + water → hydrated iron(III) oxide

Hydrated iron(III) oxide is the chemical name for rust. In this reaction, iron is oxidised, and oxygen is reduced. In every reaction in which something is oxidised, something else *must* be reduced. If something is reduced, something else *must* be oxidised. The term 'redox' is helpful – it reminds us that the two processes always happen together.

Figure 6b.1a Pieces of the element iron.

Figure 6b.1b Pieces of iron covered in hydrated iron(III) oxide, commonly known as rust.

SAQ

1. **a** Write a word equation for the rusting of iron.
   b Which substance is reduced in this reaction?
   c Which substance is oxidised in this reaction?
   d What is the chemical name for rust?

Gain and loss of electrons

When iron rusts, the iron atoms lose electrons and form Fe^{3+} ions. When the atoms in a substance lose electrons, that substance is oxidised. Iron is oxidised when it rusts because each iron atom loses electrons. The ionic half-equation for this process is:

$$Fe \rightarrow Fe^{3+} + 3e^-$$

When iron rusts, it reacts with oxygen and water. The oxygen atoms gain electrons and form O^{2-} ions. When the atoms in a substance gain electrons, that substance is reduced. Oxygen is reduced during the rusting of iron because each oxygen atom gains electrons. The ionic half-equation for this process is:

$$O + 2e^- \rightarrow O^{2-}$$

Figure 6b.2 shows a good way of remembering what happens when substances are oxidised and reduced.

OIL RIG

Oxidation	Reduction
Is	Is
Loss	Gain

Figure 6b.2 This is a very useful mnemonic to remember!

When iron rusts, it is the oxygen that takes the electrons from the iron. Oxygen therefore causes iron to be oxidised. A substance that oxidises another substance in this way is called an **oxidising agent**. Oxygen acts as an oxidising agent when iron rusts.

During rusting, it is the iron that gives the electrons to the oxygen. Iron therefore causes oxygen to be reduced. A substance that reduces another substance in this way is called a **reducing agent**. Iron acts as a reducing agent when it rusts.

C6b Redox reactions 173

SAQ

2. When iron rusts, the iron is oxidised and oxygen is reduced. Justify this statement, using ionic half-equations to help you.

3. Use the rusting of iron as an example to explain the terms *oxidising agent* and *reducing agent*.

Preventing rust

Iron is one of the most important materials in use today, particularly as an **alloy** known as **steel**. Steel is strong and keeps a sharp edge. With modern manufacturing techniques like casting, rolling, machining and pressing, it is possible to make steel objects in many different and precise shapes. Rusting is a problem, though. Rusting weakens iron and steel, and must be prevented. Figure 6b.3 illustrates several ways of doing this.

Why sacrificial protection works

When a metal is being corroded, it is reacting with something else to form a metal compound. The metal is oxidised in this process, as its atoms lose electrons. If two metals are in contact with each other, the metal that will corrode is the one that loses electrons more easily.

When zinc and iron are in contact, the zinc will corrode, not the iron. This is because zinc atoms lose electrons more easily than iron atoms do. The zinc protects the iron from rusting.

When magnesium and iron are in contact, the magnesium will corrode, not the iron. This is because magnesium atoms lose electrons more easily than iron atoms do. The magnesium protects the iron from rusting.

When tin and iron are in contact, the iron will corrode, not the tin. This is because iron atoms lose electrons more easily than tin atoms do. Tin protects iron from rusting by providing a barrier, but it cannot provide sacrificial protection.

Figure 6b.3a This chisel is being oiled. Oil, or grease, on a tool acts as a barrier, stopping oxygen or water reaching the surface of the steel.

Figure 6b.3b Painting the car bodywork stops oxygen or water reaching the steel.

Figure 6b.3c This photoframe is made of stainless steel. Stainless steel is an alloy of iron, nickel and chromium. This alloy does not rust.

Figure 6b.3d This unusual plant pot has been galvanised – that means the steel has been coated with zinc. The zinc stops oxygen or water reaching the steel. If the zinc layer is scratched, then the zinc will corrode rather than the iron. This is called sacrificial protection, as the zinc will be 'sacrificed' in place of the iron.

Figure 6b.3e Here you see several pieces of magnesium attached to the rudder and other parts of a large ship. When the ship goes to sea, the magnesium will corrode away but the steel of the ship won't be corroded. This is another example of sacrificial protection.

Figure 6b.3f This steel can is covered in a thin layer of tin. The tin stops oxygen or water reaching the steel. Unfortunately, if the tin layer is scratched it will be the steel that corrodes, not the tin.

SAQ

4 **a** Use Figure 6b.3 to describe six ways of preventing rusting.

   b How do oil, grease and paint prevent rusting?

   c Explain what is meant by *sacrificial protection*. Give two examples of how steel items can be protected in this way.

   d Why can't tin stop iron rusting by sacrificial protection?

Displacement reactions

If pieces of magnesium metal are added to a solution of iron(II) sulfate, a reaction occurs (Figure 6b.4). The products are iron metal and magnesium sulfate. The iron is said to be *displaced* from the solution of iron(II) sulfate. This reaction is called a **displacement reaction**.

The word equation for this displacement reaction is:

magnesium + iron(II) sulfate
$\rightarrow$ magnesium sulfate + iron

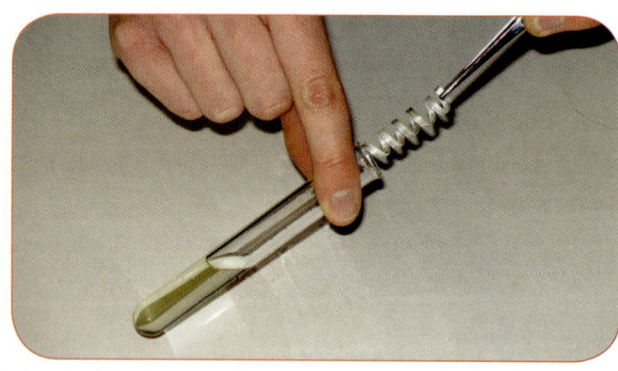

Figure 6b.4a A piece of magnesium is about to be added to some iron(II) sulfate solution.

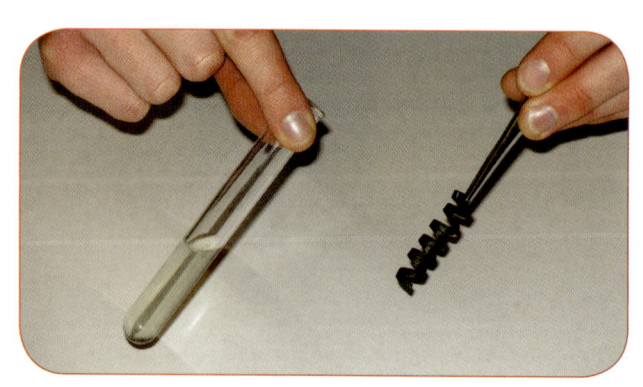

Figure 6b.4b After a little while, iron has become visible on the magnesium and in the test tube. The solution is now magnesium sulfate solution, which is colourless.

Like most displacement reactions, this reaction is exothermic, and the temperature of the solution rises.

This displacement reaction shows that magnesium is more reactive than iron. A *more* reactive metal can displace a less reactive metal from a solution of one of the less reactive metal's compounds.

If a *less* reactive metal is added to a solution of a compound of a more reactive metal, nothing happens.

SAQ

5 Magnesium is more reactive than zinc, zinc is more reactive than iron, and iron is more reactive than tin.

   a Write these four metals in order of their reactivity, starting with the most reactive.

   b What happens if:
      i zinc is added to magnesium sulfate solution?
      ii zinc is added to iron(II) sulfate solution?
      iii tin is added to zinc sulfate solution?
      iv iron is added to tin(II) sulfate solution?

6 Write a word equation for each displacement reaction that takes place in SAQ **5b**.

Balanced symbol equations

The formulae of the sulfate compounds in SAQ **5b**, including the charges on the ions, are:
- $Mg^{2+}SO_4^{2-}$ – magnesium sulfate
- $Zn^{2+}SO_4^{2-}$ – zinc sulfate
- $Fe^{2+}SO_4^{2-}$ – iron(II) sulfate
- $Sn^{2+}SO_4^{2-}$ – tin(II) sulfate.

A balanced symbol equation for the reaction between magnesium and iron(II) sulfate is:

$Mg + Fe^{2+}SO_4^{2-} \rightarrow Mg^{2+}SO_4^{2-} + Fe$

Magnesium atoms are oxidised in this reaction. The ionic half-equation is:

$Mg \rightarrow Mg^{2+} + 2e^-$

Fe^{2+} ions are reduced in the reaction. The ionic half-equation is:

$Fe^{2+} + 2e^- \rightarrow Fe$

C6b Redox reactions

SAQ

7 Write a balanced symbol equation for the displacement reaction that occurs when:

   a magnesium is added to zinc sulfate solution

   b magnesium is added to tin(II) sulfate solution

   c zinc is added to tin(II) sulfate solution.

8 a Identify the substance that is oxidised and the substance that is reduced for each reaction in SAQ 7. Include ionic half-equations.

   b Identify the substance that behaves as an oxidising agent and the substance that behaves as a reducing agent for each reaction in SAQ 7.

Displacement reactions are redox reactions

A displacement reaction occurs when a more reactive metal is added to a solution of a compound of a less reactive metal. The more reactive metal forms one of its compounds – it is oxidised. The less reactive metal starts as part of one of its compounds, but in the reaction it forms the metal itself – it is reduced. Since oxidation and reduction take place, displacement reactions are redox reactions.

Redox is everywhere

This item is called *Redox reactions*, but many of the other items in this book involve redox reactions too.

- When copper is being purified (Item C5b, *Electrolysis*), the copper is oxidised at the anode and reduced at the cathode.
- The first and second steps in the production of sulfuric acid (Item C5f, *Equilibria*) involve the oxidation of sulfur and the reduction of oxygen.
- In a fuel cell (Item C6a, *Energy transfers – fuel cells*), hydrogen is oxidised and oxygen is reduced.
- When brine is electrolysed (Item C6d, *Chemistry of sodium chloride*), hydrogen ions are reduced and chloride ions are oxidised.

Check them out. You might notice that in Item C5f, *Equilibria*, it is easier to identify the substance being oxidised by looking for the substance that gains oxygen, rather than the substance that loses electrons. When we define oxidation and reduction in this way, the substance that is reduced is the substance that loses oxygen.

Thinking of redox in terms of losing or gaining oxygen helps us to recognise that two of the most fundamental reactions for life on planet Earth are also redox reactions.

- During photosynthesis, carbon atoms are reduced and oxygen atoms are oxidised:
carbon dioxide + water → glucose + oxygen
- During the aerobic respiration of glucose, oxygen atoms are reduced and carbon atoms are oxidised (Figure 6b.5):
glucose + oxygen → carbon dioxide + water

Figure 6b.5 Oxygen atoms are reduced and carbon atoms are oxidised as aerobic respiration of glucose takes place in an athlete's muscles. Respiration is a redox reaction.

C6b Redox reactions

Summary

You should be able to:

- state that the rusting of iron is a redox reaction requiring both oxygen and water
- write a word equation for the rusting of iron
- **H** use the idea of the gain and loss of electrons to explain why the rusting of iron is a redox reaction
- recognise and use the terms *oxidising agent* and *reducing agent*
- describe methods of preventing the rusting of iron, explaining how they work
- remember the order of reactivity: magnesium, zinc, iron, tin
- interpret observations made during displacement reactions
- write word equations for displacement reactions
- **H** write balanced symbol equations for displacement reactions
- explain in terms of oxidation and reduction the inter-conversion of Fe and Fe^{2+}, Fe^{2+} and Fe^{3+}, Cl_2 and Cl^- (see Q4 on page 177)

Questions

1. Some pieces of unprotected iron are left out in the rain for several days.
   a. How would the appearance of the iron change?
   b. Name the new substance that forms on the surface of the iron.
   c. Write a word equation for the chemical reaction that takes place.
   d. Which substance is oxidised in this reaction?
   e. Which substance is reduced in this reaction?

2. Describe an experiment you have performed which shows that a piece of iron will only rust if *both* oxygen *and* water are in contact with the surface of the iron.

3. If a piece of iron is put into copper(II) sulfate solution, a displacement reaction takes place. Iron(II) sulfate and copper are formed.
   a. Describe all the observations you could make during this reaction.
   b. Which metal is more reactive, iron or copper?
   c. Which substance is reduced in this reaction?
   d. Which substance is oxidised in this reaction?
   e. Write a word equation for the reaction.

continued on next page

Questions - *continued*

H 4 The following balanced symbol equations all represent redox reactions. All ionic charges have been included.

  i $Fe + 2H^+Cl^- \rightarrow Fe^{2+}Cl^-_2 + H_2$

  ii $2Fe^{2+}Cl^-_2 + Cl_2 \rightarrow 2Fe^{3+}Cl^-_3$

  iii $2Na^+Br^- + Cl_2 \rightarrow 2Na^+Cl^- + Br_2$

  iv $Fe^{2+}SO_4^{2-} + Mg \rightarrow Mg^{2+}SO_4^{2-} + Fe$

  v $2Fe^{3+}Cl^-_3 + 2Na^+I^- \rightarrow 2Fe^{2+}Cl^-_2 + 2Na^+Cl^- + I_2$

  a Identify the substance that is *oxidised* in each reaction. You should do this by finding the substance that *loses* electrons.

  b Identify the substance that is *reduced* in each reaction. You should do this by finding the substance that *gains* electrons.

  c Identify the oxidising agent in each reaction.

  d Identify the reducing agent in each reaction.

5 The charge on a magnesium ion is 2+. The charge on a copper(II) ion is 2+. The charge on a chloride ion is 1−. Magnesium is more reactive than copper.

  a What is the formula of magnesium chloride?

  b What is the formula of copper(II) chloride?

  c Write a balanced symbol equation for the reaction between magnesium and copper(II) chloride solution.

  d Which substance is reduced in this reaction?

  e Which substance is oxidised in this reaction?

  f Use ionic half-equations to justify your answers to **d** and **e**.

C6c Alcohols

Ethanol

Alcohol is a familiar word in the early 21st century. Drinks like beers, wines and spirits all contain alcohol, and so does bread dough before it is cooked. The word 'alcohol' is a general word, however. It is the name for a whole family of different chemicals. Alcoholic drinks, and bread dough, all contain one particular member of the family. This alcohol is called **ethanol**.

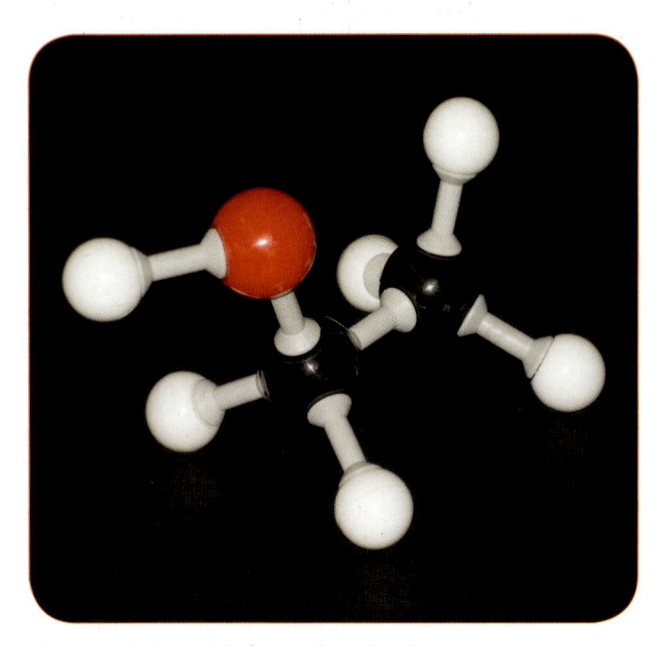

Figure 6c.1 A model of one ethanol molecule.

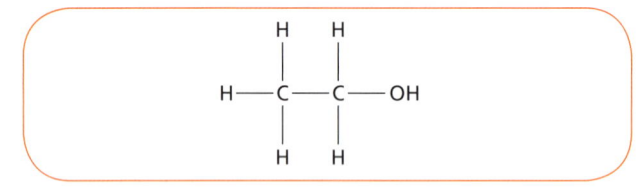

Figure 6c.2 The displayed formula of ethanol. The molecular formula of ethanol is C_2H_5OH.

SAQ

1. **a** Which three elements are combined together in ethanol?
   b How many atoms of each element are present in one molecule of ethanol?
   c How are these atoms bonded to each other in one ethanol molecule? (Hint: look at *Gateway Additional Science*, Items C3b and C3c.)

A family of alcohols

Ethanol is one member of a family of compounds called the alcohols. Every member of the alcohol family has a simple molecular structure (See *Gateway Additional Science*, pages 128 and 129). In every alcohol molecule, there is one hydrogen atom joined to an oxygen atom – called an 'OH group'. The remainder of each alcohol molecule consists of carbon atoms and hydrogen atoms. This is called the 'hydrocarbon portion' of each alcohol molecule.

Different alcohols have molecules with hydrocarbon portions of different sizes. Figure 6c.3 shows this for four alcohols with one, two, three and four carbon atoms.

Name	Displayed formula
methanol	H–C(H)(H)–OH
ethanol	H–C(H)(H)–C(H)(H)–OH
propan-1-ol	H–C(H)(H)–C(H)(H)–C(H)(H)–OH
butan-1-ol	H–C(H)(H)–C(H)(H)–C(H)(H)–C(H)(H)–OH

Figure 6c.3 The first four members of the alcohol family – each molecule has one carbon atom and two hydrogen atoms more than the last.

SAQ

2. Pentan-1-ol has a hydrocarbon portion with five carbon atoms.
   a Draw the displayed formula of pentan-1-ol.
   b What is the molecular formula of pentan-1-ol?

Every member of the alcohol family has a similar molecular formula. If the number of carbon atoms is called 'n', the number of hydrogen atoms bonded to these carbons is '2n+1'. Every molecule also has an OH group. The **general formula** of any member of the alcohol family is $C_nH_{2n+1}OH$.

Fermentation

Beer, wine and bread are all made by a process called **fermentation** (Figures 6c.4 and 6c.5). This process uses the **enzymes** in yeast to change a sugar, such as glucose, into carbon dioxide and ethanol. A word equation for fermentation is:

glucose → carbon dioxide + ethanol

Yeast does not appear in the equation. The enzymes from the yeast act as natural catalysts in this process. They are not used up.

Figure 6c.4 Bread dough consists largely of flour, water and oil. A little bit of sugar and yeast are also added. Enzymes from the yeast change the sugar into carbon dioxide and ethanol. Bubbles of carbon dioxide gas make the dough rise. The ethanol evaporates when the dough is baked in an oven. The heat of the oven will kill the yeast.

Yeast is a living organism. The enzymes in yeast will ferment sugar into carbon dioxide and ethanol under certain conditions:
- a warm temperature is needed, between 25 °C and 50 °C
- some water must be present
- there must be no oxygen present
- a sugar, such as glucose, must be present.

A temperature of 25–50 °C is ideal for the yeast – we call this the **optimum** temperature for the yeast. An optimum temperature is one that is neither too high nor too low.

Fermentation can be used to make an ethanol solution like wine or beer. Wine is typically about 12% ethanol; beer is about 5% ethanol. The rest of the wine or beer consists of water plus small amounts of colours and flavours. If either of these solutions is distilled, it is possible to produce almost pure ethanol. This ethanol can be burnt – it makes an excellent fuel.

Figure 6c.5 Making wine at home. Yeast has been added to the white grape juice in the bottle. Enzymes from the yeast are changing sugars in the juice into carbon dioxide and ethanol. The carbon dioxide is bubbling out through the airlock at the top of the bottle. The ethanol will give the wine its alcohol content. The yeast has to be filtered out before the wine is bottled.

SAQ

3 a Write a word equation for fermentation.
   b What is the role of yeast enzymes during fermentation?
   c What conditions are needed for fermentation?
   d How might a higher concentration of ethanol be made from wine?

Ethanol can be made by the fermentation of sugars. This process will not work if the conditions are wrong.
- If the temperature is too low, the yeast will remain inactive.
- If the temperature is too high, the enzymes in the yeast will be **denatured** and will not work.
- If oxygen is present, the yeast will convert the sugars into ethanoic acid instead of ethanol. Ethanoic acid is the chemical name for vinegar. This will give the drink or the bread a sour taste.

The balanced symbol equation for the fermentation of glucose is:

$C_6H_{12}O_6 \rightarrow 2C_2H_5OH + 2CO_2$

Gasohol

Gasohol is a fuel used for powering cars. It is also called 'E10'. The 'E' stands for ethanol and the '10' means 10%. Gasohol is 10% ethanol mixed with 90% petrol. It is replacing pure petrol in many countries. Ethanol made by fermenting plant material is a renewable fuel – unlike crude oil from which petrol is made. Another environmental benefit from using ethanol as a fuel is that it helps slow the rise of carbon dioxide levels in the atmosphere. This is because plants are grown in order to produce the ethanol. These plants use up carbon dioxide as they photosynthesise.

Research has not stopped at E10. Many vehicles are now being manufactured that use E85, which is 85% ethanol (Figure 6c.6). There are even some E100-powered vehicles on the roads, particularly in Brazil and other South American countries.

Figure 6c.6 This pump dispenses E85 bioethanol.

Industrial methylated spirits

Ethanol can be made industrially by reacting a gas called ethene with steam. Heated phosphoric acid is used as a catalyst in this process. Ethene (see *Gateway Science*, page 176) is made by cracking certain crude oil fractions (see *Gateway Science*, pages 169 and 170). The word equation for the reaction that makes ethanol from ethene is:

ethene + water → ethanol

This reaction is called a **hydration** reaction, as water is added to the ethanol. The ethanol is used in industry and can also be sold to the public. So that it cannot be drunk, a poisonous chemical called methanol is added to it. This mixture of ethanol and methanol is called **industrial methylated spirits**. Industrial methylated spirits is used as a fuel, and as a solvent. Ethanol is an excellent solvent, able to dissolve both greasy and non-greasy dirt.

SAQ

4 a Describe how ethanol can be made industrially.
   b Write a word equation for this process.
   c Give two uses for industrial ethanol.

H Comparing fermentation and hydration

Each method of making ethanol has its advantages and disadvantages (Table 6c.1).

Fermentation	Hydration
produces ethanol in small batches	produces large amounts of ethanol by a continuous process
produces ethanol from a renewable source – plant material	produces ethanol from a non-renewable source – crude oil
produces dilute ethanol solutions at a concentration that can be drunk – the solutions have to be distilled if the ethanol is to be used as a fuel or solvent	produces ethanol of a high concentration that can be used directly as a fuel or solvent

Table 6c.1 Fermentation and hydration compared.

SAQ

5 a Explain why the ethanol in alcoholic drinks is made by fermentation.
   b Explain why most ethanol used as a solvent is made by the hydration of ethene.
   c If ethanol is being used to power cars in 2050, how do you think it will be made? Why?

C6c Alcohols   181

Using ethanol to make ethene

If ethanol vapour is passed over heated aluminium oxide, the ethanol decomposes. The products of this reaction are ethene and water. Here is the word equation:

ethanol → ethene + water

The aluminium oxide acts as a catalyst in this process. The ethanol is being **dehydrated** – each ethanol molecule loses one water molecule in this reaction.

This dehydration reaction can be demonstrated in a school lab, as shown in Figure 6c.7. The ethene gas that is produced can be collected in inverted test tubes full of water.

SAQ

6 a Describe how ethene can be made in a school lab.
b Write a word equation for this process.
c Use *Gateway Science*, page 178, to give one important use for ethene.

H The balanced symbol equation for the dehydration of ethanol is:

$$C_2H_5OH \rightarrow C_2H_4 + H_2O$$

This is the reverse of the balanced symbol equation for the hydration of ethene, which is:

$$C_2H_4 + H_2O \rightarrow C_2H_5OH$$

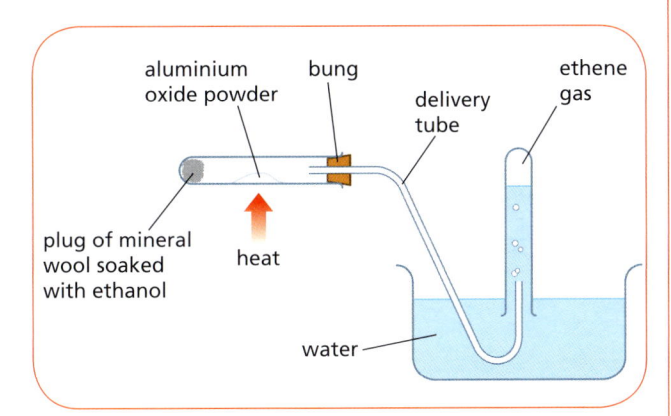

Figure 6c.7 Dehydrating ethanol.

Summary

You should be able to:

◆ state three main uses of ethanol

◆ state the molecular formula and the displayed formula of ethanol

H ◆ state the general formula of an alcohol and use it to write the molecular formulae of the alcohols with 1 to 5 carbon atoms

◆ draw the displayed formulae of the alcohols with 1 to 5 carbon atoms

◆ describe how ethanol can be made by fermentation, giving conditions and writing a word equation

H ◆ write a balanced symbol equation for fermentation and explain the conditions used in fermentation

◆ describe how ethanol can be made industrially, writing a word equation

◆ describe how ethene can be made by dehydrating ethanol, writing a word equation

H ◆ evaluate the relative merits of fermentation and hydration as ways of making ethanol

◆ write balanced symbol equations for the hydration of ethene and the dehydration of ethanol

Questions

1. **a** Describe how fermentation is used to make wine.

   b What is the typical percentage of ethanol in wine?

   c Describe how fermentation is used to make bread.

   d Why is there very little ethanol in baked bread?

2. **a** Explain what is meant by the terms *hydration reaction* and *dehydration reaction*.

   b Give details of one hydration reaction that is used to produce ethanol.

   c Give details of one dehydration reaction that uses ethanol as a reactant.

   d Write word equations for the reactions you have described in **b** and **c**.

3. What do the following terms mean?

   a *industrial methylated spirits*

   b *enzymes*

   c *optimum temperature*

H 4. Describe how you could make some ethene gas starting off with some sugar and some yeast.
   - Three processes are needed. Describe each process as fully as you can.
   - Include any necessary conditions.
   - Include a balanced symbol equation for the chemical reactions involved in the first and third processes.

   (If you cannot get started on this question, there is a hint after **Q6** below.)

5. **a** What is the general formula for an alcohol?

   b The molecular formula of ethanol is C_2H_5OH. Explain how this molecular formula fits in with the general formula you have just given.

   c What is the molecular formula of octan-1-ol, an alcohol with eight carbon atoms per molecule?

6. Look at Figure 6c.3.

   a Borrow an A-level chemistry textbook and find out what the '1' means in the names of propan-1-ol, butan-1-ol and pentan-1-ol.

   b Draw the displayed formulae of propan-2-ol and pentan-3-ol.

(Hint for **Q4** – fermentation, distillation, dehydration)

6d Chemistry of sodium chloride (NaCl)

Salt mines

One of the most important industrial areas of England stretches from Northern Cheshire up into Lancashire. There are several reasons why it is so important, but one of them is salt. There are large deposits of rock salt under Cheshire. This salt was left behind when a prehistoric sea dried up, and is now buried 50 m underground. Rock salt is impure sodium chloride.

The sodium chloride is extracted in two ways. The salt can be dug out by mining (Figure 6d.1). Alternatively, water can be pumped into the salt deposit. The salt dissolves in the water, and the resultant salt solution, called brine, is pumped back to the surface. This method is called **solution mining**.

The salt has brought industry and prosperity to the area, but it has also brought some problems. Removing underground material from so near the surface can result in the surface collapsing into the holes that are produced. This is known as **subsidence**.

Figure 6d.1 A modern salt mine.

SAQ

1 a Describe two ways of extracting salt from an underground deposit.
   b Describe a possible problem resulting from salt extraction.
   c What do we mean by *brine*?

Electrolysis of brine

The salt is used as a source of many important chemicals. The first step in producing most of these chemicals is electrolysis. The equipment used to electrolyse brine is shown in Figure 6d.2. The electrodes used have to be **inert**. Inert means unreactive. If the electrodes are not inert, then they react with the products of the electrolysis.

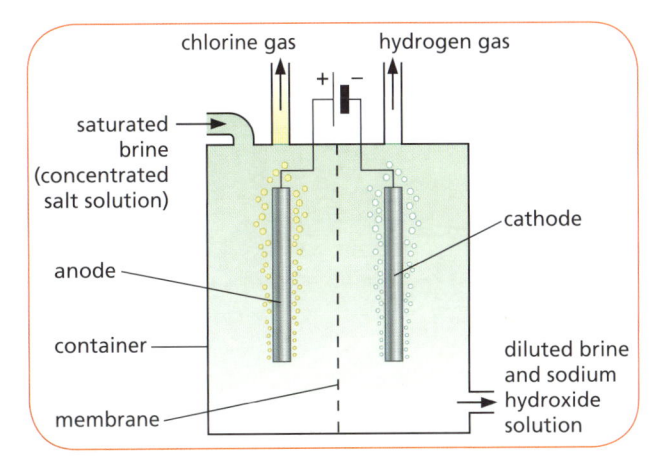

Figure 6d.2 Electrolysing brine.

SAQ

2 a What do we mean by *electrolysis*?
   b What does *inert* mean?
   c Look at the power supply in Figure 6d.2. Is it AC or DC?

Chlorine

The electrolysis of brine produces chlorine. Chlorine is used to treat drinking water and to make household bleach, plastics and solvents (Figures 6d.3 to 6d.6).

Figure 6d.3 Chlorine was used at the water treatment plant to kill germs in this drinking water.

184 C6d Chemistry of sodium chloride (NaCl)

Figure 6d.4 Chlorine was reacted with sodium hydroxide to make this household bleach.

Figure 6d.5 These plastic pipes are made of PVC. PVC is a chlorine compound.

Figure 6d.6 The solvent used by dry cleaners is a chlorine compound.

SAQ

3 a When brine is electrolysed, chlorine is collected above one of the electrodes. Which one?

   b Give four uses for the chlorine produced by electrolysing brine.

Figure 6d.7 Hydrogen was used, with vegetable oil, to make this margarine.

Hydrogen

The electrolysis of brine produces hydrogen. Hydrogen is used to make margarine (Figure 6d.7).

SAQ

4 a When brine is electrolysed, hydrogen is collected above one of the electrodes. Which one?

   b Give one use for the hydrogen produced by electrolysing brine.

Sodium hydroxide

The electrolysis of brine produces sodium hydroxide. Sodium hydroxide is used to make household bleach (Figure 6d.4) and soap (Figure 6d.8).

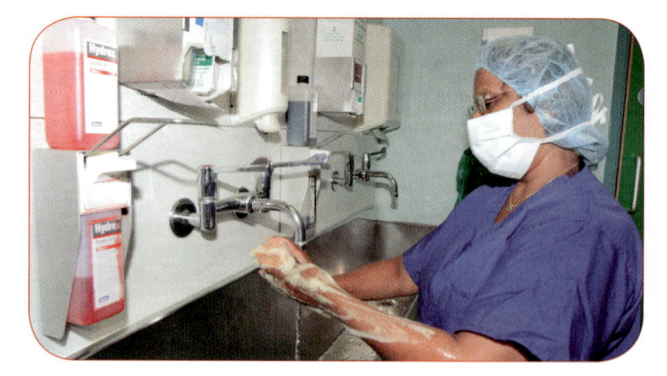

Figure 6d.8 Sodium hydroxide was used to make this soap.

SAQ

5 a When brine is electrolysed, sodium hydroxide is not collected above one of the electrodes. Where is the sodium hydroxide collected?

   b Give two uses for the sodium hydroxide produced by electrolysing brine.

Making soap

You can make soap safely and successfully at home with some readily available ingredients. You will need:

- safety glasses (DIY shops sell them)
- rubber gloves
- 930 cm³ water
- 620 cm³ olive oil
- 295 g sodium hydroxide (often labelled 'caustic soda' in DIY shops)
- a large plastic bucket and a plastic spoon for stirring
- a stainless steel saucepan (make sure you have permission to use it)
- a thermometer
- chopped up wax crayons for colouring, and essential oils like lavender or rose for perfuming, if you wish
- a mould made by lining a small cardboard box with greaseproof paper

To make your soap:

1 Put on your safety glasses and rubber gloves. Put the water into the bucket and add the sodium hydroxide, stirring all the time with the plastic spoon until it is all dissolved.

2 Put the olive oil into the saucepan and warm it until it is between 35 °C and 38 °C. Check that the sodium hydroxide solution is at a similar temperature.

3 When these two ingredients are at the correct temperatures, add the sodium hydroxide solution to the oil and stir continuously. Keep stirring until a line made on the surface of the mixture with the spoon stays there. (This normally takes over 45 minutes!)

4 Stir in a chopped up wax crayon and a few drops of essential oil if you wish.

5 Pour your soap into your mould. After 24 hours, remove it from the mould and cut it into individual bars (Figure 6d.9). Leave these bars in a warm dry place (for example, wrap them in an old towel and put them in the airing cupboard) for a month. The soap should harden over this time.

6 Before using the soap, get some universal indicator paper from school and press a wet piece of it against the soap. The pH should be between 7 and 10.

Figure 6d.9 Homemade soap.

Electrolysing molten sodium chloride

Useful chemicals can also be obtained from salt if it is melted and electrolysed as a molten liquid instead of as a solution in water. When molten salt is electrolysed, the products are sodium and chlorine.

Identifying chlorine

Chlorine gas dissolves in water to produce a solution that is a powerful bleach. If chlorine gas comes in contact with moist litmus paper, the litmus paper turns white. Chlorine gas can be identified by its ability to bleach moist litmus paper (Figure 6d.10 on page 186).

SAQ

6 A student electrolyses saturated brine. She collects a test tube full of the gas given off at the anode, and a test tube full of the gas given off at the cathode.

   a How could she identify each gas?
   b What safety precautions should she take?

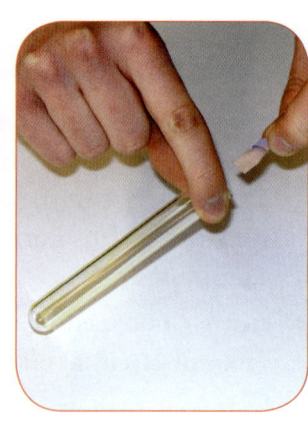

Figure 6d.10 The gas in the tube bleaches the litmus paper. The gas is chlorine.

The reactions at the electrodes

Saturated aqueous brine contains four different ions: Na^+, H^+, Cl^- and OH^-. The Na^+ ions and Cl^- ions come from the sodium chloride. The H^+ ions and OH^- ions come from the water.

When saturated brine is electrolysed, H^+ ions are discharged at the cathode:

$$2H^+ + 2e^- \rightarrow H_2$$

Cl^- ions are discharged at the anode:

$$2Cl^- - 2e^- \rightarrow Cl_2$$

The Na^+ ions and OH^- ions are not discharged – they remain in solution. A solution of NaOH (sodium hydroxide) is produced, mixed with what is left of the brine.

Chloride ions are discharged at the anode if a highly concentrated salt solution is used. If a dilute salt solution is used, the hydroxide ions are discharged at the anode instead, producing oxygen. So in industry, in order to produce chlorine gas as a product, *saturated* brine must be used in the electrolysis.

When molten sodium chloride is electrolysed, Na^+ ions are discharged at the cathode:

$$Na^+ + e^- \rightarrow Na$$

Cl^- ions are discharged at the anode:

$$2Cl^- - 2e^- \rightarrow Cl_2$$

SAQ

7 If dilute sodium chloride solution is electrolysed:
   a which ions are discharged at the anode?
   b which ions are discharged at the cathode?
   c which ions remain in solution?

Summary

You should be able to:

- describe how salt is mined, stating one problem that can result
- describe how brine is electrolysed using inert electrodes, naming the three products of the process
- state uses for the chlorine, hydrogen and sodium hydroxide that are produced
- name the products of electrolysing molten sodium chloride
- name the ions present in saturated brine and describe what happens to each ion during electrolysis, using half-equations for the electrode reactions
- describe the difference between the electrolysis of saturated brine and the electrolysis of dilute brine
- describe what happens during the electrolysis of molten sodium chloride, using half-equations for the electrode reactions

Questions

1. **a** Where are salt deposits found in England?

   b How is the salt removed from underground?

2. **a** What *three* useful products are obtained by electrolysing brine?

   b Which substance is made by reacting two of these three products?

   c Name three other things that are made from the products of electrolysing brine.

3. Useful products are obtained by electrolysing saturated brine and also by electrolysing molten sodium chloride.

   a Which process do you think operates at the higher temperature?

   b How will the higher operating temperature affect the cost of this process?

   c Which product is formed in both processes?

   d Name the electrode where sodium forms during the electrolysis of molten sodium chloride. Explain your choice.

   e Is the electrolysis of brine a continuous process or a batch process?

4. When saturated brine is electrolysed:

   a Which ions are discharged at the anode? Write a half-equation for this process.

   b Which ions are discharged at the cathode? Write a half-equation for this process.

5. Describe how your answers to Q4 would change if you had been describing the electrolysis of very dilute brine.

6. 'The electrolysis of molten sodium chloride is an example of a redox reaction.'

   Explain this statement as fully as you can. Include relevant half-equations in your answer.

C6e Depletion of the ozone layer

The ozone layer

The Earth's atmosphere changes with increasing height above the Earth's surface. At a height of 15 km from the surface of the Earth, there is a significantly higher concentration of a gas called **ozone**. This is the beginning of the **ozone layer**. The ozone layer extends from 15 km to 40 km above the Earth. This is part of the Earth's higher atmosphere called the **stratosphere**. The ozone layer is in the stratosphere.

Ozone is O_3

At the Earth's surface, 21% of the atmosphere is oxygen. The molecular formula of this oxygen is O_2. The oxygen molecules consist of two oxygen atoms held together by a double covalent bond. This is the form of oxygen that we need in order to stay alive.

Ozone is also a form of the element oxygen. The molecular formula of ozone is O_3. Each ozone molecule consists of three oxygen atoms held together by covalent bonds (Figure 6e.1). There is hardly any ozone in the atmosphere at the Earth's surface. This is fortunate, because ozone irritates our eyes and airways. The ozone layer starts at 15 km above the Earth's surface.

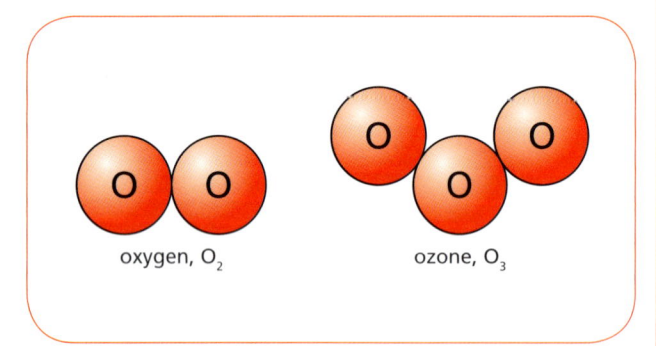

Figure 6e.1 An oxygen molecule has two oxygen atoms (O_2), while an ozone molecule has three (O_3).

SAQ

1 **a** What is ozone?
   b Where is the ozone layer?
   c Why is ozone undesirable at the Earth's surface?

The importance of the ozone layer

The electromagnetic radiation that comes to the Earth from the Sun includes many different wavelengths. As well as visible light that we can see, there is also ultraviolet light (UV) that we cannot see. UV is harmful:
- it causes sunburn
- it speeds up the ageing of skin
- it can cause skin cancers called melanomas
- it can cause eye damage, including cataracts.

Fortunately, UV is absorbed by ozone. We are protected from this harmful part of the Sun's radiation by the ozone layer.

When UV strikes a molecule of ozone in the stratosphere, the ozone molecule is split into an oxygen molecule (O_2) and an oxygen atom (O) (Figure 6e.2). UV is absorbed in this process.

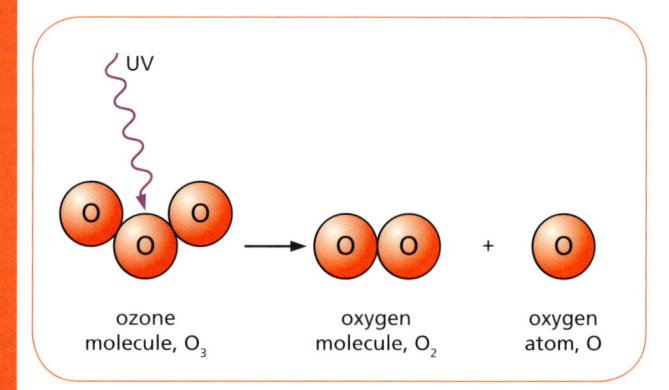

Figure 6e.2 When an ozone molecule absorbs UV, it breaks into an oxygen molecule and an oxygen atom.

Ozone depletion

In 1985, weather scientists announced that they had discovered that the amount of ozone in the stratosphere was decreasing. This decrease is known as ozone depletion. Ozone depletion was not happening equally in all parts of the world, but it was definitely happening. The problem became known as 'the hole in the ozone layer' (Figure 6e.3 on page 190).

Ozone depletion is a serious issue. Less ozone means more UV reaching the Earth's surface, which will lead to major health problems. It was soon clear to the scientists who investigated the problem that ozone depletion was due to compounds known as CFCs.

C6e Depletion of the ozone layer 189

Figure 6e.3 The hole in the ozone layer is shown in this picture as the blue patch over Antarctica.

SAQ

2 What health problems would result from ozone depletion?

CFCs

CFCs are compounds of chlorine, fluorine and carbon (Figure 6e.4). They are called chlorofluorocarbon compounds, or CFCs for short. CFCs have some very useful properties:
- they are inert – this means they are unreactive
- they have low boiling points
- they are insoluble in water.

These properties made CFCs very useful in lots of ways, but over 50% of the CFCs being used in the early 1980s were in fridges and aerosol cans.

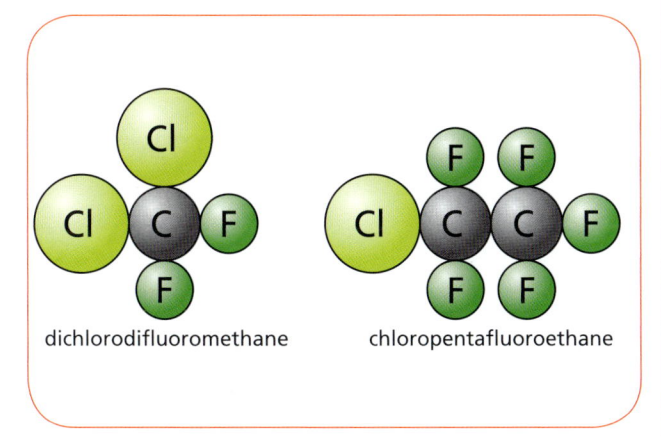

dichlorodifluoromethane   chloropentafluoroethane

Figure 6e.4 Two CFC molecules.

Refrigerants

All fridges use a liquid called a **refrigerant** that cools down the compartment inside. The refrigerant travels through pipes on the inside of the fridge, in a panel at the back. It also travels through pipes on the outside, at the back of the fridge (Figure 6e.5). CFCs were used as refrigerants.

Figure 6e.5 The refrigerant runs through these black tubes on the back of the fridge.

Aerosol propellants

Body spray is usually sold in aerosol cans (Figure 6e.6). When you press the button on the top of an aerosol, the contents start to spray out of the can. They do this because of a liquid called the **propellant**. The propellant sprays out of the can, carrying with it small amounts of perfume and other active ingredients from the body spray. The body spray dries quickly on your skin, leaving the active ingredients. CFCs made excellent aerosol propellants. In the 1970s, aerosols released huge amounts of CFCs into the atmosphere.

Figure 6e.6 This modern aerosol contains a propellant, but it is not a CFC.

SAQ

3
a What do we mean by *CFCs*?
b Describe three properties of CFCs.
c Describe two uses of CFCs.
d As CFCs are unreactive, they are also non-poisonous. Why is this important for the two uses you gave in **c**?

What CFCs do in the stratosphere

In the 1970s and 1980s, CFCs were released into the atmosphere when aerosols were used and when old fridges rusted away in dumps. The CFCs were carried into the stratosphere by natural air movement – by winds. In the stratosphere, UV light hit the CFC molecules and caused single chlorine atoms to break off (Figure 6e.7). These single chlorine atoms, also called chlorine **free radicals**, caused ozone depletion.

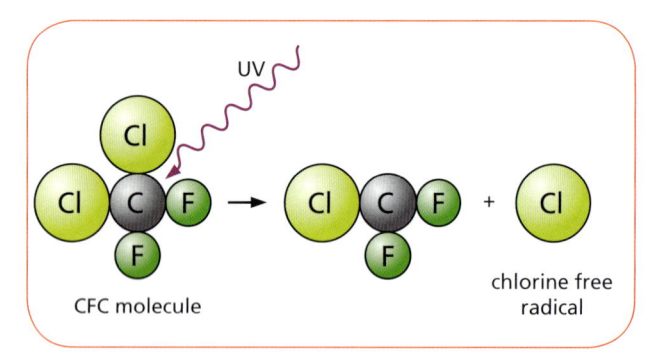

Figure 6e.7 A CFC molecule can be broken up by the action of UV. The chlorine free radical that is produced will cause ozone depletion.

At a conference in Montreal, Canada, in 1987 all the world's developed countries, including the UK, agreed to decrease CFC usage. These countries all stopped producing new CFCs by 1995. Because they are unreactive, CFC molecules remain in the stratosphere for a long time. One CFC molecule is able to cause the loss of many ozone molecules. Because of this, CFCs have continued to cause ozone depletion *since* 1995, despite the ban.

SAQ

4
a How do single chlorine atoms form from CFC molecules in the stratosphere?
b Why are these single chlorine atoms a problem?
c What are these single chlorine atoms also called?

H Free radicals

A covalent bond can break in two ways (Figures 6e.8 and 6e.9).

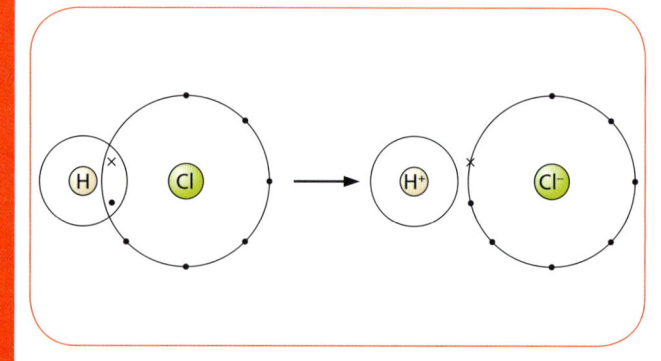

Figure 6e.8 This dot and cross diagram shows the covalent bond in an HCl molecule breaking. The chlorine atom gets both bonding electrons. An H^+ ion and a Cl^- ion are formed.

- The two electrons in the covalent bond can be shared unevenly. The two electrons in the covalent bond can both go to the same atom. In this case, the products are two ions (Figure 6e.8).

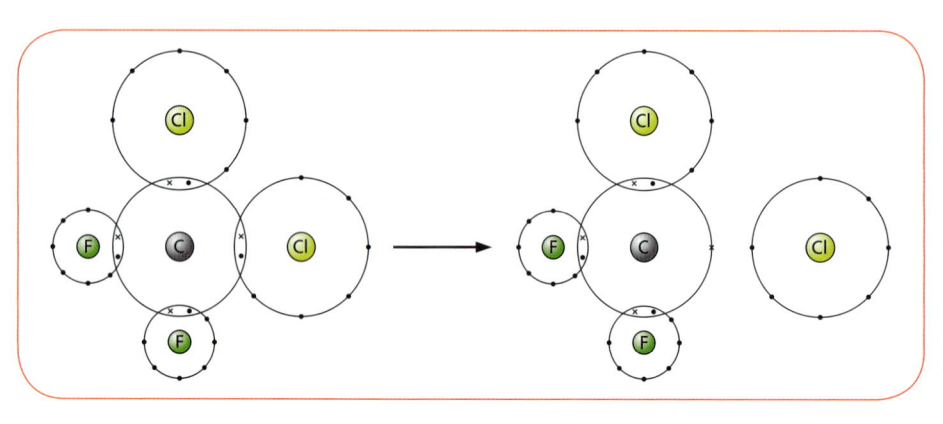

Figure 6e.9 This dot and cross diagram shows a covalent bond in a CCl_2F_2 molecule breaking. The carbon atom and the chlorine atom each get one of the bonding electrons. Free radicals are formed.

C6e Depletion of the ozone layer 191

- The two electrons in the covalent bond can be shared evenly. One electron goes to one atom and one electron goes to the other atom. In this case, the products are two free radicals (Figure 6e.9).

Free radicals are formed when a bond in a CFC molecule is broken by UV. The chlorine free radical that forms is responsible for ozone depletion. The following two reactions happen one after the other:

$$Cl + O_3 \rightarrow ClO + O_2$$
$$ClO + O_3 \rightarrow 2O_2 + Cl$$

In these two reactions, two ozone molecules are used up. A chlorine free radical is used up in the first reaction, but a chlorine free radical is re-formed in the second reaction. This chlorine free radical can then cause the reaction of another two ozone molecules, generating another chlorine free radical, and so on. This sort of continually repeating reaction is called a chain reaction.

One chlorine free radical can cause the loss of many thousands of ozone molecules by causing this pair of reactions to occur thousands of times. The chlorine free radical is acting as a catalyst – it causes the ozone breakdown to happen more quickly, but is not used up itself.

SAQ

5 a Describe how free radicals can form when a covalent bond breaks.

b Describe how ions can form when a covalent bond breaks.

c Explain why the chlorine free radical can be said to act as a catalyst.

Miracle materials?

CFCs are a good example of how scientists' and engineers' opinions of a material can change.

Initially CFCs were thought of as wonderful materials. They had a combination of properties that made them seem perfectly suited to many uses (Figure 6e.10). Their apparent inertness made them particularly useful.

After CFCs had been in use for many years, ozone depletion was detected. CFCs were found to be much less inert when they reached the stratosphere – an environment where there is a high level of UV radiation. By the time this was understood, ozone depletion had become a serious problem.

CFCs were the cause of ozone depletion. Developed countries that had previously welcomed them now banned them. Unfortunately, some of the world's newly developing countries have not reduced their use of CFCs. A further change in international attitudes is necessary before we see a final end to this particular cause of environmental damage.

The health of the ozone layer is now much better than it was in 1985. CFCs last a long time in the stratosphere, but they don't last forever. There is evidence that ozone levels are no longer falling.

Figure 6e.10 Two more uses of CFCs – they were used as the foaming agents for food packaging, and as propellants for inhalers. Their unreactiveness, and low toxicity, made them seem to be ideal for both uses.

C6e Depletion of the ozone layer

Alternatives to CFCs

Many of the jobs that were done by CFCs are now being done by chemicals known as HFCs and by hydrocarbons, such as alkanes. In some ways hydrocarbons are not as good at these jobs as CFCs were. Hydrocarbons are flammable. Now that alkanes are used as aerosol propellants it is common to find a warning on an aerosol can – 'do not spray near naked flames'. This was not a problem with CFCs, which are not flammable. The hydrocarbons have one big advantage, however. They don't damage the ozone layer.

Summary

You should be able to:

- state that ozone is a form of oxygen with the formula O_3
- describe the consequences of ozone depletion and medical problems caused by increased levels of UV light
- describe a CFC and give three properties and two uses of CFCs
- explain that CFCs are banned in many countries because they cause ozone depletion
- **H** describe how the action of UV light on CFCs produces chlorine free radicals
- describe how attitudes to CFCs have changed over time
- explain how covalent bonds can break in two different ways, producing either ions or free radicals
- use symbol equations for the reactions between chlorine free radicals and ozone
- explain why CFCs have caused ozone depletion for many years after their use was banned
- describe hydrocarbons and HFCs as alternatives to CFCs that will not cause ozone depletion

Questions

1. Explain the meaning of the following terms:

   a an oxygen molecule

   b an ozone molecule

   c an oxygen atom

2. Figure 6e.11 shows the information panel from an aerosol can of deodorant.

   a Which alkanes does it contain?

   b Why does it contain alkanes instead of CFCs?

   c How does the information on the can warn the user to be careful?

Figure 6e.11

continued on next page

Questions - continued

3 a Use the internet to find out what is meant by *HFCs*.

   b Find out why HFCs do not cause ozone depletion.

4 a Why were CFCs welcomed as excellent materials when they were first used?

   b Why have opinions about CFCs changed?

   c What has been the response to these changing opinions?

   d What could be the reasons why some countries continue to use CFCs?

5 a Explain what is meant by the term *chlorine free radical*.

   b Explain how a chlorine free radical can be produced from a CFC molecule.

   c Use equations to explain how chlorine free radicals cause ozone depletion.

C6f Hardness of water

Hard and soft water

Our tap water is not pure. It contains dissolved substances. Our tap water comes from aquifers, reservoirs and rivers. The dissolved substances come from minerals that are in the area where the water was collected. Since these minerals vary around the UK (and around the world), tap water is different in different places.

In some areas, the minerals that dissolve in the water make it difficult for the water to form a lather with soap. Water like this is called **hard water**. In particular, minerals that contain calcium ions and magnesium ions result in hard water if they dissolve.

In other areas, particularly where the minerals are quite insoluble and the water is relatively pure, the water forms a lather with soap much more easily. Water like this is called **soft water**.

Figure 6f.1 Kielder Water reservoir supplies drinking water to the north-east of England. Minerals from the rocks and stones on the bottom of the reservoir dissolve very slowly in the water.

SAQ

1. What is meant by the terms *hard water* and *soft water*?
2. Use *Gateway Additional Science*, pages 208 and 209, to explain what is meant by an *aquifer* and a *reservoir*.

Identifying hard and soft water

Figure 6f.2 shows how you can test a sample of water to see if it is hard or soft.

If you add five drops of soap solution to some distilled water in a test tube, put in a stopper and shake the tube, you will see that:
- a lather forms immediately
- the lather stays on top of the water for quite a long time before it disappears (Figure 6f.2c).

These two observations show that distilled water is soft. It lathers easily with very little soap solution. The lather is persistent.

If you add five drops of soap solution to some hard water in a test tube, put in a stopper and shake the tube, you will see that:
- no lather is formed
- small specks of white solid can be seen floating near the water surface – this white solid is called **scum** (Figure 6f.2d).

These two observations show that the water sample is hard. It does not lather easily with soap solution. It forms scum with soap solution.

Figure 6f.2a Adding soap solution to a water sample.

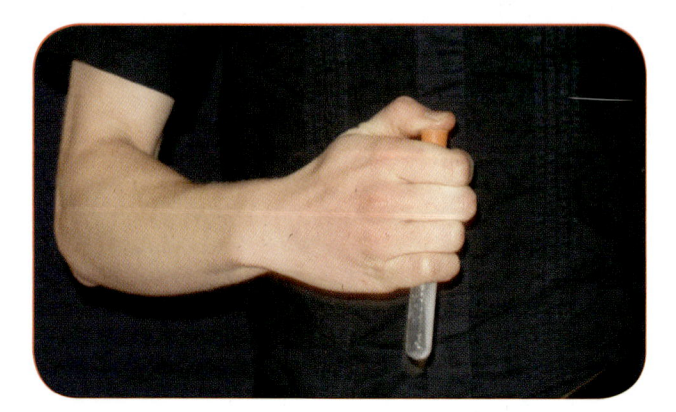

Figure 6f.2b Shaking the water and soap solution.

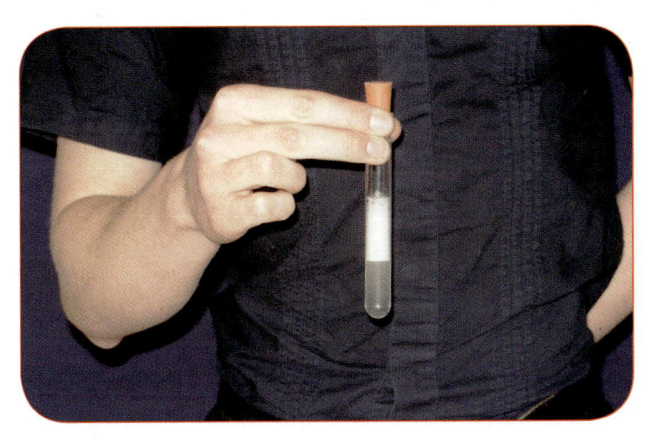

Figure 6f.2c A persistent lather forms – this sample contained soft water.

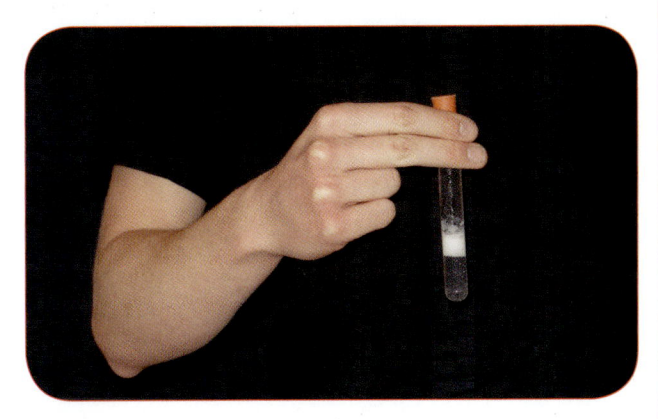

Figure 6f.2d No lather forms, and some white scum can be seen – this sample contained hard water.

SAQ

3 **a** Describe a simple experiment you could do to show the difference between a sample of hard water and a sample of soft water.

   b What observations would you expect to see in each case?

Measuring hardness

The amount of hardness in different water samples can be compared using soap solution.

- A measured volume of each water sample is put into separate test tubes.
- Two drops of soap solution are added to each water sample. The tubes are stoppered and shaken. Observations are noted.
- Two more drops of soap solution are added to each water sample. The tubes are stoppered and shaken. Observations are noted.
- The experiment is continued until a persistent lather has formed in each tube.

The harder the water is, the more soap has to be added to form a persistent lather. More scum will be seen with harder water.

The softer the water is, the less soap has to be added to form a persistent lather. Less scum will be seen with softer water.

Soapless detergents

An experiment that finds out whether a water sample is hard or soft has to be done with soap. Soap is made when a natural oil is reacted with an alkali (see Item C6d, *Chemistry of sodium chloride*, page 185). Soap has been used for many hundreds of years. More recently substances called soapless detergents have been developed and manufactured. Soapless detergents will lather well with soft water *and* with hard water (Figure 6f.3).

Figure 6f.3 The foaming in this river has been caused by pollution with a soapless detergent. The river water might be quite hard, but with a soapless detergent the lather forms anyway.

Permanent hardness

Some types of hard water become soft if you boil the water. This is called **temporary hardness**.

Other types of hard water remain hard when you boil the water. This is **permanent hardness**. It can be due to a number of different dissolved substances – dissolved calcium sulfate is a common cause of permanent hardness. Gypsum is a mineral that contains calcium sulfate.

Remember:

- temporary hardness is *removed by boiling*
- permanent hardness is *not removed by boiling*.

Removing permanent hardness

People prefer soft water to hard water. It is easier to wash yourself or your clothes if your water supply is soft, and you need less soap. It is possible to change hard water so it becomes soft. Removing hardness from water is known as **softening** the water.

Although calcium sulfate causes what we call permanent hardness, it is still possible to remove the hardness from such water supplies. The formula of calcium sulfate is $Ca^{2+}SO_4^{2-}$. In order to remove the hardness, the Ca^{2+} ions must be removed from the solution.

SAQ

4 **a** What is the difference between *temporary hardness* and *permanent hardness*?

   b Name one compound that causes permanent hardness of water supplies.

Using an ion-exchange resin

One method of softening hard water involves passing the water through a cartridge containing an **ion-exchange resin**. The resin has many sodium ions (Na^+ ions) bound onto it. As the water flows through the cartridge, calcium ions from the water bind onto the resin, while sodium ions from the resin are released into the water. The resin *exchanges* the calcium ions that caused the hardness for sodium ions. The sodium ions that are now in the water don't cause hardness.

Ion-exchange resins are a good way of softening the water used in bathrooms, but they should not be used in kitchens. We get our drinking water, and water for cooking, from our kitchen taps. Sodium ions in our drinking water are not good for us. By contrast, calcium ions in our drinking water *are* good for us – they help us to form healthy teeth and bones.

Washing soda

Washing soda is another name for sodium carbonate. If it is added to hard water, washing soda softens the water. It does this by reacting with the calcium ions in the hard water to form an insoluble precipitate of calcium carbonate. Because the calcium ions are now part of a solid precipitate, they are no longer dissolved in the water. This means the water is now soft. The word equation for the reaction between calcium sulfate and sodium carbonate is:

calcium sulfate + sodium carbonate
$\rightarrow$ calcium carbonate + sodium sulfate

As the name suggests, washing soda is a good way of softening water used for washing. It is often one of the ingredients in washing powder. With a bit of colouring and perfume added it is also sold as 'bath salts' (Figure 6f.4).

Figure 6f.4 These bath salts are mostly sodium carbonate. They act as a water softener by removing dissolved calcium ions.

SAQ

5 Name two ways permanent hard water can be softened. Explain how each method works.

Temporary hardness

In some areas, the water supply is hard, but it becomes softer if the water is boiled. This water is called temporary hard water, because the hardness is removed if the water is boiled. Temporary hard water contains dissolved calcium hydrogencarbonate.

Removing temporary hardness

When temporary hard water is boiled, the dissolved calcium hydrogencarbonate decomposes. It breaks down to form carbon dioxide, water and a precipitate of calcium carbonate. The water is no longer hard because it no longer contains dissolved calcium ions. A word equation for the chemical reaction that takes place is:

calcium hydrogencarbonate
→ carbon dioxide + water + calcium carbonate

SAQ

6 Three water samples – distilled water, sample A and sample B – were tested for hardness with soap solution. The results are shown in Table 6f.1.

Sample	Drops of soap solution needed for permanent lather
distilled water	2
A	14
boiled A	4
A + sodium carbonate	2
B	24
boiled B	22
B + sodium carbonate	2

Table 6f.1

  a Why does distilled water lather so easily?
  b Is one of the water samples temporary hard water? Explain your answer.
  c Is one of the water samples permanent hard water? Explain your answer.
  d Why does adding sodium carbonate make A and B softer?

H Calcium hydrogencarbonate is an ionic compound consisting of calcium ions and hydrogencarbonate ions. Each calcium ion has a 2+ charge. Each hydrogencarbonate ion has a 1− charge. The formula of calcium hydrogencarbonate is therefore $Ca(HCO_3)_2$. A balanced symbol equation for the chemical reaction that takes place when temporary hard water is boiled is:

$Ca(HCO_3)_2 \rightarrow CO_2 + H_2O + CaCO_3$

Limescale

In an area with temporary hard water, the water supply contains dissolved calcium hydrogencarbonate. When water is heated in a kettle or in a central heating boiler, the calcium hydrogencarbonate decomposes, forming solid calcium carbonate. This solid builds up in the kettle or boiler (Figure 6f.5). You might also see it as an unsightly deposit on showerheads and hot taps. It is called **limescale**.

Figure 6f.5 Limescale has built up on this metal pipe, which makes it much less efficient at its job.

Removing limescale

You can buy products to remove limescale. They contain weak acids. When they are put on the limescale, the weak acids react with it to make carbon dioxide, water and a soluble salt. The salt that forms can be easily washed away or wiped away.

Limescale removers usually contain *weak* acids, not strong acids. Strong acids would remove limescale effectively but they would also corrode metal parts and be more dangerous to use.

H A balanced symbol equation for the reaction between hydrochloric acid and calcium carbonate is:

$CaCO_3 + 2HCl \rightarrow CaCl_2 + H_2O + CO_2$

Hydrochloric acid is a strong acid, but most weak acids have more complicated formulae, so that the equations become more difficult to write and to interpret. Hydrochloric acid has been used here for simplicity.

SAQ

7   a   How does limescale form?
    b   What is the chemical name for limescale?
    c   How do limescale removers work?

How water becomes hard

Permanent hardness of water is caused by calcium sulfate from minerals dissolving in water. Temporary hardness is caused by a chemical reaction. This reaction is between rainwater, containing dissolved carbon dioxide, and calcium carbonate. Many minerals contain calcium carbonate. Limestone, chalk and marble are three of the best known. The carbon dioxide dissolved in the rainwater makes it slightly acidic, so it will react with the calcium carbonate. Temporary hard water is formed because the product of this reaction is calcium hydrogencarbonate. A word equation for the reaction is:

calcium carbonate + carbon dioxide + water
$\rightarrow$ calcium hydrogencarbonate

H A balanced symbol equation for the reaction is:

$CaCO_3 + CO_2 + H_2O \rightarrow Ca(HCO_3)_2$

SAQ

8   a   Name three minerals containing calcium carbonate.
    b   What property of carbon dioxide solution makes it able to react with calcium carbonate?
    c   What is the product of the reaction in **b**?
    d   Write a word equation for the reaction in **b**.

White Scar Cave

Underneath Ingleborough in the Yorkshire Dales there is grit, shale and limestone. Limestone is one form of calcium carbonate. During a period of over 100 000 years, the Yorkshire rain, including dissolved carbon dioxide, has dripped down through cracks in these rocks. Where the water flowed over and through the limestone, the rock dissolved. Very slowly, caves were formed.

This cave system was discovered in 1923, and is now known as White Scar Cave. Walking through the tunnels in the rock, it is amazing to think that it is all the product of a very slow chemical reaction. The weakly acidic carbon dioxide solution has dissolved the calcium carbonate, by a tiny amount every year.

The reverse of the chemical reaction that created White Scar Cave has also decorated it. This has produced some amazing sights. When the calcium carbonate reacted with the water and carbon dioxide, it formed calcium hydrogencarbonate solution. Elsewhere in the cave this reaction reversed, producing calcium carbonate again. This 'new' calcium carbonate has resulted in stalactites, stalagmites and other beautifully shaped deposits (Figure 6f.6).

You can read more about White Scar Cave, and the chemistry that made it, at www.whitescarcave.co.uk. (In your reading, you might come across the term 'calcium bicarbonate' – this is the old name for calcium hydrogencarbonate.) White Scar caves are well worth a visit, but do mind your head.

Figure 6f.6 A curtained stalactite at White Scar Cave.

Summary

You should be able to:

- explain what is meant by hard and soft water
- describe an experiment to measure the hardness of a water sample
- state that both hard and soft water lather well with soapless detergents
- explain what is meant by temporary and permanent hardness
- describe how permanent hardness can be removed with an ion-exchange resin or with washing soda
- describe how temporary hardness can be removed by boiling
- state that limescale is calcium carbonate, explaining how it forms and how it can be removed
- **H** construct balanced symbol equations for the formation of limescale when calcium hydrogencarbonate decomposes, and for the reaction of limescale with an acid
- explain how temporary and permanent hard water are formed
- state the word equation for the formation of calcium hydrogencarbonate from calcium carbonate, water and carbon dioxide
- **H** write a balanced symbol equation for the reaction between calcium carbonate, water and carbon dioxide

Questions

1. You are given a sample of permanent hard water and a sample of temporary hard water.

   a Describe how you could decide which was permanent and which was temporary by a series of simple experiments.

   b Name a compound which might be causing the hardness in each case.

2. a Explain how temporary and permanent hard water are formed.

   b Write a word equation for the reaction involved in the formation of temporary hard water.

H 3. a Write a balanced symbol equation for the chemical reaction that takes place when temporary hard water is heated. Include state symbols.

   b Is this chemical reaction reversible? Give a reason for your answer.

continued on next page

200 C6f Hardness of water

Questions - continued

4. Four water samples – A, B, C and D – were tested for hardness. A volume of 5 cm³ of each water sample was put into one of four separate test tubes. Soap solution was added two drops at a time. The tubes were stoppered and shaken. The observations are shown in Table 6f.2.

Drops of soap solution added	A	B	C	D
2	no lather forms, specks of scum visible	permanent lather forms, no scum seen	no lather forms, specks of scum visible	no lather forms, specks of scum visible
4	no lather forms, specks of scum visible	permanent lather forms, no scum seen	no lather forms, specks of scum visible	no lather forms, specks of scum visible
6	permanent lather forms, specks of scum visible	permanent lather forms, no scum seen	no lather forms, specks of scum visible	no lather forms, specks of scum visible
8	permanent lather forms, specks of scum visible	permanent lather forms, no scum seen	no lather forms, many specks of scum visible	no lather forms, many specks of scum visible
10	permanent lather forms, specks of scum visible	permanent lather forms, no scum seen	no lather forms, many specks of scum visible	no lather forms, many specks of scum visible
12	permanent lather forms, specks of scum visible	permanent lather forms, no scum seen	no lather forms, many specks of scum visible	permanent lather forms, many specks of scum visible
14	permanent lather forms, specks of scum visible	permanent lather forms, no scum seen	no lather forms, heavy scum visible	permanent lather forms, many specks of scum visible
16	permanent lather forms, specks of scum visible	permanent lather forms, no scum seen	no lather forms, heavy scum visible	permanent lather forms, many specks of scum visible

Table 6f.2

a. How was fair testing ensured?

b. Rank the four water samples from hardest to softest. Explain your decision.

c. The four water samples were:
- tap water from an area of the UK where the water is comparatively soft
- tap water from an area of the UK where the water is comparatively hard
- distilled water
- calcium sulfate solution (very hard)

Decide which of these water samples was sample A, B, C and D. Explain your decision.

6g Natural fats and oils

Fats and oils are esters

Fats and oils are naturally occurring compounds of carbon, hydrogen and oxygen. Fats are solid at room temperature, while oils are liquid at room temperature. Fats and oils have molecules with very similar structures. An example of one of these molecules is drawn in Figure 6g.1.

Look at Figure 6g.1. The coloured backgrounds are there to help you to identify three sections of the molecule.

- The part of the molecule with a green background has three carbon atoms. This part is called a glycerol residue.
- The part of the molecule with a blue background consists of three long chains of carbon atoms with hydrogen atoms attached. These are called hydrocarbon chains.
- The hydrocarbon chains are joined to the glycerol residue by linkages. These three linkages have a yellow background in the diagram. Each linkage is made of one carbon atom and two oxygen atoms. The linkages are called **ester linkages**.

Fats and oils are classified as **esters**. Esters are compounds with molecules that have two or more distinct sections, held together by ester linkages.

Figure 6g.2 Butter is an animal fat, made from cow's milk. Butter is an ester. It is called a fat because it is solid at room temperature.

Figure 6g.3 Sunflower oil is a vegetable oil, made from sunflower seeds. Sunflower oil is an ester. It is called an oil because it is liquid at room temperature.

SAQ

1. a What is the difference between a fat and an oil?
   b Give one example of an animal fat.
   c Give one example of a vegetable oil.
2. a What is meant by an *ester linkage*?
   b Why are fats and oils classified as esters?

Saturation and unsaturation

A typical fat or oil has three long hydrocarbon chains (Figure 6g.1). Each of these is a chain of carbon atoms with hydrogen atoms bonded to it. If all the carbon atoms in each chain are linked by single bonds, as in Figure 6g.1, then we say the

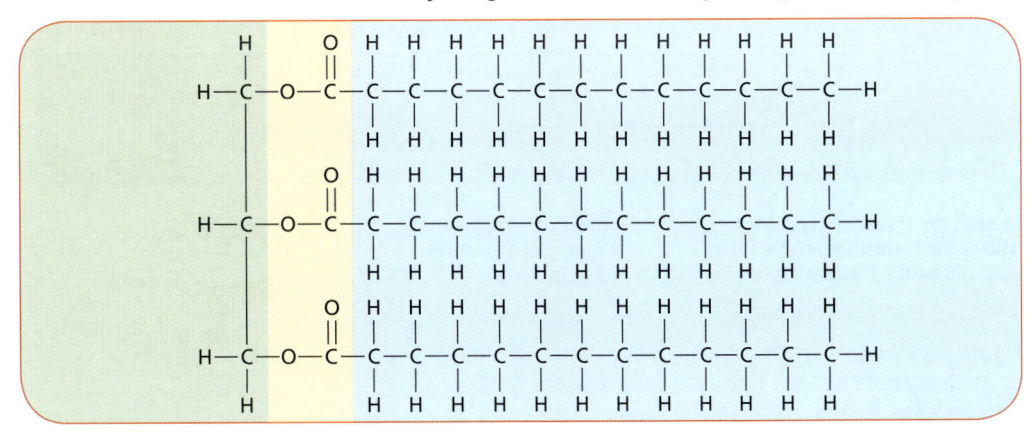

Figure 6g.1 One molecule of a typical fat.

Figure 6g.4 An unsaturated fat molecule.

fat or oil is **saturated**. If at least one of the chains has two carbon atoms linked by a double bond, as in Figure 6g.4, then the fat or oil is **unsaturated**.

Testing a fat with bromine water

Bromine water is an orange colour. It can be used to find out whether a fat or oil is saturated or unsaturated. A little bromine water should be added to some of the fat or oil, and the two should then be mashed or shaken together. If the fat or oil is unsaturated, the bromine water will lose its colour. If it is saturated, the orange colour will remain.

SAQ

3 a What is meant by an *unsaturated fat*?

   b What is meant by a *saturated fat*?

   c How can you tell a saturated fat from an unsaturated fat?

How the bromine test works

If bromine water is shaken with an unsaturated fat or oil, it reacts with the carbon–carbon double bonds. The bromine water has an orange colour, but the products of this reaction are colourless (Figure 6g.5). This is why bromine water goes colourless when it is shaken with an unsaturated fat or oil.

Making margarine

If there are several double bonds in the hydrocarbon chains, then the substance will be a liquid – an oil. If there are few double bonds in the hydrocarbon chains, or none at all, then the substance will be a solid – a fat.

An oil can be changed into a fat if the oil is reacted with hydrogen in the presence of a nickel catalyst. The hydrogen reacts with the double bonds so that only single bonds are left. This makes the oil more saturated, so that it becomes a fat.

This reaction is used to make margarine. A vegetable oil with many double bonds is reacted with hydrogen, producing a solid fat with fewer double bonds. This solid fat is called margarine.

a bromine molecule approaches the carbon–carbon double bond in a molecule of unsaturated fat

a reaction occurs and this product forms – it is colourless

Figure 6g.5 Bromine reacts with the carbon–carbon double bonds in unsaturated fats. If the fat is saturated, this reaction can't happen, and the bromine water stays orange.

Figure 6g.6 If this vegetable oil is reacted with hydrogen in the presence of a nickel catalyst, a solid fat like this margarine can be produced.

Is margarine healthier than butter?

Our livers use saturated fat to make a substance called cholesterol. Cholesterol is necessary for healthy body function (see *Gateway Additional Science*, page 22). Unfortunately, cholesterol can also form bulges in artery walls, called plaques (Figure 6g.7). These plaques can help to cause heart disease, or a stroke. Eating saturated fats and oils has therefore been identified as 'unhealthy'.

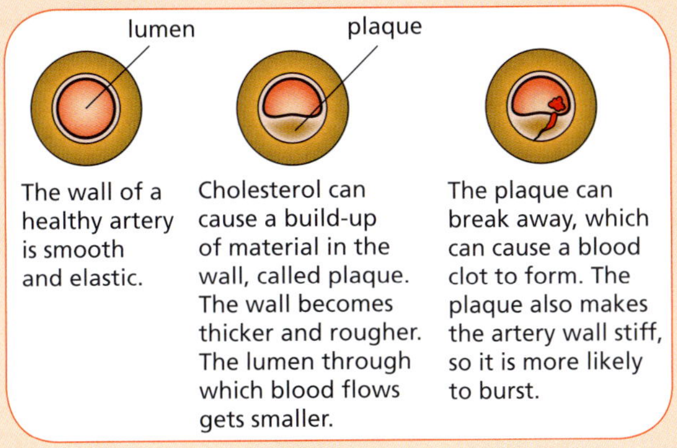

Figure 6g.7 Too much cholesterol can lead to unhealthy arteries.

- The wall of a healthy artery is smooth and elastic.
- Cholesterol can cause a build-up of material in the wall, called plaque. The wall becomes thicker and rougher. The lumen through which blood flows gets smaller.
- The plaque can break away, which can cause a blood clot to form. The plaque also makes the artery wall stiff, so it is more likely to burst.

Animal fats and oils are more often saturated, while vegetable fats and oils are more often unsaturated. A high intake of animal fats and oils has therefore become linked with poor health. Some people are reducing their intake of animal fat. They do this by eating margarine rather than butter, and by trimming the fat off meat before eating it. Of course, vegetarians eat very little animal fat, and vegans eat no animal fat – often for ethical as well as health reasons.

Since vegetable fats and oils are more often unsaturated, they have become identified as healthier. Current scientific evidence suggests, though, that it is not as clear cut as this. Firstly, the process of hydrogenating vegetable oils to make margarine can also produce other chemicals. Some of these can be bad for us. Secondly, the human liver makes cholesterol from other starting materials as well as saturated fats. The amount of cholesterol in our blood is not simply due to the saturated fat we eat.

SAQ

4 a Which has one or more carbon–carbon double bonds, a saturated fat or an unsaturated fat?
   b How can the number of double bonds in a fat be reduced?

Fats and oils in industry

Fats and oils are a part of everybody's diet, but they are also important raw materials in the chemical industry.

Vegetable oils can be used to make bio-diesel. Diesel-engined cars and lorries can run on bio-diesel. Unlike diesel made from crude oil, bio-diesel is a renewable fuel.

Fats and oils are also used to make soap. Vegetable oils in particular are used as starting materials by soap manufacturers. This is reflected in the names of some soaps. 'Palmolive' soap is made using oils from palm trees and olives as two of its ingredients.

Making soap

Soap is made by boiling a fat or oil with sodium hydroxide solution (see Item C6d, *Chemistry of sodium chloride*, page 185). A chemical reaction takes place. In this reaction each oil molecule splits apart, producing four product molecules. One of these product molecules is called **glycerol**. The other three molecules are all molecules of soap. This process is called **saponification**. Saponification is illustrated in Figure 6g.8 on page 204.

SAQ

5 a Describe how soap is made.
   b What is the molecular formula of glycerol?
   c Use Figure 6g.8 to write a word equation for saponification.

Figure 6g.8 Saponification.

Saponification

When a fat or oil is saponified, the ester linkages in it are broken. They break when they react with sodium hydroxide. Water will also react with ester linkages, breaking them in a similar fashion. This type of reaction is called a **hydrolysis** reaction (*hydro* means water, *lysis* means breaking). Because saponification involves a similar type of reaction, it is also described as a hydrolysis reaction.

Emulsions

A fat or oil will not normally mix with, or dissolve in, water. However, a fat or oil can form an **emulsion** with water. In an emulsion, one liquid is finely dispersed in the other. If you could look at an emulsion under a microscope, you would see tiny droplets of the substance that is present in the smaller amount.

Simply putting some oil and some water in a bottle and shaking it vigorously can form an emulsion. This emulsion will separate out, but it is possible to stabilise an emulsion so it does not do this. Salad cream is an example of a stable emulsion. It is made from a vegetable oil and vinegar, which is mostly water. A substance called an **emulsifier** enables the oil and water to mix, stabilising the emulsion.

Oil-in-water emulsions

Typically, whole milk is around 4% fats and oils, 88% water and 8% other components such as carbohydrates, proteins and minerals. Semi-skimmed milk contains less than 2% fats and oils, and skimmed milk contains only 0.1%. These fats and oils are dispersed in the water as tiny droplets, which can be seen under a microscope. Milk is an oil-in-water emulsion.

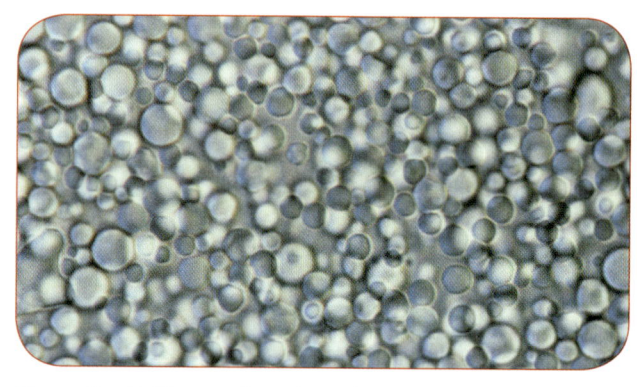

Figure 6g.9 Milk seen through a microscope. The circular shapes are the droplets of fat and oil.

Water-in-oil emulsions

Butter consists of over 80% fat. The rest is mostly water. The water forms tiny droplets that are dispersed in the fat. Butter is a water-in-oil emulsion.

Summary

You should be able to:

◆ state examples of animal and vegetable fats and oils, distinguishing between fats as solids and oils as liquids

continued on next page

Summary - *continued*

- state that animal and vegetable fats and oils are esters
- describe a saturated fat and an unsaturated fat, and a method of distinguishing one from the other
- **H** state that vegetable fats and oils are more likely to be unsaturated than animal fats and oils, and explain why this makes them healthier
- describe how soap is made from vegetable oils, using a word equation and the term *saponification*
- **H** explain that saponification is a hydrolysis reaction
- state the importance of natural fats and oils to the chemical industry, including the importance of bio-diesel
- describe the formation of an emulsion when oil and water are shaken together
- describe milk as an oil-in-water emulsion and butter as a water-in-oil emulsion

Questions

H 1 Read the box called *Is margarine healthier than butter?* (page 203) and answer the following questions.

   a Describe the changes that may be caused to a healthy artery by high levels of cholesterol in the blood.

   b What diseases are associated with these changes?

   c Why are vegetable fats and oils thought to be better for us than animal fats and oils?

2 Soap is made by a saponification reaction between a fat or oil and sodium hydroxide. One soap formula unit consists of one positive ion and one negative ion.

   a What happens to the fat or oil molecule in a *saponification* reaction?

   b What is the positive ion in the soap in Figure 6g.8?

   c What is the negative ion in the soap in Figure 6g.8?

3 A fat or oil can be *saturated* or *unsaturated*.

   a Explain the meanings of these two terms.

   b Describe a simple experiment to distinguish between a saturated oil and an unsaturated oil.

   c Which is more likely to be unsaturated, a vegetable fat or a vegetable oil?

4 a What is meant by an *emulsion*?

   b Explain why milk and butter are both described as emulsions.

   c How do milk and butter differ as emulsions?

C6h Analgesics

Drugs

Three important chemicals produced in large amounts every year are **aspirin**, **ibuprofen**, and **paracetamol**. These chemicals are all **drugs**.

A drug is a substance that is externally administered to the body, and which affects the way the body works in some way. *Externally administered* means we take drugs into our bodies by swallowing, injecting or inhaling them, or in some other manner. Once a drug is inside the body, it modifies or affects one or more chemical reactions in the body.

SAQ

1. **a** What is meant by a drug?
   b Give three ways of taking a drug.
   c What does a drug do to the human body?

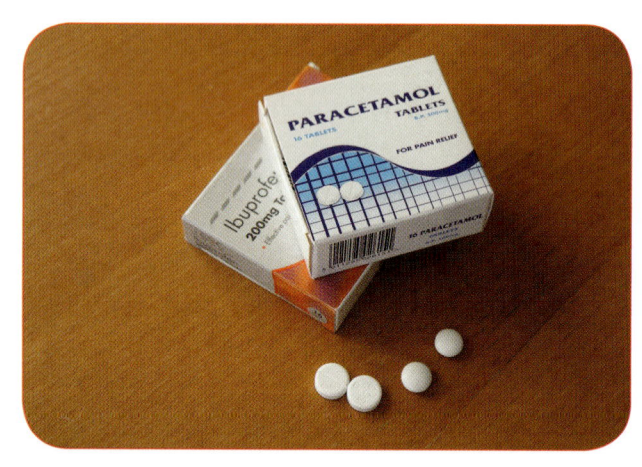

Figure 6h.1 Paracetamol and ibuprofen are drugs that are sold as tablets, which are easy to swallow.

Analgesics reduce pain

Aspirin, ibuprofen and paracetamol are **analgesics**, or painkillers. This means they are drugs that reduce pain. To take them, we usually swallow them in tablet form. They are transported around the body in the bloodstream.

Drugs such as analgesics, that benefit the person taking them, are also known as **medicines**. Medicines can be obtained in a number of places. If you visit your doctor, he or she may decide you need medicine. The doctor gives you a slip of paper called a prescription, which has the name of the medicine written on it, as well as the amount of medicine you need to take each day. If you take this to a chemist's shop, or to a hospital, a trained worker called a **pharmacist** gives you the medicine. Pharmacists are highly trained, so that they know and understand the medicines they supply.

Figure 6h.2 A pharmacist at work.

SAQ

2. What is a pharmacist?

Analgesics and safety

You do not need a prescription to buy aspirin, ibuprofen or paracetamol. They can be bought at a chemist's shop, or at many other shops including newsagents and supermarkets. Although they are widely available, we must not assume they are 'safe'. They would be unsafe if they were not made properly, or if they were not used correctly.

These drugs are made by the chemical industry. They use extremely pure, high quality starting materials. The processes they use to make the drugs are carefully and precisely controlled. In this way, they produce high-quality drugs that are safe, if used correctly.

They are often sold in restricted amounts, such as a maximum of two to four packets, and are not sold to people under 16 years old. This is because, although these medicines can have a beneficial effect, taking too much can be dangerous. This is known as taking an **overdose**. An overdose of paracetamol, for example, can kill.

SAQ

3 a Give two ways the manufacturers of analgesics maintain the quality of the product.
 b Give two ways the suppliers of analgesics can help to ensure safety.
 c What is an overdose?

Analgesic molecules

Aspirin, ibuprofen and paracetamol are covalent compounds. Aspirin and ibuprofen are compounds of carbon, hydrogen and oxygen. Paracetamol is a compound of carbon, hydrogen, oxygen and nitrogen. They all have simple molecular structures. Their structural formulae are shown in Figure 6h.3.

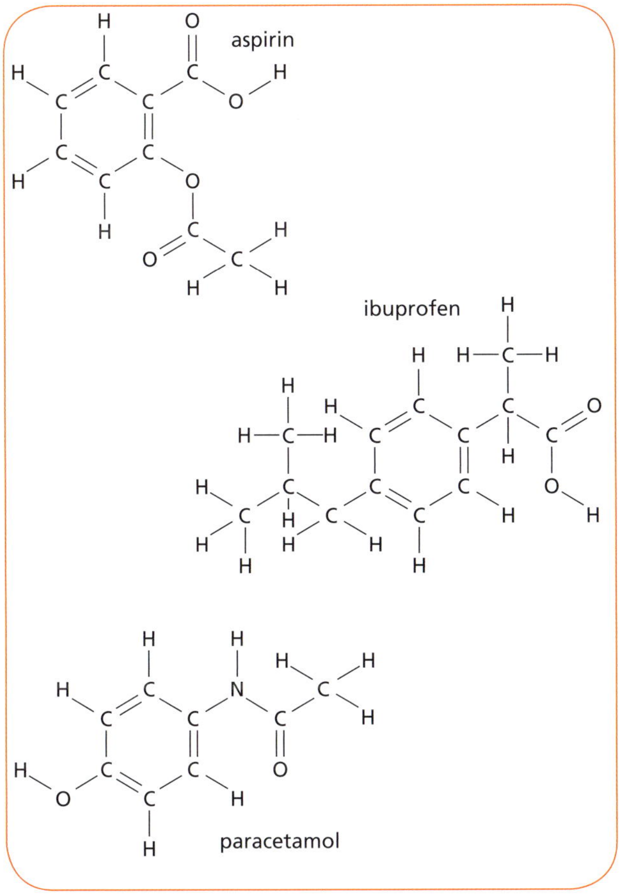

Figure 6h.3 An aspirin molecule, an ibuprofen molecule and a paracetamol molecule.

The aspirin molecule shown in Figure 6h.3 consists of nine carbon atoms, eight hydrogen atoms, and four oxygen atoms. The molecular formula of aspirin is $C_9H_8O_4$.

SAQ

4 a What is the molecular formula of ibuprofen?
 b What is the molecular formula of paracetamol?

Structure and function

Figure 6h.3 shows the structural formulae of aspirin, ibuprofen and paracetamol. These three structures have some similarities and some differences. The structures of aspirin and ibuprofen are the most similar. Both have a COOH group of atoms, and a group of six carbon atoms, which are shown held together by three single bonds and three double bonds. This group of six carbon atoms is called a **benzene ring**.

Aspirin and ibuprofen both block pain in the same way. They are able to do this because of their similar structures.

A paracetamol molecule also has a benzene ring, but that is its only major similarity with aspirin and ibuprofen molecules. Structurally, a paracetamol molecule is different in many ways from an aspirin or an ibuprofen molecule. Because of these structural differences, paracetamol is a different type of drug from aspirin and ibuprofen. It is still an analgesic, but it blocks pain in a different way.

SAQ

5 Copy the structures of aspirin, ibuprofen and paracetamol from Figure 6h.3. Label the following:
 a three benzene rings
 b two COOH groups
 c a feature of the paracetamol molecule that is not found in either of the other molecules.

Aspirin

Aspirin has been used as a painkiller for a long time. It is found naturally in willow bark, but is now manufactured synthetically by the chemical industry. It has other beneficial effects as well as reducing pain. It lowers the body temperature rapidly, thus reducing fever. It 'thins' the blood, making it less likely to clot. This reduces the risk of heart disease or a stroke.

208 C6h Analgesics

Figure 6h.4 Salicylic acid and ethanoic anhydride react to make aspirin and ethanoic acid.

salicylic acid + ethanoic anhydride → aspirin + ethanoic acid

Manufacturing aspirin

Aspirin can be made from an acid called salicylic acid. If this acid is made to react with a chemical called ethanoic anhydride, aspirin is formed (Figure 6h.4).

SAQ

6 What is the molecular formula of ethanoic anhydride?

Soluble aspirin

Aspirin itself is not very soluble in water. It can be converted into a form known as **soluble aspirin**. In this form, it dissolves in water. Soluble aspirin can therefore be dissolved and drunk, rather than swallowed as a solid tablet. This is better for people with sore or swollen throats. The soluble aspirin also dissolves quickly in the blood. It is then transported swiftly around the body, causing more rapid pain relief.

SAQ

7 a Give three beneficial effects of swallowing aspirin.
  b What does *soluble* mean?
  c Give two advantages of soluble aspirin.

Soluble aspirin is produced by reacting aspirin with sodium hydroxide (Figure 6h.6). The soluble aspirin produced in this reaction is an ionic salt. Ionic compounds are often more soluble in water than covalent compounds are. The electric charges on the positive and negative ions in the salt make it more soluble in water than aspirin is.

SAQ

8 Why is soluble aspirin more soluble than ordinary aspirin?

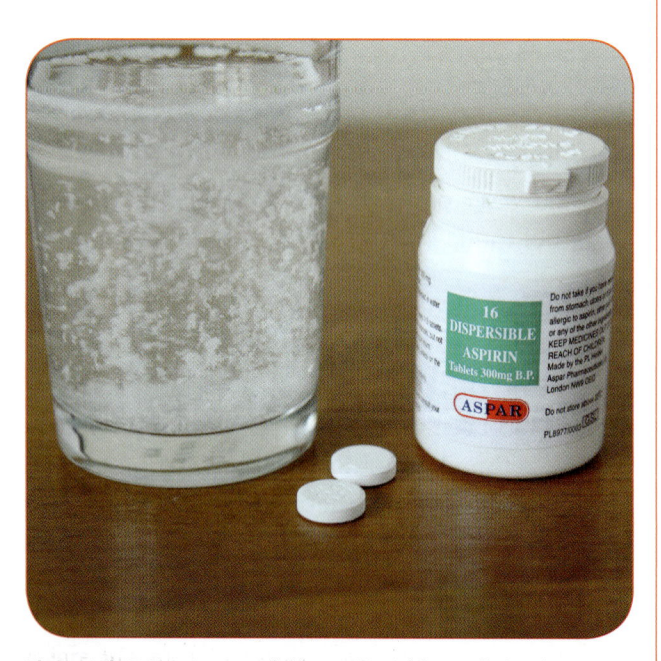

Figure 6h.5 These two soluble aspirin tablets will dissolve in water to produce this pain-relieving solution.

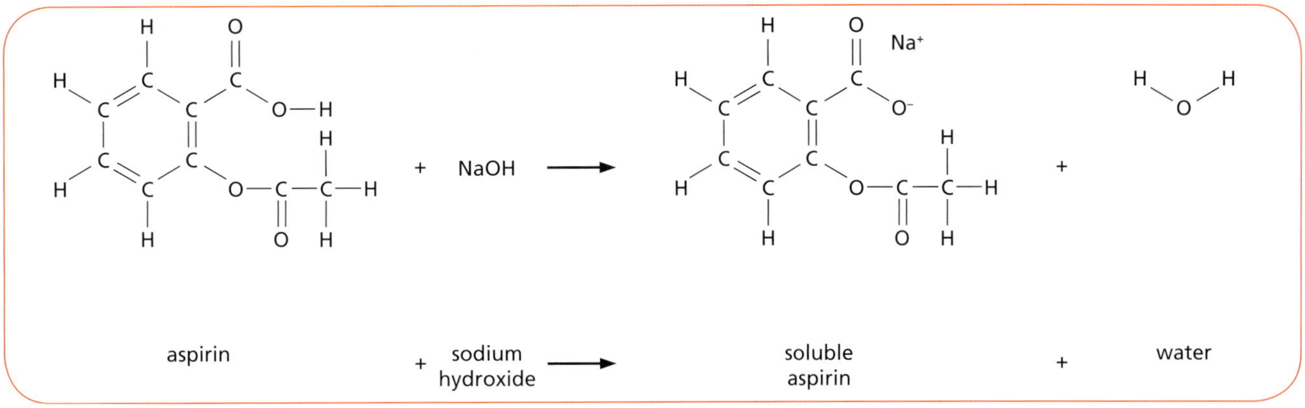

Figure 6h.6 Aspirin and sodium hydroxide react to make soluble aspirin and water.

Side-effects

Analgesics have side-effects. Aspirin causes an upset stomach in 6% of users, and can also cause an allergic reaction. It should not be given to children under sixteen. Because aspirin thins the blood, it has been known to cause bleeding within the stomach. This particular side-effect, blood thinning, has also been responsible for a new beneficial use of this drug. A person who is diagnosed as having an increased risk of a heart attack often takes a small amount of aspirin every day. This may be as little as half a tablet, but it helps to prevent the blood-clotting that puts them at risk.

In general, paracetamol has fewer side-effects than aspirin. Because of this, it is safer to use with small children. The flavoured painkilling liquids administered to sick children contain paracetamol. Unfortunately, paracetamol is very toxic – much more so than other analgesics – if an overdose is taken. This is because an overdose of paracetamol will cause liver damage. Swallowing too many paracetamol tablets can be fatal – the warnings on the packets are deadly serious.

Ibuprofen has not been around for as long as aspirin or paracetamol, and in many ways it is the best of the three analgesics. It has a very low toxicity, much less than paracetamol. It causes side-effects similar to those caused by aspirin. This is hardly surprising, since they are quite similar drugs. Ibuprofen, however, is only half as likely as aspirin to cause any particular side-effect.

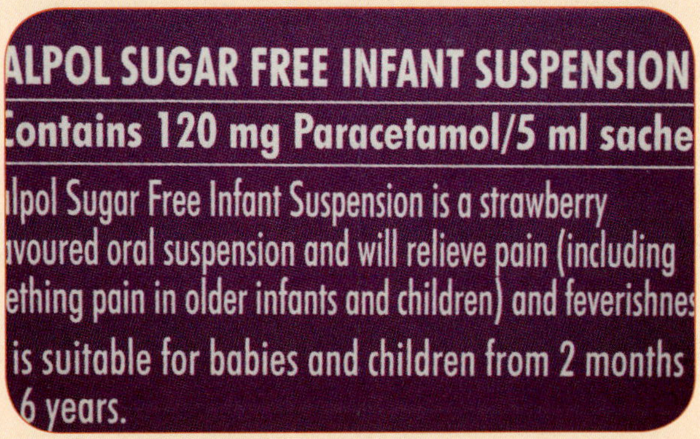

Figure 6h.7 Paracetamol suspension is given to children to relieve pain or fever.

Summary

You should be able to:

- state that analgesics are painkillers and name three common painkillers

continued on next page

Summary - continued

- describe what we mean by a drug
- state that most medicines are sold in a shop where there is a trained pharmacist
- explain that the chemicals used to make analgesics must be very pure
- use a displayed formula to work out a molecular formula
- **H** find similarities and differences in the displayed formulae of aspirin, ibuprofen and paracetamol
- state that aspirin was first discovered in willow bark but is now manufactured synthetically
- describe the beneficial effects of aspirin
- describe dangers linked to aspirin and paracetamol, particularly when an overdose is taken
- describe the advantages of soluble aspirin
- **H** describe how aspirin is manufactured from salicylic acid
- explain the solubility of soluble aspirin with reference to its structural formula

Questions

1. **a** What is an *analgesic*?

   b Name three analgesics, and give their molecular formulae.

2. Read the box called *Side-effects* (page 209).

   a What side-effects does aspirin have?

   b How do aspirin and paracetamol compare for side-effects?

   c How do aspirin and ibuprofen compare for side-effects?

3. Analgesics are *synthetic* materials, made by *batch processes*.

   a Explain the meanings of the two terms in italics (if you need help, use the glossary of *Gateway Additional Science*).

   b Why are medicines commonly made by batch processes?

4. **H** **a** How is aspirin made from salicylic acid?

   b How is soluble aspirin made from aspirin?

   c How do the molecular formulae of aspirin and soluble aspirin differ?

   d In what ways is soluble aspirin a better medicine than aspirin?

5a Satellites, gravity and circular motion

Space tourist

To see the Earth from space must be a marvellous thing. The first astronauts, who went into space in the early 1960s, were in orbits less than 200 km above the Earth's surface. This meant that they could only see a part of the Earth. However, when astronauts reached the Moon, they could look back and see the whole of the Earth, a light blue ball in the darkness of space, flecked with white clouds, and with the brown and green shapes of the continental landmasses (Figure 5a.1).

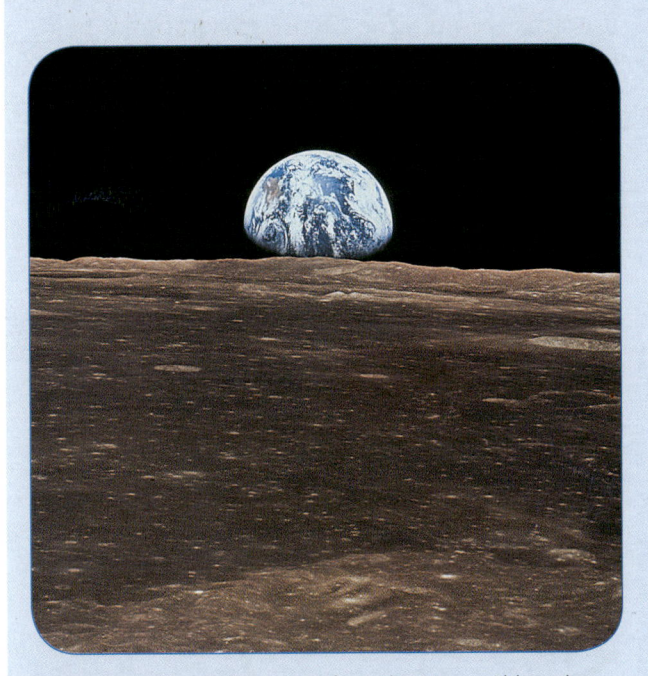

Figure 5a.1 The Earth, as seen from the Moon. Although such images are familiar today, they had a dramatic impact on people's view of the Earth when they were first published. This picture was taken from *Apollo 11* by NASA astronauts who visited the Moon in 1969.

Today, if you have the money, you can become a space tourist. In 2006, Anousheh Ansari spent two weeks in the International Space Station, in orbit around the Earth (Figure 5a.2). She had previously undergone rigorous physical training and was required to carry out experiments in the weightless conditions she experienced in the space station. At the end of her stay, she said, 'I am drowning in the sadness of my departure.'

Would space be a desirable holiday destination? In the 1990s, the first tickets were sold to would-be space holidaymakers. For tens of thousands of pounds, they were promised a trip in a shuttle and a stay in a space hotel in low Earth orbit. Although the first flight was fully booked, the tickets showed no departure date.

Figure 5a.2 Anousheh Ansari, an Iranian, became the first woman space tourist in 2006.

Satellites

Sometimes the term *satellite* is used to describe any spacecraft. This is not strictly accurate: for a spacecraft to be a satellite, it must be in **orbit** around the Earth or the Moon, or some other object in space.

The Moon itself is a satellite because it orbits the Earth, and the Earth (and the other planets) are satellites of the Sun. Notice that the object in orbit is smaller than the object it is orbiting around.

We can divide satellites into two types:
- **natural satellites** such as the Moon, Earth, Jupiter's moons and so on
- **artificial satellites**, which are manufactured spacecraft, and which may be in orbit around the Earth, Moon or some other object in space.

SAQ

1. **a** For each of the following satellites, state whether it is natural or artificial, and name the object it is in orbit around:

    Mars, International Space Station, the Moon

   b Name another satellite of each type.

Staying in orbit

When a spacecraft orbits the Earth, it does not need to use its rocket motors to keep it in orbit. So why doesn't it simply disappear off into space? The answer is that the satellite is held in its orbit by the pull of the Earth's gravity.

Gravity is the most important force in space. A spacecraft is attracted to the Earth by gravity, because both the spacecraft and the Earth have mass. Without the Earth's gravitational pull on the satellite, it would fly off into space (Figure 5a.3).

All objects attract each other because all objects have mass. Your mass attracts the mass of a ball, but the effect is too small to notice because the two masses are small. However, if the mass of one of the objects is large, like the Earth or the Moon, then you do notice the attraction. That is why a ball falls to the ground if you let go of it.

Any two objects with mass attract each other because of their masses; for this reason, gravity is described as a *universal* force of attraction between masses. Your mass pulls on the most distant star, but your effect is tiny. The planets pull on you, but their effects are tiny, too. That is one of the reasons why physicists find it hard to accept the claims of astrologers (people who write horoscopes) that we are influenced by the positions of the stars and planets.

SAQ

2. What property must two objects have if there is to be a gravitational attraction between them?

Going up

As a space rocket lifts off, it starts its climb upwards, away from the Earth. It is moving upwards through the Earth's gravitational field. We picture this as the region around the Earth in which its gravity acts. We can represent it as shown in Figure 5a.4, by lines of force, rather like the lines of force of a magnetic field. The arrows show the direction in which the Earth's gravity pulls. This diagram shows us two things about the Earth's gravitational field.

1. As the rocket gets higher, the pull of the Earth's gravity gets weaker. This is shown by the lines of force becoming further apart at greater distances from the Earth. In fact, if the rocket doubles its distance (measured from the centre of the Earth), the pull of gravity decreases to one-quarter of its previous value.

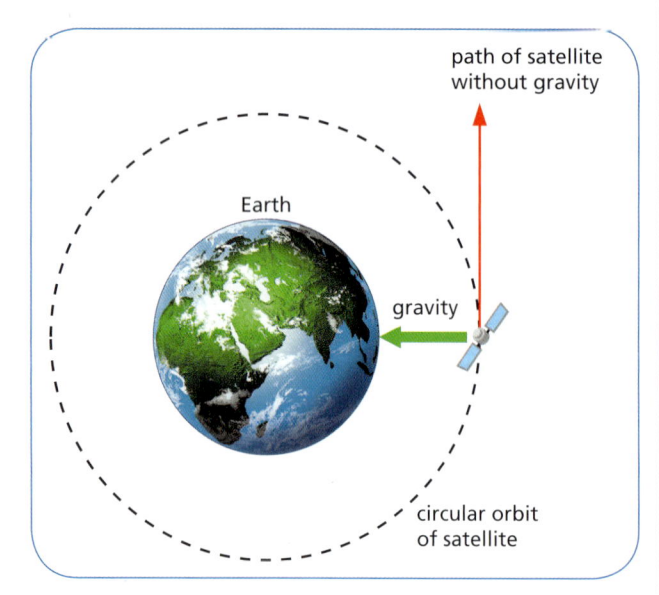

Figure 5a.3 Gravity keeps a satellite in orbit; without gravity, the satellite would fly off into space.

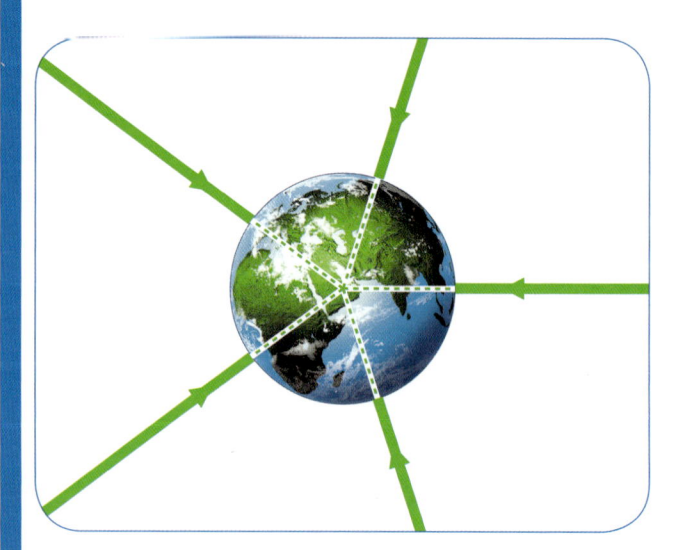

Figure 5a.4 The Earth exerts a gravitational pull on other objects. This is represented by its gravitational field, shown by lines of force radiating outwards. The field gets weaker the higher you are above the Earth's surface.

At three times the distance, gravity has one-ninth of the strength. *The gravitational force is inversely proportional to the square of the distance.*

2 The lines of force extend outwards forever. This tells us that the Earth's gravity extends to the most distant objects in space. For objects beyond the solar system, the Earth's pull is very tiny indeed. However, we know that the Earth's pull on the Sun must be significant, even though the Sun is 150 million kilometres away. This force is equal and opposite to the Sun's pull on the Earth, without which we would not continue in our orbit around the Sun.

SAQ

3 If the distance between two masses is doubled, their gravitational attraction is one-quarter of its previous value. What will happen if the distance between them is increased by a factor of 4?

Into orbit

Isaac Newton used a *thought experiment* to explain how gravity could hold an object in orbit around the Earth. Picture a very high mountain on the Earth. At the top of the mountain, above the Earth's atmosphere, stands a large cannon. The cannon can fire a shell horizontally. When a shell is fired, its path depends on its initial speed. Figure 5a.5 shows four typical tracks.

1 Shell 1 is the slowest. It follows a curved path down to the ground.

2 Shell 2 is faster than shell 1, so that it travels further before it reaches the ground. Notice that, because of the curvature of the Earth, it travels a greater distance than if the Earth had been flat.

3 Shell 3 is faster still. In fact, it is fired at just the right speed so that its curved path follows the curve of the Earth. Gravity is constantly pulling it towards the centre of the Earth but, at the same time, the surface of the Earth is curving away below it. As a consequence, it continues along a circular path around the Earth – it is in orbit.

4 Shell 4 is the fastest. Gravity pulls on it so that its path is slightly curved. However, because it is travelling so fast, gravity is not strong enough to prevent it from flying off into space.

If you could do this experiment in practice, you would find that a shell with an initial speed of 8 km/s would stay in a circular orbit, just above the Earth's surface (assuming there was no air resistance to slow it down). Spacecraft that travel a few hundred kilometres above the Earth move at a speed of 8 km/s. They have an **orbital period** – the time taken to travel once round the Earth – of about 90 minutes.

A satellite at a greater distance above the Earth travels more slowly and has further to go, so it takes longer. The higher a satellite is above the Earth's surface, the greater its orbital period.

SAQ

4 How many orbits will a satellite complete in a day if its orbital period is 90 minutes?

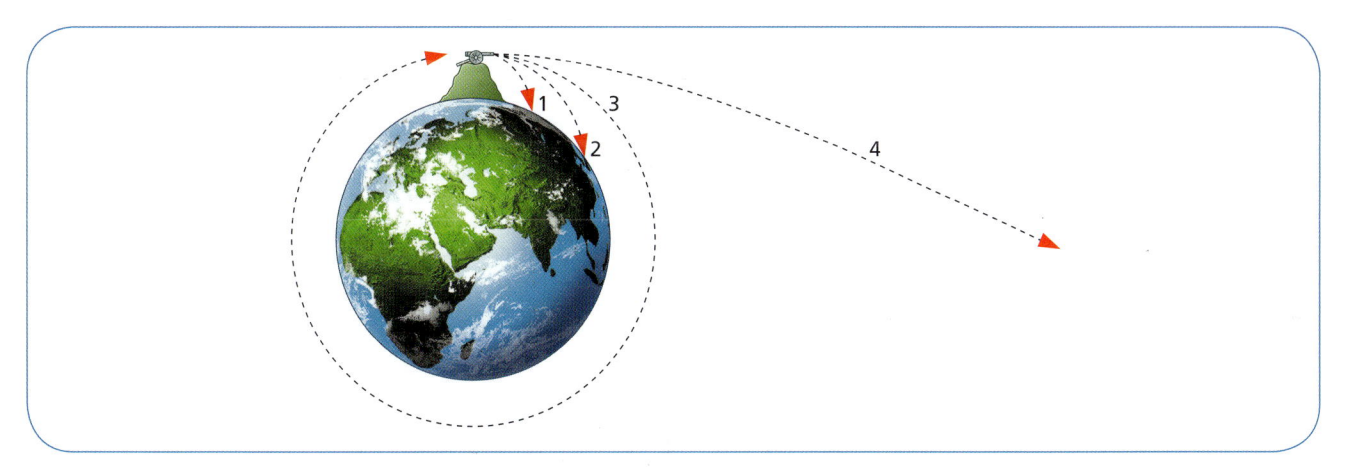

Figure 5a.5 Newton's thought experiment. He wanted to explain how gravity could hold an object in a circular orbit around the Earth. An object moving too slowly will fall to the ground; an object moving too fast will fly off into space.

214 P5a Satellites, gravity and circular motion

Going round in circles

Many spacecraft orbit the Earth along circular paths. Gravity keeps them in their orbits – in fact, gravity is the only force acting on them.

Figure 5a.6 shows that, as a spacecraft travels around the Earth:

- its weight (the pull of gravity) is always directed towards the centre of the Earth
- at any instant, it is moving at right angles to its weight.

Notice that, because there is only one force acting on the spacecraft, it is not in equilibrium. If the forces on it were balanced, it would travel in a straight line.

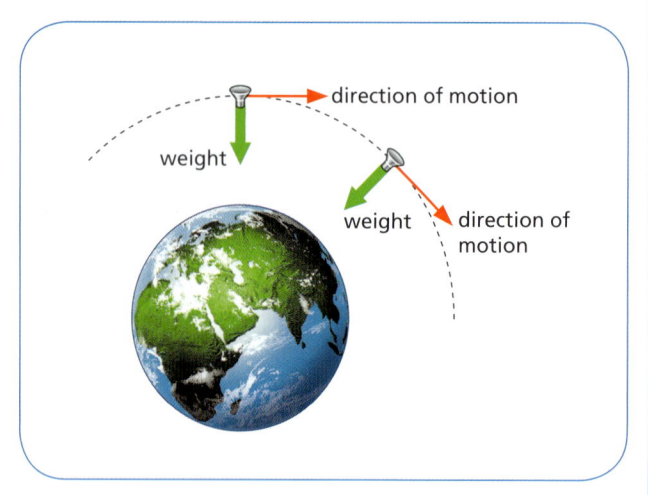

Figure 5a.6 The only force acting on a spacecraft in a circular orbit around the Earth is its weight – the pull of the Earth's gravity on it. Its direction of motion is always at 90° to this force.

Circular motion

Any object moving in a circle has an unbalanced force acting on it (otherwise it would travel in a straight line). The force is always at right angles to the direction in which the object is moving, and so is directed towards the centre of the circle. A force that points towards the centre of a circle is called a **centripetal force**. Here are some examples (see Figure 5a.7).

- The Moon orbits the Earth; it is pulled on by the Earth's gravity.
- If you whirl an apple around on the end of a piece of string, the tension in the string keeps pulling on the apple. If the string breaks, the apple flies off at a tangent.
- A motorcyclist changes direction by leaning into the corner. The frictional force of the

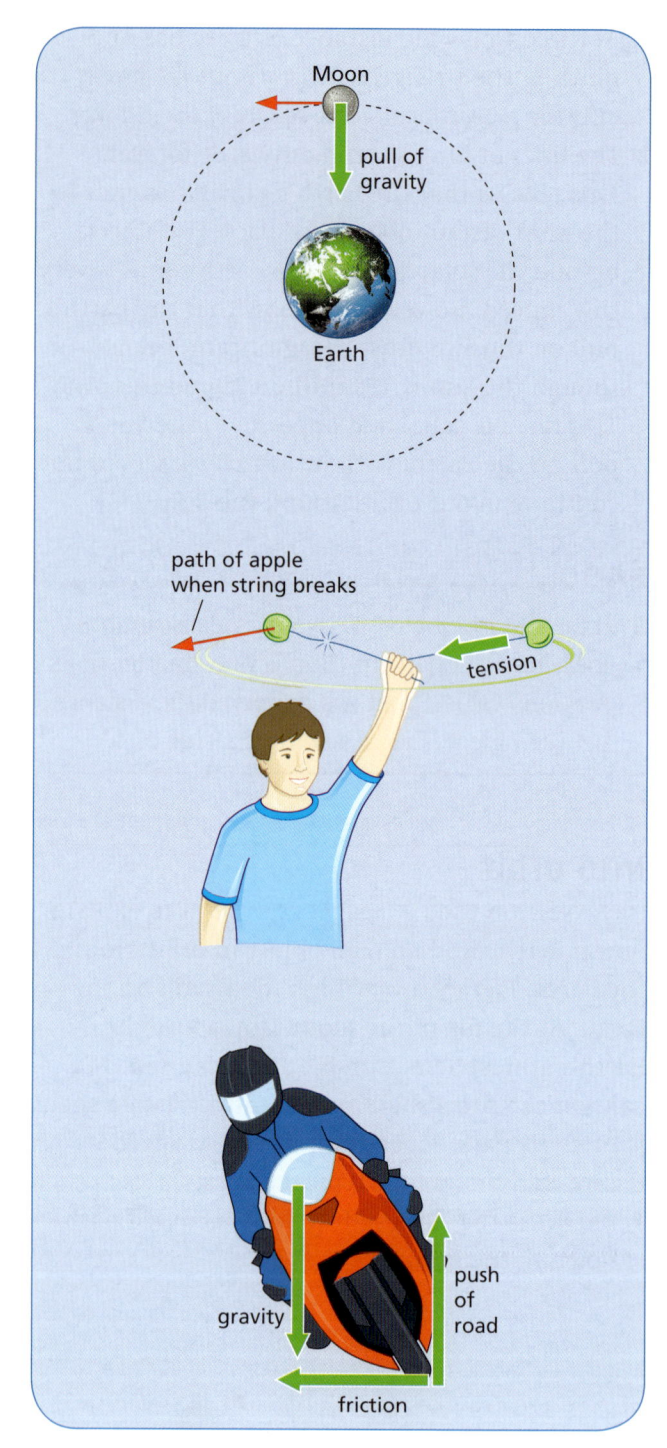

Figure 5a.7 Examples of circular motion. In each case, there is a sideways force which holds the object in its circular path; these forces are called centripetal forces.

road provides the sideways push. (If the road is slippery, friction is insufficient and the bike slides from underneath the motorcyclist.)

SAQ

5 Which force is the centripetal force that keeps the Earth in its orbit around the Sun?

Force and acceleration

An artificial satellite in a circular orbit around the Earth is acted on continuously by the force of the Earth's gravity. This is an unbalanced force, so it makes the satellite accelerate. The acceleration is towards the centre of the circle (because that is the direction of the force).

However, this doesn't make it go any faster. The force is just enough to ensure that the satellite follows a curved path; its tangential motion keeps it moving in an approximately circular orbit.

From Newton's thought experiment, you can see that there is a 'right speed' at which a satellite must move if it is to remain at a particular height above the Earth. For satellites close to the Earth's surface, such as those in low polar orbits, this is about 8 km/s. Any faster and they move further out into space; any slower and they drop downwards and crash into the Earth. At this speed, it takes about 90 minutes for a single orbit.

Further into space, gravity is weaker and the 'right speed' is smaller. The geostationary satellites used for satellite TV (see below) are at a height of about 36 000 km, and travel at 2.6 km/s; this level is chosen so that they complete one orbit in exactly 24 hours. At the distance of the Moon (400 000 km), the 'right speed' is just 1 km/s. The great distance and the relatively low speed explain why the Moon takes a month to complete one orbit.

Mercury is the closest planet to the Sun. Here, the Sun's gravitational field is strong, and the planet must travel quickly. It takes just 88 days to complete one orbit (a 'year'). The furthest planet is Neptune, almost 100 times as distant from the Sun. It travels at one-tenth of the speed of Mercury, and its year lasts 165 Earth years.

Not all satellites follow circular paths. A comet, for example, follows a highly elliptical path around the Sun (Figure 5a.8).

- When the comet is far from the Sun, the gravitational pull on it is weak, and it travels slowly.
- As it 'falls' towards the Sun, it speeds up. The force acting on it increases, and it has to travel faster to stay in its orbit.

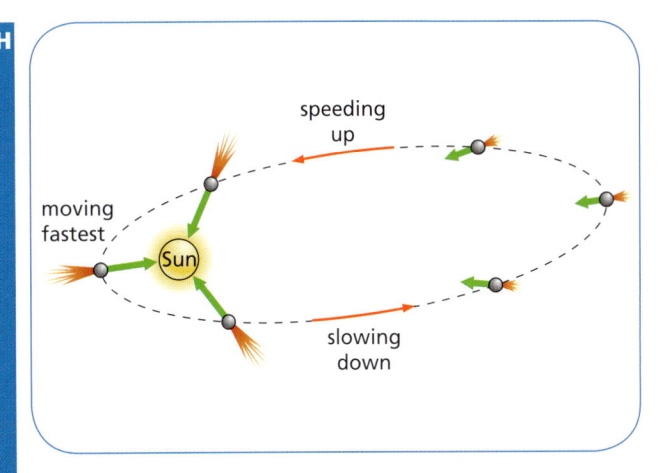

Figure 5a.8 A comet's orbit around the Sun is far from circular. The pull of gravity increases as the comet approaches the Sun, and its speed increases.

SAQ

6 Some telecommunications satellites are in low Earth orbit. Although the atmosphere at this height is very thin, it still causes a slight frictional force on these satellites. Explain how this will affect the orbit of the satellite.

Spacecraft at work

Spacecraft have many different uses, and they have made a big difference to our everyday lives. Here are some uses of spacecraft.

Geostationary satellites

Hundreds of communications satellites are in orbit above the equator. These are used for beaming television broadcasts down to Earth; individual consumers fit a dish receiver to the wall or roof of their home to collect the electromagnetic waves transmitted by a satellite.

Such satellites are described as **geostationary**. This means that they orbit the Earth once each day. As they orbit, the Earth turns below them so that they stay above the same point on the Earth's surface all the time (Figure 5a.9). The receiving dish does not have to alter its position to track the satellite. The radius of a geostationary orbit is about six times the radius of the Earth, so they are a long way off in space.

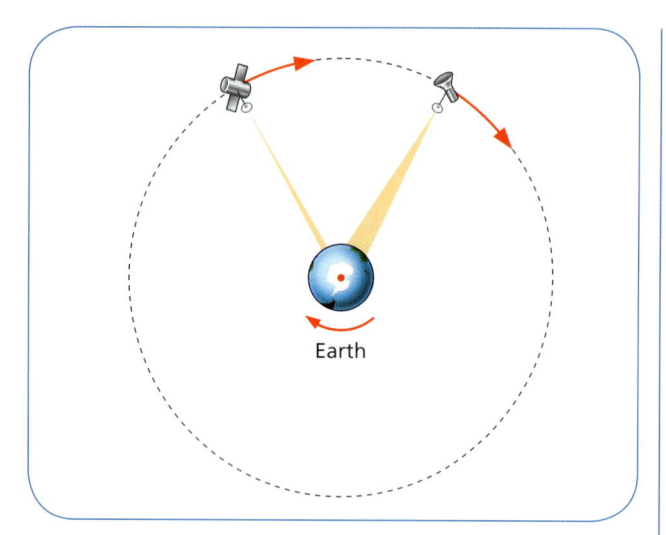

Figure 5a.9 Geostationary satellites are used for satellite TV transmissions. They travel once around the Earth in 24 hours, remaining above the same point on the equator.

Only a limited number of geostationary satellites (about 400) can be put in place. This is because, if two were close together, a dish receiver would collect signals from both, and the signals would interfere with each other. Some geostationary satellites also contribute to weather forecasting because they give a complete view of half the Earth's surface.

Polar-orbiting satellites

As well as satellites whose orbits are over the equator, there are satellites that orbit the Earth from pole to pole (Figure 5a.10). It takes such a satellite about 90 minutes to complete one orbit of the Earth if it is just a few hundred kilometres above the surface. As the Earth turns beneath it, it can scan the Earth's surface. In the course of a day, it can send back images of a large part of the Earth. Such monitoring satellites have many uses:

- taking weather measurements – cloud cover, pressure, wind speed and direction, temperature, and so on, for use by meteorologists
- observing environmental features of the Earth, such as crops and other vegetation, rock formations for mineral prospecting, and changes in sea levels
- making astronomical observations (Figure 5a.11) – ground-based telescopes produce poor images because the atmosphere absorbs or distorts much of the radiation
- for military purposes, including spying and possibly carrying weapons.

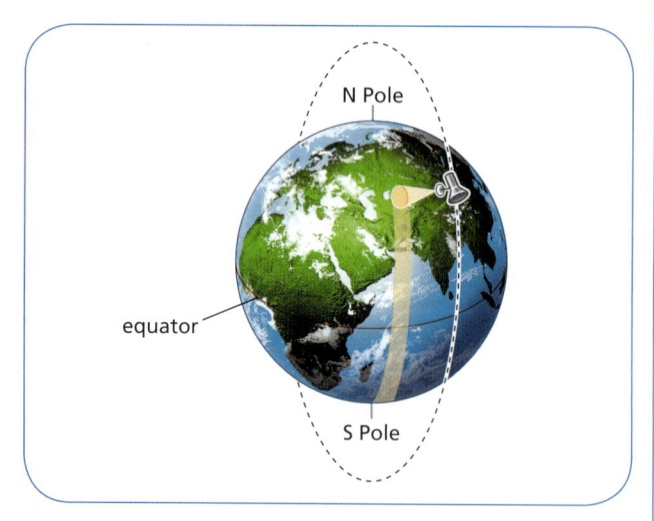

Figure 5a.10 A polar-orbiting satellite travels around the Earth in an orbit that takes it directly over the North and South Poles. On each orbit, it can monitor a strip of the Earth's surface.

Figure 5a.11 The Hubble Space Telescope has produced many images to excite astronomers. Because it is positioned above the Earth's atmosphere, it has a much clearer view of space than ground-based telescopes. The stars it sees do not 'twinkle', because their light has not been distorted by irregularities in the atmosphere.

At night, you may be able to see polar-orbiting satellites – they look like bright stars, moving steadily across the sky, heading north or south and taking perhaps 20 minutes to cross your field of view. There are astronomy websites that give details of when you may see such a satellite like the International Space Station.

The Global Positioning System (GPS) uses a network of satellites orbiting the Earth. Each satellite transmits radio signals that can be detected on Earth. At any one time, four or five satellites are above the horizon for an observer on Earth. Someone who requires an accurate measurement of their own position uses a receiver that picks up the signals from these satellites. A tiny computer measures the time it has taken each signal to reach the receiver and works out the distance of each satellite. From this, it can calculate the receiver's position on the Earth to within a fraction of a metre. GPS receivers have proved to be a great benefit to people who enjoy sailing and wilderness walking, but they also have many commercial uses, including navigation systems in vehicles and aircraft.

SAQ

7 Some weather satellites are in low polar orbits; others are in geostationary orbits. Explain why one type of satellite gives a detailed view of a small area of the Earth's surface whereas the other gives a less detailed view of a larger area.

Summary

You should be able to:

- state that gravity is a universal force of attraction between bodies that have mass
- **H** explain that gravitational attraction decreases with distance – it is inversely proportional to the square of the distance
- state that circular motion requires a centripetal force that acts towards the centre of the circle
- describe a satellite as an object that orbits a larger object in space, held in its orbit by the centripetal force of gravity
- state that the orbital period of a satellite is greater the higher it is above the Earth's surface
- **H** explain that the lower the orbit of a satellite, the faster it must travel to remain in orbit
- state that satellites may follow low orbits passing over the Earth's poles or geostationary orbits high above the equator with a period of 24 hours
- describe some uses of artificial satellites, for example communications, weather forecasting, military, scientific and GPS

218 P5a Satellites, gravity and circular motion

Questions

1. Any object travelling in a circle requires a centripetal force to keep it in its orbit.

   a. Figure 5a.12 shows a satellite at three points in its orbit around the Earth. Copy the diagram and add an arrow to show the centripetal force on the satellite at each point.

   b. What force keeps the satellite in its orbit?

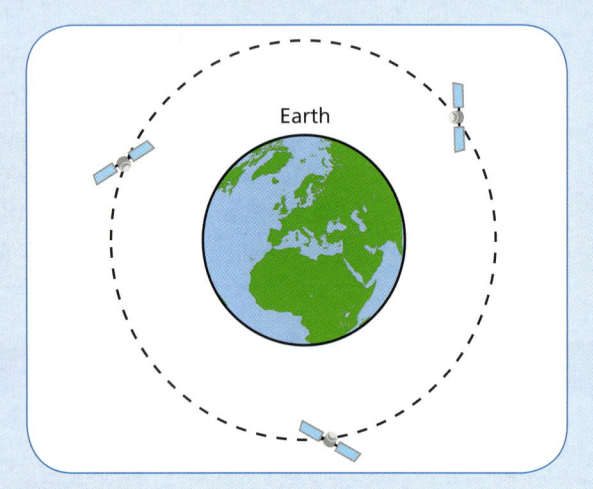

Figure 5a.12

2. Explain why the Earth and the Moon can both be described as *natural satellites*. What forces keep them in their orbits?

3. A weather satellite may be positioned in a low polar orbit or a geostationary orbit. Draw diagrams to show the differences between these two types of orbit.

4. Copy and complete Table 5a.1 to show as many uses of artificial satellites as you can.

Type of satellite	Uses
communications	
military	
scientific research	
weather forecasting	
GPS	

Table 5a.1

H 5. Every geostationary satellite orbits the Earth once every 24 hours. The radius of its orbit is 42 300 km.

   a. Calculate the speed of a geostationary satellite.

   b. Is a geostationary satellite in equilibrium? Explain your answer.

   c. Explain why a geostationary satellite remains above the same point on the Earth's surface as it orbits.

continued on next page

Questions - continued

H 6 What benefits can come from having an astronomical telescope on board a spacecraft orbiting the Earth?

7 A satellite orbits the Earth at a constant speed and at a constant height above the Earth's surface.

   a Explain why a force is needed to keep it in its orbit.

   b What is this force?

   c Explain whether the satellite is in equilibrium.

   d Which of the following remain constant as the satellite orbits?
- speed
- acceleration
- kinetic energy
- gravitational potential energy

8 Venus orbits the Sun. Its orbit is closer to the Sun than the Earth's orbit.

   a Which travels at a greater speed around the Sun – Venus or the Earth?

   b Which takes longer to orbit the Sun?

9 The Moon follows a roughly circular path around the Earth. Explain how the Moon's path would change if:

   a its speed suddenly halved;

   b its speed suddenly doubled.

P5b Vectors and equations of motion

Speed and velocity

Although the words *speed* and *velocity* have the same meaning in everyday speech, there is an important distinction between them in Physics. **Speed** tells us about how fast something is moving; **velocity** tells us how fast it is moving *in a particular direction*. So, if you are cycling along a road, your speed might be 10 m/s, but your velocity might be 10 m/s *towards the east*. To state your velocity completely, it is essential to include the direction in which you are moving.

Velocity is an example of a **vector quantity**. Vector quantities have both magnitude (size) and direction. Another example of a vector quantity is force. If you are lifting a heavy load, you might say that the force you use is 200 newtons *vertically upwards*. You need to state its direction if someone else is to know the full details of your lifting force.

In diagrams, vector quantities are often represented by arrows. The direction of the arrow shows the direction of the vector quantity.

The alternative to a vector quantity is a **scalar quantity**. This is a quantity that has magnitude but no direction. Speed is a scalar quantity, because we do not need to give a direction when we state an object's speed. Another scalar quantity is mass – it makes no sense to talk of an object having a mass of 50 kg in a particular direction. However, weight does have a direction – your weight is a force that acts towards the centre of the Earth.

SAQ

1. State whether each of the following quantities is a vector or a scalar:

   mass, speed, weight, force, velocity

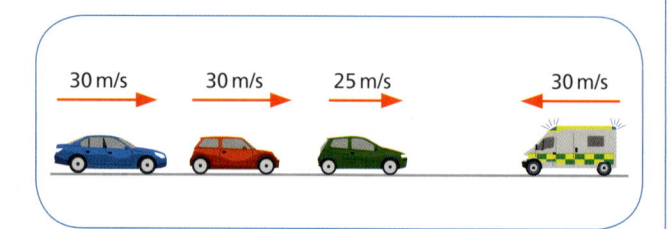

Figure 5b.1 If we know the speed and direction of each car, we can work out their speeds relative to each other.

Relative velocities

Figure 5b.1 shows two cars (red and blue) travelling along at 30 m/s. Although each car is moving at 30 m/s relative to the road, their speed relative to each other is 0 m/s, because the distance between them is not changing.

The red and blue cars are catching up on the green car. The green car is moving at 25 m/s, so their relative speed is 5 m/s. Because the cars are all travelling in the same direction, their **relative speed** is smaller than their individual speeds.

The ambulance is approaching from the opposite direction at 30 m/s. Its speed relative to the green car is 30 m/s + 25 m/s = 55 m/s. So, for vehicles approaching one another, their relative speed is greater than their individual speeds.

SAQ

2. What is the speed of the ambulance relative to the blue car?

Adding velocities, adding forces

Imagine that you are on a train which is moving at 50 m/s. You walk quickly towards the front of the train at 5 m/s. Your velocity relative to the track is now 55 m/s. If you turn round and walk back in the opposite direction at 5 m/s, your velocity relative to the track will be 45 m/s.

Figure 5b.2 shows how we can work this out, using arrows to represent the quantities involved. It helps to draw longer arrows for bigger quantities.

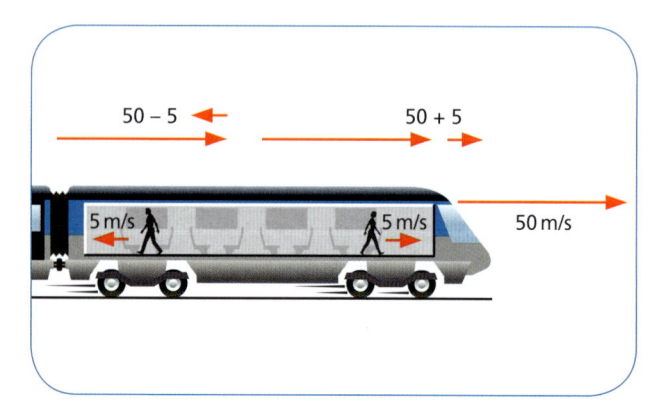

Figure 5b.2 Velocities are vector quantities, so we have to take account of their directions when adding them.

If an object is acted on by two or more forces, their directions must be taken into account because forces are vectors (see Figure 5b.3).
- If the forces are pointing in the same direction, they add up.
- If the forces are pointing in opposite directions, they subtract.

When forces act together like this, their combined effect is called the **resultant force**.

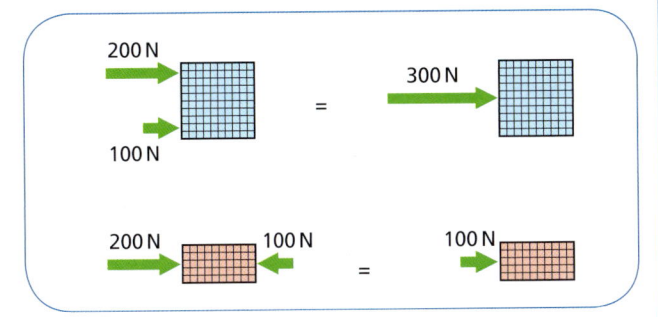

Figure 5b.3 When adding forces, their directions must be taken into account.

Worked example 1

A car has a mass of 800 kg. Its engine provides a forward force of 400 N. There is a frictional force of 160 N, acting to oppose the car's motion. What is the resultant force acting on the car? What is its acceleration?

Step 1: the first step in a problem like this is to draw a diagram, to represent all of the information we have – see Figure 5b.4. The two forces are represented by arrows. The car's mass is written by the car, but it has no arrow – it has no direction.

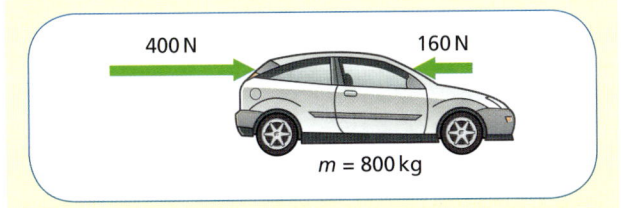

Figure 5b.4 This summarises the information in the question.

Step 2: it is helpful to consider forces in one direction as positive – forces to the right in this case.

We have two forces, 400 N to the right (+400 N), and 160 N to the left (−160 N). We can thus find the resultant force:

resultant force = +400 N − 160 N = +240 N

The resultant force is positive, meaning that it acts to the right. (This makes sense because the forward force is greater than the frictional force.)

Step 3: now we can calculate the car's acceleration.

$$\text{acceleration } a = \frac{\text{resultant force}}{\text{mass}} = \frac{+240\,\text{N}}{800\,\text{kg}}$$

$$= +0.3\,\text{m/s}^2$$

So, the car's acceleration is $+0.3\,\text{m/s}^2$.

We have kept the positive sign from the resultant force. The acceleration is also positive, and this tells us that the car's acceleration is towards the right, as you would have expected.

SAQ

3 A parachutist is falling through the air at high speed. Two forces act on her:
- her weight, 800 N, acting downwards
- air resistance, 1250 N, acting upwards.

What is the resultant force acting on her? Give its size and direction. What effect will the resultant force have on how she is moving?

Vectors at right angles

Imagine that you dive into a fast-flowing river and attempt to swim directly across to the opposite bank. As you swim forwards, the current carries you sideways. The effect is that you reach the opposite bank at a point downstream of where you intended.

In effect, you have two velocities, which are at right angles to each other: the velocity you are swimming with directly across the river, and the velocity of the river at right angles to your swimming – see Figure 5b.5. To find the **resultant velocity**, we draw a **vector triangle**.

222 P5b Vectors and equations of motion

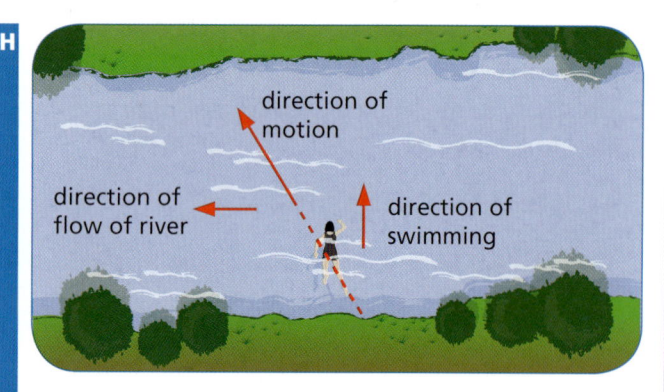

Figure 5b.5 This swimmer has two velocities; her resultant velocity is at an angle to the river bank.

Worked example 2 shows how this is done. (Forces at right angles to each other can be added in the same way.)

Worked example 2

A small boat moves through the water at 4 m/s. It is directed straight across a river which is flowing at 3 m/s. At what speed will it cross the river? And in what direction?

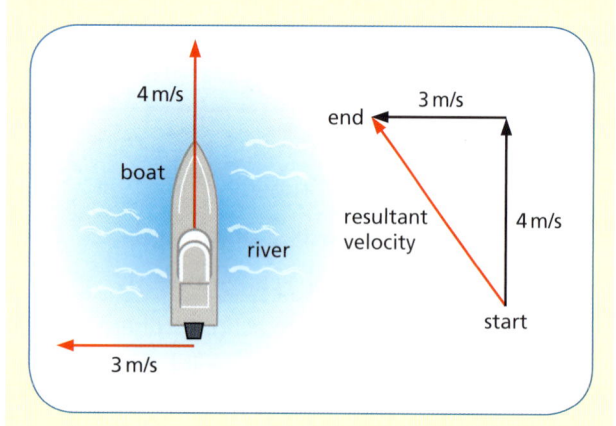

Figure 5b.6 Combining two velocities using a vector triangle.

Step 1: Figure 5b.6 shows the situation. It also shows how we can represent the two velocities by arrows, drawn to scale.

- The length of the arrow represents the size of the velocity.
- The direction of the arrow represents the direction of the velocity.

To construct the triangle, we first draw an arrow to represent one of the velocities. Then, *starting at the end of the first arrow,* we draw a second arrow to represent the other velocity.

Step 2: to find the resultant, we go back to the start of the first vector in the vector triangle and draw an arrow from there to the end of the second vector. This line represents the resultant velocity, and there are two ways to find its length:

- by scale drawing
- by Pythagoras' theorem.

Since we have a right-angled triangle in this case, we can use Pythagoras' theorem:

(resultant velocity)^2 = 3^2 + 4^2 = 25

resultant velocity = 5 m/s

So, the boat crosses the river at 5 m/s. You could achieve the same answer by accurate scale drawing. The drawing also shows the direction in which the boat will move.

SAQ

4 Look at Figure 5b.6. The vector triangle was drawn starting with the velocity of the boat (4 m/s). Then the velocity due to the current was added. Draw a second vector triangle to find the resultant, but starting with the velocity of the current. Show that the resultant velocity is the same as calculated in Worked example 2.

The equations of motion

In order to estimate how fast a car was moving when it became involved in an accident, or how far it might have moved, the police make use of a set of equations known as the **equations of motion**. These relate five important quantities, which describe an object's motion. These are shown in Table 5b.1.

Quantity	Symbol	Unit
distance travelled	s	m
time taken	t	s
initial velocity	u	m/s
final velocity	v	m/s
acceleration	a	m/s^2

Table 5b.1 These five important quantities give us information about the way an object is moving.

Explaining accidents

After a road accident, traffic police may be sent to examine the scene (Figure 5b.7). They look for evidence that may show whether any of the drivers involved were breaking the law. They may look for skid marks on the road, or anything that might have obstructed the view. They will have taken statements from those involved, and now want to check whether what witnesses have said matches up with the situation on the ground.

How can traffic police reconstruct the events leading up to an accident? Suppose they find long skid marks, left on the road when one of the cars decelerated suddenly. Can they deduce how fast the car must have been travelling when the driver braked? Can they work out how much time a driver had to brake when a problem arose ahead?

In this sort of situation, the police carry out experiments to test the road surface. They drag a device called a sledge along the road, to see what sort of acceleration they can achieve without slipping or skidding. They may obtain a car of the same make and model as the one involved in the accident and test it on the same stretch of road. They measure the road and make a map of the accident situation.

Once they know the distance the car skidded, its acceleration and so on, they can work back to calculate whether the driver might have been travelling too fast or whether he or she was an innocent victim of a genuine accident.

Figure 5b.7 Traffic police carrying out investigations at the scene of an accident.

There are equations that link the quantities in Table 5b.1. Note that they only apply in situations in which an object is moving with constant acceleration – when its velocity is increasing or decreasing at a steady rate.

Equations 1 and 2

Here is the equation that defines what we mean by acceleration:

$$a = \frac{(v - u)}{t}$$

It can be rearranged like this:

$v = u + at$ **Equation 1**

This tells us that the final velocity v of an object is equal to its initial velocity u plus the amount it increases by when it moves with acceleration a for time t:

$$\text{final velocity} = \text{initial velocity} + \text{change in velocity}$$

The second equation comes from the definition of average velocity:

$$s = \frac{(v + u)}{2} \times t \quad \textbf{Equation 2}$$

This tells us that the distance travelled s is the average velocity $(v + u)/2$ multiplied by the time taken t:

distance travelled = average velocity × time

Worked example 3 (over the page) shows how to use these equations.

Worked example 3

A car is travelling at 10 m/s. It speeds up to 26 m/s in a time of 20 s. How far does it travel in this time?

Step 1: write down what you know, and what you want to know.

$u = 10$ m/s
$v = 26$ m/s
$t = 20$ s
$s = ?$

(This first step is crucial in helping you to decide which equation to use.)

Step 2: choose the equation of motion that links these quantities.

$$s = \frac{(v + u)}{2} \times t$$

Step 3: substitute in values and solve.

$$s = \frac{(26 + 10)}{2} \times 20$$

$$= 18 \times 20$$

$$= 360 \text{ m}$$

So, the car travels 360 m in this time.

SAQ

5. A train, moving at 5 m/s, speeds up with an acceleration of 0.4 m/s^2.
   a. How fast will it be moving after 30 s?
   b. How far will it travel in this time?

Equations 3 and 4

If we do not know an object's final velocity, we can use a different equation to work out how far it has travelled:

$s = ut + \frac{1}{2}at^2$ **Equation 3**

Note that, if the object was not accelerating ($a = 0$), this would simply be
$s = ut$, or distance = speed × time

Finally, if we know the distance over which an object is accelerating, rather than the time during which it is accelerating, we can find its final velocity using this equation:

$v^2 = u^2 + 2as$ **Equation 4**

Again, for zero acceleration ($a = 0$), this would simply give us $v^2 = u^2$ so that the final velocity would equal the initial velocity.

The four equations of motion are collected together in Table 5b.2. Note that you may need to rearrange an equation in order to solve problems. Worked examples 4 and 5 illustrate this.

Equation 1	$v = u + at$
Equation 2	$s = \frac{(v + u)}{2} \times t$
Equation 3	$s = ut + \frac{1}{2}at^2$
Equation 4	$v^2 = u^2 + 2as$

Table 5b.2 The four equations of motion, for an object moving with constant acceleration.

Worked example 4

A car driver is travelling at 20 m/s along a road. 50 m ahead, a child runs out into the road, and the driver presses the brake pedal. What must the car's acceleration be if the car is to stop before it reaches the child? The police have estimated that no vehicle could achieve an acceleration of greater magnitude (size) than 5 m/s^2. Should the driver be able to stop in time?

Step 1: write down what you know, and what you want to know.

$u = 20$ m/s
$v = 0$ m/s
$s = 50$ m
$a = ?$

Step 2: choose the equation of motion that links these quantities, and rearrange it to make the unknown quantity (a) the subject.

$$v^2 = u^2 + 2as$$

$$a = \frac{(v^2 - u^2)}{2s}$$

continued on next page

Worked example 4 - continued

Step 3: substitute in values and solve.

$$a = \frac{-(20\,\text{m/s})^2}{2 \times 50\,\text{m}}$$

$$= \frac{-400\,\text{m}^2/\text{s}^2}{100\,\text{m}}$$

$$= -4\,\text{m/s}^2$$

So, the car must accelerate at $-4\,\text{m/s}^2$. Note that the acceleration is negative because the car is slowing down (decelerating). The magnitude of the acceleration ($4\,\text{m/s}^2$) is less than the maximum estimated by the police, so the driver should be able to stop in time.

Worked example 5

A car joins a motorway travelling at 15 m/s. It accelerates at $2\,\text{m/s}^2$ for 8 s. How far will it travel in this time?

Step 1: as before, write down what you know, and what you want to know.

$u = 15\,\text{m/s} \qquad t = 8\,\text{s}$
$a = 2\,\text{m/s}^2 \qquad s = ?$

Step 2: choose the equation of motion that links these quantities. In this case, there is no need to rearrange it to make the unknown quantity (s) the subject.

$$s = ut + \tfrac{1}{2}at^2$$

Step 3: substitute in values and solve.

$$s = 15\,\text{m/s} \times 8\,\text{s} + \tfrac{1}{2} \times 2\,\text{m/s}^2 \times (8\,\text{s})^2$$

$$s = 120\,\text{m} + 64\,\text{m}$$

$$= 184\,\text{m}$$

So, the car will travel 184 m in this time.

SAQ

6 A stone is dropped from the top of a high cliff. It falls with an acceleration of $9.8\,\text{m/s}^2$, and its speed is 24 m/s when it reaches the foot of the cliff. How high is the cliff?

Summary

You should be able to:

♦ state that vector quantities have both size and direction, while scalar quantities do not have direction

♦ explain that when vector quantities are combined, their directions must be taken into account

♦ use the equations of motion, which apply to an object moving with constant acceleration:

$v = u + at$

$s = \dfrac{(v + u)}{2} \times t$

$s = ut + \tfrac{1}{2}at^2$

$v^2 = u^2 + 2as$

Questions

1. Name three scalar quantities, and three vector quantities.

2. An aircraft flies through the air at a speed of 300 m/s. There is a headwind (a wind blowing in the opposite direction) with a speed of 40 m/s. What is the aircraft's speed relative to the ground? If it flew in the opposite direction, what would be its speed relative to the ground?

3. Figure 5b.8 shows the forces acting on various objects. What resultant forces act on objects **a** and **b**?

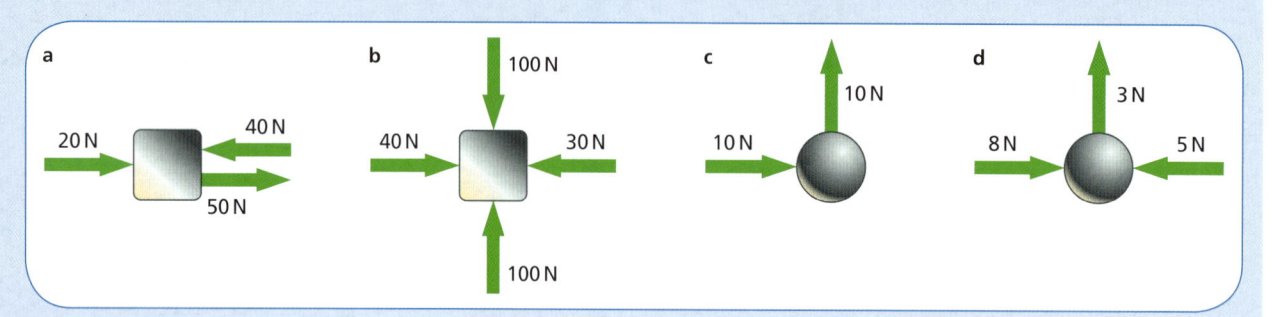

Figure 5b.8

4. What resultant forces act on objects **c** and **d** in Figure 5b.8?

5. If you are going to use any of the four equations of motion to calculate some aspect of the motion of an object, what must be true about the object's acceleration?

6. A car is moving at 20 m/s. The driver applies the brakes and slows down, with an acceleration of $-2\,m/s^2$. If the brakes are applied for 5 s, what speed will the car have now?

7. An aircraft starts to move down the runway. Its initial speed is 0 m/s. After 20 s, it is moving at 100 m/s. It then takes off.

    a What is its average speed along the runway?

    b How far does it travel down the runway before it takes off?

8. A sprinter starts from stationary. He can maintain an acceleration of $12\,m/s^2$ for 0.8 s.

    a What speed will he reach?

    b How far does he travel whilst accelerating?

9. Which of the equations of motion follows from the definition of acceleration (acceleration = change in velocity ÷ time taken)?

10. A motorcyclist joins a motorway travelling at 18 m/s. She accelerates at $4\,m/s^2$ for 5 s. How far does she travel in this time?

11. How long will it take an aircraft to accelerate from 200 m/s to 260 m/s if its acceleration is $4\,m/s^2$?

12. A car is initially moving at 6 m/s. Over a distance of 32 m it accelerates at $1\,m/s^2$. What speed does it reach?

13. A plane landing on an aircraft carrier must come to a halt in a distance of 120 m. If it touches down at a speed of 80 m/s, what acceleration is required to bring it to a halt in this distance?

5c Projectile motion

Flying through the air

Footballers can do some amazing things when they kick a ball. It might follow a long arc through the air to land at a team-mate's feet, for example, or curve past a wall of defenders. A skilful footballer can kick the ball so that it flies off at just the right angle, spinning in such a way that it follows the desired path through the air.

Today we have cameras to show us the path of a ball through the air. In the past, people found it hard to picture how objects moved when they were thrown or fired. Figure 5c.1 shows one idea from a 15th-century book on warfare. A cannon ball is fired into the air; it follows a straight path upwards until it suddenly drops vertically to the ground. Why was it drawn like this?

Before the work of Galileo and Newton, people said that the ball was given a 'force' by the hand that threw it. This force got weaker as the ball went upwards. Eventually, the ball ran out of force and fell back down to the ground. Today, we do not imagine that the force of our hand continues to have an effect once the ball has left it. Once the ball is free, gravity is the only force on it (apart from air resistance).

Figure 5c.2 shows clearly the path of a ball through the air. Once it leaves the thrower's hand, it follows a curved trajectory, slowing down as it goes upwards and speeding up as it comes back down.

Figure 5c.1 A medieval illustration showing the path of a cannon ball, as imagined by the artist.

Figure 5c.2 This photo was made using a stroboscopic flash light. Once the ball has left the thrower's hand, the thrower has no influence over the path it follows.

Projectile paths

A ball thrown into the air is an example of a **projectile** – it has been *projected* into the air. Cannon balls, bullets, darts – these are all projectiles. Even a high-jumper or long-jumper is a projectile, whilst they are in the air. The point about a projectile is that, once it has been thrown or fired, it doesn't have any force pushing it along. So a rocket is not a projectile, because the rocket's engines keep firing after it has left the ground.

The path of a projectile is called its **trajectory**. In Figure 5c.2, you can see the curved shape of a projectile's trajectory. If there is no air resistance to affect the projectile's motion, its trajectory has the shape of a **parabola**.

SAQ

1 A sky-rocket is a popular type of firework. It flies up into the air; when it has burnt out, it falls back to Earth. Is it correct to describe a sky-rocket as a projectile? Explain your answer.

Gravity at work

When a ball flies through the air, only two forces act on it: **gravity** and **air resistance**. Gravity is the more important of these. It always acts downwards on the ball. (Air resistance acts in the opposite direction to a projectile's velocity. From here on, we will ignore the effects of air resistance.)

Figure 5c.3 shows a ball that has been projected (thrown or fired) horizontally. Its path has the shape of a parabola. The arrows show the force of gravity acting on it (its weight).

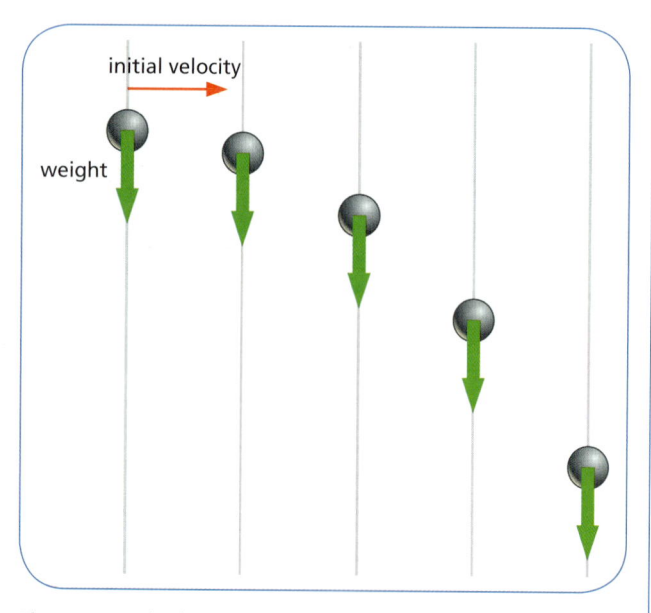

Figure 5c.3 The force of gravity always acts downwards on a projectile.

The images of the ball are shown at equal intervals of time. To understand this motion, we think separately about what is happening vertically and horizontally.
- **Vertically:** Gravity is pulling downwards on the ball, making it accelerate. Its **vertical velocity** is increasing.
- **Horizontally:** Gravity cannot affect the horizontal motion of the ball. It moves at a steady rate towards the right. It has a constant **horizontal velocity**.

(Note that, here, we have ignored air resistance. This force is usually small for a ball; however, it is much more important for an object like a shuttlecock. Because of air resistance, a shuttlecock will follow a path similar to that of the cannon ball shown in Figure 5c.1.)

SAQ

2 When a projectile moves through the air, are the forces on it balanced or unbalanced? In which direction does it accelerate?

At an angle

When a projectile is fired into the air at an angle, its trajectory is still a parabola. The only force acting on it is gravity (downwards), and so its horizontal velocity is unaffected.

You can see this in Figure 5c.4. The images of the bouncing ball are equally spaced horizontally, because it moves steadily from left to right. However, the images become closer together as the ball rises upwards, because gravity is slowing it down; they become further apart as it falls, because gravity is causing it to accelerate.

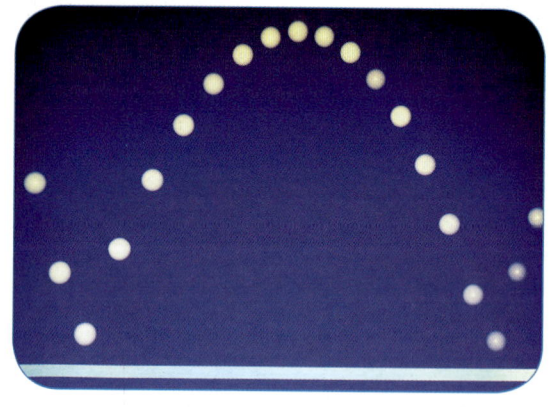

Figure 5c.4 A bouncing ball is a projectile (except when it is in contact with the ground). The pattern of the images shows that it is only acted on by a vertical force (the pull of the Earth's gravity).

Thinking vectors

We can think of a projectile as having two separate velocities.
- Its horizontal velocity is constant, because it is unaffected by gravity.
- Its vertical velocity changes, because gravity causes it to accelerate downwards.

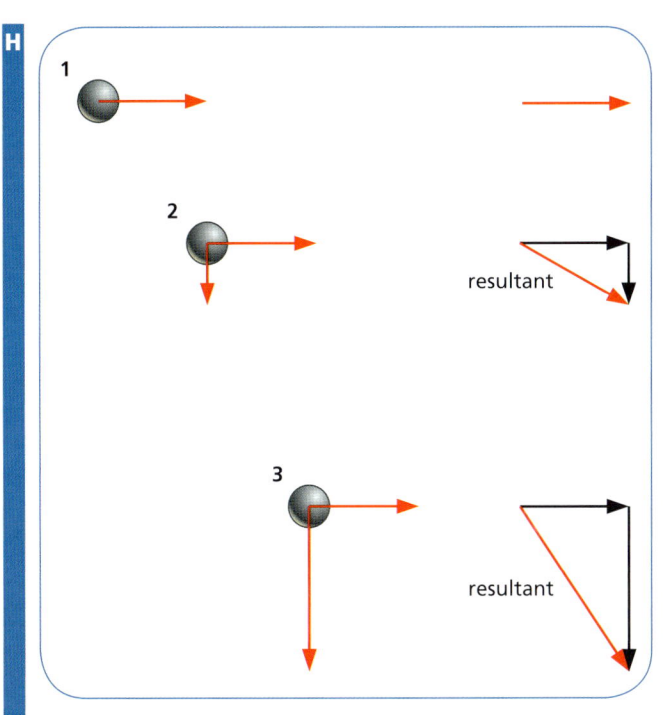

Figure 5c.5 Using vectors to determine the resultant velocity of a projectile. In this case, the projectile was projected horizontally.

To find the resultant velocity of a projectile, we must calculate the **vector sum** of these two. Figure 5c.5 shows how this is done.

1 Initially, the ball has only a horizontal velocity.
2 After a short while, it has a small vertical velocity. Its horizontal velocity remains constant. Its resultant velocity is at an angle downwards.
3 Later, its vertical velocity has increased further. Its resultant velocity is greater, and at a steeper angle.

SAQ

3 At one point in its trajectory, a projectile has a vertical velocity of 4 m/s upwards, and a horizontal velocity of 5 m/s to the right. Draw a vector diagram to show this, and use it to calculate the resultant velocity of the projectile.

Using the equations of motion

Can we use the equations of motion (Item P5b, *Vectors and equations of motion*) to calculate the motion of a projectile? Recall that, if we are to use the equations, the object's acceleration must be constant. This is the case for a projectile because it has a constant downward acceleration (caused by gravity), and zero horizontal acceleration. So, provided we treat horizontal and vertical motions separately, we can use the equations.

Any object falling close to the Earth's surface has an acceleration of about $10\,\text{m/s}^2$ downwards. This quantity is known as the **acceleration due to gravity**, and is given the symbol g.

acceleration due to gravity = $g = 10\,\text{m/s}^2$

(A closer value is $9.8\,\text{m/s}^2$, but it is convenient to use $10\,\text{m/s}^2$ in calculations.)

Worked example 1 shows how to apply the equations of motion to an object projected horizontally.

Worked example 1

A stone is thrown horizontally from the top of a 20 m high cliff. Its initial velocity is 8 m/s.

- How long will it take to reach the foot of the cliff?
- How far from the foot of the cliff will it land?

Figure 5c.6 shows the situation. Remember that we can treat vertical and horizontal motions separately, because the stone will keep moving horizontally at a constant speed whilst accelerating downwards.

Step 1: we have to find the time taken for the stone to fall 20 m vertically. Initially, its vertical velocity is 0 m/s. So we have:

$s = 20\,\text{m}$
$u = 0\,\text{m/s}$
$a = g = 10\,\text{m/s}^2$
$t = ?$

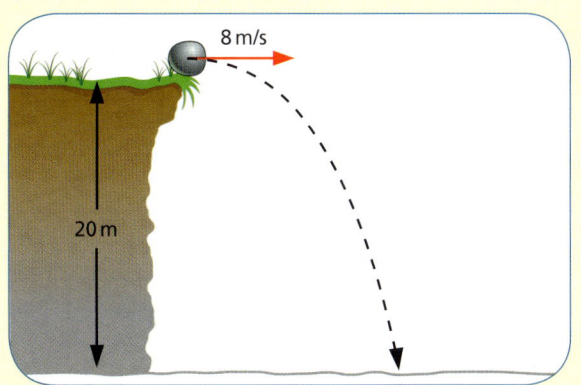

Figure 5c.6

continued on next page

Worked example 1 – continued

Step 2: select the equation of motion that contains s, u, a and t.

$s = ut + \frac{1}{2}at^2$

Step 3: substitute values and solve.

$20 = 0 \times t + \frac{1}{2} \times 10 \times t^2$

$5t^2 = 20$

$t^2 = 4$

$t = 2$

So the stone takes 2 s to reach the foot of the cliff.

Step 4: now it is easy to find how far the stone has travelled horizontally. It travels horizontally at 8 m/s for 2 s, so the distance from the foot of the cliff is:

distance = 8 m/s × 2 s = 16 m

Note that the equations of motion can be applied in more complex situations – for example, where a projectile is fired at an angle to the horizontal. They can be used to calculate the horizontal and vertical velocities of a projectile, and these can then be added (as vectors) to find the projectile's resultant velocity.

SAQ

4 A stone is projected horizontally from the top of a high cliff with an initial velocity of 5 m/s. It lands 15 m from the foot of the cliff. How high is the cliff?

Summary

You should be able to:

- state that the trajectory of a projectile is a parabola

- explain that the only force (ignoring air resistance) acting on a projectile is gravity, which causes it to accelerate towards the ground

- explain that an object projected horizontally has a constant horizontal velocity and an increasing vertical velocity

- explain that the horizontal and vertical velocities of a projectile are vectors, and that their vector sum is the resultant velocity of the projectile

- use the equations of motion applied to a projectile (horizontal acceleration = 0; vertical acceleration = g).

Questions

1. Explain why a tennis ball or a high-jumper could be regarded as a projectile but a parachutist could not.

2. What word means *the path of a projectile*? What word describes the shape of such a path?

3. Figure 5c.7 shows the paths of two projectiles. One ball has been dropped vertically; the other has been projected horizontally at the same instant.

   a Name the force that causes both balls to accelerate downwards.

   b Explain why the second ball moves horizontally at a steady speed.

   c Explain why both balls reach the ground at the same time.

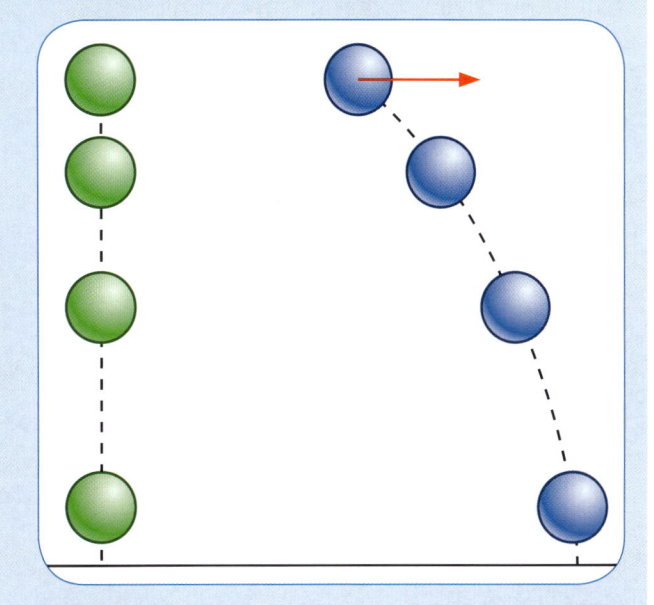

Figure 5c.7

H 4 A ball is thrown with a horizontal velocity of 10 m/s. After 1 s, its vertical velocity is also 10 m/s. Calculate its resultant velocity. (Include a diagram to show how you arrive at your answer.)

5 A snooker ball rolls off a table, 0.8 m high. Its initial horizontal velocity is 1 m/s. How long will it take the ball to reach the floor? How far from the table will it land?

6 A stone is thrown horizontally from the top of a 45 m high cliff. It lands 18 m from the foot of the cliff. What was the stone's initial velocity?

P5d Momentum

Mass, speed and sport

Physics can give you an insight into sport. For example, in tennis, your opponent may hit the ball towards you so that it is travelling very quickly. If your return stroke is weak, you may not give the ball enough velocity for it to get over the net.

If you play tennis in wet weather, the ball may absorb water and become waterlogged. Then it is more difficult to strike it so that it travels a long way. Its mass has increased, so the effect of your hit is reduced.

Mass and velocity are important in many other sports. For example, in rugby, it is difficult to stop an opposing player who is large and moving fast (Figure 5d.1). It is much easier to tackle a small, slow-moving player.

In hockey or cricket, if you strike the ball very hard, you can make it move very fast. It is easier to make the ball move fast if you strike it as it goes past, so that you increase its velocity.

It is much harder to hit it head-on, so that it stops and flies off in the opposite direction.

If you play a sport, you develop an intuitive grasp of these ideas. Your experience allows you to predict where a ball will go, or where a player's movement will take them. Physics can help us to explain these and other observations of how things move.

Figure 5d.1 Jonah Lomu, the New Zealand rugby player, has a large mass and can run fast. This makes it very difficult for his opponents to stop him as he runs for the line. It is much easier to tackle a lightweight opponent who is running slowly.

Collisions

Here are two situations that show how, in Physics, we can think about what happens when two objects collide.

- In a game of snooker, the white ball runs across the table and strikes a red ball. The collision is head-on, rather than a glancing blow. What happens? The white ball stops dead, while the red ball moves off. It moves in the same direction as the white ball was originally moving, and with the same speed. The 'movement' of the white ball has been transferred to the red ball.
- Twin brothers (with equal masses) are skating on an ice-rink. One is stationary in the middle of the rink. The other skates towards him, and they collide. They move off together, but at half the speed of the brother who was originally moving. His 'movement' has been shared with his brother.

Collisions can be much more complicated than these, with objects bouncing off one another, striking each other at glancing angles, and so on. To understand how to give a scientific description of what is going on, and to be able to predict what will happen, we can start with two simple experiments. (These correspond to the two collisions described above.)

In the first experiment (Figure 5d.2a), a trolley of mass m, moving with velocity v, collides with an identical stationary trolley. The collision must be springy, so the spring-load of the moving trolley must be released. When it hits the stationary trolley, it stops moving and the second trolley moves off. Its velocity v is equal to that of the first trolley. This is just how the snooker balls behaved. Before the collision, we have a single trolley of mass m moving with velocity v; afterwards, we again have mass m moving with velocity v.

In the second experiment (Figure 5d.2b), the collision is sticky, rather than springy.

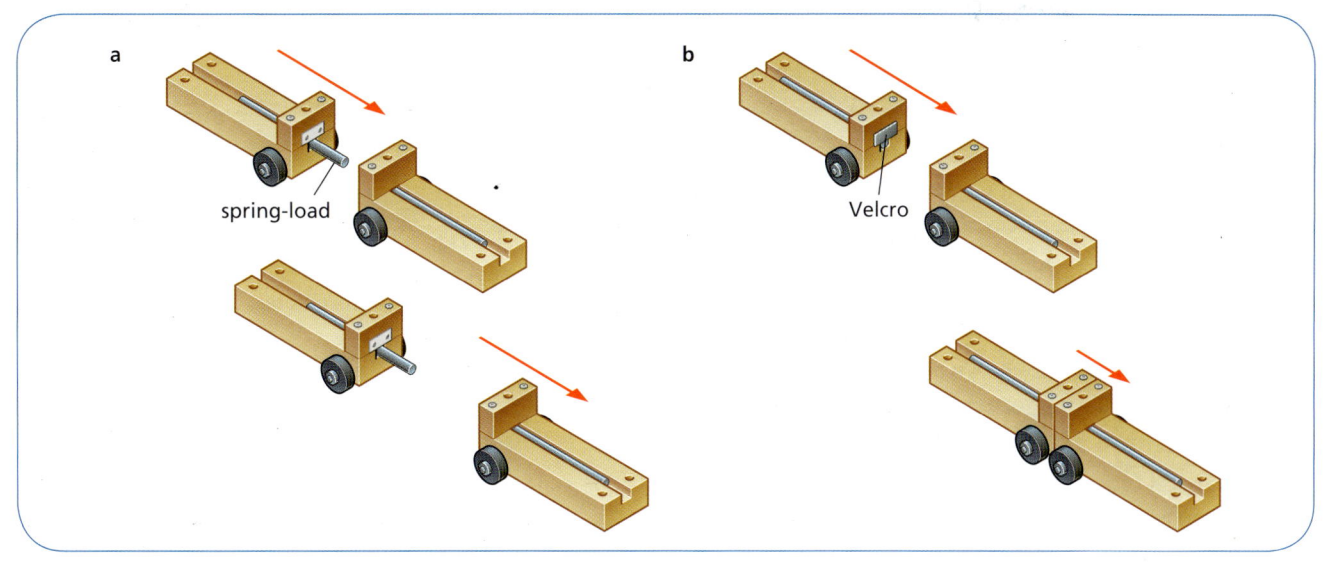

Figure 5d.2 In each of these collisions, a moving trolley runs into a stationary one. **a** The collision is springy; the second trolley moves off, leaving the first one stationary. **b** The collision is sticky; the trolleys stick together and move off together after the collision.

The first trolley collides with the second, and they stick together and move off, just like the ice-skating twins. Measurements of the trolleys' velocities show something that you might have guessed: if the first trolley's velocity is v, then the velocity of the two trolleys together is $v/2$. After the collision, the mass that is moving has doubled, but its velocity has halved.

Predictability

These experiments suggest that, in order to understand collisions, we need to think about both mass m and velocity v. There is a very important quantity that combines both of these; this quantity is called **momentum**. To calculate the momentum of a moving body, we multiply together its mass and velocity:

momentum = mass × velocity
momentum = mv

In the springy collision, the first trolley starts off with momentum mv; in the collision, it transfers all of its momentum to the second trolley. In the sticky collision, the first trolley again has momentum mv; this time, after the collision, we have a double trolley (mass = $2m$) moving more slowly (velocity = $v/2$). The momentum of the double trolley is $2m \times v/2 = mv$. So the first trolley has shared its momentum with the second.

In the discussion of the snooker and ice-skating collisions above, we said that 'movement' was being transferred or shared. Now we can see that the correct quantity to describe this is momentum. This is a very important understanding, because it helps us to predict what will happen in the event of a collision, and predicting the outcome of events is one of the main aims of science.

Worked example 1 shows how to calculate momentum. Momentum is calculated as mass (in kg) × velocity (in m/s), so its units are kg m/s.

Worked example 1

In a rugby match, a forward is about to tackle the fly half of the opposing team. Who has more momentum, the forward (mass 125 kg, running at 8 m/s) or the fly half (mass 80 kg, running at 10 m/s)?

In each case, we can calculate the momentum using momentum = mass × velocity.

momentum of forward = 125 kg × 8 m/s
    = 1000 kg m/s

momentum of fly half = 80 kg × 10 m/s
    = 800 kg m/s

So the forward has more momentum. He is moving a little more slowly than the fly half, but his mass is much greater.

SAQ

1. Calculate the momentum of a runner of mass 75 kg running at 8 m/s.
2. Calculate the momentum of a bird of mass 200 g which flies 100 m in 20 s.

Momentum is conserved

The trolley experiments above suggest something very important about momentum: there is as much momentum after a collision as there is before. This is true in any situation, not just in collisions. The total amount of momentum is always the same before and after a collision, an explosion or any other event. We say that *momentum is conserved*. This is the **principle of conservation of momentum**:

> When two objects interact, the total amount of momentum remains constant.

(This is only true if there are no forces acting on the objects from the outside. The only forces must be the ones the objects exert on each other.)

This principle can allow us to calculate the outcomes of events, as is shown in Worked example 2.

Worked example 2

A trolley of mass 2 kg is moving at 5 m/s. It collides with a second trolley of mass 8 kg, and it bounces back with a velocity of 3 m/s. With what velocity does the second trolley move off?

In problems like this, it is best to draw a 'before and after' diagram to show the information provided in the question – see Figure 5d.3. The top half of the diagram shows the situation before the collision; the lower half shows the situation after the collision. The trolleys are marked with their masses and velocities. The only unknown quantity is v, the velocity of the second trolley after the collision.

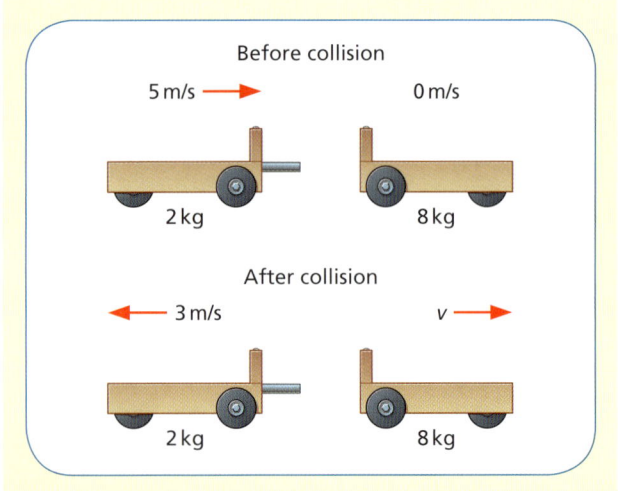

Figure 5d.3 This shows the situation before and after the collision of the two trolleys. Velocities are shown with arrows, because their direction is important.

Step 1: we calculate the total momentum of the two trolleys before the collision. (Note that the velocity of the second trolley is zero.)

momentum = 2 kg × 5 m/s + 8 kg × 0 m/s

= 10 kg m/s

Step 2: we write down the momentum of the trolleys after the collision. In this case, the first trolley is now moving backwards, so its velocity is negative.

momentum = 2 kg × (−3 m/s) + 8 kg × v

= 8 kg × v − 6 kg m/s

Step 3: because we know that momentum is conserved, we can put this equal to the total momentum calculated in Step 1.

8 kg × v − 6 kg m/s = 10 kg m/s

Rearranging this gives:

8 kg × v = 10 kg m/s + 6 kg m/s

8 kg × v = 16 kg m/s

$$v = \frac{16 \text{ kg m/s}}{8 \text{ kg}}$$

= 2 m/s

So the second trolley moves off at a speed of 2 m/s.

continued on next page

Worked example 2 - continued

Checking: In calculations like this, it is a good idea to check the final answer by calculating the total momentum before and after the collision. We need to be sure that no momentum has appeared or disappeared.

As we calculated above:

momentum before collision
$$= 10 \text{ kg m/s}$$

momentum after collision
$$= 2 \text{ kg} \times (-3 \text{ m/s}) + 8 \text{ kg} \times v$$
$$= 2 \text{ kg} \times (-3 \text{ m/s}) + 8 \text{ kg} \times 2 \text{ m/s}$$
$$= -6 \text{ kg m/s} + 16 \text{ kg m/s}$$
$$= 10 \text{ kg m/s}$$

Hence total momentum before collision = total momentum after collision, as required by the principle of conservation of momentum.

A useful formula

You can also do this type of calculation using a formula. We imagine object 1 colliding with object 2. The formula is:

$$m_1 u_1 + m_2 u_2 = m_1 v_1 + m_2 v_2$$

Here, m_1 is the mass of object 1, u_1 is its initial velocity and v_1 is its final velocity. The subscript 2 shows the same quantities for object 2.

SAQ

3 A trolley of mass 5 kg is moving at a speed of 0.4 m/s. It collides with a stationary trolley of mass 15 kg. They move off together at 0.1 m/s. Calculate the momentum of the first trolley before the collision, and the combined momentum of the two trolleys after the collision. Is momentum conserved?

4 A ball of mass 500 g, moving at 2 m/s, collides with an identical stationary ball. They move off in the same direction. The first ball moves at 0.5 m/s. How fast does the second ball move?

Momentum is a vector quantity

The idea of momentum can also help us to predict what happens in an explosion. Figure 5d.4 shows the explosion of a firework fixed to the top of a tall tower. Before the explosion, the firework is stationary; its momentum is zero. Afterwards, material flies off in all directions. Does this mean that momentum has been created out of nothing? No. The clue lies in the words *in all directions*. Some material is moving to the left; an equal mass is moving to the right. Some is moving upwards, while some is moving downwards. The momentum of the material moving to the left is equal and opposite to the momentum of the material moving to the right, and cancels it out. The momentum of the material moving upwards is equal and opposite to the momentum of the material moving downwards.

Because momentum involves *velocity* (rather than *speed*), it follows that momentum is a vector quantity. We must always take account of its direction. (We did this in Worked example 2, where the trolley rebounded with a negative velocity, so that its momentum was −6 kg m/s.)

Figure 5d.4 A firework explodes. The sparks have momentum; the momentum of particles moving to the left is balanced by the momentum of particles moving to the right. It would be very surprising if all the sparks flew off to the left, with nothing moving to the right.

SAQ

5 A ball is moving to the right with momentum 5 kg m/s. It bounces off a wall, and moves in the opposite direction at the same speed as before.

   a What is the ball's new momentum?

   b By how much has the ball's momentum changed? (Hint: remember that momentum is a vector quantity.)

   c How much momentum has been transferred to the wall?

Explaining explosions

When a bullet is fired from a gun, it flies off at high speed. The person firing the gun must be prepared for its recoil. The bullet is an object of small mass moving at high speed; the gun has a much greater mass, so it moves backwards at a much smaller speed. You may have seen the same effect in a historical drama where soldiers are firing a cannon (Figure 5d.5). They ignite the fuse and jump aside; the cannon ball flies off, and the cannon jerks backwards. Similarly, cannon-fire from early naval ships had to be carefully controlled. As the cannon recoiled, strong ropes attached to the ship's timbers brought it to a halt. This transferred its momentum to the ship, which moved back slightly. If all the cannons on one side of the ship were fired at the same time, it was in danger of capsizing.

Rockets and jet engines also rely on recoil for their motion. A rocket is a controlled explosion. As its fuel burns, hot gases rush out of the back end. They have momentum backwards; the rocket is given an equal amount of momentum in the opposite direction. Similarly, jet engines blast hot gases backwards to give an aircraft momentum forwards.

Figure 5d.5 Recoil can be dangerous. When the cannon is fired, it leaps backwards. In the process of giving momentum to the cannon ball, the cannon is given an equal amount of momentum in the opposite direction. Because the mass of the cannon is much greater than that of the ball, its velocity is much smaller.

Momentum and force

Cats in New York live a dangerous life. Falling from a skyscraper is no fun. When some New York vets investigated their records, they were surprised to find that cats that fell from higher than the third floor were more likely to survive than those that fell from lower down. How could this be?

When a cat lands on the ground, it uses its legs to break its fall. It is advisable for people to do the same thing – allowing your legs to bend as you land results in a smaller force and a smaller chance of a broken limb. The vets found that, if a cat fell from high up, it had time to twist its body round and have its legs ready to break its fall.

When an object falls to the ground, it has momentum. It loses its momentum when it hits the ground; if this process is spread over a longer period of time, the force will be smaller. A cat's 'crumple zone' is its legs, which it bends as it lands on the ground. Dogs can't do this, so don't drop a dog.

Athletes have to think about this too. A sudden halt can result in a large force, which can cause serious injury. The pole-vaulter in Figure 5d.6 has a large, soft mat to land on – that's essential when falling from a height of 5 m.

Figure 5d.6 A pole-vaulter lands on a soft mat. This reduces the force of impact.

A tennis racket strikes a ball; the ball slows down, stops for an instant, and then flies back in the opposite direction. This is an example of a collision. How can we use the idea of momentum to understand what is going on here?

When the ball is struck, its momentum is changed. The force of the racket changes the ball's momentum. The bigger the force, and the longer it acts for, the greater its effect will be. This idea can help us to understand how car safety features such as crumple zones work (Item P3f, *Crumple zones* in *Gateway Additional Science*).

If you are in a car that is involved in a collision, you come to a sudden halt. Your momentum has been reduced very rapidly to zero by the force acting on you. The bigger the force that acts on you, the more your body can be damaged. Modern cars are designed to avoid the occupants coming to a sudden halt in the event of a collision; rather, they should come gradually to a halt, so that the forces they experience are as small as possible.

- A crumple zone ensures that the car doesn't bounce backwards. Instead, the car halts relatively gradually, as its front end collapses.
- A seat belt has a certain amount of 'give' in it.
- An airbag also helps to prevent the passenger from coming to an abrupt halt.

These devices all help to ensure that the momentum of the car's occupants decreases slowly to zero.

SAQ

6 Why should a high-jumper bend her legs on landing? Use the words *momentum* and *force* in your answer.

Momentum, force and time

An unbalanced force is needed to change an object's momentum. The bigger the force and the longer it acts, the greater the change in momentum. We can write this as an equation:

force × time = change in momentum

This equation shows that an alternative unit for momentum is the newton second (Ns). Worked example 3 shows how we can use this equation to calculate a force.

Worked example 3

The Ariane rocket travels into space carrying liquid oxygen and hydrogen as its fuel supply (Figure 5d.7). Its third stage burns fuel at the rate of 14.2 kg/s, and the exhaust gases leave the rocket at a speed of 4440 m/s. What thrust does the rocket provide?

Step 1: re-writing the equation for change in momentum, we have:

$$\text{force} = \frac{\text{change in momentum}}{\text{time}}$$

Step 2: in 1 s, 14.2 kg of fuel is given a speed of 4440 m/s. So the momentum gained by the fuel in 1 s is:

change in momentum = 14.2 kg × 4440 m/s
= 63 048 kg m/s

Step 3: the force provided by the rocket is thus:

$$\text{thrust} = \frac{\text{change in momentum}}{\text{time}}$$

$$= \frac{63\,048\,\text{kg m/s}}{1\,\text{s}}$$

$$= 63\,048\,\text{N}$$

So the rocket provides a thrust of about 63 kN (kilonewtons).

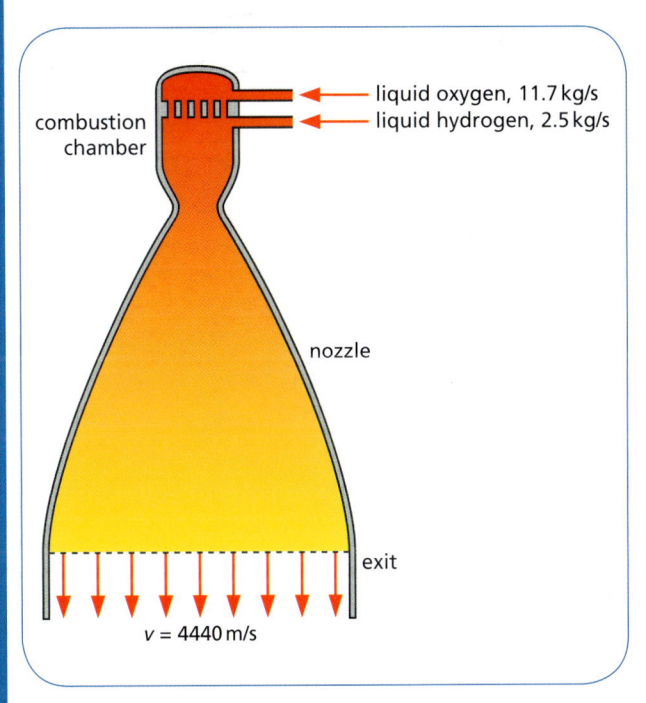

Figure 5d.7 The third stage of an Ariane rocket. Because this rocket travels well above the Earth's atmosphere, it has to carry its own oxygen supply. The calculation of Worked example 3 shows that it provides a thrust of about 63 kN.

238 P5d Momentum

SAQ

7 Cars have crumple zones to reduce the effects of a sudden impact. A more rigid car would bounce back in the event of a collision. The crumple zone also ensures that the impact is spread over a longer time. Use the equation *force × time = change in momentum* to explain how both of these effects help to reduce the force of the impact.

Pairs of forces

Imagine being stationary in a swimming pool. If you can swim, imagine what you do to get moving (Figure 5d.8). You use your hands and legs to push the water backwards. This provides the force to move you forwards. In momentum terms, you give the water some momentum backwards; it gives you an equal amount of momentum forwards.

Here we have two forces: your push on the water, and the water's push on you. These forces are equal in size, and opposite in direction. One is exerted by you, and acts on the water. The other is exerted by the water, and acts on you. Without the force of the water on you, you would not start moving forwards.

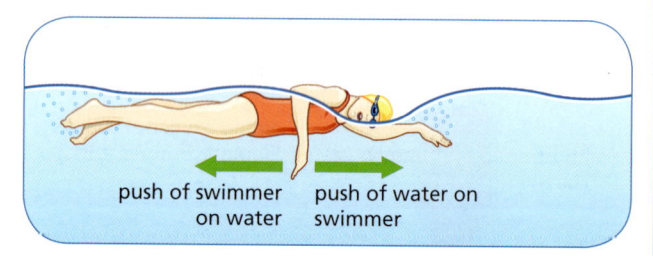

Figure 5d.8 Forces are created in pairs. The swimmer pushes backwards on the water, and the water pushes forwards on the swimmer. These forces are equal and opposite.

Isaac Newton realised that forces are *always* created in pairs. A single force cannot be created out of nothing. Let us look more closely at four more examples of these pairs of equal and opposite forces.

1 If you hold two bar magnets so that the north pole of one faces the south pole of the other, they will attract one another (Figure 5d.9). You will feel that they attract each other equally strongly. Even if one magnet is more strongly magnetised than the other, you will feel that they are pulled equally. Reverse one of the magnets, and you will feel that they now repel each other with equal forces.

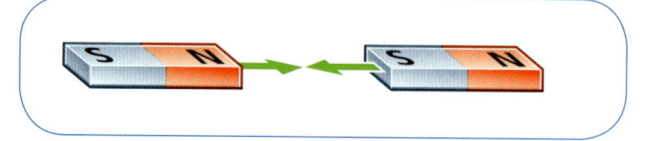

Figure 5d.9 Magnets attract and repel one another with equal and opposite forces. Even if one magnet was replaced by a piece of unmagnetised iron, the forces of attraction would still be equal and opposite.

2 The force of friction is necessary for walking. To walk, you push backwards on the ground. Your foot is tending to slide backwards on the ground, so a frictional force pushes forwards to oppose you. There are thus two frictional forces acting: one that acts backwards on the ground, and the other that acts forwards on your foot. It is this second force that pushes you forwards.

3 We are used to the idea that the Earth's gravity pulls on us. If we fall down, it is because of gravity. The force of gravity on us is our weight – say, 500 N. But at the same time, we pull on the Earth. We have a gravitational pull, too. We pull upwards on the Earth with a force of about 500 N. However, because the Earth's mass is so great, our pull has very little effect on it.

4 An important example of a pair of equal and opposite forces is shown in Figure 5d.10a. When two objects are touching one another, there is a contact force between them. It is important to realise that contact forces, like all forces, come in pairs. If you stand on the ground, you push down on the ground and the ground pushes up on you. If we concentrate on the point where your feet touch the ground, Figure 5d.10a shows this pair of forces. (If you don't believe that they exist, imagine putting your hand between your foot and the ground. It would be squashed between the upward and downward forces.) Now let's look at your body as a whole (Figure 5d.10b), and consider only the forces acting on *you*. These are gravity downwards and the contact force upwards. If these forces are equal, you are in equilibrium. (If the contact force is smaller than gravity, you will accelerate downwards through the floor!) These forces are

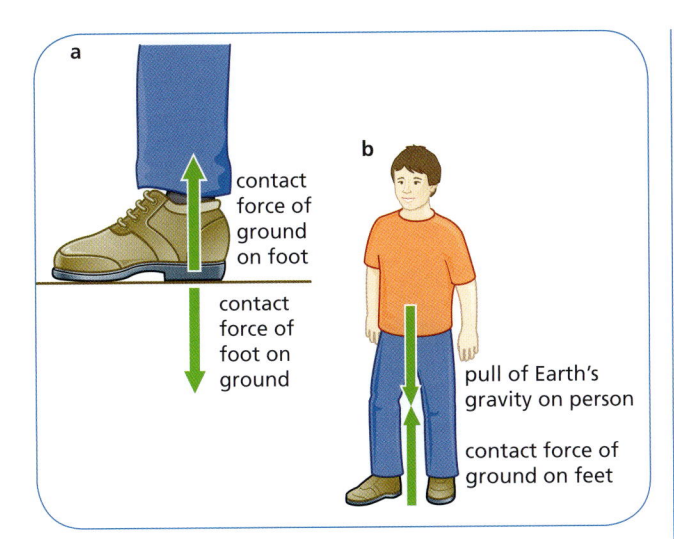

Figure 5d.10 a When any two objects touch, each exerts a contact force on the other. These forces are an equal and opposite pair. **b** This diagram shows the forces acting on the person; it doesn't show the forces exerted by the person on anything else. The forces are balanced – but they are different types, so they do not make a 'pair' like the contact forces do.

balanced, but they are not a 'pair' of the kind Newton described (below), because they are not of the same type.

Newton's third law

The idea that forces are always created in pairs is known as **Newton's third law of motion**:

When bodies interact, they exert equal and opposite forces on each other.

The forces in such a pair must be:
- equal in size
- opposite in direction
- of the same type.

Of the same type means both forces are either gravitational, or contact, or frictional, or magnetic, and so on. In addition, one force acts on one of the bodies; the other acts on the other body. (They don't act on the same body.)

Why is this a law of motion? Newton realised that forces make things speed up or slow down – they cause acceleration. Thus forces change momentum. Whenever an object's momentum is changing, there must be an unbalanced force acting on it (Newton's second law of motion). The principle of conservation of momentum tells us that momentum cannot be created out of nothing – it is always created in equal and opposite amounts, by equal and opposite forces.

You may sometimes hear a pair of forces like this referred to as *action* and *reaction*. The idea is that one force, the action, always results in a second force, the reaction. So the third law is sometimes stated as *For every action, there is an equal and opposite reaction*. However, you need to be careful that you understand what the terms *action* and *reaction* mean.

SAQ

8 You may have carried out experiments on static electricity, in which toy balloons are rubbed so that they become charged. Draw a diagram to show two such balloons placed close to each other, each having a positive charge. Show the force each exerts on the other. What can you say about the sizes and directions of these forces?

Summary

You should be able to:

- state that the momentum of an object is the product of its mass and velocity:

    momentum = mass × velocity = mv

- **[H]** state that momentum is conserved

- describe how, to reduce the force on an object in a collision, the length of time the collision takes should be increased, and give some examples of car safety features that use this idea

- **[H]** explain that a force can cause a change in momentum, depending on the time for which it acts:

    $$\text{force} = \frac{\text{change in momentum}}{\text{time}}$$

- state Newton's third law of motion: when bodies interact, they exert equal and opposite forces on each other

Questions

1. What is the momentum of a child of mass 40 kg running at 4 m/s?

2. Which has more momentum, spacecraft A (mass 500 kg, velocity 1000 m/s) or spacecraft B (mass 1000 kg, velocity 500 m/s)?

3. When you stand on the floor, your feet and the floor exert equal and opposite forces on each other. What phrase is used to describe such a pair of forces?

4. A girl stands on the floor. She exerts a downward force of 500 N on the floor. What force does the floor exert on her?

5. Your weight is the force of the Earth's gravity on you. What is the equal and opposite reaction to this force?

6. **[H]** Sketch a diagram to show how a jet engine provides a force to move an aircraft forwards.

7. An astronaut of mass 100 kg is floating in space. He throws a hammer of mass 2 kg with a velocity of 10 m/s. The astronaut moves off in the opposite direction. What will be the astronaut's velocity?

8. A trolley of mass 2 kg and moving at 4.5 m/s collides with a stationary trolley of mass 1 kg. They stick together. At what speed do they move?

9. A trolley of mass 20 kg is moving at 1.2 m/s. A bag of sand is placed on the trolley; it slows down to 1.0 m/s. What is the mass of the bag of sand?

10. A bus driver applies the brakes. A force of 5000 N acts for 4 s. By how much does the bus's momentum change?

11. A car of mass 800 kg is moving at 20 m/s. What force is needed to bring it to a halt in 16 s?

5e Satellite communication

Tsunami telephone

Communications systems using geostationary satellites can be invaluable. In December 2004, a tsunami struck in the Indian Ocean. Several remote island groups, including the Maldives and the Andaman Islands, were swamped. However, news of their predicament could be spread rapidly by means of satellite telephone systems (Figure 5e.1). These rely on an electricity supply, which can be provided using solar cells. They do not require expensive transmitters and long cables, so they can stand up to adverse conditions.

After the tsunami, a warning system was established for the area. In the future, advice of an approaching tsunami will be transmitted over satellite links, giving people a better chance of escape.

Figure 5e.1 These satellite communication dishes on Raa island, one of the Maldives, send and receive signals. They are hi-tech devices, but they are simple to operate and prove invaluable to the islanders.

Keeping in touch

Today, many hundreds of artificial satellites orbit the Earth. Each operational satellite is in touch with one or more ground stations, receiving instructions and other data, and sending signals back down to Earth. As we saw in Item P5a, *Satellites, gravity and circular motion*, communications satellites beam TV broadcasts and other signals down to receivers on Earth. In this item, we will look at the way ground stations and satellites exchange signals, and at other ways in which radio and TV signals are sent around the world.

Transmit and receive

A satellite ground station usually has a large 'dish' aerial, which can be moved around to direct its signal at a particular satellite (Figure 5e.2, over the page). Positioned above the dish is the transmitter, which sends out the microwave signals. These reflect off the dish to form a narrow beam, which passes up through the atmosphere to the satellite.

Such a dish can both send and receive signals. Incoming signals are focused on the detector, which is next to the transmitter. A cable carries the electrical signals to a receiver, which decodes them.

The signals are carried by microwaves, a type of electromagnetic radiation similar to radio waves. The signals are in digital form. This allows a lot of information to be transmitted in a short time. It also means that the signals can be 'cleaned up' easily if they gather noise (Item P1e, *Communicating with infra-red radiation*).

The orbiting satellite has a similar, smaller dish that collects the signals. The two dishes must be carefully lined up so that the narrow beam of radiation reaches its target. The frequencies used are:

- between 100 MHz and 2 GHz for low polar orbit satellites
- between 6 GHz and 24 GHz for geostationary satellites.

(1 MHz = 1 megahertz = 1 million waves per second; 1 GHz = 1 gigahertz = 1 billion waves per second.)

SAQ

1. Many homes in the UK have a satellite dish (aerial) to pick up television broadcasts. Explain why the dish must be carefully aligned, so that it points at the right part of the sky.

P5e Satellite communication

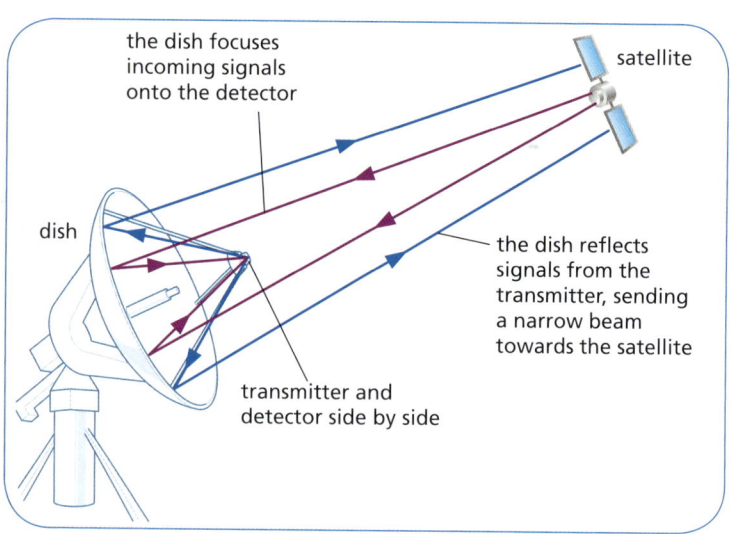

Figure 5e.2 a A satellite dish receives and transmits microwave signals. These three dishes are at a NASA ground station in Australia. Each is pointing at a different satellite. **b** The transmitter sends out the microwave signals, which reflect off the dish to form a narrow beam. The dish also focuses incoming signals onto the detector, which is next to the transmitter.

Through the atmosphere

The microwave signals sent up to satellites must pass through the Earth's atmosphere. Similarly, satellite television broadcasts must pass down through the atmosphere if the millions of small receiver dishes on Earth are to pick them up.

To us, the atmosphere looks clear (unless it is cloudy). This is because visible light passes straight through without being absorbed. However, the situation is more complicated for radio waves. Only certain wavelengths can pass through the atmosphere without being absorbed or reflected.

Radio waves and microwaves are part of the electromagnetic spectrum. The spectrum (Figure 5e.3) ranges from gamma rays to radio waves. Microwaves are part of the radio wave section of the spectrum.

Figure 5e.3 shows the parts of the spectrum that can pass through the atmosphere, and the parts that are absorbed. In broad terms, there are two 'windows': one lets through visible light and infra-red, while the other lets through parts of the microwave and radio wave region.

Transmit, reflect, absorb

The radio 'window' in the atmosphere extends from about 30 MHz to about 30 GHz. Radio waves with frequencies within this range are transmitted through the atmosphere. This means they are suitable for communicating with spacecraft (and for observing radio waves coming from far out in space).

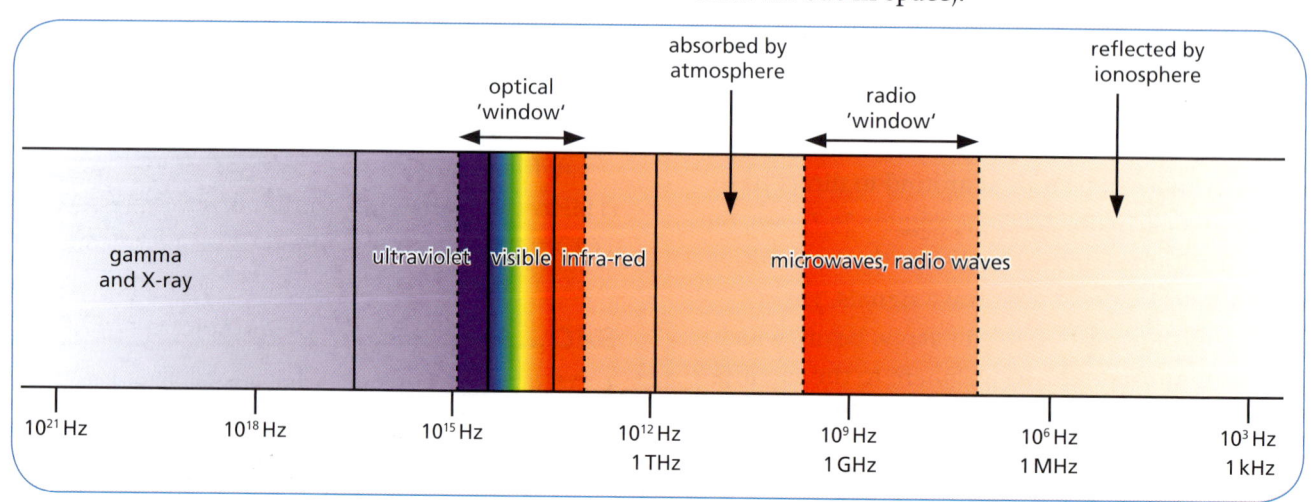

Figure 5e.3 The electromagnetic spectrum, showing the regions that can pass through the Earth's atmosphere.

What happens to radio waves with frequencies outside this range?

- Radio waves with frequencies below 30 MHz are reflected back down to Earth (Figure 5e.4). They reach the ionosphere, a layer of charged particles high in the Earth's upper atmosphere, and are reflected back towards the ground.

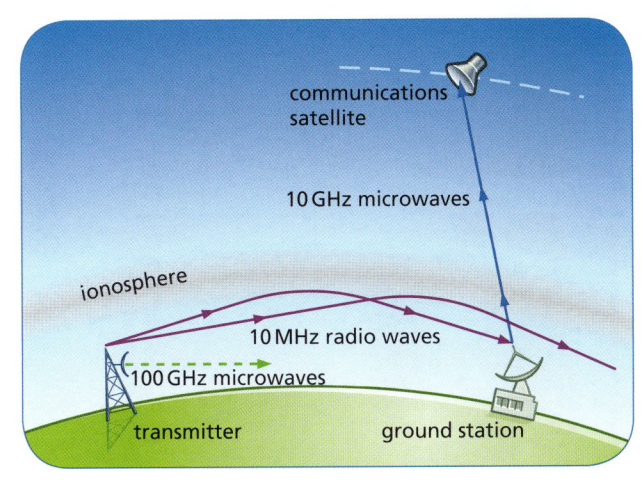

Figure 5e.4 Only radio waves with frequencies in the radio 'window' can pass through the Earth's atmosphere.

- Radio waves (microwaves) with frequencies greater than 30 GHz are absorbed and scattered by dust, rain and water vapour in the atmosphere. This means that they are unsuitable for communicating with satellites. (However, it does mean that these frequencies can be used to detect clouds and rainfall, so they are used in weather-forecasting.)

SAQ

2 What will happen to radio waves with the following frequencies when transmitted upwards from the Earth's surface?

   a 10 MHz
   b 100 MHz
   c 10 GHz
   d 100 GHz

3 Explain why an orbiting spacecraft could not use radio waves of frequency 10 MHz to communicate with a ground station.

Diffraction effects

Radio waves are not just used for communicating with spacecraft. They are also used to broadcast radio and television programmes. A radio set like the one shown in Figure 5e.5 can pick up radio signals broadcast by stations many hundreds of kilometres away. It has two aerials, one external and one internal, which collect radio signals. Then its electronic circuits separate the different stations so that the user can tune in to any one of many stations.

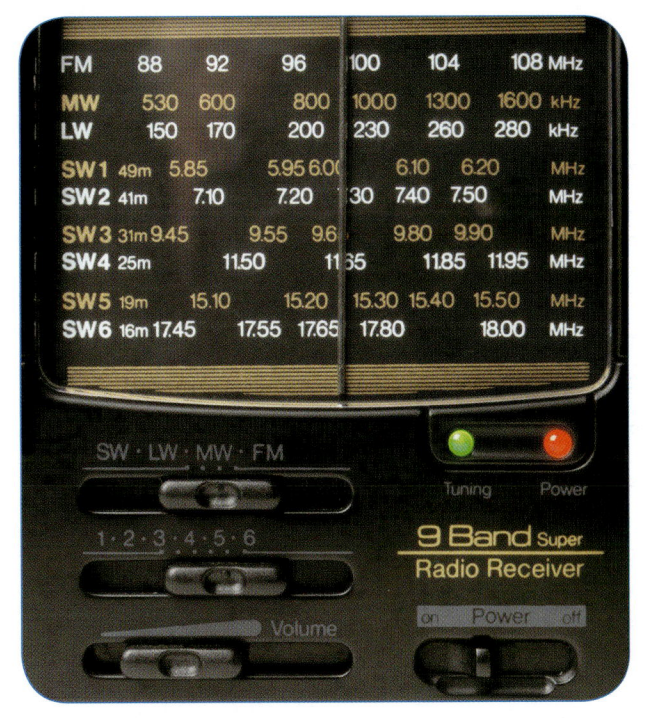

Figure 5e.5 This radio set can receive signals broadcast by stations in many different countries.

The Earth is spherical. So how can a radio set pick up signals from distant stations? You might expect it to be essential for the receiving aerial to be in *line-of-sight* with the transmitter. However, this is not so. There are two effects which make it possible for radio waves to be picked up even when the transmitter is out of 'sight' (Figure 5e.6, overleaf).

a Radio waves can be reflected by the ionosphere. This allows them to travel around the curve of the Earth's surface.

b Radio waves curve around obstacles such as tall buildings or hills. This also means that they can curve over the horizon. This curving or bending of waves around an obstacle is called **diffraction**.

244 P5e Satellite communication

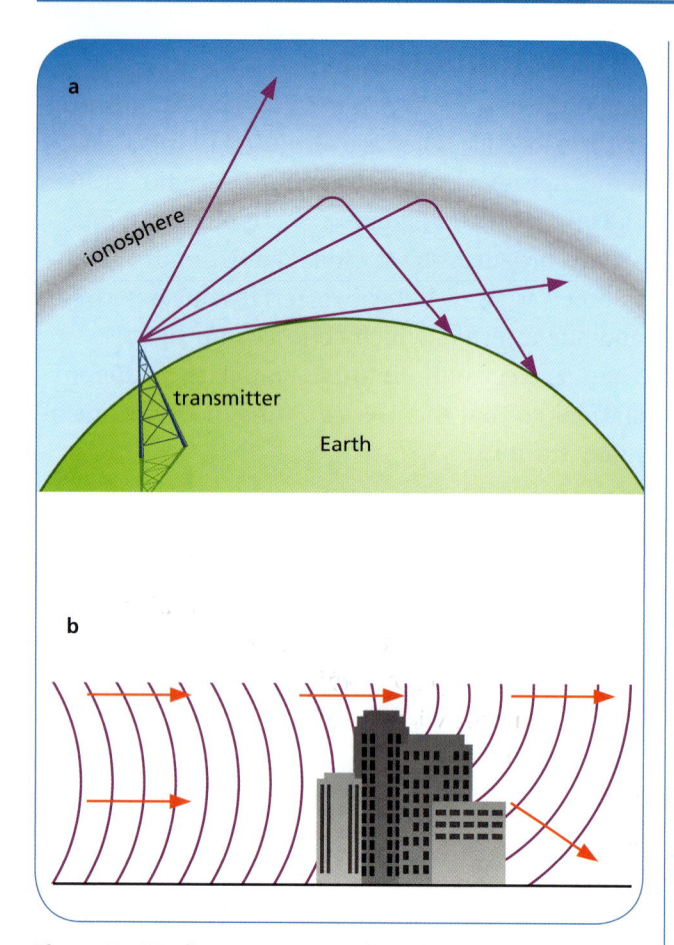

Figure 5e.6 Radio waves can travel over long distances around the Earth. **a** They are reflected by the ionosphere. **b** They are diffracted around large obstacles.

SAQ

4 Look at Figure 5e.6, showing how radio waves travel around the Earth. Use it to explain why one listener may be unable to pick up a particular radio station, while someone further away may be able to do so.

Observing diffraction

A ripple tank is a useful device for observing the behaviour of waves. The tank contains shallow water, and a vibrating bar disturbs the surface, sending out parallel waves known as **ripples**. Figure 5e.7 shows what happens when a barrier blocks the path of the ripples.

a Ripples passing the edge of a barrier curve 'round the corner', into the space behind the barrier.

b Ripples passing through a gap in a barrier spread out into the space beyond the barrier.

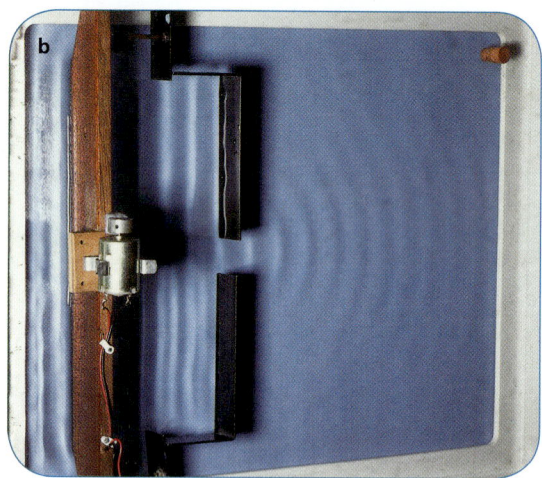

Figure 5e.7 Ripples are diffracted: **a** as they pass the edge of a barrier; or **b** as they pass through a gap in a barrier. In both cases, they spread into the space behind the barrier.

These are examples of diffraction. Diffraction can be observed with all types of waves. For example, if someone is talking in the next room, you may be able to hear them through an open doorway even if you cannot see them. The sound waves of their voice are diffracted through the doorway to fill the room beyond. Another example: you might notice water waves being diffracted in a harbour. The waves enter the harbour mouth and spread around corners, so that no part of the harbour is entirely undisturbed. Boats bob up and down on the diffracted waves.

Figure 5e.8 shows the patterns observed when plane (straight) waves pass through gaps of different sizes. If the gap is wide, there is little diffraction. A medium gap gives a lot of diffraction. If the gap is very narrow, the waves cannot pass through it.

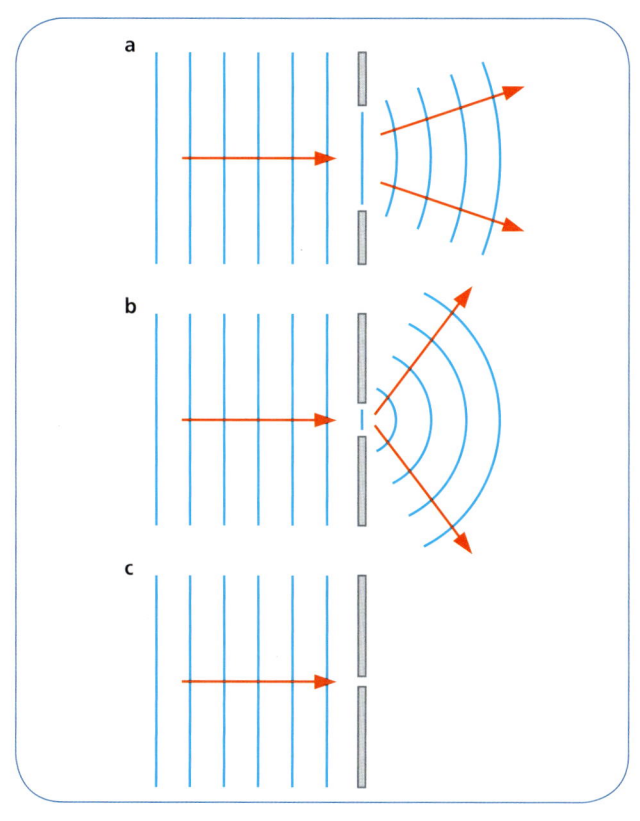

Figure 5e.8 Diffraction is greatest when the width of the gap is similar to the wavelength of the waves being diffracted. When the gap is much smaller than the wavelength, the waves do not pass through at all.

Waves are diffracted most when their wavelength is similar to the size of the obstacles that they encounter. The radio waves used for broadcasting have wavelengths of a few metres (FM) up to several hundred metres (AM). Hence they are diffracted a lot when they encounter large obstacles such as ranges of hills; this is useful because it means that there is no 'shadow' beyond the hills where the signal cannot be received.

SAQ

5 Draw a diagram to show how radio waves are diffracted as they pass a range of hills.

Wavelength and gap size

From Figure 5e.8, you can see that the biggest diffraction effect occurs when the width of the gap is equal to the wavelength of the waves (part b of the diagram). If the gap is much wider than the wavelength, the diffraction effect is less pronounced. If the gap is much narrower than the wavelength, the ripples cannot pass through.

SAQ

6 Look at Figure 5e.8. Use a ruler to measure the wavelength of the ripples (the distance between adjacent blue lines).

   a Does the wavelength change when the ripples have been diffracted?

   b In part b of the diagram, is the size of the gap equal to the wavelength?

Practical considerations

A microwave 'dish' aerial is used to send microwaves up to an orbiting spacecraft. The microwaves have a wavelength of about 20 cm, while the dish is several metres across. The dish is effectively a gap from which the microwaves emerge, and the width of the gap is many times greater than the wavelength of the microwaves. Hence there is little diffraction of the waves, and they therefore form a narrow beam. This is good, because it means that the energy of the beam is not spread out over a wide area. However, it also means that the transmitting and receiving dishes must be carefully aligned to ensure that the beam from the transmitting dish does not miss the receiving dish.

The signals used to communicate with spacecraft are digital; that is, they are a series of *ons* and *offs*, rather like Morse code. Digital signals are also used for satellite TV transmissions, and increasingly for terrestrial radio and TV broadcasts (*terrestrial* means Earth-based).

However, longwave radio transmissions are analogue signals. To carry the signal, the wave height (amplitude) is varied to match the amplitude of the sound wave it represents (see Figure 5e.9). This is called **amplitude modulation**, or AM for short. Because these waves have wavelengths of hundreds or thousands of metres, they are diffracted around large obstacles and even over the Earth's horizon.

246 P5e Satellite communication

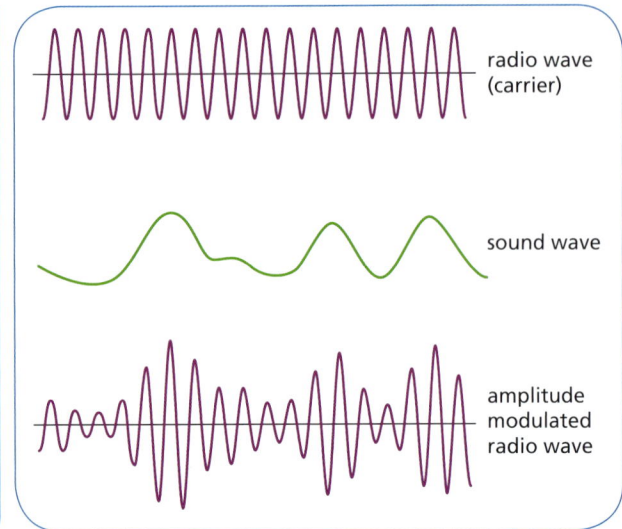

Figure 5e.9 Longwave radio transmissions make use of amplitude modulation; the amplitude (height) of the carrier wave varies with the shape of the original sound wave. One alternative to AM is frequency modulation (FM); this is used for terrestrial radio and TV broadcasts in the VHF and UHF wavebands.

Summary

You should be able to:

- describe how microwaves are used to transmit signals between spacecraft and ground stations

- explain that the atmosphere is transparent to radio waves with frequencies between 30 MHz and 30 GHz; waves with lower frequencies are reflected by the ionosphere, while those with higher frequencies are absorbed by dust, rain and water vapour in the atmosphere.

- describe how waves are diffracted when they pass through a gap or around the edge of an obstacle

- explain how diffraction causes radio waves to curve over hills and over the horizon

- state that the diffraction effect is greatest when the width of the gap is equal to the wavelength of the waves

- describe how longwave radio waves are used to carry signals by amplitude modulation

Questions

1 From which part of the electromagnetic spectrum are the waves that are used to communicate with orbiting spacecraft?

2 Here is a list of frequencies that lie in the radio wave section of the electromagnetic spectrum:

  50 GHz, 50 MHz, 5 MHz, 5 GHz

  a Which of these will be reflected in the upper atmosphere?

  b Which will be absorbed by substances in the atmosphere?

  c Which would be suitable for communicating with a spacecraft?

continued on next page

Questions - continued

3. Radio waves can spread around large objects; this is diffraction.

   a. Explain why the diffraction of radio waves is useful in broadcasting.

   b. Give another example of waves spreading out from a gap.

4. Figure 5e.10 shows waves of two different wavelengths approaching a gap in a barrier. Which waves will be diffracted more as they pass through the gap? Copy the diagram and complete it to show the effect you would expect to see.

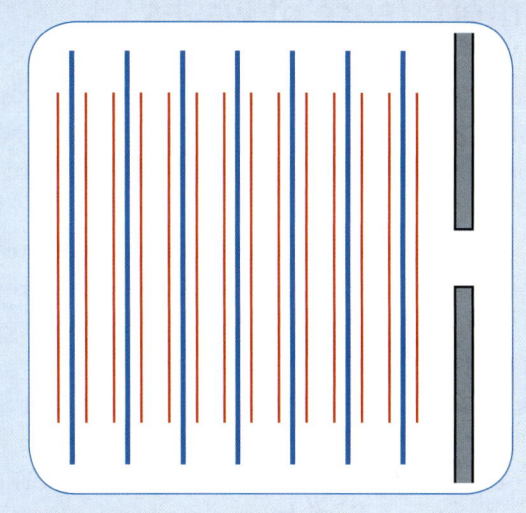

Figure 5e.10

5. The ionosphere is a series of electrically charged layers in the upper atmosphere. Figure 5e.11 shows that some of these layers disappear at night. Copy the diagram and add to it, to explain why it may be possible to receive a signal from a distant radio station at night, while this is impossible during the day.

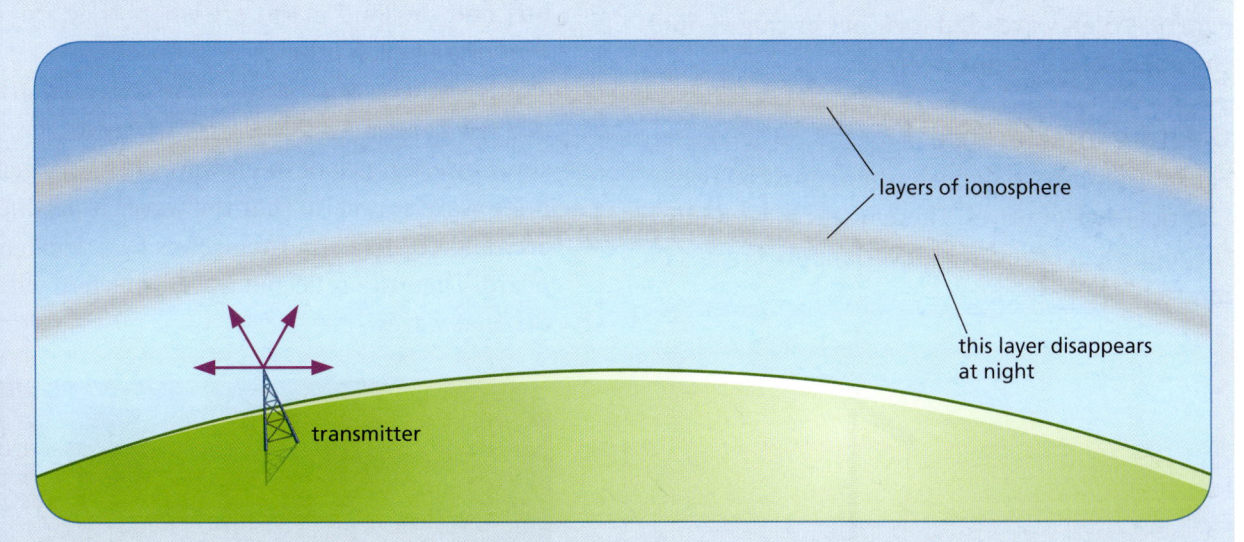

Figure 5e.11 Some layers of the ionosphere disappear at night.

6. Different frequencies of microwaves are used to communicate with different types of satellite.

   a. Which use higher frequencies, geostationary or polar-orbit satellites?

   b. Which use shorter wavelengths?

   c. Which will be diffracted more as they leave the transmitting dish?

P5f The nature of waves

Interference of waves

In Item P5e, *Satellite communication*, you saw how waves can be diffracted. Diffraction is a property of waves; particles cannot be diffracted. A beam of particles passes through a gap and continues in its original direction.

Figure 5f.1 shows another example of wave behaviour. In this case, two sets of circular ripples are created in a ripple tank. Where they overlap, something unusual is observed.

- At some points, the surface of the water is greatly disturbed; it goes up and down a lot because the waves are adding to each other.
- At other points, the surface of the water is almost still, despite the fact that two sets of ripples are passing by. The waves are cancelling each other out.

These points form a sort of grid on the surface of the water. The pattern is known as an *interference pattern*. When waves add together or cancel out, the effect is called **interference**.

Creating interference

For interference to happen, there must be two identical sets of waves. These waves meet and pass through each other. Their overlapping creates an interference pattern.

- At some points, the two waves reinforce each other (they add together), producing a large disturbance.
- At points in between, the two waves cancel each other (they subtract), so that there is no disturbance.

Interference can be shown with ripples on the surface of water, and also with other types of waves such as light (Figure 5f.7) and sound (Figure 5f.2).

Sound waves: Figure 5f.2 shows two loudspeakers connected to a single signal generator. The speakers produce identical sound waves, and an interference pattern can be detected in the space beyond them. What is this interference pattern like? If you move around in the space in front of the speakers, you will find that there are points where the sound is very soft, and other points where it is loud. The soft places are where the two sets of waves are cancelling out, and the loud places are where they are reinforcing each other.

Microwaves: Microwaves have a wavelength of a few centimetres. The easiest way to show interference of microwaves is to place a microwave source so that the waves it produces are reflected back on themselves by a wooden board. Then the reflected waves interfere with the original waves.

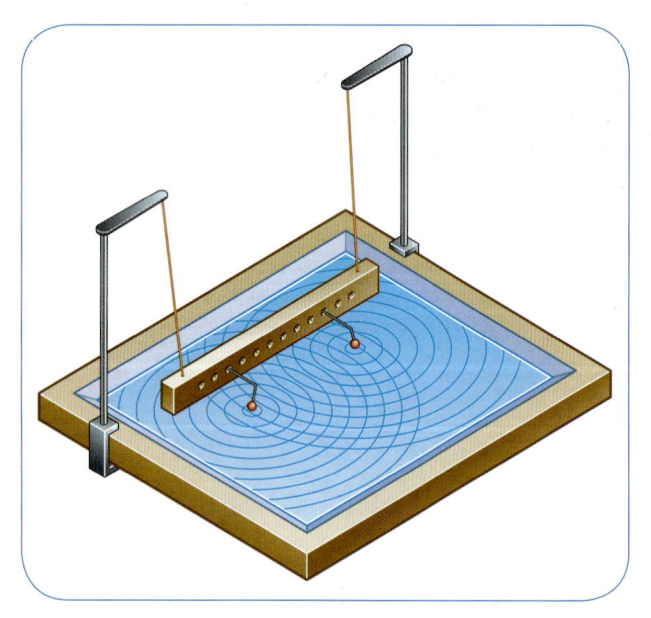

Figure 5f.1 Interference in a ripple tank. **a** Each dipper produces a set of circular ripples. **b** Together, the ripples produce an interference pattern.

P5f The nature of waves   249

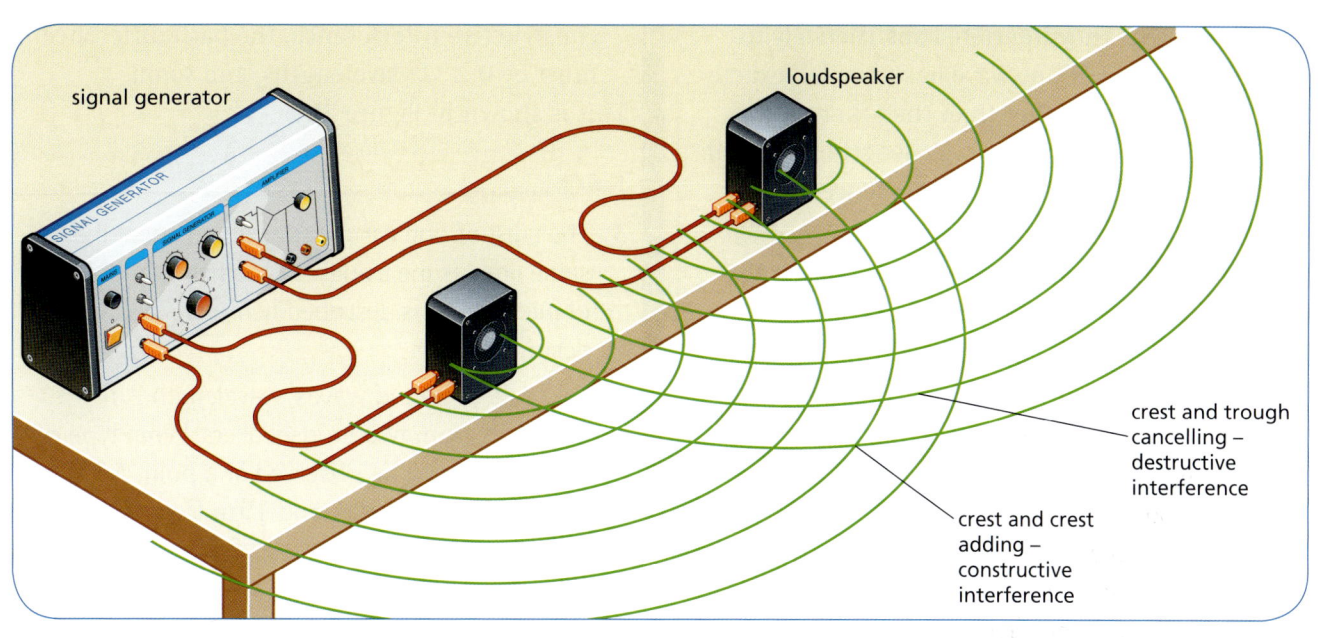

Figure 5f.2 How to observe an interference pattern for sound. It is easier to detect the pattern if you put a finger in one ear so that you are only listening with one ear.

SAQ

1 In an experiment to observe interference of sound waves, a pattern is produced with some loud points and some soft points.

   a At which points are the sound waves reinforcing each other?

   b At which points are they cancelling out?

2 For interference to occur, the two sets of waves must have the same frequency (so that they have the same wavelength).

   a Explain why the two sets of ripples in the ripple tank (Figure 5f.1) have the same frequency.

   b Explain why the two sets of sound waves in the sound experiment (Figure 5f.2) have the same frequency.

Explaining interference

Think back to the example of interference in the ripple tank. Each point in the ripple tank receives one set of ripples from one dipper, and another set from the other dipper. Whether the waves add up or cancel out depends on whether the two sets are in step or not (see Figure 5f.3).

a If the ripples are in step, they combine to give a bigger ripple with twice the amplitude. This is **constructive interference**.

b If the ripples are out of step, they cancel to give no ripple. This is **destructive interference**. (At most points, the ripples are neither exactly in step nor exactly out of step, and they may not have exactly the same amplitude, so the outcome is somewhere in between the two extremes shown in the diagram.)

Why are waves sometimes in step, and sometimes out of step? Think about the experiment showing how sound waves interfere (Figure 5f.2). The waves set off in step with one another from the loudspeakers. However, they may have to travel different distances to reach your ear. We say that there is a **path difference** between them. If the two waves have travelled exactly the same distance to your ear, the path difference is zero and the waves are in step when they meet.

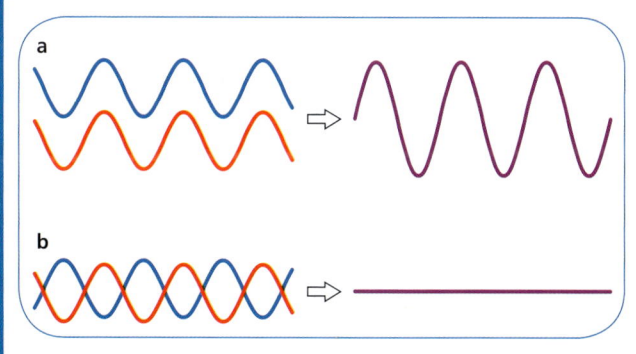

Figure 5f.3 a Identical waves that are exactly in step show constructive interference. **b** If the waves are exactly out of step, they show destructive interference.

250 P5f The nature of waves

This means that they interfere constructively to give a loud sound. If one wave has travelled exactly half a wavelength further than the other (that is, the path difference is half a wavelength), they will interfere destructively to produce silence.

- If the path difference is an odd number of half-wavelengths, the waves will be out of step and they will interfere destructively. This means that, for destructive interference, the path difference must end in half a wavelength: $\frac{1}{2}$, $1\frac{1}{2}$, $2\frac{1}{2}$ wavelengths, and so on.
- If the path difference is an even number of half-wavelengths, the waves will be in step and they will interfere constructively. So, for constructive interference, the path difference must be 0, 1, 2 wavelengths, and so on.

This is shown in Figure 5f.4.

SAQ

3. The crest of one wave arrives at a point at the same time as the trough of another. What type of interference is described here, constructive or destructive?

4. Two sets of sound waves, wavelength 2 m, travel out from two loudspeakers. What type of interference will be observed at a point 3.5 m from one loudspeaker and 1.5 m from the other?

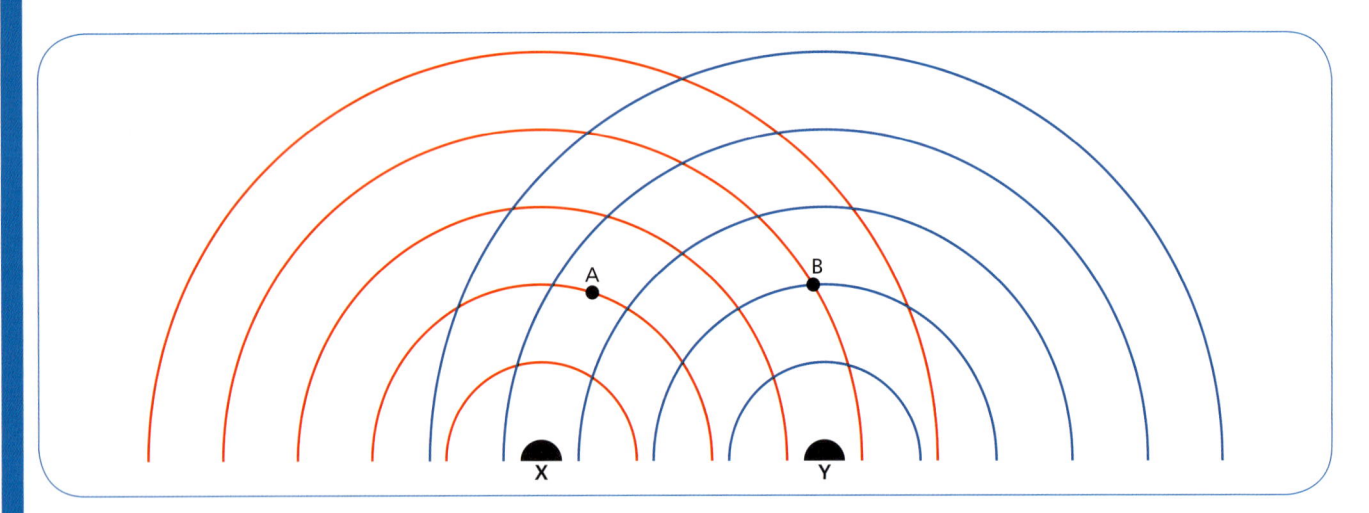

Figure 5f.4 At A, the path difference between the waves is $1\frac{1}{2}$ wavelengths ($3\frac{1}{2}$ wavelengths from source **Y**, and 2 from source **X**), so they interfere destructively. At B, the path difference is 2 wavelengths (4 wavelengths from source **X**, and 2 from source **Y**), so they interfere constructively.

Light: waves or particles?

For two centuries, roughly from 1600 to 1800, physicists argued about the nature of light. Does light travel in the form of waves, or as a stream of particles? No-one doubted that sound travels as a wave. You could see or feel the vibrations of a drum or a guitar string, and imagine the molecules of air being pushed back and forth as the sound waves travelled towards your ears. But what about light?

Experiments with ripple tanks show that water waves are reflected in a similar way to light. However, this does not prove that light is a wave. Particles also reflect off hard surfaces in the same way – think about how a ball bounces off the side of the table in snooker or pool, so that its angle of incidence equals its angle of reflection. So reflection cannot tell us whether light travels as waves or particles. Perhaps refraction can tell us? (Refraction is the bending of a ray of light when it passes from one material into another.)

Isaac Newton refused to accept the idea that light travels as waves. He believed that light travels in the form of tiny particles. He had his own explanation for the refraction of light, shown in Figure 5f.5. He pictured a small particle (such as a marble) rolling down a

continued on next page

Light: waves or particles? – continued

sloping board. When the board gets steeper, the particle moves faster. It also changes direction. If we look down on this model from above, we see that the particle turns towards the normal as it speeds up. This is the opposite to what we observe for ripples: they bend towards the normal when they slow down.

Hence Newton's particle theory predicted that light speeds up when it enters glass. When you try to explain refraction using waves, you find that you can do it if light slows down when it enters glass. Unfortunately, In Newton's time there was no way of doing an experiment to find which idea was right, so the question was unresolved. Eventually, measurements showed that light travels more slowly in glass than in air, so Newton's idea was wrong. However, his influence was so strong that his particle theory of light was favoured by physicists for more than a century after his death.

In Item P5e, *Satellite communication*, we saw that waves can be diffracted – they spread out when they pass the edge of an obstacle, or through a gap. In this item, we have seen that waves can produce interference patterns when they overlap. So a test of the question, *Does light travel as waves or particles?* is to see if light can be diffracted, and if it can produce interference patterns. The answer, as we shall now see, is *Yes, it can*. Figure 5f.6 shows the sort of effect that photographers can achieve using the diffraction and interference of light.

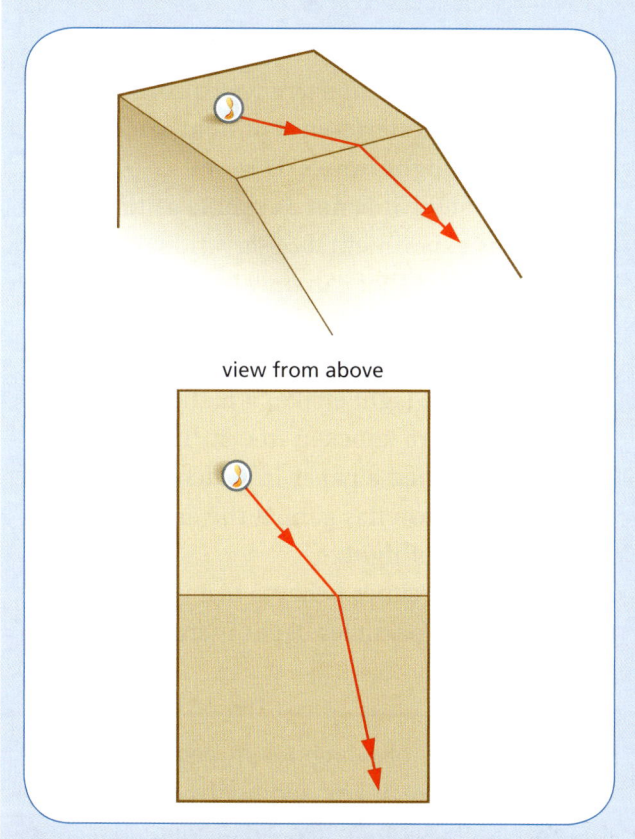

view from above

Figure 5f.5 Newton's idea of refraction. The marble *speeds up* and changes direction as the slope of the board changes. The pattern is the same as we see when a ray of light is refracted, as it goes from air into glass, for example. However, later experiments showed that light *slows down* when it goes from air into glass. So, for once, Newton was wrong!

Figure 5f.6 Photographers use the diffraction of light to produce some striking effects.

Diffraction and interference of light

When rays of light pass the edge of a solid object, we are used to seeing shadows. Shadows arise because light travels in straight lines, so they have sharp edges. This might suggest that light is not diffracted. However, to know where to look for diffraction effects with light, we need to think about the connection between wavelength and the size of the gap that causes diffraction.

To have the maximum effect, the gap should be similar to the wavelength of light. Light has a very small wavelength – less than a thousandth of a millimetre – so we need to look for some very tiny gaps.

Figure 5f.7 shows what happens when light from a laser shines through a narrow vertical slit between two straight-sided pieces of metal. When the light reaches a screen, it has spread out sideways to form a diffraction pattern. So light can be shown to be diffracted.

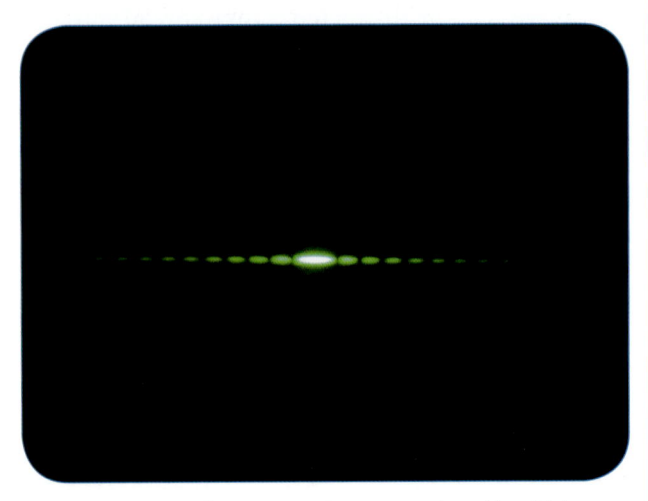

Figure 5f.7 This diffraction pattern was produced by shining green laser light through a narrow slit. The light spreads out beyond the slit, proving that light can be diffracted if the gap through which it passes is narrow enough.

You may have seen diffraction effects with light (Figure 5f.8). For example, if you see a streetlight on a misty night, you may notice a glowing ring of light around it. This is light that has been diffracted by the tiny droplets of water that make up the mist. (The droplets are a similar size to the wavelength of light.) Think about how the light has reached your eyes: it has travelled outwards from the lamp, and has then been bent towards you by the mist, so that it doesn't appear to be coming directly from the light. (You may see a similar effect looking through a steamed-up car windscreen at the headlights of cars coming towards you.)

Patterns like this are produced by a combination of diffraction (bending) and interference (reinforcing and cancelling). Bright areas are where light waves are reinforcing each other; darker areas are where they are cancelling out.

Figure 5f.8 A diffraction effect – light from the Moon has been diffracted by pollen grains in the atmosphere. Not a good sign if you suffer from hay fever.

Now we know that light can be diffracted. It can also produce interference patterns. So diffraction and interference of light are both evidence that light travels as waves.

Unfortunately, it is not as simple as that. There is a famous experimental observation called the *photoelectric effect* that can only be explained by saying that light travels as particles, not waves. The effect is made use of in devices such as digital cameras. Light strikes a metal surface and electrons are knocked free. In 1905, Albert Einstein published a paper in which he showed that each electron has gained the energy of a single particle of light.

So light sometimes behaves like waves and sometimes like particles. Light is strange stuff.

SAQ

5 Which of the following are properties that are *only* shown by waves, and not by particles?
reflection, refraction, diffraction, interference

Interference patterns

One way to produce an interference pattern with light in the laboratory is to use a double-slit. This is made from a piece of metal or dark glass, with two tall, narrow slits in it, side by side. The slits need to be a fraction of a millimetre wide and about a millimetre apart. Light from a laser is shone through the two slits, and an interference pattern appears on a screen beyond them (Figure 5f.9).

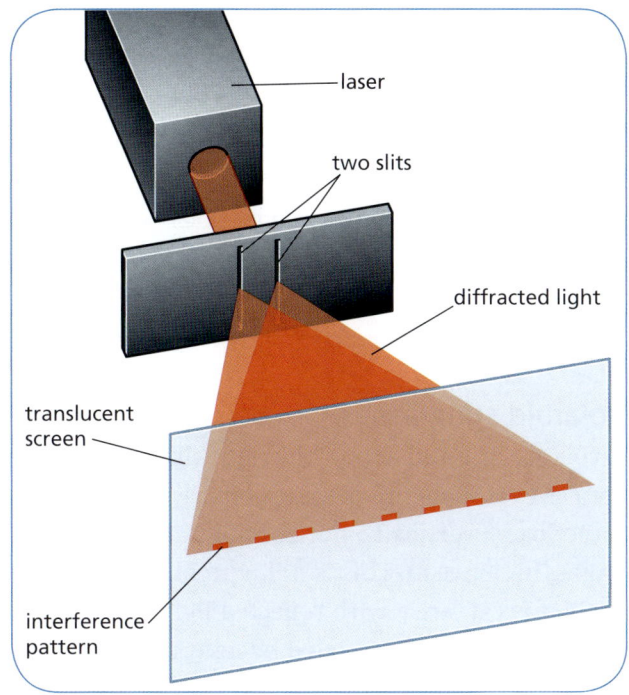

Figure 5f.9 How to produce an interference pattern using light. The two slits act as two sources of light waves. The waves interfere on the screen.

Both diffraction and interference contribute to the effect we see.
- Light passing through a slit is diffracted; it spreads out into the space beyond. So each slit acts as the source of a widening area of light.
- In the space beyond the slits, these two spreading beams of light overlap, so that interference can occur. Each point on the screen receives two light waves, one from each slit. If the waves are in step, constructive interference produces a bright spot. If they are out of step, destructive interference is seen. At the central point on the screen, directly opposite the two slits, there is a bright spot. This is because the waves from the two slits have travelled exactly the same distance to reach this spot, so their path difference is zero. This gives constructive interference.

SAQ

6 Explain why, in the experiment shown in Figure 5f.9:

   a two slits are needed

   b the slits must be narrow.

Polarisation

You should recall from Item P4d *Ultrasound* in *Gateway Additional Science* that there are two types of wave, **longitudinal** and **transverse**.

Sound waves are longitudinal. Think about how they are produced: a vibrating source pushes molecules of air back and forth, so that they vibrate along the direction in which the wave travels.

Light waves (and other electromagnetic waves) are transverse. It's not so easy to see how we know this.

Figure 5f.10 shows how to produce a transverse wave on a stretched spring. In the drawing, the free end of the spring is being moved from side to side, at right angles to the direction in which the wave moves. But there is another way to produce a transverse wave on a spring: move the end up and down. (This won't work on a table.)

We say that a transverse wave can be *polarised*. The wave in Figure 5f.10 is polarised in the horizontal plane; if the spring is moved up and down, the wave is vertically polarised.

A longitudinal wave cannot be polarised, because there is only one way to move the end of the spring to produce a longitudinal wave. So **polarisation** is something that only transverse waves can show. Figure 5f.11 shows one consequence of polarisation: the waves on the string can only pass through the slot if they are polarised in the same direction as the slot.

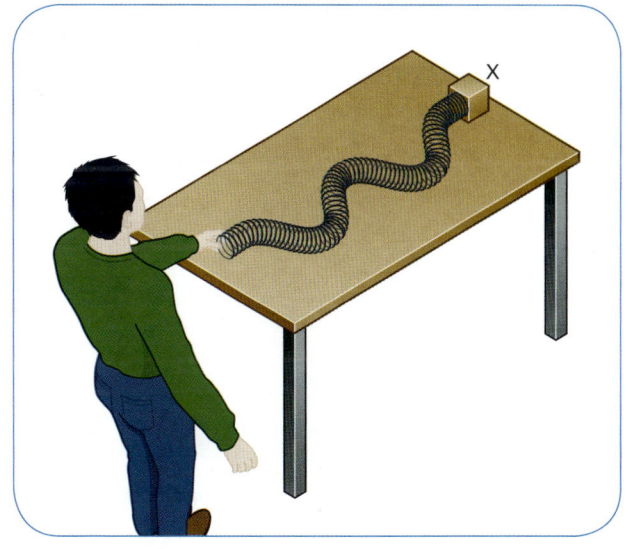

Figure 5f.10 Moving the end of the spring from side to side produces a transverse wave. An alternative would be to move the spring up and down (or at any angle in between).

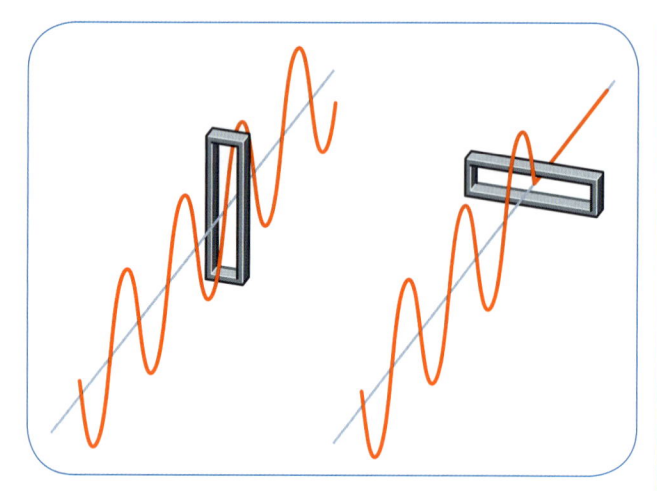

Figure 5f.11 Waves polarised vertically can pass through the vertical slot. When the slot is horizontal, the waves are blocked.

Polarising light

Can light be polarised? That is what Polaroid film does (Figure 5f.12). A piece of Polaroid only allows through light that is polarised in one direction – say, vertically. If this light then meets a piece of Polaroid that is at right angles to the first one, it cannot pass through. The result is darkness.

Figure 5f.12 Polaroid allows through light that is polarised in one direction only; *crossed polars* completely prevent light from passing through, because one blocks vertically polarised light and the other blocks horizontally polarised light.

So light and other types of electromagnetic wave can be polarised. This shows that they are transverse waves.

SAQ

7 **a** Which type of wave can be polarised?
   b Light can be polarised. What does this tell you?

Polaroid sunglasses

If you are out and about, the light that reaches your eyes is made up of waves polarised in all directions – vertically, horizontally, and at all angles in between. This is not a problem. However, problems can arise with reflected light. Motorists, for example, can be dazzled by light reflecting off a wet road surface. How can this be overcome?

The trick is to make use of the fact that reflected light is polarised. When light strikes a horizontal surface, only the horizontally polarised part is reflected. (The vertically polarised waves are absorbed by the reflecting surface.) If a motorist wears Polaroid sunglasses with lenses that only let through vertically polarised light, the annoying reflection will be absorbed.

SAQ

8 Imagine that you are looking at light reflected from a pond while wearing Polaroid sunglasses. You take off the glasses and rotate them slowly in front of your eyes (or tip your head on its side). What would you observe? Use the idea of polarisation to explain your idea.

Summary

You should be able to:

- state that interference occurs when two waves overlap, so that they reinforce each other or cancel out
- describe how an interference pattern for light waves has light and dark areas, while for sound it has louder and quieter areas
- **H** explain that when the path difference between two waves is an even number of half-wavelengths, the waves are in step, and this gives constructive interference
- explain that when the path difference between two waves is an odd number of half-wavelengths, the waves are out of step, and this gives destructive interference
- explain that when light passes through two slits, it is diffracted and an interference pattern of light and dark areas is produced
- state that light can show both diffraction and interference, which is evidence for its wave nature
- state that light can be polarised, which is evidence that it travels as transverse waves

Questions

1. Complete Table 5f.1, by writing *Yes* or *No*, to show the properties of some different types of wave.

Type of wave	Electromagnetic?	Transverse?	Can be diffracted?	Can show interference?	Can be polarised?
sound					
light					
microwaves					
ripples on water					

Table 5f.1

2. Here are five properties of light waves: reflection, refraction, diffraction, interference, polarisation

    a Which *three* of these can result in a light wave changing direction?

    b Which *three* of these are evidence that light travels as waves (rather than as particles)?

    c Which *one* of these is evidence that light travels as transverse waves?

3. a Draw a diagram to show an experiment in which interference of sound waves can be observed.

    b Describe what you would expect to observe.

H 4. Sound waves of wavelength 0.4 m are produced by a loudspeaker. Some reach a listener directly, at a distance of 3 m from the loudspeaker. Others reflect off a wall and reach the listener after travelling 3.8 m. Explain what you would expect the listener to hear. What would they notice if they moved their head a short distance to one side?

P5g Refraction of waves

Light direction

If you should ever become a professional astronomer, you might have the pleasure of working at a telescope set high on an exotic island – in the Canaries or Hawaii, perhaps (Figure 5g.1). Your nights would be spent working at an advanced telescope, but on your days off you could drive down the mountainside and enjoy a swim in a sunlit pool, surrounded by tropical plants. Idly floating in the pool, you might notice the ripple pattern that the Sun's rays cast on the bottom of the pool. Being a physicist, you might wonder *How is this pattern formed? And anyway, why is the telescope placed in such an unusual location?* These two questions have a shared answer.

Astronomers use telescopes to gather light. The lens of the telescope is much bigger than the lens of your eye, so it can gather more light and allow you to see fainter objects in the night sky. Light from distant stars travels for millions of years across empty space, carrying with it a record of the star's existence. In empty space there is nothing to interfere with the light, so it travels on and on, unimpeded. Then, just a few kilometres before it gets to your telescope, it reaches the Earth's atmosphere, and that's where the trouble starts. The atmosphere is rather irregular. There are small patches that are denser, others that are less dense. As the starlight passes through these patches, it changes direction slightly. It moves from side to side, and that is why stars appear to twinkle in the night sky. Using a telescope positioned at sea level, astronomers get a very poor view of the faintest stars. Looking at stars from sea level is rather like trying to recognise a visitor when you see their face through a frosted glass panel in your front door.

The obvious answer is to get above the atmosphere. There are two ways to do this. You could put your telescope in space (like the Hubble Space Telescope), but this is very expensive. Alternatively, you can build your telescope on a high mountain. If the mountain is, say, 3 km high, you will be above most of the Earth's atmosphere, and you will get a much clearer view. Choose a tropical location where you won't have to worry about snow in the winter, and you have the ideal site for an observatory.

And why do we see patterns of shadowy ripples on the bottom of a swimming pool? The surface of the water is irregular; there are always small disturbances on the water, and these cause the rays of sunlight to change direction, just like the irregularities in the atmosphere. Where the pattern is darker, rays of light have been deflected away, producing a sort of shadow. This bending of rays of light when the material they are moving through changes is called *refraction*, and it is the subject of this item.

As you drive back up the mountain to start your next night's work, you might notice that the scenery around you looks blurred, rather like a mirage seen in the desert. There is a heat haze as warm air rises from the ground after a day in the baking sunshine. The warm air is less dense than the colder air it is moving through. It isn't difficult to explain why things look blurred through this irregular air.

Figure 5g.1 This astronomical observatory is on La Palma, one of the Canary Islands, off the west coast of Africa. It is at a height of about 2400 m above sea level, so that the telescopes are sited above most of the Earth's atmosphere. This gives them a much clearer view of the stars, without the twinkling effects produced by irregularities in the atmosphere.

Refraction effects

There are many effects caused by the refraction of light – the sparkling of diamonds, the way your eye produces an image of the world around you, the distorted images you see when looking through textured glass in a bathroom window or a bulls-eye window in an old house. The 'broken stick' effect (Figure 5g.2) is another consequence of refraction. The word **refraction** is related to the word *fractured*, meaning broken. Picture a bent and fractured bone and you will remember the meaning of refraction.

Figure 5g.2 The pencil is partly immersed in water. Because light from the underwater part of the pencil is refracted, the pencil appears broken.

Refraction occurs when a ray of light travels from one **medium** (or material) into another. The ray of light may change direction. You can investigate this using a ray box and a block of glass or Perspex, as shown in Figure 5g.3. Note that the ray travels in a straight line when it is in the air outside the block, and when it is inside the block. It only bends at the point where it enters or leaves the block, so it is the *change* of material that causes the bending.

From Figure 5g.3, you will notice that the direction in which the ray bends depends on whether it is entering or leaving the glass.

- The ray bends towards the normal when entering the glass.
- The ray bends away from the normal when leaving the glass.

One consequence of this is that, when a ray passes through a parallel-sided glass block, it returns to its original direction of travel, although it is shifted to one side. When we look at the world through a window, we are looking through a parallel-sided sheet of glass. We do not see a distorted image because the rays of light reach us travelling in their original direction, although they are shifted slightly as they pass through the glass.

Figure 5g.3 Demonstrating the refraction of a ray of light when it passes through a rectangular glass block. The ray bends as it enters the glass; as it leaves, it bends back to its original direction.

Changing direction

As with reflection, we define angles relative to the normal. Figure 5g.4a (over the page) shows the terms used.

The **incident ray** strikes the block. The **angle of incidence** i is measured from the normal to the ray. The **refracted ray** travels on at the **angle of refraction** r, measured relative to the normal. (You may have previously used r for the angle of reflection; here it stands for the angle of refraction.)

A ray of light may strike a surface head-on, so that its angle of incidence is 0°, as shown in Figure 5g.4b. In this case, it does not bend; it simply passes straight through and carries on in the same direction. Usually we say that *refraction is the bending of light when it passes from one material to another*. But we should bear in mind that, when the light is travelling perpendicular to the boundary between the two materials, there is no bending.

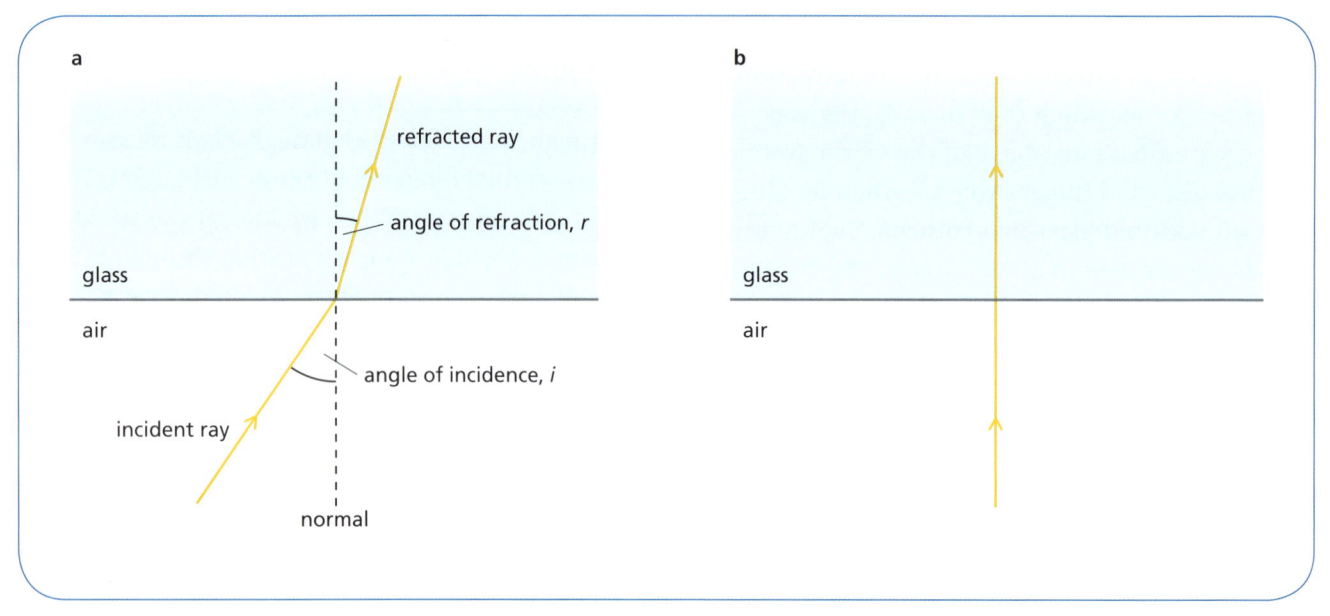

Figure 5g.4 a Refraction: defining terms. The normal is drawn perpendicular to the surface at the point where the ray passes from one material to another. The angles of incidence and refraction are measured relative to the normal. **b** When a ray strikes the glass at 90°, it carries straight on without being deflected.

SAQ

1 Look at Figure 5g.4a. Which is greater, the angle of incidence or the angle of refraction? Copy the correct equation: $i > r$ or $r > i$.

Explaining refraction

Why does light change direction when it passes from one material to another? The answer lies in the way its speed changes. Light travels fastest in a vacuum (empty space). Table 5g.1 shows that light travels very slightly slower in air, but much more slowly in transparent solid materials. Of all the materials shown in the table, light travels most slowly in diamond, where its speed is less than half of that in a vacuum.

Material	Speed of light in m/s	Speed in vacuum / Speed in medium
vacuum	2.998×10^8	1 exactly
air	2.997×10^8	1.0003
water	2.3×10^8	1.33
glass	$1.8–2.0 \times 10^8$	1.5–1.7
Perspex	2.0×10^8	1.5
diamond	1.25×10^8	2.4

Table 5g.1 The speed of light in some transparent materials. (The value for a vacuum is shown, for comparison.) Note that the values are only approximate. The third column shows the factor by which the light is slowed down.

One way to explain why a change in speed leads to a change in direction is shown in Figure 5g.5. A truck is driving along a road across the desert. The driver is careless, and allows the wheels on the left to drift off the road onto the sand. Here, they spin around, so that the left-hand side of the truck moves more slowly. The right-hand side is still in contact with the road and keeps moving quickly, so that the truck starts to turn to the left.

The boundary between the two materials is the edge of the road; the normal is at right angles to the road. The truck has veered to the left, so its direction has moved towards the normal. Thus we would expect a ray of light to move towards the normal when it enters a material where it moves more slowly, and this is indeed what we see with glass (Figure 5g.3). Light travels more slowly in glass than in air, so it bends towards the normal as it enters glass.

(Compare this with Newton's explanation of refraction, shown in Figure 5f.5. He thought that light bends towards the normal because it speeds up.)

When light slows down, it bends towards the normal. The quantity that describes how much the light is slowed down is the **refractive index**. The more the light slows down, the greater the refractive index of the medium and the more the light bends.

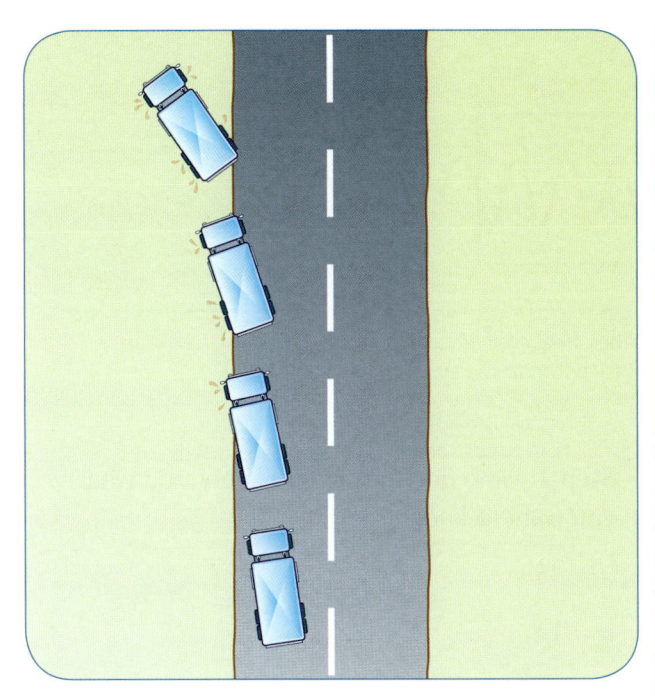

Figure 5g.5 To explain why a change in speed causes the bending we see in refraction, we picture a truck whose wheels slip off the road into the sand. The truck veers to the side because it cannot move so quickly through sand.

When light leaves a medium, it speeds up again. Then it bends away from the normal. The more it speeds up, the greater the bending. (Note that a ray of light only bends at the *boundary* between two materials.)

SAQ

2 Draw a labelled diagram similar to Figure 5g.4a, showing a ray of light travelling from glass into air.

3 The refractive index of glass depends on its composition. Here is the speed of light in two different types of glass:

Glass A: 1.8×10^8 m/s

Glass B: 1.9×10^8 m/s

  a Which type of glass has the greater refractive index?
  b Which type will cause a ray of light to bend more on entering the glass?
  c Which type will cause a ray of light to bend more on leaving the glass?

Calculating refractive index

If the speed of light is halved when it enters a material, the refractive index is 2, and so on. There is an equation for the refractive index n of a medium:

$$\text{refractive index } n = \frac{\text{speed of light in vacuum}}{\text{speed of light in medium}}$$

Water has the refractive index $n = 1.33$. This means that light travels 1.33 times as fast in a vacuum, compared to its speed in water. For calculations, it is usually satisfactory to take the speed of light to be 3×10^8 m/s in a vacuum.

It is a *change* in speed that causes rays of light to bend; the bigger the refractive index, the bigger the change in speed and the greater the effect. If one part of the atmosphere is warmer than another, it will be less dense and its refractive index will be smaller. This is what causes problems for astronomers as they try to look out through the atmosphere towards the distant stars.

SAQ

4 Look at Table 5g.1. What alternative heading could we give to the third column?

5 The speed of light in a particular type of transparent polymer is 1.8×10^8 m/s. What is the refractive index of this material?

Changing material, changing speed

When waves travel from one material into another, the frequency of the waves remains unchanged. Because the speed of the waves changes, their wavelength must also change (because speed, frequency and wavelength are connected by the equation speed = frequency × wavelength). This is illustrated in Figure 5g.6 (over the page), which shows light waves travelling quickly through air. They reach some glass and slow down; their wavelength decreases. When they leave the glass again, they speed up, and their wavelength increases again.

260 P5g Refraction of waves

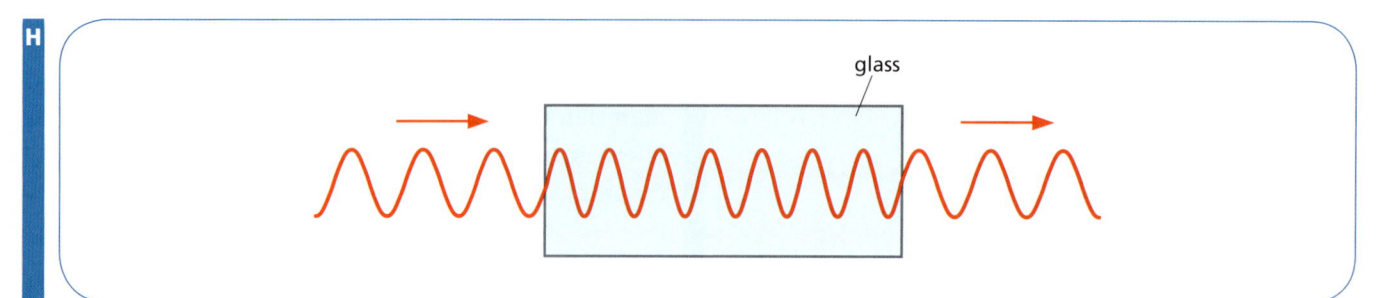

Figure 5g.6 Waves change their wavelength when their speed changes. Their frequency remains constant. Here, light waves slow down when they enter glass and speed up when they return to the air.

Snell's law

There is a law that relates the size of the angle of refraction r to the angle of incidence i. This is **Snell's law**. It also involves the refractive index n, since the greater the refractive index, the more a ray is bent.

The law is written in the form of an equation:

$$\frac{\sin i}{\sin r} = n$$

Worked example 1 shows how to use this equation to find the angle through which a ray is refracted.

Worked example 1

A ray of light strikes a glass block with an angle of incidence of 45°. The refractive index of the glass is 1.6. What will be the angle of refraction?

The situation is shown in Figure 5g.7.

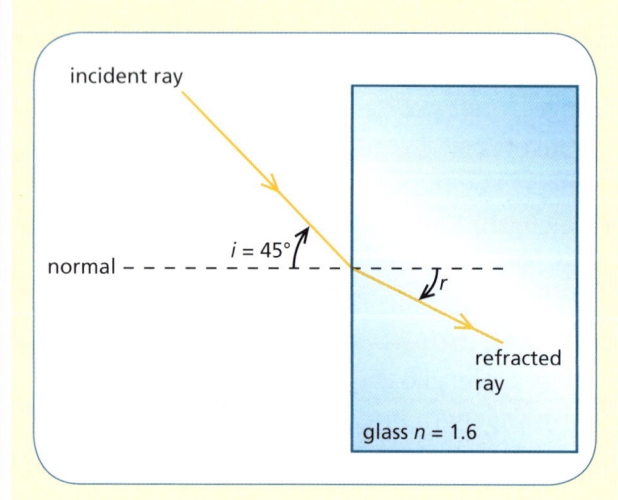

Figure 5g.7

Step 1: write down what you know, and what you want to know.

$i = 45°$

$n = 1.6$

$r = ?$

Step 2: write down the equation for Snell's law. Since we want to know r, rearrange it to make $\sin r$ the subject.

$$\frac{\sin i}{\sin r} = n$$

$$\sin r = \frac{\sin i}{n}$$

Step 3: substitute values and calculate $\sin r$.

$$\sin r = \frac{\sin 45°}{1.6}$$

$$= 0.442$$

Step 4: use the $\sin^{-1}$ function on your calculator to find r. (This will tell you the angle whose sine is 0.442.)

$r = \sin^{-1} 0.442$

$= 26.2°$

So the angle of refraction is 26.2°. You can see that Snell's law correctly predicts that the ray will be deflected towards the normal.

Snell's law can also be used to find the value of the refractive index of a material: simply measure values of i and r and substitute them in the equation.

SAQ

6 Careful measurements of the refraction of a ray of light entering a glass block gave the following results:

angle of incidence = 60°

angle of refraction = 28°

Calculate the refractive index of the glass.

7 A ray of light entered a tank of water with an angle of incidence of 80°. What will be its angle of refraction? (Refractive index of water = 1.33.)

Dispersion of light

Diamonds are attractive because they sparkle. As you turn a cut diamond, light flashes from its different surfaces. You may also notice that you can see all sorts of colours in the diamond, even though the diamond itself is likely to be colourless. (Some diamonds are slightly yellow, because they contain impurities.) Where do these varying colours come from?

Cut glass is a lot cheaper than diamonds, and has many more uses (Figure 5g.8). It is used for chandeliers, which move gently in the air. It is also used for glass ornaments; your eye is caught by the changing colours as you walk past. Again, the glass itself is colourless, so where do these colours come from?

Figure 5g.8 Cut glass is used for ornaments and chandeliers, because it shows all the colours of the rainbow as it moves in the light.

The underlying principle is shown in Figure 5g.9. A ray of white light is shone at a prism; it is refracted as it enters and leaves the glass. At the same time, it is split into a **spectrum** of colours. Notice that the colours merge into one another, and that they are not all of equal widths in the spectrum.

Figure 5g.9 A spectrum can be produced by shining a ray of white light through a glass prism. The light is split up into a spectrum.

Traditionally, we say that there are seven colours in the spectrum. The number seven was chosen because it had a mystical significance in the 17th century. It is very hard to distinguish between indigo and violet at the end of the spectrum, so you might say that there are really only six colours. Alternatively, you might suggest that there are many shades of red present, and of each of the other colours, so the spectrum shows many more than seven colours. The standard list is as follows:

red, orange, yellow, green, blue, indigo, violet

There are different ways of remembering this list. One simple way is to remember the sequence of initial letters in the form of someone's name: Roy G Biv.

Explaining the spectrum

This splitting up of white light into a spectrum is known as **dispersion** (spreading out). Isaac Newton set out to explain how it happens. It had been suggested that light is coloured by passing it through a prism. Newton showed that this was the wrong idea by arranging for the spectrum to be passed back through another prism. The colours recombined to form white light again. He concluded that white light is a mixture of all the different colours of the spectrum.

So what happens in a prism to produce a spectrum? As the white light enters the prism, it slows down. We say that it is refracted and, as we have seen, its direction changes. Dispersion occurs because each colour is refracted by a different amount. Violet light slows down the most, and so it is refracted the most. Red light is the least affected.

A rainbow (Figure 5g.10) is a naturally occurring spectrum. White light from the Sun is dispersed as it enters and leaves droplets of water in the air. It is also reflected back to the viewer by total internal reflection (see below), which is why you must have the Sun behind you to observe a rainbow.

Laser light is not dispersed by a prism. It is refracted so that it changes direction, but it is not split up into a spectrum. This is because it is light of a single colour; it described as monochromatic (*mono* = one, *chromatic* = coloured).

Figure 5g.10 A rainbow shows all the colours of the spectrum. White light from the Sun is split up into its constituent colours as it is refracted by water droplets in the air.

SAQ

8 Light at the blue/violet end of the spectrum is bent most on entering a glass prism.
   a What does this tell you about the speed of violet light in glass, compared to red light?
   b Which colour of light will be bent most on leaving glass?

H 9 Different colours of light are bent to different extents on entering and leaving glass (or another transparent material). We can explain this by saying that the refractive index of glass is different for different colours. For which colour is the refractive index of glass greatest?

Total internal reflection

If you have carried out a careful investigation of refraction using a ray box and a transparent block, you may have noticed something extra that happens when a ray strikes a block. Not only is the ray refracted; in addition, a reflected ray also appears. You can see this in Figure 5g.3, but it has been ignored in Figure 5g.4. When the ray strikes the block, some of the light passes into the block and is refracted; some is reflected. When the light leaves the block, some is again reflected. These reflected rays obey the law of reflection: angle of incidence = angle of reflection.

These reflected rays can be a nuisance. If you try to look downwards into a pond or river to see if there are any fish there, your view may be spoilt by light reflected from the surface of the water. You see a reflected image of the sky, or of yourself, rather than what is in the water. On a sunny day, reflected light from windows or water can be a hazard to drivers.

To see how we can make use of reflected rays, you can use the apparatus shown in Figure 5g.11. A ray box shines a ray of light at a semicircular glass block. The ray is always directed at the curved edge of the block, along the radius. This means that it enters the block along the normal, so that it is not bent by refraction. Inside the glass, the ray strikes the midpoint of the flat side, point X.

What happens next? This depends on the angle of incidence of the ray at point X. The various possibilities are shown in Figure 5g.12.

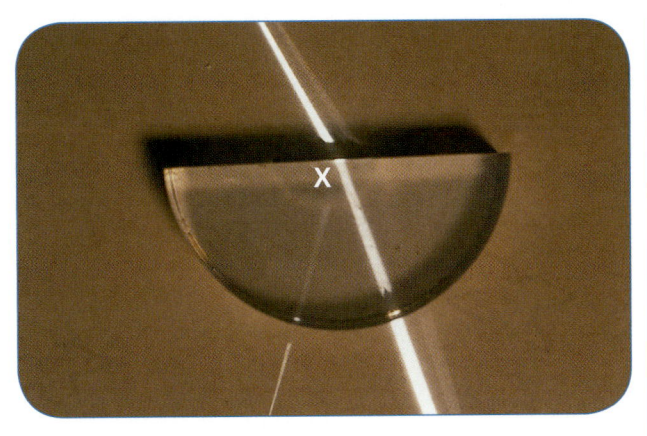

Figure 5g.11 Using a ray box to investigate reflection when a ray of light strikes a glass or Perspex block. The ray enters the block without bending, because it is directed along the radius of the block.

a If the angle of incidence is small, most of the light emerges from the block. There is a faint reflected ray inside the glass block. The refracted ray bends away from the normal, because the air is less dense than the glass.

b If the angle of incidence is increased, more light is reflected inside the block. The refracted ray bends even further away from the normal.

c Eventually, as the angle of incidence is increased, the refracted ray emerges parallel to the surface of the block. Most of the light is reflected inside the block.

d Now, at an even greater angle of incidence, all of the light is reflected inside the block. No refracted ray emerges from point X.

We have been looking at how light is reflected *inside* a glass block. We have seen that, if the angle of incidence is greater than a particular value, the light is entirely reflected inside the glass. This phenomenon is known as **total internal reflection**, or TIR:

- *total*, because 100% of the light is reflected
- *internal*, because it happens inside the glass
- *reflection*, because the ray is entirely reflected.

For total internal reflection to happen, the angle of incidence of the ray must be greater than a particular value, known as the **critical angle** c. The value of c depends on the material being used. For glass, the critical angle is about 42° (though this depends on the composition of the glass.) For water, which has a lower refractive index, the critical angle is greater, about 49°. For diamond, with a very high refractive index, the

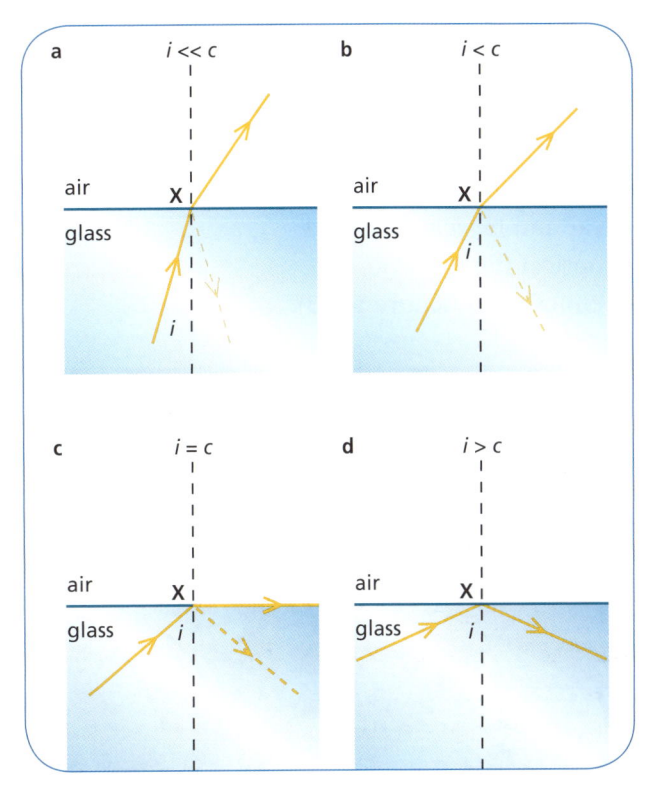

Figure 5g.12 The way a ray of light is refracted or reflected inside a glass block depends on the angle of incidence.
a, b For angles less than the critical angle, some of the light is reflected and some is refracted. **c** At the critical angle, the angle of refraction is 90°. **d** At angles of incidence greater than the critical angle, the light is totally internally reflected; there is no refracted ray.

critical angle is small, about 25°. Hence rays of light that enter a diamond are very likely to be totally internally reflected, so they bounce around inside, eventually emerging from one of the diamond's cut faces. That explains why diamonds are such sparkly jewels.

SAQ

10 Sketch a graph to show how the fraction of light reflected internally changes as the angle of incidence i increases. What fraction do you think will be reflected when the ray strikes the surface along the normal ($i = 0$)? What fraction will be transmitted?

Using TIR

It is the fact that total internal reflection is total – that is, that 100% of the light is reflected – that makes it so useful. No light is lost. This means that suitably arranged prisms can be more effective than mirrors for some uses.

A typical mirror reflects perhaps 90% of the light that falls on it. The rest of the light is absorbed, so that the image is rather dim.

TIR is used in:
- prismatic binoculars, telescopes and periscopes
- car and bicycle reflectors
- optical fibres
- endoscopes (Figure 5g.13).

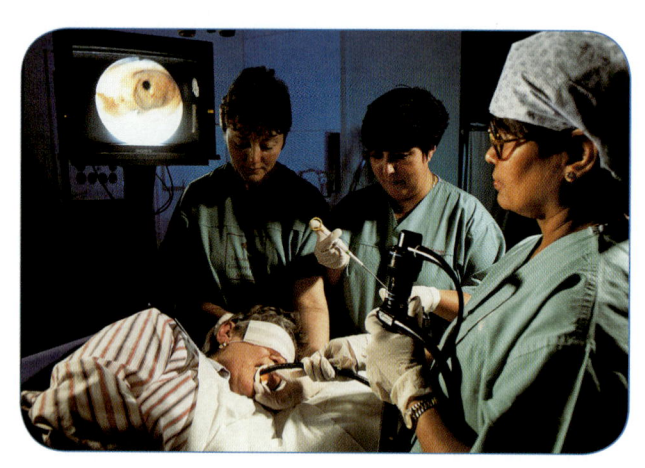

Figure 5g.13 This medical team is looking at the patient's stomach using an endoscope. Light is shone down into the body along optical fibres, making use of total internal reflection. A tiny camera provides images of the patient's insides, which are shown on a monitor. The endoscope may also include small tools that the surgeon can manipulate.

Calculating c

So far, we have looked at TIR when a ray of light is trying to get out of a material such as glass, with a vacuum (or air) beyond. Can we have TIR with a different material outside the glass – water, for example?

Figure 5g.14 shows the situation. A ray of light strikes the inner surface of a block of glass; there is water on the other side. TIR *can* occur, because the refractive index of glass is greater than that of water. However, the critical angle is much greater than if the water was replaced by air. This is because the refractive index of water is quite similar to that of glass.

The critical angle c depends on two refractive indexes:

n_i = the refractive index of the *internal* medium – the one through which the light is travelling

n_r = the refractive index of the *reflecting* medium – the one which reflects the light

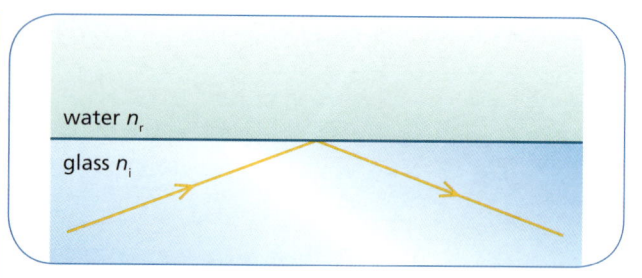

Figure 5g.14 TIR can occur at the boundary between two different transparent materials, provided the outer one has a lower refractive index.

Then we have:

$$\sin c = \frac{n_r}{n_i}$$

(If you remember Figure 5g.14, you will recall that the reflecting medium is 'on top of' the internal material.)

Worked example 2

A ray of light is travelling through a block of glass of refractive index 1.40. It strikes the surface of the glass with an angle of incidence of 80°. The surface of the glass is covered with water (refractive index 1.33). Will it undergo total internal reflection, or will it refract out of the glass?

Step 1: the situation is as in Figure 5g.14. To know whether TIR will occur, we need to know the critical angle.

We know: $n_i = 1.40$ $n_r = 1.33$

Step 2: then we have:

$$\sin c = \frac{n_r}{n_i} = \frac{1.33}{1.40} = 0.95$$

$$c = 71.8°$$

So the critical angle is 71.8°. The angle of incidence is 80°, and this is greater than c, so the ray will be totally internally reflected.

SAQ

11 Use the equation $\sin c = n_r / n_i$ to show that the critical angle of water is about 49°. (Refractive index of water = 1.33.) (Assume that the reflecting medium is air, refractive index = 1.00)

Summary

You should be able to:

- explain that refraction of light occurs at the boundary between two mediums because the speed of light changes as it passes from one medium to another

- explain that when the wave speed decreases, the ray bends towards the normal

- state that the greater the change in wave speed and in refractive index, the greater the bending

- **H** use the formula: refractive index $n = \dfrac{\text{speed of light in vacuum}}{\text{speed of light in medium}}$

- use Snell's law: $\dfrac{\sin i}{\sin r} = n$

- describe how dispersion to form a spectrum (ROYGBIV) occurs because different colours of light have different speeds in a medium

- explain that total internal reflection occurs when light strikes the inner surface of a medium at an angle greater than the critical angle c

- explain that at smaller angles of incidence, the ray is partly refracted and partly reflected

- **H** use the formula $\sin c = \dfrac{n_r}{n_i}$ to calculate the critical angle

Questions

1. Draw a diagram to show what we mean by the *angle of incidence* and the *angle of refraction* for a refracted ray of light.

2. A ray of light passes from air into a block of glass. Does it bend *towards* the normal, or *away* from it?

3. Draw a diagram to show how a ray of light passes through a parallel-sided Perspex block. What can you say about its final direction of travel?

4. A vertical ray of light strikes the horizontal surface of some water. What is its angle of incidence? What is its angle of refraction?

5. When a ray of light passes from air to glass, is the angle of refraction greater than, or less than, the angle of incidence?

6. Why do we see a distorted view of the world when we look through a window which is covered with raindrops?

7. Light travels more quickly through water than through glass. If a ray passes from glass into water, which way will it bend: towards or away from the normal?

continued on next page

Questions – *continued*

8. Explain the meaning of the words *total* and *internal* in the expression *total internal reflection*.

9. The critical angle for water is 49°. If a ray of light strikes the upper surface of a pond at an angle of incidence of 45°, will it be totally internally reflected?

H 10. Look back at Table 5g.1 on page 258. What is the value of the refractive index of diamond?

11. The speed of light in quartz is 1.94×10^8 m/s.

    a. What is the refractive index of quartz? (Speed of light in vacuum = 3×10^8 m/s)

    b. What is the critical angle for quartz?

12. Write down an equation to represent Snell's law. Explain the symbols used. Include a diagram to show the angles involved.

13. A ray of light strikes the surface of water with an angle of incidence of 40°. Calculate the angle of refraction. (Refractive index of water = 1.33.)

14. A ray of light is directed internally within a block of glass (refractive index 1.47) towards the surface, which has a coating of polymer (refractive index 1.45). Calculate the critical angle at the boundary between the two mediums. Explain why the critical angle is close to 90°.

15. Snell's law can be used to determine the refractive index of a transparent medium. A ray of light is shone into the medium at different angles of incidence, and the angle of refraction is measured. Table 5g.2 shows the results of such an experiment, using a block of transparent polymer.

    Copy and complete the table; draw a graph of sin *i* (*y*-axis) against sin *r* (*x*-axis).

    Calculate the gradient of the graph; this is the refractive index of the polymer.

i	*r*	sin *i*	sin *r*
20°	13°		
34°	22°		
45°	30°		
52°	32°		
60°	36°		
68°	39°		

Table 5g.2

5h Optics

Revolutionary lenses

We are all familiar with lenses in everyday life – in spectacles and cameras, for example. The development of high quality lenses has had a profound effect on science. In 1609, using the newly invented telescope, Galileo discovered the moons of Jupiter and triggered a revolution in astronomy. In those days, scientists had to grind their own lenses starting from blocks of glass, and Galileo's skill at this was a major factor in his discovery.

Later in the 17th century, a Dutch merchant called Anton van Leeuwenhoek managed to make microscope lenses that gave a magnification of 200 times. One of his microscopes is shown in Figure 5h.1. The left-hand picture shows the side through which the user looked; on the right you can see the spike on which a specimen was mounted. Turning the screw moved the specimen in front of the lens.

Van Leeuwenhoek used his microscopes (which were really no more than powerful magnifying glasses) to look at a great variety of things and was amazed to find a wealth of tiny micro-organisms, including bacteria, that were invisible to the naked eye (Figure 5h.2). This provided later scientists with a clue to how infectious diseases might be spread. At that time, people thought infections were carried by smells or by mysterious 'vapours'. Van Leeuwenhoek's microscope helped sow the seeds of a revolution in medicine.

The work of Galileo and van Leeuwenhoek shows how newly invented instruments can open up whole new fields of study for scientists. Because they were able to make better instruments than their competitors, they were the individuals who made the new discoveries.

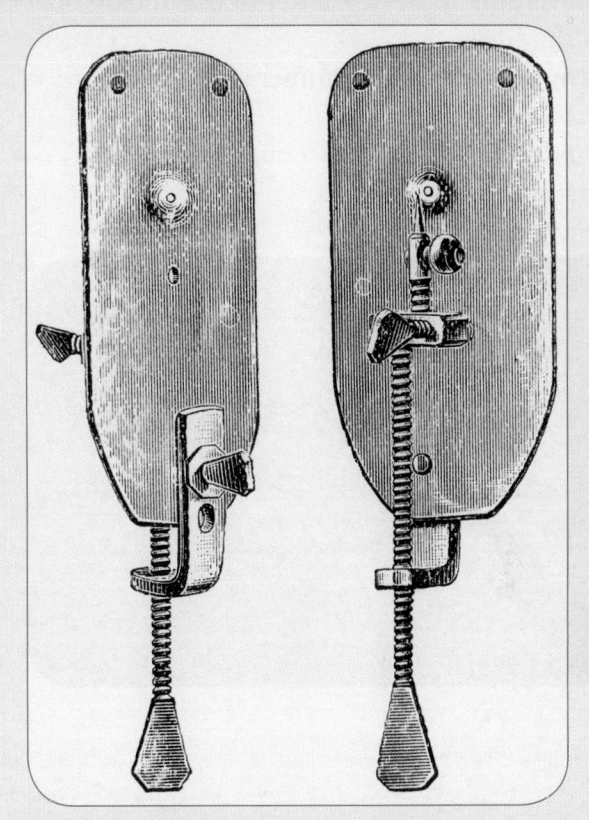

Figure 5h.1 Van Leeuwenhoek's microscope. The tiny lens is mounted in a hole in the metal plate, and the specimen is attached to the tip of the spike.

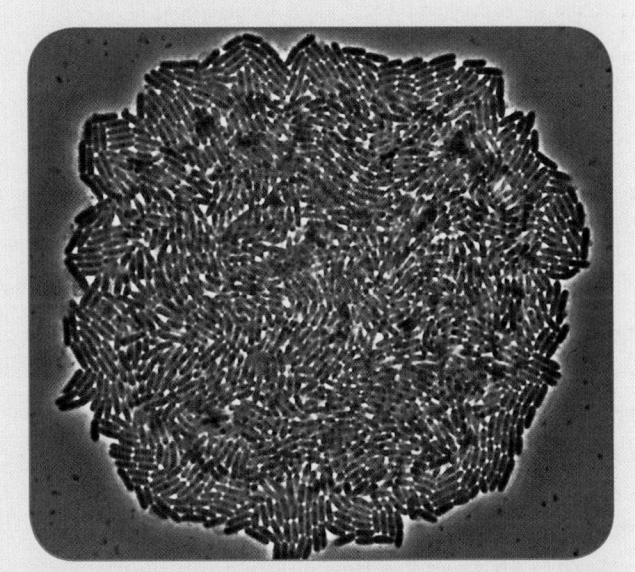

Figure 5h.2 Bacteria that cause infections cannot be seen with the naked eye. This photograph, taken using a modern microscope, shows *E. coli* bacteria which are responsible for food poisoning. The microscope uses two lenses to give an image which is 2000 times the size of the object being viewed.

268 P5h Optics

Converging lenses

Lenses can be divided into two types, according to their shape (Figure 5h.3):
- **converging lenses** are fatter in the middle than at the edges
- **diverging lenses** are thinner in the middle than at the edges.

In this item, we will restrict ourselves to converging (convex) lenses.

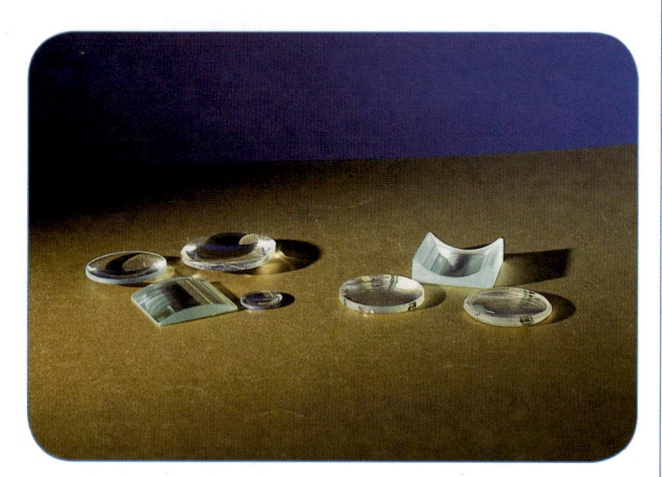

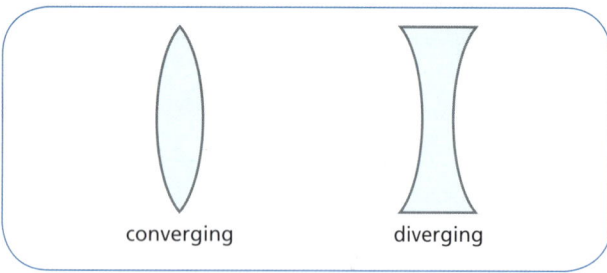

Figure 5h.3 The lenses on the left are converging (convex) lenses; they are fattest at the middle. On the right are diverging lenses, thinnest at the middle. They are given these names because of their effect on parallel rays of light. Usually we simply draw the cross-section of the lens, to indicate which type we are considering.

You have probably used a magnifying glass to look at small objects. This is a converging lens. You may even have used a magnifying glass to focus the rays of the Sun onto a piece of paper, to set fire to it. (Over a thousand years ago, an Arab scientist described how people used lenses for starting fires.) This gives a clue to the name *converging*.

Figure 5h.4a shows how a converging lens focuses parallel rays of light (like the rays we receive from the Sun). The rays are all parallel to the **axis** of the lens (the line passing symmetrically through the centre of the lens). On one side of the lens, the rays are parallel; after

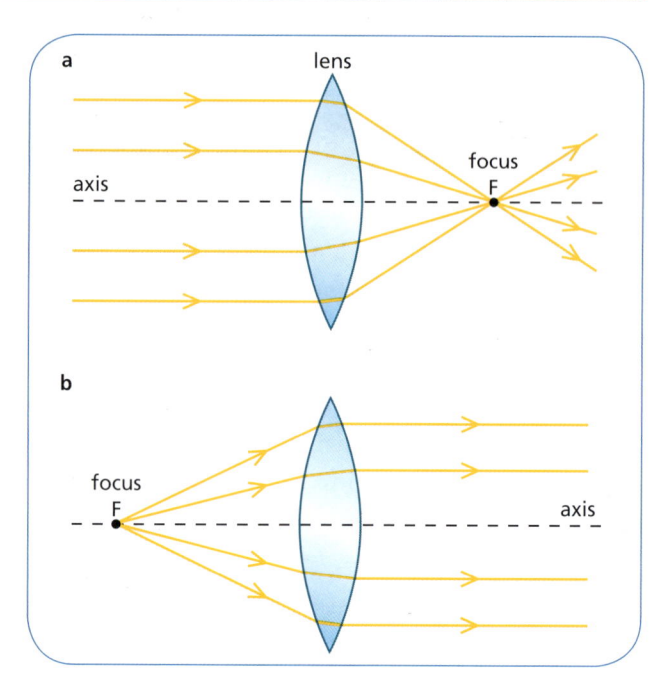

Figure 5h.4 The effect of converging lenses on rays of light. **a** A converging lens makes rays parallel to the axis converge at the focus. **b** Rays from the focus of a converging lens are turned into a parallel beam of light.

they pass through the lens, they converge on a single point, the **focus** or **focal point**. After they have passed through the focus, they spread out again. So a converging lens is so-called because it makes parallel rays of light converge. The focus is the point where rays parallel to the axis are concentrated together – and where a piece of paper needs to be placed if it is to be burned.

The **focal length** is the distance from the centre of the lens to the focus. The 'fatter' the lens, the closer the focal point is to the lens, and so fatter lenses have shorter focal lengths.

A converging lens can be used in reverse to produce a beam of parallel rays. If a source of light such as a small light bulb is placed at the focus, the rays that reach the lens are bent so that they become a beam parallel to the axis (Figure 5h.4b). This diagram is the same as Figure 5h.4a, but in reverse.

Lenses work by refracting light. When a ray strikes the surface of the lens, it is refracted towards the normal because it is entering a denser material. When it leaves the glass of the lens, it bends away from the normal. The clever thing about the shape of a converging lens is that it bends all rays parallel to the axis just enough for them to meet at the focus.

SAQ

1 Sketch the shapes of two converging lenses, one fatter than the other. Label them *shorter focal length* and *longer focal length*.
2 Copy Figure 5h.4a and mark the focal length of the lens.

Cameras and projectors

A camera has a converging lens that is used to produce an **image** on the film (or on the electronic sensor in a digital camera). Figure 5h.5a shows how this works. Rays of light from a distant object are bent by the lens so that they produce a focused image on the film. Notice that the image formed on the film is inverted (upside down).

If the object is close to the camera, the image will be formed further back, so that it is out of focus on the film (Figure 5h.5b). To overcome this, the lens must be moved forwards. You can observe this happening if you have an auto-focus camera. Point it at a distant object, and then turn it towards a nearby object – the lens moves outwards to produce a focused image.

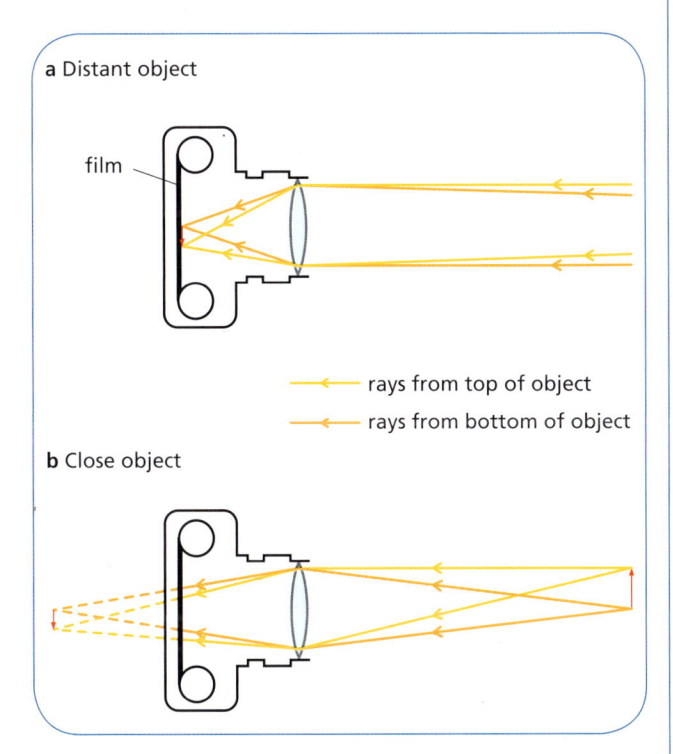

Figure 5h.5 How an image is formed on the film in a camera. **a** The single lens focuses rays onto the film. **b** If the object is close up, the image will form further from the lens, so the lens must be moved away from the film. This is what happens when we 'focus' a camera before taking a picture.

A projector shines light through a transparent slide to form a large image on a screen. This is rather like a camera in reverse. The bright lamp illuminates the slide; light spreads out from the slide, and a converging lens focuses the light on the screen. The lens must be moved back and forth until the image is in focus. As with a camera, the image is inverted, so the slide must be put in upside down if the image is to be the right way up.

SAQ

3 A data projector is used with a computer to project an image onto a screen. What type of lens is used at the front of the projector? Explain why it is important that the position of this lens can be adjusted backwards and forwards.

Magnifying glass

A magnifying glass is a converging lens. You hold it close to a small object and look through it to see a magnified image. Figure 5h.6 shows how a magnifying glass can help to magnify print for someone with poor eyesight.

If you move a magnifying glass away from the object you are looking at, the image will get bigger. After a certain distance, the lens stops working as a magnifying glass and you see a small, inverted image.

SAQ

4 Give another word that means the same as *enlarged*.

Figure 5h.6 This long converging lens is designed to help people to read. It produces a magnified image of a line of print. The user simply slides it down the page.

Forming a real image

When the Sun's rays are focused onto a piece of paper, a tiny image of the Sun is created. It is easier to see how a converging lens makes an image by focusing an image of a light bulb or a distant window onto a piece of white paper. The paper acts as a screen to catch the image. Figure 5h.7 shows an experiment in which an image of a light bulb (the object) is formed by a converging lens. Things to note about the image:
- it is *inverted* (upside down)
- it is *reduced* (smaller than the object)
- it is *nearer* to the lens than the object
- it is *real*.

We say that the image is **real**, because light really does fall on the screen to make the image. A magnifying glass does not produce a real, inverted image like this; its image is upright (the right way up), and appears to be behind the object you are looking at. Light *appears* to be coming from behind the object; an image like this is described as **virtual**. (The image in a plane mirror is also virtual. When you look at yourself in a mirror, your image appears to be behind it, although no rays of light actually come from behind the mirror.)

The size of the image depends on how fat or thin the lens is – that is, on its focal length. It also depends on the distance of the object from the lens.

SAQ

5 What words are the opposites of *real*, *inverted*, and *reduced*?

Figure 5h.7 Forming a real image of a light bulb using a converging lens.

Constructing a ray diagram

We can explain the formation of this real image using a **ray diagram**, as shown in Figure 5h.8. Here are the steps in drawing this diagram.
1. Draw the lens – a simple outline shape will do – and add the horizontal axis through the middle of it.
2. Mark the positions of the focal points F on either side, at equal distances from the lens. Mark the position of the object O, an arrow standing on the axis.
3. Draw ray 1, a straight line from the top of the arrow and passing undeflected through the middle of the lens.
4. Draw ray 2, from the top of the arrow parallel to the axis. As it passes through the lens, it is deflected down through the focal point. Look for the point where the two rays cross. This is the position of the top of the image I.

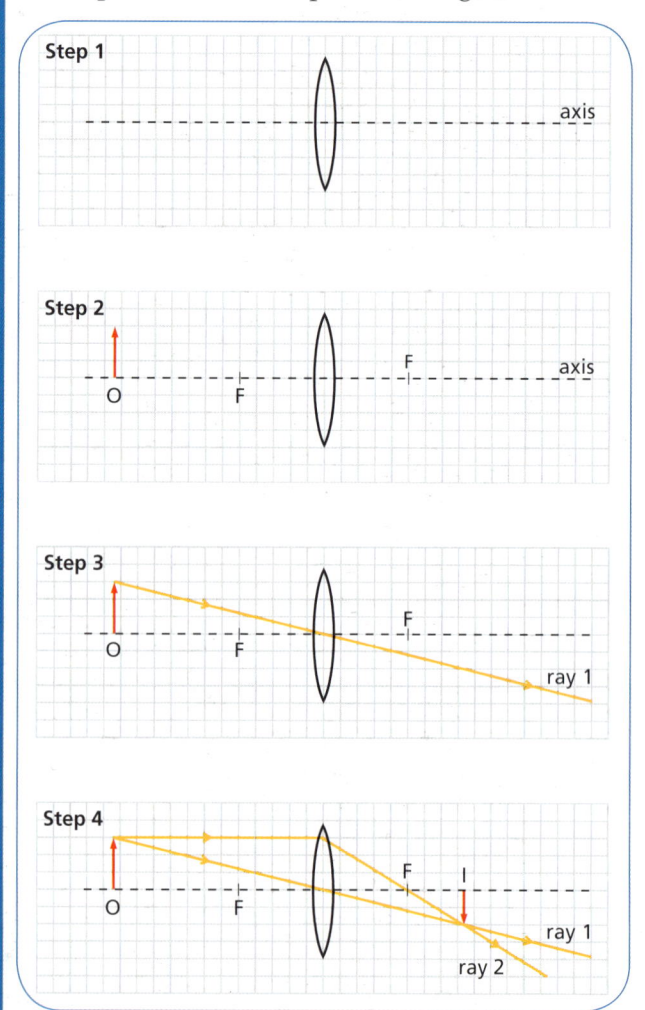

Figure 5h.8 A ray diagram can be used to show how an image is formed by a converging lens. The steps in drawing this diagram are given in the text.

With an accurately drawn ray diagram, you can see that the image is inverted, reduced and real.

It is helpful to understand why we draw rays 1 and 2 as shown in Figure 5h.8. Ray 1 passes through the middle of the lens. Here, the sides of the lens are parallel, so the ray emerges undeflected. Ray 2 bends twice, at the two surfaces of the lens. It is easier to show it bending once in the middle of the lens, though this is not a correct representation of what really happens.

SAQ

6 Explain why ray 2 passes through the focal point F.

7 Ray 1 passes through the centre of the lens, where the opposite faces of the lens are parallel to each other. Explain why ray 1 does not change direction as it passes through the lens.

Worked example

A converging lens has a focal length of 10 cm. An object is placed on the axis of the lens, at a distance of 15 cm from the lens. Draw an accurate ray diagram and use it to determine the distance from the lens of the image formed.

Figure 5h.9 shows the ray diagram required, constructed as in Figure 5h.8. Measuring the diagram shows that the image is 30 cm from the lens.

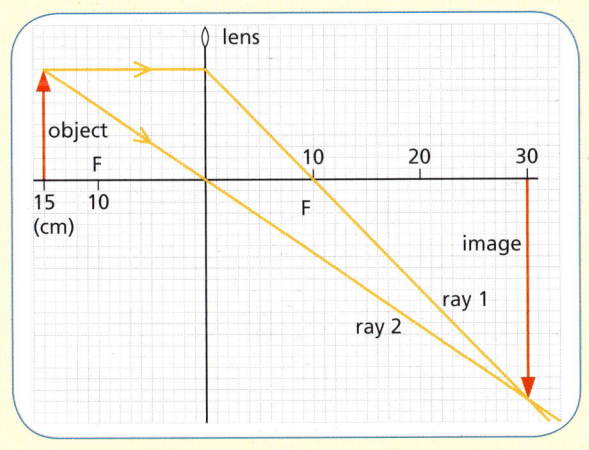

Figure 5h.9 The lens is shown as a vertical line, with a small drawing above it to show the shape of the lens.

Magnification

The image produced by a converging lens can be larger or smaller than the object. In other words, it can be magnified or reduced. You can see that the image in Figure 5h.9 is twice the size of the object. We say that the **magnification** produced by the lens in this situation is 2.

Here is how we define magnification:

$$\text{magnification} = \frac{\text{image size}}{\text{object size}}$$

A magnifying glass is obviously designed to produce a magnified image, so its magnification is greater than 1. However, a camera lens produces a reduced image, so its magnification is less than 1.

SAQ

8 Look at the ray diagram shown in Step 4 of Figure 5h.8.

  a How many squares high is the object O?

  b How many squares high is the image I?

  c What is the magnification produced by the lens in this diagram?

P5h Optics

Summary

You should be able to:

- describe how a converging (convex) lens causes rays of light parallel to the axis to converge at the focus, and causes rays of light diverging from the focus to form a parallel beam of light
- explain that the focal length of a lens is measured from the centre of the lens to the focus
- describe how converging lenses are used as magnifying glasses, and in cameras and projectors
- **H** state that real images can be projected onto a screen, while virtual images cannot
- use a ray diagram to find the position and size of the real image formed by a converging lens
- state that the image formed by a converging lens can be real and inverted, or virtual and the right way up
- use the formula: magnification = $\frac{\text{image size}}{\text{object size}}$

Questions

1. Draw a diagram to show the difference in shape between a converging lens and a diverging lens.
2. Draw a ray diagram to show how a converging lens focuses parallel rays of light.
3. How would you alter your diagram in question 2 to show how a converging lens can produce a beam of parallel rays of light?
4. What is meant by the focus (or focal point) of a converging lens?
5. Between which two points is the focal length of a converging lens measured?
6. A camera is focused on a distant scene. How must its lens be adjusted to focus on a nearby object?
7. **H** What type of image is produced by the converging lens of a camera – real or virtual?
8. What is the difference between a real image and a virtual image?
9. Look at the ray diagram shown in Figure 5h.8. How does it show that the image formed by a converging lens is inverted?
10. Draw an accurate ray diagram for this situation: An object is placed at a distance of 10 cm from a converging lens of focal length 5 cm. How far is the image from the lens? What can you say about the magnification produced by the lens?
11. A photographic slide is 24 mm high. When projected, the image on the screen is 1.2 m high.

    a What magnification is produced by the projector?

    b The slide is 36 mm wide. How wide will the image be?

6a Resisting

Georg Ohm and electrical resistance

Georg Ohm (Figure 6a.1) was a mathematician and physicist. He made a famous contribution to Physics known as Ohm's law.

Figure 6a.1 Georg Ohm published his famous law in 1827.

Ohm was working in Berlin in the 1820s. His experiments on electrical circuits showed that the greater the voltage across a wire, the greater the current that flowed through it. In fact, he found that current was proportional to voltage ($I \propto V$). He even developed a theory to explain *why* this should be so.

You can easily check Ohm's ideas in a school laboratory (Figure 6a.2) – you just need a voltmeter and an ammeter. These instruments, which are generally quite accurate, cost a few pounds each but it wasn't so easy for Ohm, working 200 years ago. He had to devise his own methods for determining voltage and current. It is often the case that scientists who are working at the forefront of a new field have to devise their own equipment and techniques. They are working at the extreme limits of the sensitivity of the apparatus which they are using.

To provide the voltage in his experiments, he used voltaic piles, a type of battery that had been invented a few decades previously. The more piles he connected in series, the greater the voltage. There were no ammeters to measure the current. Instead, Ohm used the fact that an electric current flowing in a wire produces a magnetic field (this is how electromagnets work). A compass needle placed near a current would deflect in the field and the greater the current, the greater the deflection. Ohm was also obliged to make his own wires of different lengths and thicknesses by drawing out metal rods to the required dimensions.

Despite these difficulties, he was able to publish results that demonstrated his law: that current is proportional to voltage. At the time, many of Ohm's fellow scientists were not convinced of the importance of his findings. However, he was able to adapt a theory of heat conduction through solids so that it would explain electrical conduction. Once he had a theoretical explanation of his law, it was taken much more seriously.

Figure 6a.2 Checking Ohm's law in a school lab. It's easy with modern equipment but it was much harder in Ohm's day.

Current under control

You should recall from Item P4c *Safe electricals* that **resistors** can be used to control the current flowing in a circuit. The greater the resistance in the circuit, the smaller the current that will flow. A **variable resistor** allows the resistance in a circuit to be altered without changing any of the components.

Figure 6a.3 shows a circuit diagram in which a variable resistor is used to vary the brightness of a bulb. (You should recognise the following circuit symbols: a cell; a bulb; a variable resistor.) If the resistor's slider is moved to the right, there will be more resistance in the circuit and the bulb will be dimmer. A variable resistor can also be used to control the speed of an electric motor.

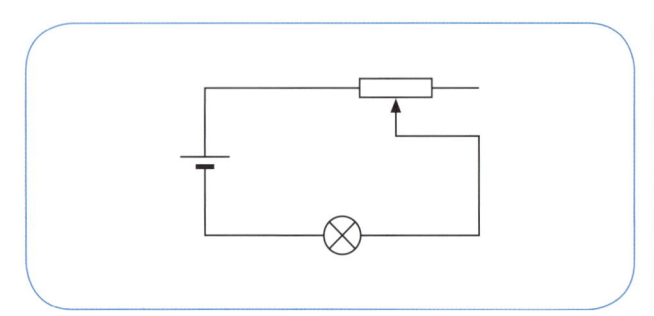

Figure 6a.3 A circuit used to control the brightness of a bulb.

SAQ

1. Figure 6a.4 shows three more circuit symbols. Using the appropriate symbols, draw a diagram to show a circuit in which a battery of cells is connected to a bulb; include a fixed resistor in series with the bulb, and a variable resistor connected so that the brightness of the bulb can be varied.

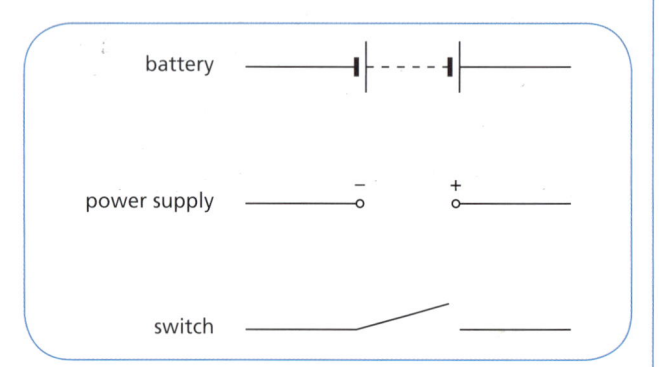

Figure 6a.4 Some circuit symbols.

Inside a variable resistor

Figure 6a.5 shows an inside view of a variable resistor. You can see the three connectors (at the top) – compare these with the circuit symbol. The outer connectors connect to the ends of the shiny resistance wire. The central connector connects to the metal slider, which you can see touching the resistance wire at the 4 o'clock position. Turning the knob of the resistor moves the slider to a different position.

Imagine that connections are made to the left-hand and central connectors. Current entering at the left-hand connector must flow through more than half of the resistance wire and out through the central connector.

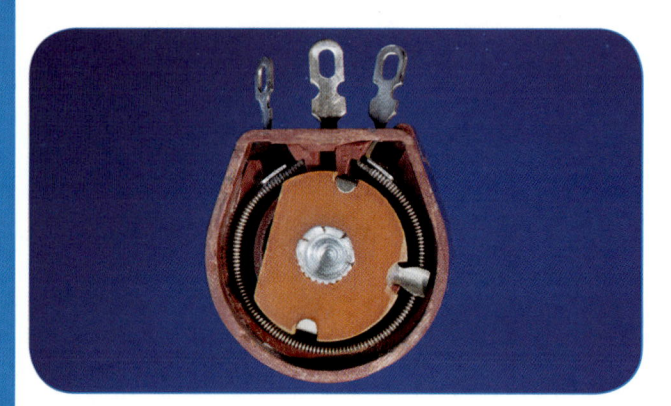

Figure 6a.5 A variable resistor with the back cover removed. You can see the coil of metallic resistance wire bent round to form an incomplete circle.

SAQ

2. Look at the variable resistor shown in Figure 6a.5. Which two connectors would you join:
   a. to obtain the greatest resistance?
   b. to obtain the least resistance?

Measuring resistance

To find the resistance R of a resistor, you need to connect it to a cell and measure two quantities:
- the current I that flows through it
- the p.d. (potential difference or voltage) V across it.

Then you can calculate the resistance R using the equation:

$$\text{resistance} = \frac{\text{voltage}}{\text{current}} \qquad R = \frac{V}{I}$$

However, this method only provides a single value each for the p.d. and the current. A better technique is shown in Figure 6a.6. In place of the cell is a **power supply** that can be adjusted to give several different values of p.d. (Notice the circuit symbol for a power supply.) For each value, the current is measured to give results like those shown in Table 6a.1.

The last column in the table shows calculated values for R. These could be averaged but a better approach is to use a graph (Figure 6a.7). The axes of the graph show p.d. and current. It is clear that the points all fall close to a straight line. A graphical method has the advantage that it is easier to spot results that do not fit the pattern. Also, the points on the graph are inevitably slightly scattered about a straight line and a 'best fit' line can be drawn to smooth out this scatter.

There are two ways to find the resistance from a graph like this. A simple method is to select one point on the line, to note the current and p.d. and then to use these values to calculate the resistance. For example, the line passes through the point (10 V, 0.4 A), and so we have:

$R = V \div I$
$ = 10 \text{ V} \div 0.4 \text{ A}$
$ = 25 \, \Omega$

A better method for finding the value of the resistance is to use the line of best fit (Worked example 1).

Potential difference V (V)	Current I (A)	Resistance R (Ω)
2.0	0.08	25.0
4.0	0.17	23.5
6.0	0.34	17.6
8.0	0.31	25.8
10.0	0.40	25.0
12.0	0.49	24.5

Table 6a.1 Typical results for an experimental measurement of resistance. The values of resistance are calculated using $R = V \div I$.

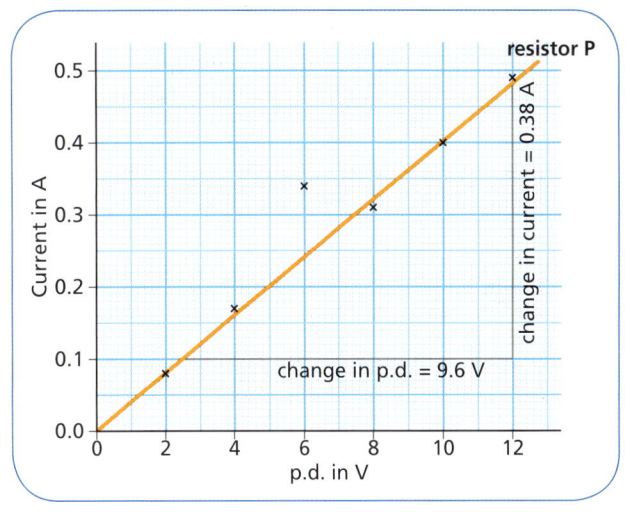

Figure 6a.7 A graph of current against p.d. for the data shown in Table 6a.1. See Worked example 1.

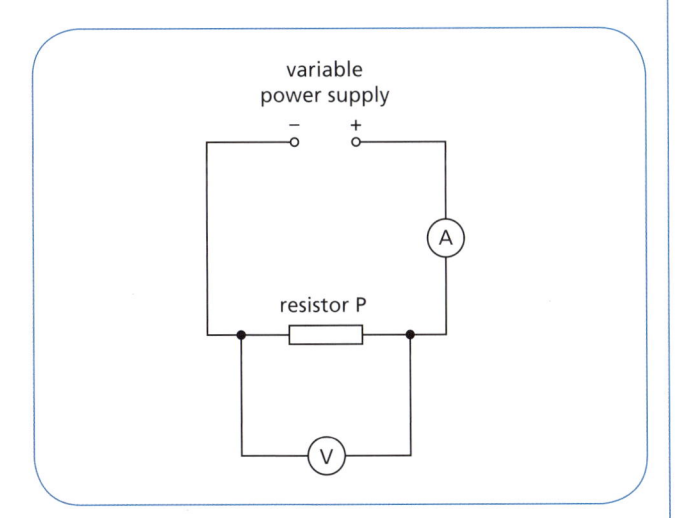

Figure 6a.6 A circuit for investigating the current–voltage characteristics of a resistor. The power supply can be adjusted to give a range of values of p.d. (typically from 0 V to 12 V). For each value of p.d., the current is recorded.

Worked example 1

Use a graphical method to find the resistance of the resistor using the results shown in Table 6a.1.

Step 1: draw a graph to show the data. Plot p.d. V on the x-axis and current I on the y-axis (Figure 6a.7).

Step 2: draw a line of best fit through the points. (Identify any points that clearly do not lie on the line and ignore them.) The line should pass close to the origin.

The line of best fit is shown in Figure 6a.7. The point at 6.0 V has been ignored. (You might guess that the value of current has been incorrectly recorded. It should have been 0.24 A but you can't assume this; you would need to repeat this measurement.)

continued on next page

Worked example 1 - continued

Step 3: draw a large triangle to find values for change in p.d. and change in current.

Change in p.d. = 12.0 V − 2.4 V = 9.6 V

Change in current = 0.48 A − 0.10 A = 0.38 A

Step 4: calculate the resistance R
= change in p.d. ÷ change in current

(This is $\dfrac{1}{\text{gradient of the line}}$.)

Resistance R = change in p.d. ÷ change in current

= 9.6 V ÷ 0.38 A = 25.3 Ω

So the resistance of the resistor is 25.3 Ω.

SAQ

3 One way to find the resistance from Table 6a.1 would be to average the values of R in the last column.

   a Why would this give an inaccurate answer?

   b Explain why using the gradient of the line of best fit is likely to give a more accurate value for R than averaging several values or selecting a single pair of values of I and V.

Ohm's law

A graph like that shown in Figure 6a.7 is known as a **current–voltage characteristic graph**. It shows how the current through a component depends on the p.d. across it. For the resistor whose current–voltage characteristic is shown in Figure 6a.7, the graph is a straight line passing through the origin. So, we can say that the current that flows through it increases in proportion to the p.d. across it.

Another way of summarising this graph is to say that the resistance of the resistor is constant and does not depend on the p.d. across it. This is the relationship now known as Ohm's law.

- A component obeys **Ohm's law** if the current that flows through it is proportional to the potential difference across it.

Usually, for a component to obey Ohm's law, its temperature must remain constant. An **ohmic conductor** is one whose resistance does not depend on the current flowing through it; in other words, it obeys Ohm's law.

SAQ

4 A current of 0.2 A flows through a resistor when the p.d. across it is 8 V. When the p.d. is reduced to 3 V, the current drops to 0.075 A. Is the resistor an ohmic conductor? Can you be sure of your answer?

Positive and negative

Figure 6a.8a shows a smaller version of Figure 6a.7, for resistor P. Now imagine reversing the contacts to the resistor shown in Figure 6a.6. The p.d. across P would be reversed and the current would flow through P in the opposite direction. The current and p.d. would now have negative values. This is shown by the line in Figure 6a.8b. We can combine

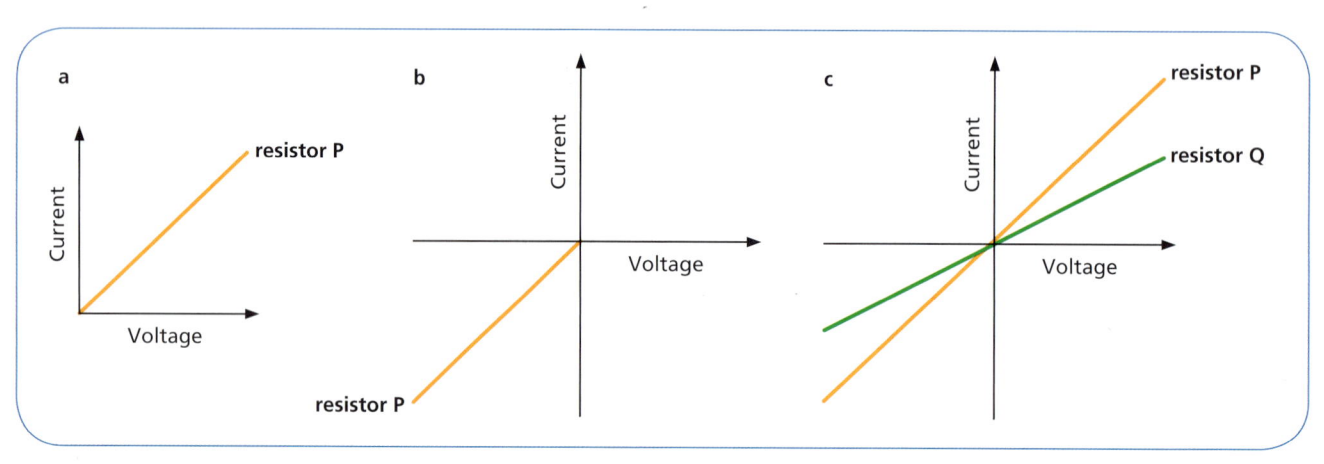

Figure 6a.8 a,b A resistor has the same resistance in both directions. **c** Current–voltage characteristic graphs for two resistors, P and Q.

Figures 6a.8a and 6a.8b on a single graph, and this is shown as the line labelled 'resistor P' in Figure 6a.8c. The line is straight and passes through the origin, showing that the resistor has the same resistance in both directions.

All ohmic conductors have a current–voltage characteristic graph that is a straight line passing through the origin.

Figure 6a.8c also shows the characteristic graph for another resistor, labelled 'resistor Q'. The graph is again a straight line passing through the origin, showing that resistor Q is an ohmic conductor. Resistor Q has a higher resistance than resistor P; you can tell this because, for any given p.d., it lets through less current.

Another characteristic graph

A filament lamp has a metal filament – a coil of fine wire through which a current flows (Figure 6a.9). The greater the current flowing through the filament, the hotter it gets and the more brightly it glows.

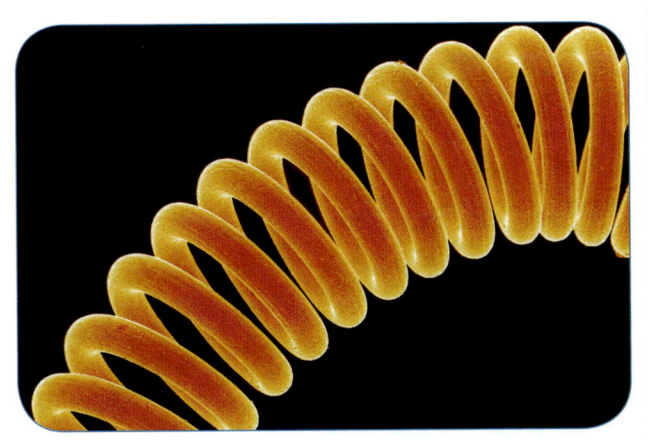

Figure 6a.9 The glowing filament of a light bulb. The resistance of the filament increases as it gets hotter.

Figure 6a.10 shows the results of an experiment to investigate the current–voltage characteristic graph of a filament lamp. The graph starts off from the origin as a roughly straight line, but it gradually curves over. A filament lamp is not an ohmic conductor.

Why does the graph curve in this way? The current passing through the metal filament causes it to get hot and this increases its resistance. As a consequence, the current flowing through the filament does not increase as rapidly as if its temperature remained constant. The result is a curved characteristic graph.

SAQ

5 Look at Figure 6a.10. It is suggested that a filament lamp is an ohmic conductor if the voltage across it is small. Is this idea supported by the graph?

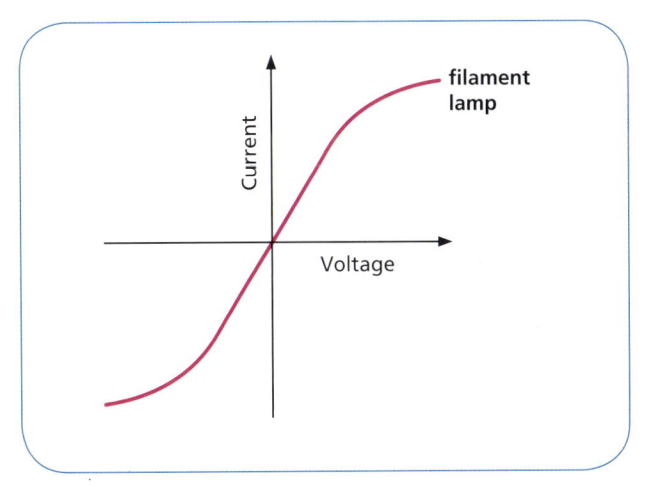

Figure 6a.10 The current–voltage characteristic graph for a filament lamp. Although the line passes through the origin, it is curved, showing that a lamp is a non-ohmic conductor.

Summary

You should be able to:

◆ state that a variable resistor can be used in a circuit to control the current flowing, and hence that it can be used to vary the brightness of a lamp or the speed of a motor

◆ calculate resistance using the equation:

$$\text{resistance } (\Omega) = \frac{\text{voltage (V)}}{\text{current (A)}} \qquad R = \frac{V}{I}$$

continued on next page

Summary - continued

- state that the current-voltage characteristic graph for an ohmic conductor is a straight line passing through the origin

- state that the resistance of an ohmic conductor is:

$$\frac{1}{\text{gradient of the current-voltage characteristic graph}}$$

- state that a filament lamp is a non-ohmic conductor and that its current-voltage characteristic graph is curved because the resistance of the filament increases as it gets hotter

Questions

1. Draw the circuit symbols for the following components:

   resistor, variable resistor, bulb, cell, battery, switch, power supply.

2. What is the resistance of a resistor if a p.d. of 20 V across it causes a current of 2 A to flow through it?

3. A resistor allows a current of 0.25 A to flow when the p.d. across it is 4 V. What is its resistance?

4. How does the resistance of a metal change if its temperature decreases?

5. Look at the graphs shown in Figure 6a.8c. How can you tell from the graphs that the two resistors obey Ohm's law?

6. Draw a diagram to show the circuit you would use to investigate the current-voltage characteristic graph of a filament lamp. (Hint: Figure 6a.6 will help you.)

7. Table 6a.2 shows the results of an experiment to investigate the current-voltage characteristic graph of a filament lamp.

   a. Copy and complete the table by adding values to the final column.

Potential difference V (V)	Current I (A)	Resistance R (Ω)
2.0	0.42	
4.0	0.80	
6.0	1.12	
8.0	1.38	
10.0	1.61	
12.0	1.73	

   Table 6a.2

continued on next page

Questions - continued

    b Explain how you can tell from the figures in the final column that the lamp is not an ohmic conductor.

    c What current would you expect to flow through the lamp if the connections to the lamp were reversed and the p.d. across it was 4.0 V?

    d On graph paper, draw the current–voltage characteristic graph for the lamp.

8 Look at the graph in Figure 6a.10. How can you tell from the graph that the filament lamp's resistance increases as the p.d. across it increases?

H 9 What p.d. is needed to make a current of 2 A flow through a 20 Ω resistor?

10 What current will flow through a 50 Ω resistor when the p.d. across it is 6 V?

11 Table 6a.3 shows the results of an experiment to investigate the current–voltage characteristic graph of a metal wire. Draw a graph to show these results and use it to deduce the resistance of the wire.

Potential difference V (V)	Current I (A)
2.5	0.15
4.7	0.27
6.0	0.35
8.7	0.53
10.0	0.60
11.9	0.71

Table 6a.3

12 A thermistor is a type of resistor whose resistance changes rapidly when it is heated. For a thermistor whose resistance decreases as it gets hotter:

    a explain why it is not an ohmic conductor

    b sketch the current–voltage characteristic graph you would expect it to have.

P6b Sharing

Automatic response

Automatic weather stations in Antarctica use several devices to detect weather conditions. Some of these sensors depend on the changing electrical resistance of components such as light-dependent resistors and thermistors. (You may have used some of these components in Technology lessons.)

On top of the station shown in Figure 6b.1 is a wind vane. It turns as the wind changes direction. As it turns, it rotates the spindle of a variable resistor, so the resistance depends on the direction of the wind.

There is also an electronic thermometer whose resistance changes with temperature and an instrument for measuring the hours of daylight. Even the sensors that detect changes in atmospheric pressure and humidity rely on changing electrical resistance. This is because a changing resistance can be readily converted to a changing voltage, and this can be used to trigger an electronic circuit. The result can then be transmitted to a recording station far away.

Figure 6b.1 An automatic weather station on Alexander Island in Antarctica. Most of the time, these stations are unmanned but here you can see one being checked by a researcher from the British Antarctic Survey.

Changing resistance

Weather conditions are recorded by automated weather stations in Antarctica. However, you don't have to experience extreme conditions like those to make use of such sensors. They are all around us, for example controlling the water temperature in a washing machine or sensing the presence of burglars. A changing electrical resistance is used to sense some factor of interest and then an electronic circuit can respond automatically.

We will look at two components whose resistance changes in response to changes in their environment, and then we will look at how these changes in resistance can be converted to a change in voltage.

Light-dependent resistors

A **light-dependent resistor** (LDR) is a type of variable resistor. When light shines on it, its resistance falls (Figure 6b.2). In the dark, an LDR has a high resistance, often over 1 MΩ. In bright light, its resistance may fall to 400 Ω.

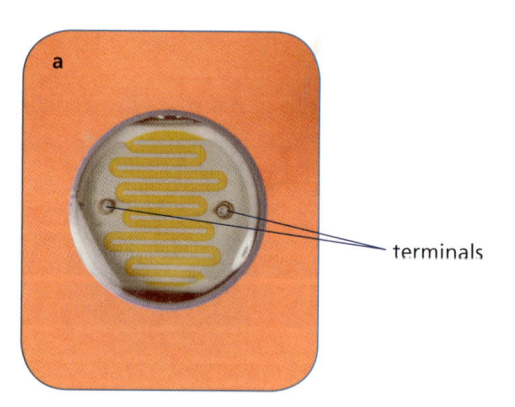

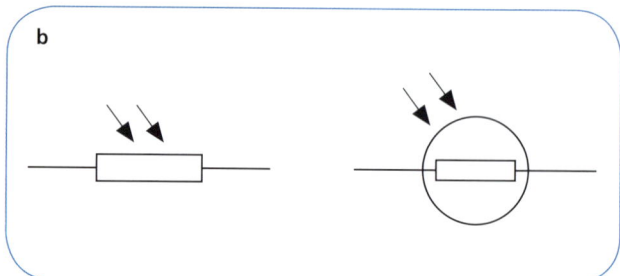

Figure 6b.2 a A light-dependent resistor. The terminals are where the current enters and leaves the resistor. In between is the resistive material (orange). When light falls on it, its resistance decreases. **b** Two alternative circuit symbols for an LDR – the circle is optional. The arrows represent light shining on the LDR.

An LDR is made of a material that does not normally conduct well, because its electrons are bound to their atoms. However, light can provide the energy needed for some electrons to break free; now there are free conduction electrons that can move through the resistor, and this means that a current can flow much more easily. Figure 6b.3 shows how the resistance of an LDR depends on the brightness of the light falling on it.

LDRs are used in circuits to detect the level of light – for example, in security lights that switch on automatically at night. Some digital clocks and radios have one fitted. When the room is brightly lit, the display is automatically brightened so that it can be seen against its bright surroundings. In a darkened room, the display need only be dim.

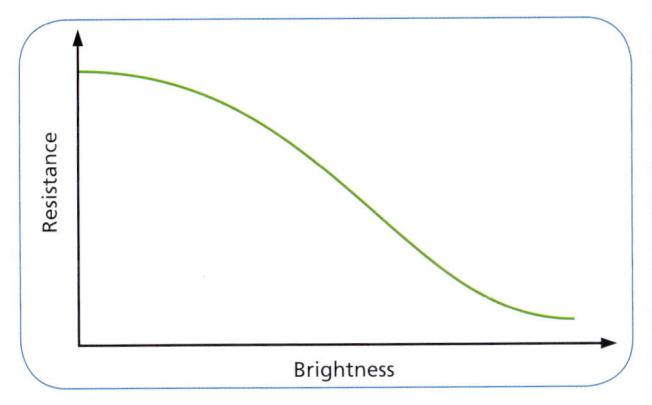

Figure 6b.3 The resistance of an LDR decreases as the brightness of the light falling on it increases.

SAQ

1 Suggest some other places where LDRs might be used.

Thermistors

A **thermistor** (Figure 6b.4) is another type of variable resistor whose resistance depends on its environment. In this case, its resistance depends on its temperature. The resistance changes by a large amount over a narrow range of temperatures. For some thermistors, the resistance decreases as they are heated – perhaps from 2 kΩ at room temperature to 20 Ω at 100 °C. These thermistors are useful as temperature probes and thermometers.

Some thermistors are made so that their resistance *increases* when the temperature increases. These are included in circuits in which we want to prevent overheating. If the current flowing is too high, components might burn out. With a thermistor in the circuit, the resistance increases as the temperature rises and the high current is reduced.

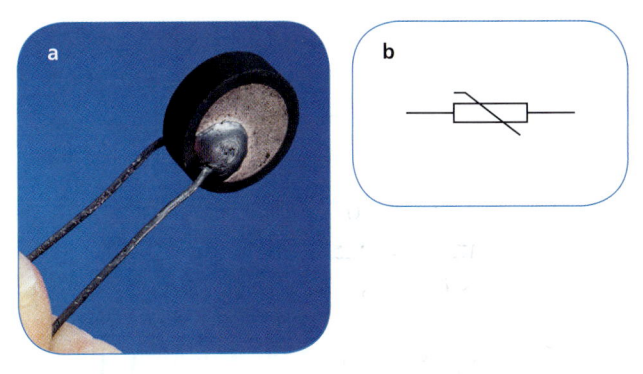

Figure 6b.4 a A thermistor. **b** The circuit symbol for a thermistor. The line through the resistor indicates that its resistance varies.

SAQ

2 Look at Figure 6b.5.
   a Over what range of temperatures does the resistance of the thermistor drop most rapidly?
   b Between what values does the resistance change over this range?
   c By what *fraction* does the resistance change over this range?

(Your answers can only be approximate.)

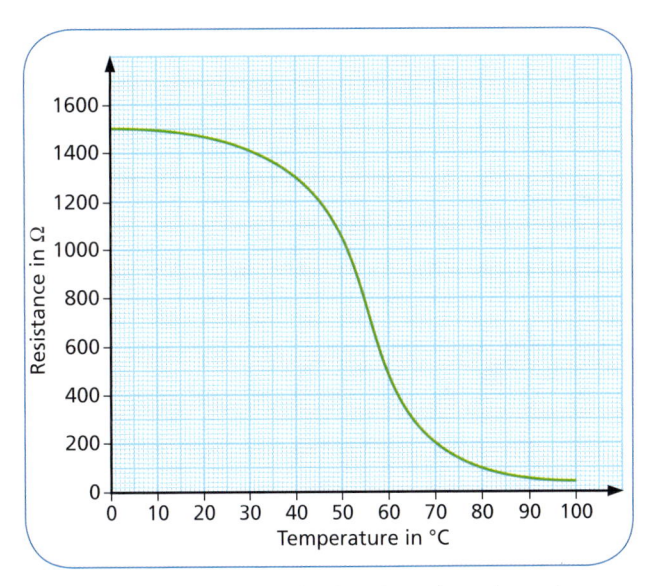

Figure 6b.5 The resistance of a thermistor depends on the temperature. In this case, its resistance drops a lot as the temperature increases by a small amount.

Potential divider circuits

Power supplies and batteries usually provide a fixed potential difference. There is a simple circuit called a **potential divider** that you can use to get a smaller p.d. from the power supply or battery. Figure 6b.6 shows the simplest form of potential divider.

In Figure 6b.6a, two identical 10 kΩ resistors are connected in series across a 6 V supply. The voltage across each of the resistors is 3 V: because the resistors have equal values, the voltage of the supply is divided equally between them.

In Figure 6b.6b, two resistors R_1 and R_2 are connected in series across the 6 V supply. The p.d. across the pair is thus 6 V; we call this the *input voltage* V_{in}. (It helps to think of the bottom line as representing 0 V and the top line as representing 6 V.) What is the p.d. at point X, between the two resistors?

The input voltage V_{in} is divided between the two resistors; the bigger resistor gets the bigger share. Suppose that R_1 = 10 kΩ and R_2 = 20 kΩ. Because R_2 is twice as big as R_1, its share of the p.d. is twice as great. Hence the p.d. across R_2 is 4 V and the *output voltage* V_{out} is 4 V. By choosing the values of the two resistors carefully, we can get any output voltage we want, from 0 V to 6 V.

Variable output

Some electronic circuits work from a fixed voltage. Suppose a radio works from a 9 V battery but one of its components needs 4.5 V. The solution is simple: use a potential divider with two equal resistors. Then the input voltage is divided in two and the output voltage is 4.5 V.

Other components may need a variable voltage. For example, you need to be able to vary the voltage to a loudspeaker so that the volume changes. Figure 6b.7 shows how to do this. The upper fixed resistor R_1 is replaced by a variable resistor R_V. When the variable resistor is altered, its share of the input voltage changes and so the output voltage changes.

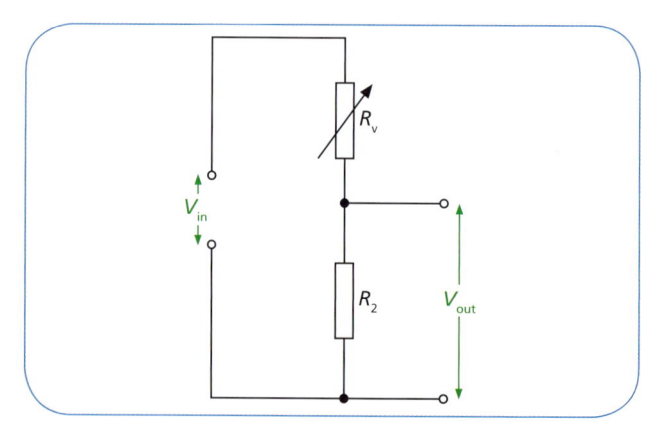

Figure 6b.7 Using a variable resistor to give a variable output voltage from a potential divider.

Recognising a potential divider

You should now be able to spot the following features of a potential divider circuit.
- Two resistors are connected in series.
- The input voltage is connected across both resistors.
- The output voltage is across the lower resistor.
- The size of the output voltage depends on the resistors; the bigger resistor gets a bigger share of the input voltage.

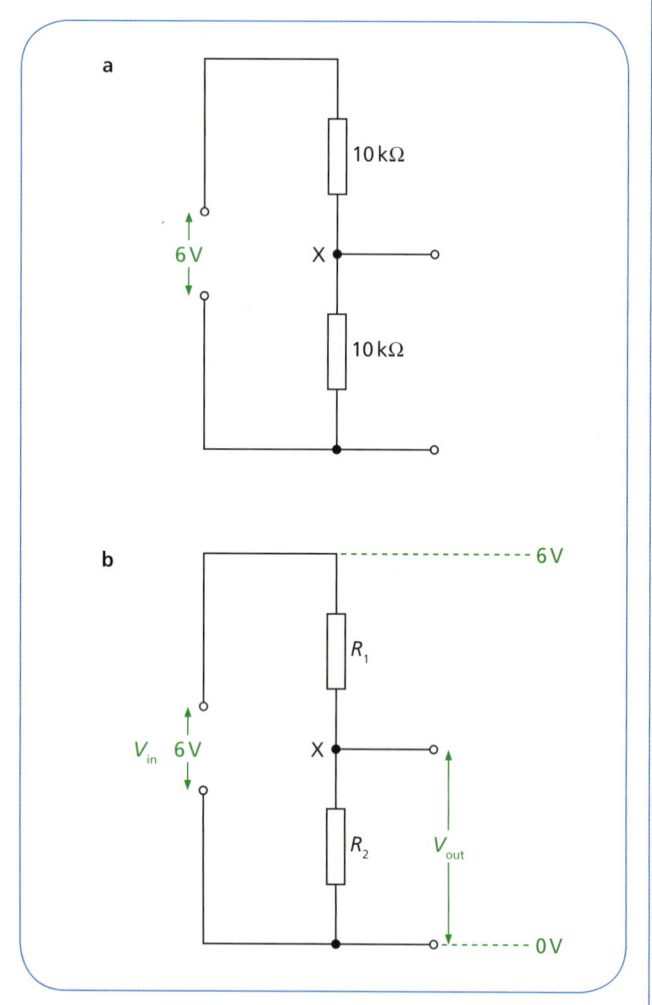

Figure 6b.6 A simple potential divider circuit. The output voltage is a fraction of the input voltage. The input voltage is divided according to the relative values of the two resistors.

SAQ

3 A potential divider circuit has an input voltage of 12 V and an output voltage of 6 V. What can you say about the sizes of the two resistors?

4 Draw a potential divider circuit formed of a 50 Ω resistor and a 100 Ω resistor. The input voltage is 6 V and the output voltage should be 2 V.

Calculating V_{out}

We can write an equation for calculating V_{out}:

$$V_{out} = \frac{V_{in} \times R_2}{(R_1 + R_2)}$$

Worked example 1

What is the output voltage for the potential divider circuit shown in Figure 6b.8?

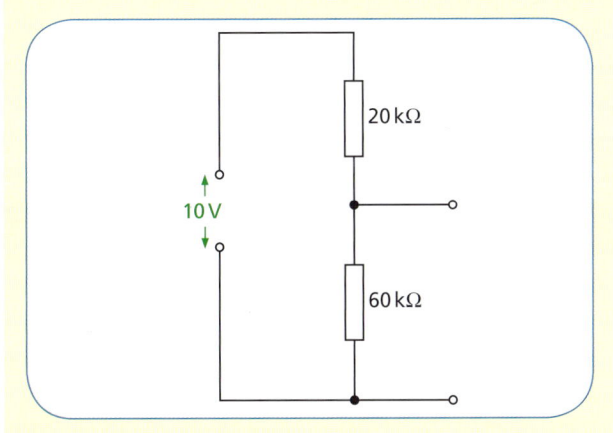

Figure 6b.8

We have $V_{in} = 10\,V$

$R_1 = 20\,k\Omega$

$R_2 = 60\,k\Omega$

Substituting in the equation gives:

$$V_{out} = \frac{V_{in} \times R_2}{(R_1 + R_2)}$$

$$= \frac{10 \times 60}{(60 + 20)}$$

$$= \frac{600}{80}$$

$$= 7.5$$

So the output voltage is 7.5 V.

(We can ignore the fact that the resistances are in kΩ, rather than Ω, because the 'kilo' parts 'cancel out'.)

SAQ

5 If the two resistors in Worked example 1 were swapped over, what would the output voltage become?

Variable output

Suppose the upper resistor R_1 in Worked example 1 is now replaced with a variable resistor whose resistance can be varied between 0 Ω and 500 Ω. How will this affect the output voltage?

When the resistor is set at 0 Ω, it will have no voltage across it and so the output voltage will be the full 10 V supplied by the input.

At the other end of its range, R_1 will be 500 Ω and the output voltage will be:

$$10\,V \times \frac{60\,k\Omega}{560\,k\Omega} = 1.1\,V$$

Notice that the output voltage cannot go all the way down to zero in this arrangement because, no matter how big R_1 is, R_2 will always have a share of the input voltage. So there is a 'threshold voltage' when a potential divider is used like this to divide a supply voltage.

Using potential dividers in circuits

A potential divider can be used as part of a light-sensing circuit. Figure 6b.9 (page 000) shows how. One of the resistors, R_1, is replaced by a light-dependent resistor. When light falls on the LDR, its resistance decreases. This means that its share of V_{in} decreases and so the p.d. across the fixed resistor R_2 increases. Hence the output p.d. increases when the light level increases, and this can then be used to trigger an electronic circuit that responds in some way. If the circuit has detected a burglar's torchlight, it might sound an alarm. If it has detected the morning sun, it might switch off a streetlight.

If the two resistors in this circuit are swapped over, the output p.d. will *increase* when the light level decreases. This could be used in a similar way to trigger an electronic circuit.

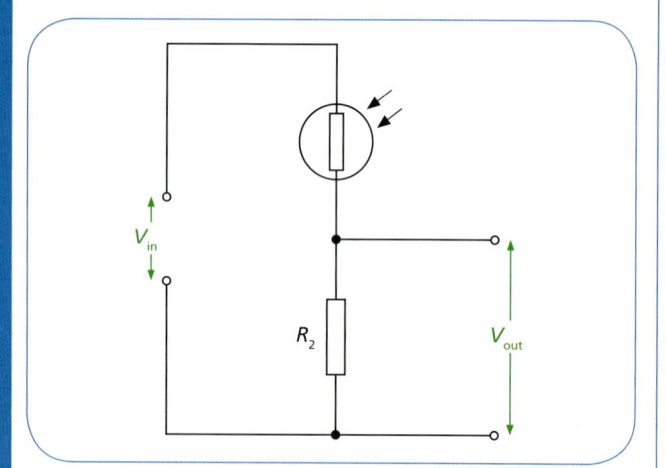

Figure 6b.9 Using a potential divider circuit to produce a p.d. that changes when the light level changes. The resistance of the LDR depends on the intensity of the light and so the output p.d. changes with the light level.

The circuit could also be adapted by using a variable resistor in place of the fixed one. Then the light level at which the electronic circuit is triggered could be set by adjusting the variable resistor.

For a circuit that detects changes in temperature, a thermistor could be used in place of the LDR. A change in temperature would make the thermistor's resistance change and so V_{out} would change.

SAQ

6 The LDR in the circuit of Figure 6b.9 is replaced by a thermistor whose resistance decreases when it is heated. How will the output voltage V_{out} change as the temperature rises?

Summary

You should be able to:

- state that the resistance of a light-dependent resistor decreases as the brightness of the light falling on it increases

- state that the resistance of a thermistor changes rapidly over a narrow temperature range when it is heated

- state that, in a potential divider circuit, two resistors are connected in series across the input p.d., which is divided in proportion to their resistances

- state that the output p.d. V_{out} is supplied at the point between the two resistors

- calculate the output p.d. of a potential divider using the equation:

$$V_{out} = \frac{V_{in} \times R_2}{(R_1 + R_2)}$$

- state that a potential divider circuit can include an LDR or a thermistor, so that the output p.d. depends on the light level or temperature, respectively

Questions

1. What components are described here?

   a Its resistance decreases when it is placed in a freezer.

   b Its resistance decreases when it is placed in bright sunlight.

2. Draw the circuit symbols for each of the components in question 1.

3. A battery supplies a voltage of 10 V.

   a Draw a circuit to show how a potential divider can be used to give a smaller voltage than this.

   b Suppose that the potential divider is required to supply a p.d. of 5 V. On your diagram, indicate suitable values of the resistances and show two points between which the output voltage can be measured.

   c How should the circuit be altered to give a variable output voltage?

H 4 Calculate the output voltage of the circuit shown in Figure 6b.10.

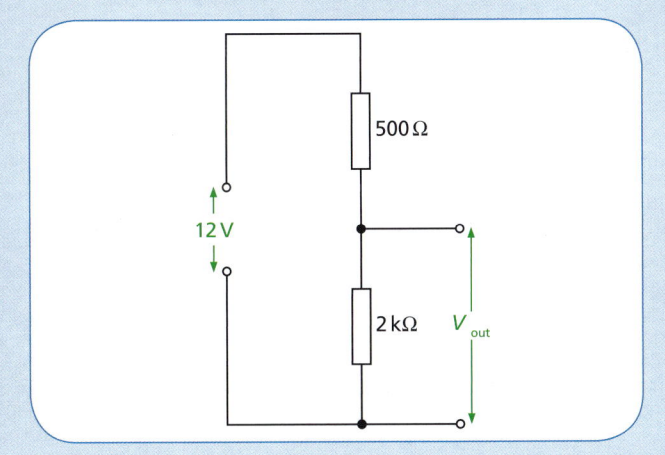

Figure 6b.10

5. A thermistor has a resistance of 20 kΩ at 30 °C and 200 Ω at 60 °C.

   a Draw a circuit diagram to show how the thermistor can be connected as a potential divider with a 100 Ω resistor.

   b If the input voltage is 12 V, what is the output voltage at 30 °C and at 60 °C?

P6c Motoring

Making the most of motors

We make good use of electric motors in our everyday lives. They spin our CDs and computer disk drives. They turn our washing machines and food processors. They blow the hot air out of our hair dryers and they cool us (and our computers) by turning fans.

There are many other electric motors at work, making our highly technological lives possible. An electric motor (the starter motor) is needed to get a petrol-driven car started (Figure 6c.1). They turn the wheels of many trains, and of the electric cars that can help to reduce air pollution in cities (Figure 6c.2). Giant electric motors operate the pumps that bring us fresh water and take away our sewage. Some of the electricity generated by a power station is needed by the power station itself, to power motors that pump cooling water through the condensers.

SAQ

1 Which of the following make use of electric motors? Which do not?

washing machine   electric heater
CD player   MP3 player   electric drill
electric lawnmower   television set

Figure 6c.1 Most cars have a motor that is a type of internal combustion engine, fuelled by petrol. It will continue to operate once it has started turning. Today, cars have electric starter motors to give them their initial spin, but in the early days of motoring drivers had to use a crank handle to start the motor. You may have seen this in old movies. An electric motor has the convenience that it will start to turn as soon as a current flows through it.

Figure 6c.2 This electric car is designed to be used in towns and cities. It has an electric motor to turn each wheel, as well as motors that operate the windscreen wipers and the door locks. Its roof has an array of solar cells that supplement the energy stored in its battery. Electric cars like this can help to reduce both noise and air pollution in city centres.

Magnetism from electricity

Electric motors use electromagnets. A typical electromagnet is made from a coil of copper wire. When a current flows through the wire, there is a magnetic field around the coil (Figure 6c.3 on page 287). A coil like this is sometimes called a **solenoid**. Copper wire is often used. The coil does not have to be made from a magnetic material – the point is that it is the electric current that produces the magnetic field.

From Figure 6c.3, you can see that the magnetic field around a solenoid is similar to that around a bar magnet. One end of the coil is a north pole and the other is a south pole. Notice that the field lines go all the way through the centre of the coil.

The field around a current

If you uncoil a solenoid, you will have a straight wire. When an electric current flows through a wire, a magnetic field is produced around

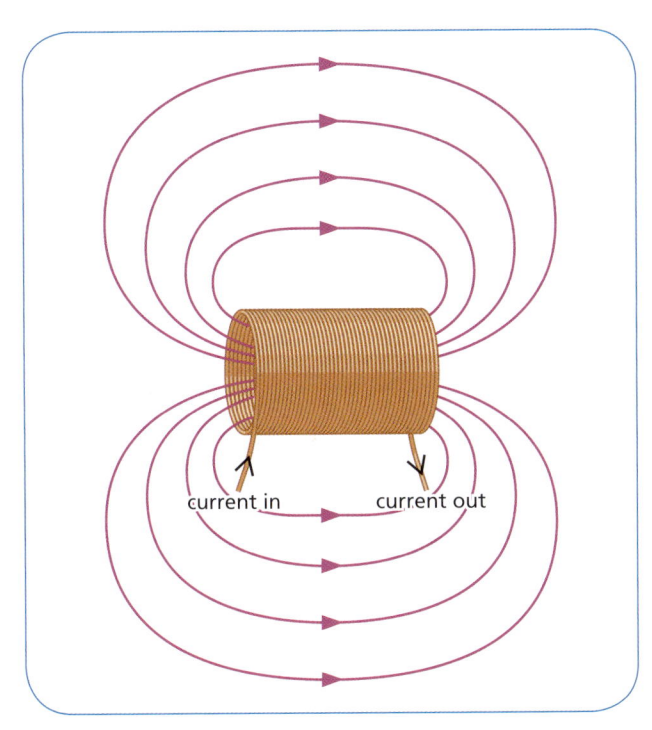

Figure 6c.3 A solenoid: when a current flows through the wire, a magnetic field is produced. The field is similar in shape to that of a bar magnet.

it, as shown in Figure 6c.4a. The field lines are concentric circles around the current. The field is weaker than that of the solenoid – winding the wire into a coil is a way of concentrating the field around the current.

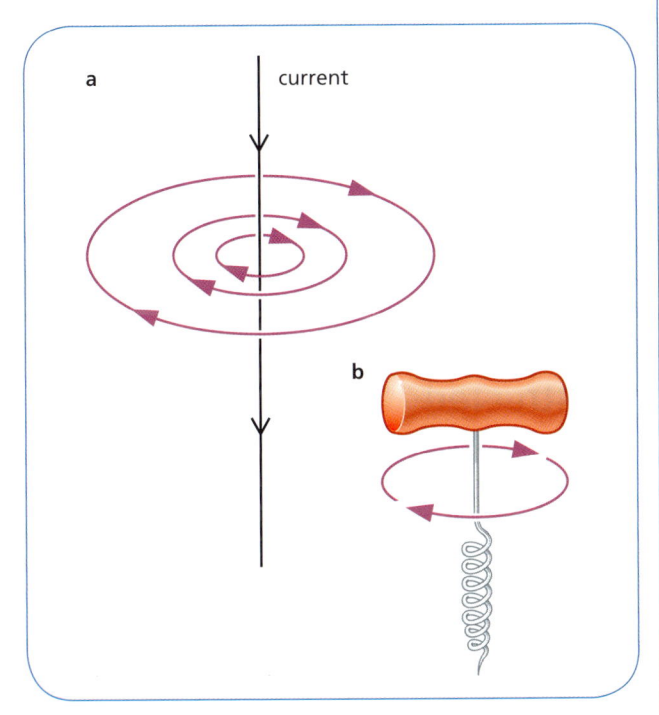

Figure 6c.4 The magnetic field around a current in a straight wire. The field lines are circles around the wire. The further you go away from the wire, the weaker the field is.

The **corkscrew rule** (or pencil-sharpener rule) tells you the direction of the field lines. Imagine screwing a corkscrew in the direction of the current (Figure 6c.4b); the direction you turn it in is the direction of the magnetic lines of force. Alternatively, use the **right-hand-grip rule**. Imagine gripping the wire by curling the fingers of your right hand around it, with your thumb pointing along it in the direction of the current. Then your fingers show the direction of the field lines around the wire.

SAQ

2 A current flows downwards in a wire that passes vertically through a table top. Will the magnetic field lines around it go clockwise or anticlockwise?

3 Look at the magnetic field pattern shown in Figure 6c.4a. How can you tell from the pattern that the field gets weaker as you get further from the wire?

The motor effect

An electric motor makes use of the magnetic field around a coil produced by an electric current. However, it is not essential to have a coil to produce movement. The basic requirements are:
- a magnetic field
- a current flowing *across* the magnetic field.

Figure 6c.5 shows a way of demonstrating this in the laboratory. The copper rod is free to roll along the two aluminium support rods. The current

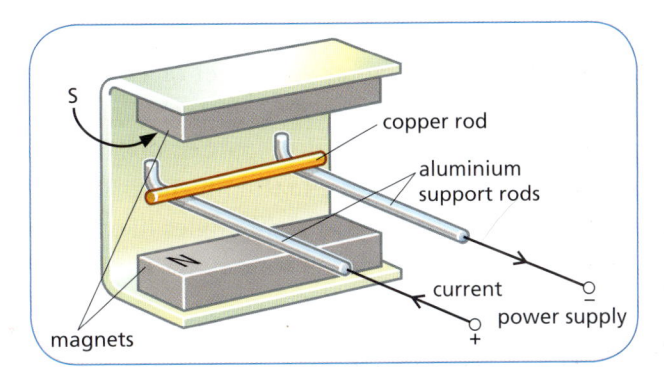

Figure 6c.5 Demonstrating the motor effect. There is a magnetic field around the current in the copper rod. This interacts with the field of the magnets and the result is a horizontal force on the rod. A copper rod is used because it is a non-magnetic material. (A steel rod would be attracted to the magnets.)

from the power supply flows along one support rod, through the copper rod and out through the other support rod. The two magnets provide a vertical magnetic field.

What happens when the current starts to flow? The copper rod rolls horizontally along the support rods. It is pushed by a horizontal force. The force comes about because the magnetic field around the current is repelled by the magnetic field of the permanent magnets. The force can be increased in two ways:
- by increasing the current
- by using magnets with a stronger magnetic field.

This force, which is the force that makes an electric motor turn, is known as the **motor effect**.

If the current is reversed, then the rod is pushed in the opposite direction because the force is reversed. If the magnets are turned upside down, the magnetic field is reversed and, again, the rod is pushed in the opposite direction because the force is reversed.

SAQ

4 Look at Figure 6c.5. Predict what will happen if the magnets are reversed and the connections to the power supply are reversed.

Fleming's left-hand rule

In Figure 6c.5, you can see that there are three things with direction and that they are at right angles to each other. These are:
- the magnetic field
- the current
- the force.

From Figure 6c.6a, you can see that they are arranged like the edges at the corner of a cube. To remember how they are arranged, physicists use **Fleming's left-hand rule** (Figure 6c.6b). It is worth practising holding the thumb and first two fingers of your left hand at right angles like this. Then learn what each finger represents, as shown in the diagram.

SAQ

5 Check that Fleming's left-hand rule correctly predicts the direction of the force on the current in Figure 6c.5.

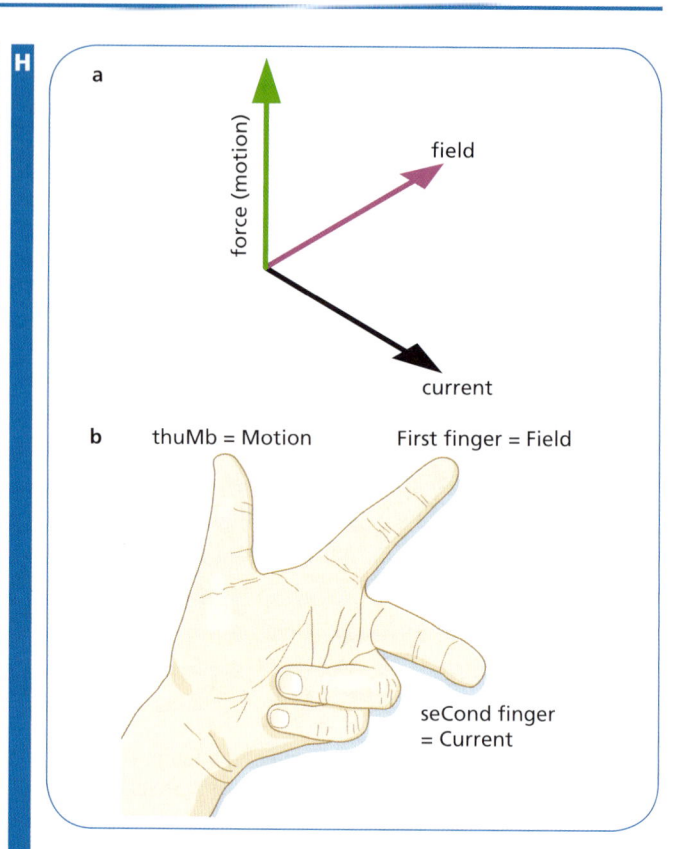

Figure 6c.6 a Force, field and current are at right angles to each other. **b** Fleming's left-hand rule.

If you know the directions of the current and the magnetic field then you can use Fleming's left-hand rule to predict the direction of the force on a current-carrying conductor in a magnetic field. By keeping your thumb and fingers rigidly at right angles to each other, you can check that reversing the direction of either the current or the magnetic field will also reverse the direction of the force. (Don't try changing the directions of individual fingers. You have to twist your whole hand around at the wrist and elbow.)

How electric motors are constructed

The idea of an electric motor is this: there is a magnetic field around an electric current that can be attracted or repelled by another magnetic field, and this can produce movement. It isn't obvious how to do this so that continuous movement is produced. If you put two magnets together so that they repel, they move apart and stop. Electric motors are cleverly designed to produce movement that continues for as long as the current flows.

You may have constructed a model electric motor like the one shown in Figure 6c.7. Its essential features are shown in Figure 6c.8:

- a **coil of wire**, which acts as an electromagnet when a direct current flows through it
- two **magnets**, to provide a steady magnetic field passing through the coil
- a **split-ring commutator** and two **brushes** through which current reaches the coil. The brushes are springy wires that press against the two metal sections of the commutator.

Here is our first explanation of how an electric motor works. A current flows in through the right-hand brush, around the coil and out through the other brush. While the current is flowing, the coil is an electromagnet. Suppose that the top of the coil becomes a north magnetic pole and the bottom a south pole. The north pole is attracted to the south pole of the permanent magnet on the left, and so the coil starts to turn to the left (anticlockwise).

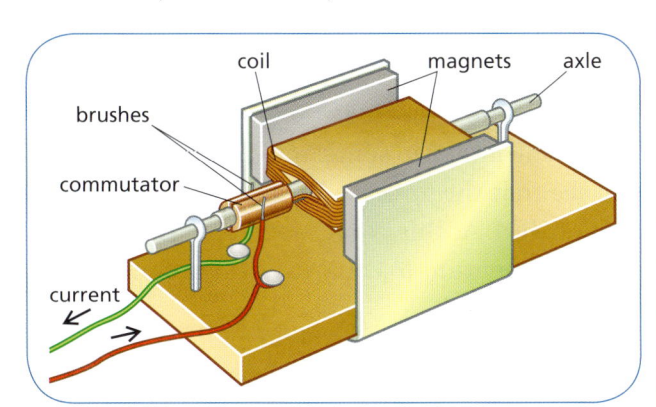

Figure 6c.7 A model of how an electric motor works.

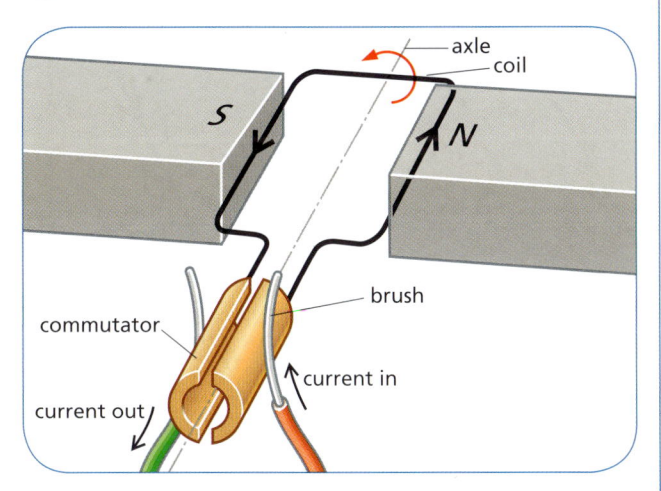

Figure 6c.8 A spinning electric motor. The coil is an electromagnet that is attracted round by the permanent magnets. Every half turn, the commutator reverses the current flowing through the coil so that it keeps turning.

These are the main features of a DC electric motor, which works with direct current. (AC motors, which work with alternating current, are built slightly differently.)

SAQ

6 Explain why the motor would start to turn clockwise if the current flowed the opposite way round to that shown in Figure 6c.8.

The role of the commutator

The commutator is there to make sure that the current always flows the same way round the coil, so that it keeps spinning. Here is how it works.

When the current flows, the coil is attracted round by the two permanent magnets. Its momentum carries it past the horizontal position. Now the brush connections to the two halves of the commutator are reversed. The current flows the opposite way around the coil, which means that we again have a north magnetic pole on the top of the coil, so it turns another 180° anticlockwise.

Without the commutator, the coil would simply turn until it was vertical and then stop. The commutator reverses the current through the coil every half turn so that the coil keeps on turning. In Figure 6c.8, the commutator ensures that the current always flows into the right-hand side of the coil and out of the left-hand side.

If you have made a model like the one shown in Figure 6c.7, you might have noticed electrical sparks flashing around the commutator. These happen as the contact between the brush and one commutator segment is broken, and as it makes contact with the other segment.

SAQ

7 Explain why the coil in Figure 6c.8 would not turn if it was vertical.

Practical electric motors

The model electric motor is a good design if you want to understand how an electric motor works, but it doesn't function very well as a practical motor. One way in which the design can be improved is by changing the shape of the magnets. Magnets with curved faces can be used. The magnetic field lines are then at right angles to the poles, so that they form a 'radial field'. As the coil turns, the field lines are always at 90° to the coil, so it feels the maximum force as it turns.

In many motors, the permanent magnets are replaced by electromagnets. Figure 6c.9 shows an electric motor that has been partially dismantled to show the rotor coils that turn inside the fixed (stator) coils.

Figure 6c.9 This electric drill has been partly disassembled to show the construction of its motor. The rotor is on the right (with cooling fan blades attached); you can see the brass commutator with its many segments. The stator is inside the body of the drill. The stator coils provide the magnetic field that causes the rotor to spin when a current flows through it.

Electric motors revisited

So far, we have said that an electric motor turns because its coil is an electromagnet that is attracted around by permanent magnets. That is correct as far as it goes, but we can also explain how a motor works in a different way, by thinking of the forces that act on the current flowing round the coil.

Figure 6c.10a shows a simple electric motor with its coil horizontal in a horizontal magnetic field. The coil is rectangular. What forces act on each of its four sides?

Side AB – the current flows from A to B, across the magnetic field. Fleming's left-hand rule shows that a force acts on it, vertically upwards.

Side CD – the current is flowing in the opposite direction to AB, so the force on CD is downwards.

Sides BC and DA – the current here is parallel to the field. Because it does not cross the field, there is no force on these sides.

Figure 6c.10b shows a simplified view of the coil. The two forces acting on it are shown; they cause the coil to turn anticlockwise. The two forces provide a turning effect (or torque) that causes the motor to spin. From Figure 6c.10c, you can see that the forces will not turn the coil when it is vertical. At this point, we have to rely on the coil's momentum to carry it further round.

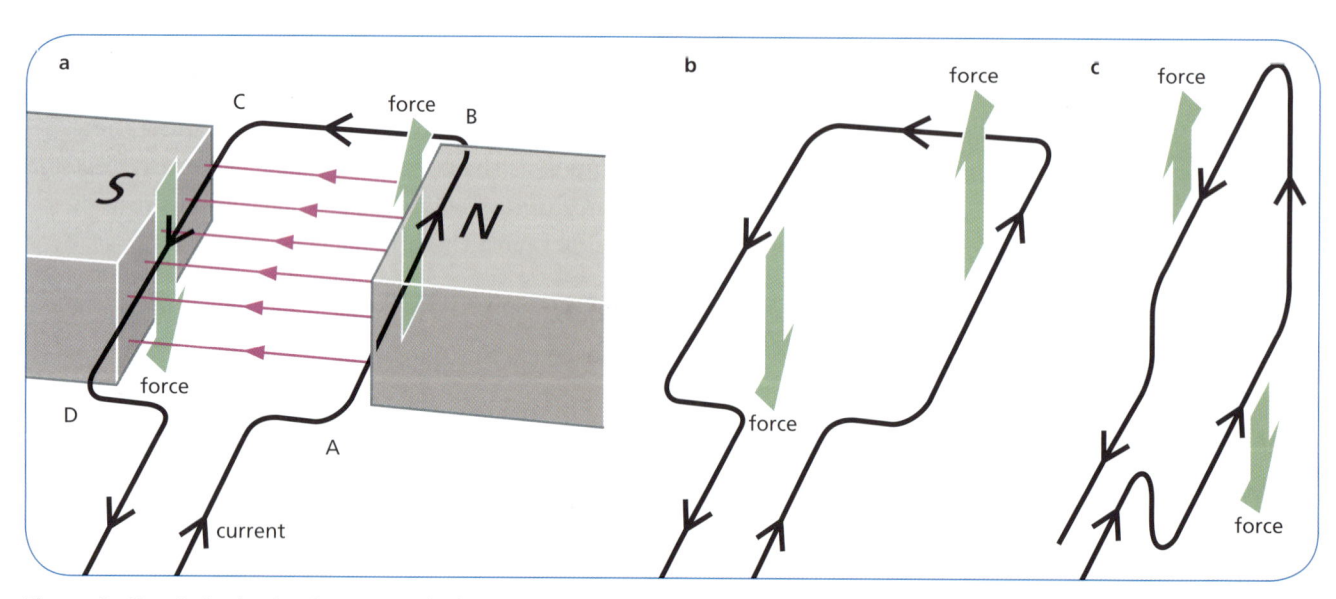

Figure 6c.10 a A simple electric motor. Only the two longer sides experience a force, because their currents cut across the magnetic field. **b** The two forces provide the turning effect needed to make the coil rotate. **c** When the coil is in the vertical position, the forces have no turning effect.

SAQ

8 Look at Figure 6c.10b. Explain how this diagram would change if the direction of the current through the coil was reversed. What effect would this have on the motor?

A more powerful motor

The diagrams in this chapter show the coil as if it was a single turn of wire. In practice, though, the coil might have hundreds of turns of wire, resulting in forces hundreds of times as great. In a coil, the current flows across the magnetic field many times, and each time it does so it feels a force. A coil is simply a way of multiplying the effect that would be experienced using a single length of wire.

Here are three ways to increase the motor's turning effect:
- supply a bigger current
- use stronger magnets
- wind more turns of wire onto the coil.

Summary

You should be able to:

- state that, when a current flows through a coil of wire, it becomes magnetised and that its magnetic field is similar to that of a bar magnet

- state that there is a magnetic field around any current and that the field lines are circles centred on the current

- state that a force is exerted on any current-carrying wire that crosses a magnetic field, and that reversing the current or the field reverses the force

- **H** understand that the relative directions of force, field and current are given by Fleming's left-hand rule

- understand that an electric motor is a current-carrying coil in a magnetic field, and that the forces on opposite sides of the coil cause the motor to turn

- **H** state that, in a DC motor, the split-ring commutator automatically reverses the current in the coil twice in each rotation to ensure that the coil keeps spinning

- state that the turning force can be increased by increasing the strength of the magnetic field, the number of turns on the coil, or the current

Questions

1. Sketch a diagram of the magnetic field pattern of a solenoid. How would the pattern change if the current through the solenoid was reversed?

2. Describe the magnetic field produced by a current flowing in a straight wire. Draw a diagram to support your answer.

3. A current-carrying conductor experiences a force when it crosses a magnetic field.

    a List two ways to increase the force.

    b List two ways to reverse the direction of the force.

4. Give three everyday uses for electric motors.

5. Look at the motor shown in Figure 6c.10 and the explanation of how it works. Suppose that the two magnets were turned round so that there was a north magnetic pole on the left. Explain how the coil would move.

6. List three ways in which the turning effect of a DC motor can be increased.

7. Which three things are at 90° to each other, in Fleming's left-hand rule? Which finger represents each of them?

8. What is the force on a current-carrying conductor that is *parallel* to a magnetic field?

9. In a DC motor, why must the current to the rotor coil be reversed twice during each rotation? What device reverses the current?

10. Explain why the forces shown in Figure 6c.10c have no turning effect on the coil.

11. Use your understanding of the moment (turning effect) of a force to explain why increasing the current through the coil of an electric motor will increase the turning effect it provides.

12. What effect would using a coil of bigger area have on the moment of the forces? Draw a diagram like that shown in Figure 6c.10b to support your answer.

13. How would the moment (turning effect) on a current-carrying coil be affected if stronger magnets were used?

6d Generating

Power plant

Modern societies depend on electricity. However, we usually don't have to think about it: we plug in a computer or switch on a light and they just work. Often, we have no idea where the electricity we use is generated.

Things can be different in a developing nation. Figure 6d.1 shows how electricity is generated and used in the Kenyan village of Tungu-Kabiri, on the slopes of Mount Kenya.

This is a micro-hydroelectric scheme. Water is fed by a pipe to a turbine, which causes a generator to spin. This generates electricity at the rate of 14 kW – not a lot, but enough to keep several enterprises working, including a metal workshop, a hairdresser's and a food shop.

Local, environmentally friendly schemes like this can show the way forward for a developing country like Kenya.

Figure 6d.1 a Water from a dam is fed through the yellow pipe to the turbine on the left; a rubber belt transfers the rotation to the generator at the top. The operator is opening the valve to control the flow of water. **b** Welding equipment in use in the workshop.

Generating electricity

A motor is a device for transforming electrical energy into mechanical (kinetic) energy. To generate electricity, we need a device that will do the opposite: it must transform mechanical energy into electrical energy. Fortunately, we can simply use an electric motor in reverse. If you connect up an electric motor to a voltmeter and spin its axle, the meter will show that you have generated a voltage (Figure 6d.2). Inside the motor, the coil is spinning around in the magnetic field provided by the permanent magnets; the result is that a current flows in the coil, and this is shown by the meter. We say that the current has been **induced** and that the motor is acting as a **generator**.

There are many different designs of generator, just as there are many different designs of electric motor. Some generate direct current,

Figure 6d.2 A motor can act as a generator. Spin the motor and the voltmeter shows that an induced current flows around the circuit.

others generate alternating current. Some use permanent magnets, whereas others use electromagnets. The power station generators shown in Figure 6d.3 generate alternating current at a voltage of about 25 kV. The turbines are made to spin by the high-pressure steam from the boiler. The generator is on the same axle as the turbine, so it spins too. The axle is connected to an electromagnet coil inside the generator, which spins around inside some fixed coils. A large current is then induced in the fixed coils, and this is the current which the power station supplies to consumers. A small proportion of it is used to supply the electromagnets of the generator itself.

Figure 6d.3 The turbine and generator in the generating hall of a Canadian nuclear power station. At the back are the turbines, fed by high-pressure steam in pipes; the generator is in the centre.

Cycle power

Old-fashioned bicycles made use of generators of a different design – a **dynamo** – for powering the lights. In a dynamo, the coil is fixed and a permanent magnet spins around inside it (Figure 6d.4). The magnet is on an axle that is made to turn by a knurled wheel that rubs on the bicycle's tyre. Look at the coil of wire on the left. First, the magnet's north pole moves past it and the induced current flows in one direction. Then the south pole moves past and the induced current flows the other way. In other words, an alternating current has been induced in the coil, and it is this current that makes the lamps light up.

Bicycle dynamos are designed like this because they have to be robust. A permanent magnet spinning around is more robust than a spinning coil, and the problem of making electrical connections to a spinning coil has been avoided. Nowadays, few bicycles make use of dynamos; instead, they often have LED lights powered by batteries. However, some wind generators use spinning permanent magnets to induce currents in fixed coils.

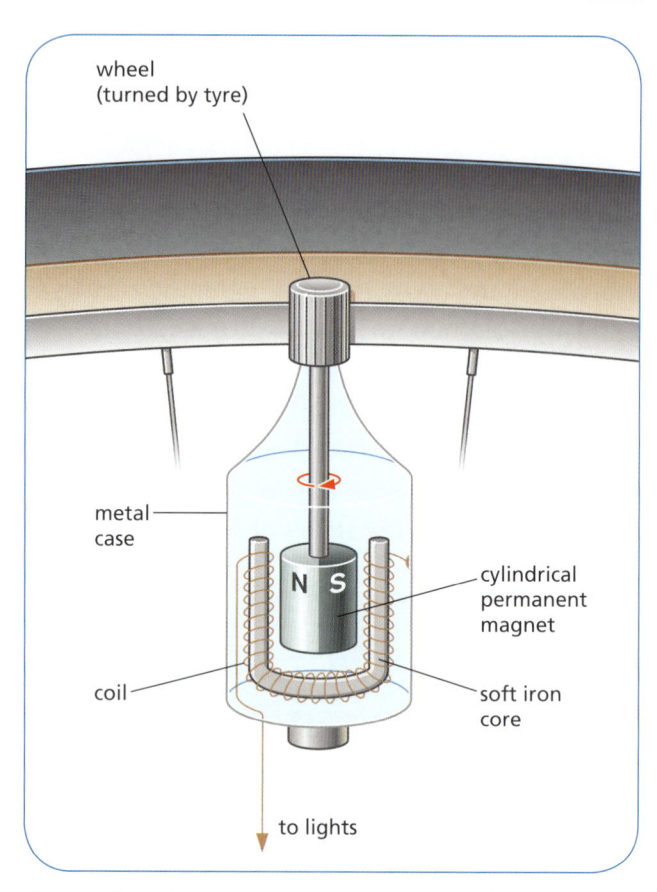

Figure 6d.4 A bicycle dynamo generates an alternating current that powers the bicycle's lights.

Generators

All of these generators have three things in common:
- a magnetic field (provided by magnets or electromagnets)
- a coil of wire (fixed or moving)
- movement (the coil and magnetic field move relative to one another).

When the coil and the magnetic field move relative to each other, a current flows in the coil if it is part of a complete circuit. This is known as an **induced current**. If the generator is not connected up to a circuit, there will be an **induced voltage** across its ends, ready to make a current flow around a circuit.

SAQ

1 In this section, what alternative word is used to mean 'generator'?
2 Explain how a coil and a magnetic field are made to move relative to each other in a bicycle dynamo (Figure 6d.4).

The principles of electromagnetic induction

Electromagnetic induction is the process of generating electricity from motion. The science of electromagnetism was largely developed by Michael Faraday (Figure 6d.5). He invented the idea of the magnetic field and drew field lines to represent it. He also invented the first electric motor. Then he extended his studies to show how the motor effect could work in reverse to generate electricity. In this section, we will look at the principles of electromagnetic induction that Faraday discovered.

As we have seen, a coil of wire and a magnet moving relative to each other are needed to induce a voltage across the ends of a wire. This is called the **dynamo effect**. If the coil is part of a complete circuit, the induced voltage will make an induced current flow around the circuit.

In fact, a single wire is enough (see Figure 6d.6a). The wire is connected to a sensitive meter to show when a current is flowing. Move one pole of the magnet downwards past the wire and a current flows. Move the magnet back upwards and a current flows in the opposite direction. Alternatively, the magnet can be stationary and the wire can be moved up and down next to it. You can see similar effects using a magnet and a coil (Figure 6d.6b). Pushing the magnet in and out of the coil induces a current that flows back and forth in the coil.

Here are two further observations.
- Reverse the magnet to use the opposite pole and the current flows in the opposite direction.

Figure 6d.5 Michael Faraday delivering a Christmas lecture at the Royal Institution in London on 27 December 1855. He was a great populariser of science and his lectures attracted many famous people. The artist, Alexander Blaikley, has included several members of the royal family in the audience, as well as famous scientists including Charles Darwin, although it is unlikely that they were all present at this lecture. The Christmas lectures started in 1826 and continue to this day. They are presented in the same lecture theatre; you might have seen them on television, because they are broadcast around the world.

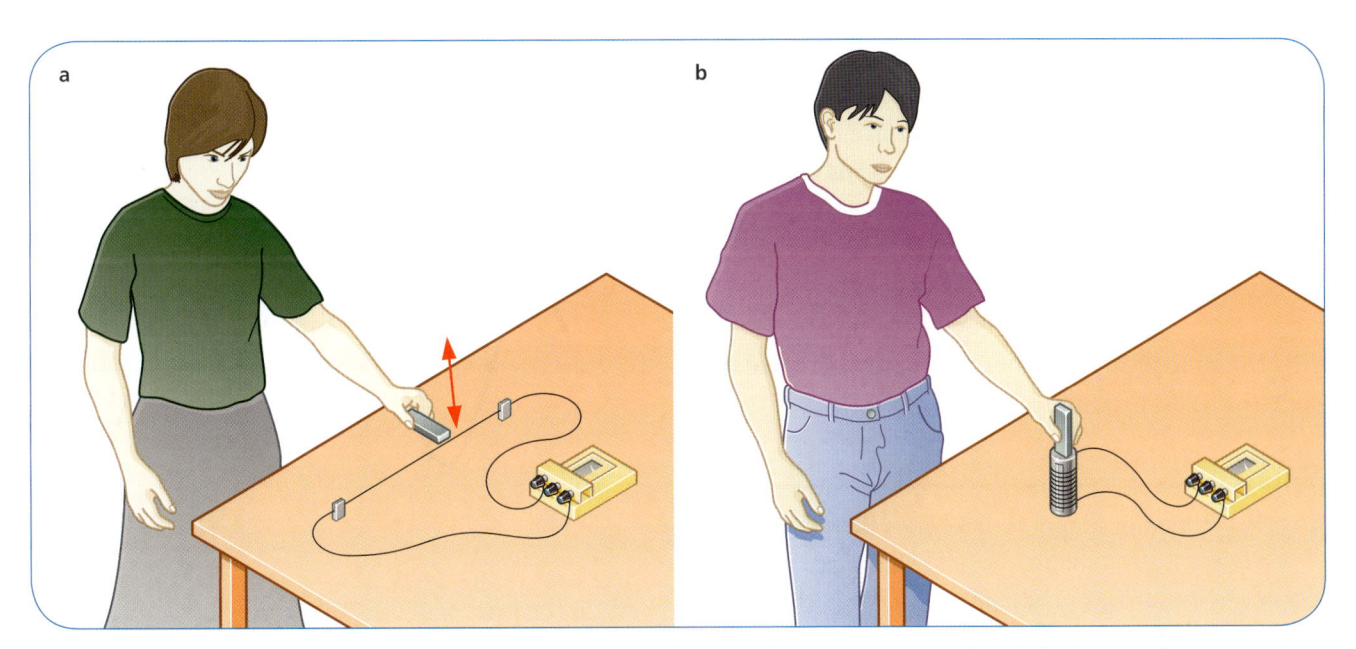

Figure 6d.6 a Move a magnet up and down next to a stationary wire and an induced current will flow. **b** Similarly, moving a magnet into and out of a coil of wire induces a current. Michael Faraday first did experiments like this in 1831.

- Hold the magnet stationary next to the wire or coil and no current flows. They must move relative to each other or nothing will happen. (This provides a good test of how steady your hand is. Hold a strong magnet next to a coil of wire; if your hand trembles, the meter will show that a current is flowing in the wire.)

SAQ

3. Look at Figure 6d.6b. When the students moves the north pole of the magnet down into the coil, a current flows in the wire. State two ways in which the student could make an induced current flow in the opposite direction.

Induction and magnetic field lines

Here are two further observations that help to show how to increase the size of the induced voltage when a magnet is moved near a wire or coil.
- With the magnet further from the wire or coil, the induced current is smaller.
- If the magnet or the wire is moved more slowly, the induced current is smaller.

Here is how Faraday explained electromagnetic induction using his idea of magnetic field lines. Picture the magnetic field lines coming out of the poles of the magnet shown in Figure 6d.6a. As the magnet is moved, the magnetic field lines are cut by the wire, and it is this cutting of the magnetic field lines that induces the current. If the magnet is further from the wire then the magnetic field lines are further apart and so fewer are cut, giving a smaller current. If the magnet is moved quickly, the lines are cut more quickly and a bigger current flows.

As usual, a coil gives a bigger effect than a single wire. This is because each turn of wire cuts the magnetic field lines and each therefore contributes to the induced current.

SAQ

4. Use the idea of magnetic field lines being cut to explain why, if a magnet is held stationary next to a wire, no current is induced.

An AC generator

Faraday's discovery of electromagnetic induction led to the development of the electricity supply industry. In particular, it allowed engineers to design generators that could supply electricity. At first, this was only done on a small scale but generators gradually got bigger and bigger until, like the ones shown in Figure 6d.3, they were capable of supplying the electricity demands of thousands of homes.

Inside a generator like this, a large electromagnet rotates inside fixed coils of wire. A voltage is induced in the fixed coils, and it is this voltage that is transmitted to the homes, offices and other workplaces of consumers.

A generator of this type produces **alternating current** (AC). This means that the current is not direct current (DC), which always flows in the same direction. Instead, it flows back and forth. Figure 6d.7 shows a graph of this. Half of the time, the current flows in the positive direction. Then it flows in the opposite direction. The **frequency** of an AC supply is the number of cycles it produces each second.

There are two ways to increase the voltage produced by a generator like this.
- Use an electromagnet with more turns of wire.
- Increase the speed at which the electromagnet rotates.

Increasing the speed has another effect: it increases the *frequency* of the alternating voltage

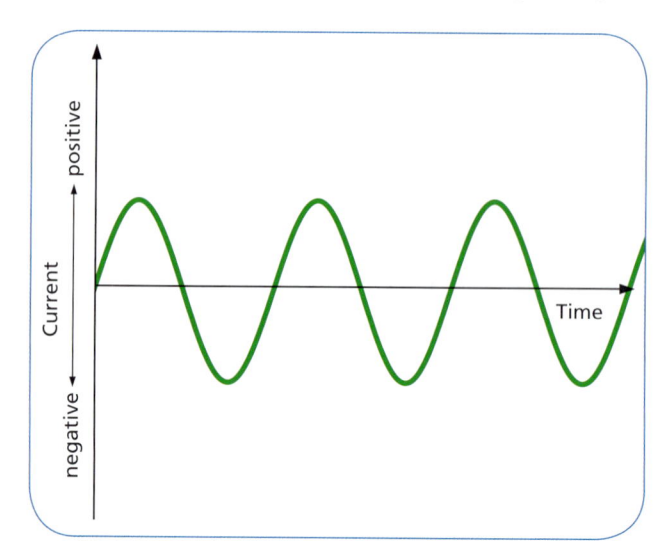

Figure 6d.7 A graph to represent an alternating current. For the first half of a cycle, the current flows one way. Then it goes into reverse.

produced. In the UK, mains electricity has a frequency of 50 Hz (50 cycles per second).

SAQ

5 Sketch a copy of the graph shown in Figure 6d.7. Add a second line to show how the graph would change if the generator was turned at twice the speed. (Both the amplitude and the frequency will change.)

Getting the current out

Figure 6d.8 shows a simple generator that produces alternating current. In principle, this is like a DC motor working in reverse: the axle is made to turn so that the coil spins around in the magnetic field, and a current is induced. The other difference is in the way the coil is connected to the circuit beyond. A DC motor uses a split-ring commutator, whereas an AC generator uses **slip rings**.

As the coil rotates, each side passes first the north magnetic pole and then the south magnetic pole. This means that the induced current flows first one way and then the other. In other words, the current in the coil is alternating. If we used a split-ring commutator, we would get direct current out because the commutator would automatically reverse the connection every half revolution. However, AC has certain advantages (see Item P6e *Transforming*) and so it is useful to be able to extract AC from the generator. Brushes are again used but, this time, they press against slip rings. Each ring is connected to one end of the coil and so the alternating current flows out through the brushes.

Turning the coil more rapidly or using a coil with more turns of wire has the effect of increasing the rate at which magnetic field lines are cut, so the induced voltage is greater.

For the AC generator shown in Figure 6d.8, each revolution of the coil generates one cycle of alternating current. Spin the coil 50 times each second and the AC generated has a frequency of 50 Hz.

SAQ

6 In power stations in the USA, generators spin 3600 times per minute. What is the frequency of the alternating voltage they produce?

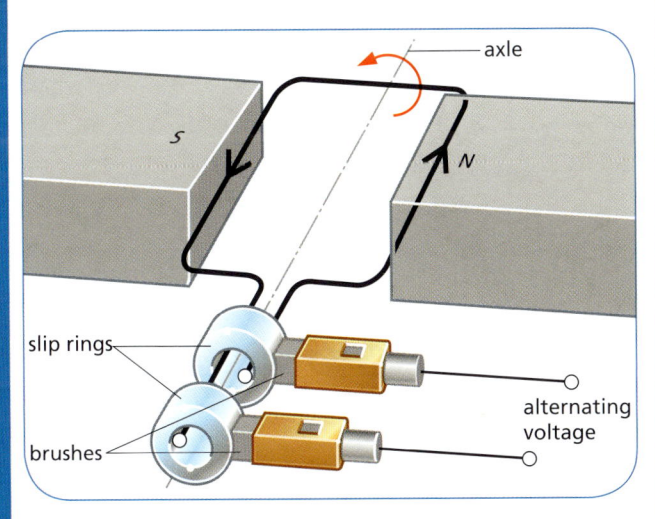

Figure 6d.8 A simple AC generator works like a motor in reverse. The slip rings and brushes are used to connect the alternating current to the external circuit.

Summary

You should be able to:

- state that electricity is generated by moving a magnet near a wire (or moving a wire near a magnet), which induces a voltage across the ends of the wire

- state that, similarly, a voltage is induced across a coil when the magnetic field within it changes and that reversing the field reverses the voltage

continued on next page

298 P6d Generating

Summary - continued

- (H) ◆ understand that, the faster the magnetic field changes, the greater the induced voltage
- ◆ state that an alternating current is generated when a magnet rotates inside a coil
- ◆ state that the rotating magnet in a power station is an electromagnet
- ◆ understand that a coil with more turns generates a bigger voltage and that, the faster the coil is rotated, the greater the voltage and the higher the frequency
- (H) ◆ understand that slip rings and brushes connect the rotating coil to the external circuit so that a current can flow

Questions

1. Draw a diagram to show the energy transformations in an electric motor and a generator. Remember that neither is 100% efficient.

2. If you hold a coil of wire next to a magnet, no current will flow. What else is needed to induce a current?

3. Look at Figure 6d.9, which shows a simple DC generator. Name the parts labelled A, B and C.

4. The north pole of a magnet is moved towards a coil of wire, as shown in Figure 6d.6b, so that an induced current flows. How would the current change if the magnet's south pole was used instead?

5. State two ways in which the current induced in the coil in Figure 6d.6b could be increased.

6. A magnet lies inside a coil of wire. Why does no induced current flow?

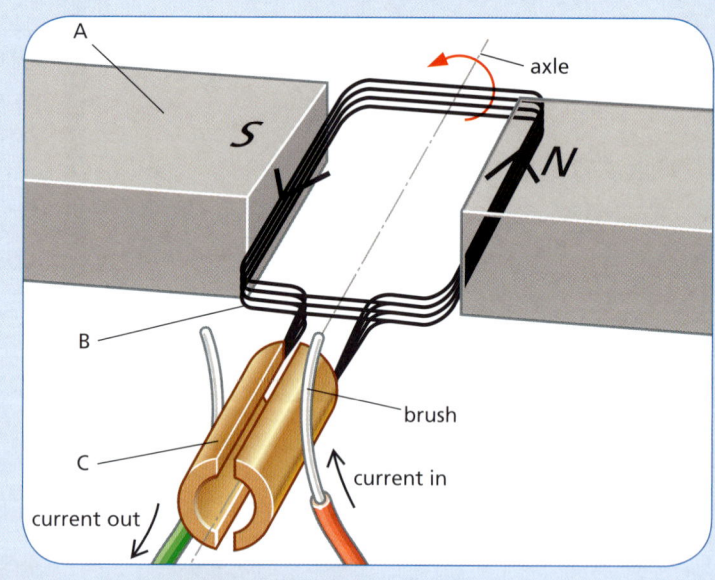

Figure 6d.9 A DC generator.

(H) 7. Look at the bicycle dynamo in Figure 6d.4. Use Faraday's idea of *cutting magnetic field lines* to explain how it works.

8. When you cycle faster, the bicycle's dynamo spins faster. In what two ways will the induced alternating current supplied by the dynamo change?

9. List the features of a large AC generator from a power station (Figure 6d.3) that make it capable of generating a higher voltage than the model AC generator shown in Figure 6d.8.

6e Transforming

From power station to you

Power stations may be 100 km or more from the places that use the electricity they generate. They are built close to a reliable source of fuel and where cooling water is readily available – on the banks of a river or lake, or near the sea. You might have noticed the cooling towers, which often have clouds of water vapour pouring out into the sky.

The electrical power generated at a power station must then be distributed around the country. High-voltage electricity leaves the power station. Its voltage may be as much as a million volts. To avoid danger to people, it is usually carried in cables slung high above the ground between tall pylons. Lines of pylons stride across the countryside, heading for the urban and industrial areas that need the power (Figure 6e.1).

When the power lines approach the area where the power is to be used, they enter a local distribution centre. Here, the voltage is reduced to a less hazardous level and the power is sent through more cables (overhead or underground) to local substations. Wherever you live, there is likely to be a substation in the neighbourhood. It might be in a securely locked building, or the electrical equipment might be surrounded by fencing that carries notices warning of the hazard (Figure 6e.2).

From the substation, electricity is distributed around the neighbouring houses. In some countries, the power is carried in cables buried underground; other countries use tall 'poles' that hold the cables above the level of traffic in the street to distribute the power. Overhead power lines and cables can be an eyesore, but the cost of burying cables underground can be ten or a hundred times as great.

Figure 6e.1 Electricity is usually generated at a distance from where it is used. If you look on a map, you may be able to trace the power lines that bring electrical power to your neighbourhood.

Figure 6e.2 An electricity substation like this has warning signs to indicate the extreme hazard of entering the substation.

Why use high voltages?

The high voltages used to transmit electrical power around the country are dangerous. That is why the cables that carry the power are supported high above people, traffic and buildings on tall pylons. Sometimes the cables are buried underground, but this is much more expensive and the cables must be safely insulated. There is a good reason for using high voltages: it means that the current flowing in the cables is relatively low, and this wastes less energy. We can understand this as follows.

When a current flows in a wire or cable, some of the energy it is carrying is lost because of the cable's resistance – the cables get warm. A small current wastes less energy than a high current.

Electrical engineers do everything they can to reduce the energy losses in the cables. If they can reduce the current to half its value (by doubling the voltage), the losses will be one quarter of their previous level. This is because power losses in cables are proportional to the *square* of the current flowing in the cables:

- double the current gives four times the losses
- three times the current gives nine times the losses.

SAQ

1. When a current of 1000 A flows in a particular cable, energy is lost at the rate of 50 kW.
   a. What will be the rate of loss if the current is doubled to 2000 A?
   b. What will be the rate of loss if the current is halved to 500 A?

Calculating current

To transmit a certain power P, we can use a small current I if we transmit the power at high voltage V. This follows from the equation for electrical power, $P = IV$ (see Item P2c *Fuels for power*).

Worked example 1

Suppose that a power station generates 500 MW of power. What current will flow from the power station if it transmits this power at 50 kV? What current will flow if it transmits it at 1 MV?

Step 1: rearranging $P = IV$, we have the equation we need to use.

$$I = \frac{P}{V}$$

Step 2: substituting values for the first case ($P = 500\,\text{MW} = 500 \times 10^6\,\text{W}$, $V = 50\,\text{kV} = 50 \times 10^3\,\text{V}$) gives the current as

$$I = \frac{500 \times 10^6\,\text{W}}{50 \times 10^3\,\text{V}} = 10\,000\,\text{A}$$

Step 3: now consider the second case, when the power is transmitted at 1 MV (10^6 V) – the operating voltage of some national grids. The current is now given by

$$I = \frac{500 \times 10^6\,\text{W}}{10^6\,\text{V}} = 500\,\text{A}$$

Energy saving

Worked example 1 shows that increasing the voltage by 20 times reduces the current by 20 times. This means that the power lost in the cables is reduced by 400 times (because 400 is 20^2). This reduction in power loss means that the cables will not get as hot and so thinner cables can safely be used – reducing the cost of installing and maintaining the network.

The current flowing in the cables is a flow of coulombs of charge. At high voltage, we have fewer coulombs flowing but each coulomb carries more energy with it.

SAQ

2. a. A power distribution system transmits 200 MW of power at a current of 500 A. At what voltage is the power distributed? Give your answer in kV.
   b. Engineers propose to double the distribution voltage. What current will now flow in the cables?
   c. If power losses in the existing system are 6 MW, what will they be if the higher-voltage system is adopted?

Transformers

Transformers are the devices used to increase or decrease the voltage of the electricity supply. They are designed to be as efficient as possible (up to 99.9% efficient). This is because the electricity we use might have passed through as many as ten transformers before it reaches us from the power station. A loss of 1% of energy in each transformer would represent a total waste of 10% of the energy leaving the power station.

Power stations typically generate electricity at 25 kV. This has to be converted to the grid voltage – say 400 kV – using transformers. For these voltages, we say that the voltage is 'stepped up' by a factor of 16. Figure 6e.3 shows the construction of a simple step-up transformer. Every transformer has three parts:

- a **primary coil** across which the incoming voltage V_p is connected
- a **secondary coil** that provides the voltage V_s to the external circuit
- an **iron core** that links the two coils.

There is *no electrical connection* between the two coils: they are linked together only by the iron core. Also, it is important to realise that the voltages are both alternating voltages – a transformer does not change AC to DC or anything of the sort. It changes the size of an alternating voltage (transformers do not work on DC). Also, if the voltage is stepped up, the current must be stepped down, and vice versa.

To step the input voltage up by a factor of 16, there must be 16 times as many turns on the secondary coil as on the primary. That is, the **turns ratio** tells us the factor by which the voltage will be changed.

- A **step-up transformer** increases the voltage; there are more turns on the secondary than on the primary.
- A **step-down transformer** reduces the voltage; there are fewer turns on the secondary than on the primary.

This means that we can write an equation that relates the two voltages V_p and V_s to the numbers of turns on each coil, N_p and N_s. This is known as the **transformer equation**:

$$\frac{\text{voltage across primary coil}}{\text{voltage across secondary coil}} = \frac{\text{number of turns on primary}}{\text{number of turns on secondary}}$$

$$\frac{V_p}{V_s} = \frac{N_p}{N_s}$$

It will help you to recall this equation if you remember that the coil with more turns has the higher voltage.

Worked example 2

A transformer is needed to step down the 230 V mains supply to 6 V. If the primary coil has 1000 turns, how many must the secondary have?

Step 1: draw a transformer symbol and mark on it the information from the question (Figure 6e.4).

Step 2: write down the transformer equation.

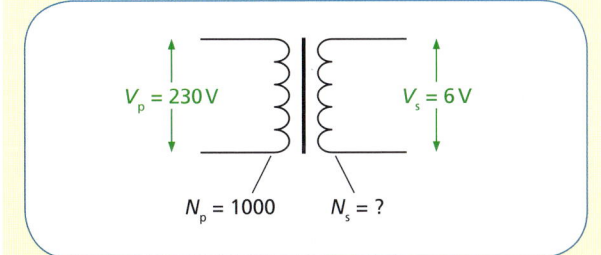

$$\frac{V_p}{V_s} = \frac{N_p}{N_s}$$

$V_p = 230\,\text{V}$, $V_s = 6\,\text{V}$, $N_p = 1000$, $N_s = ?$

Figure 6e.4

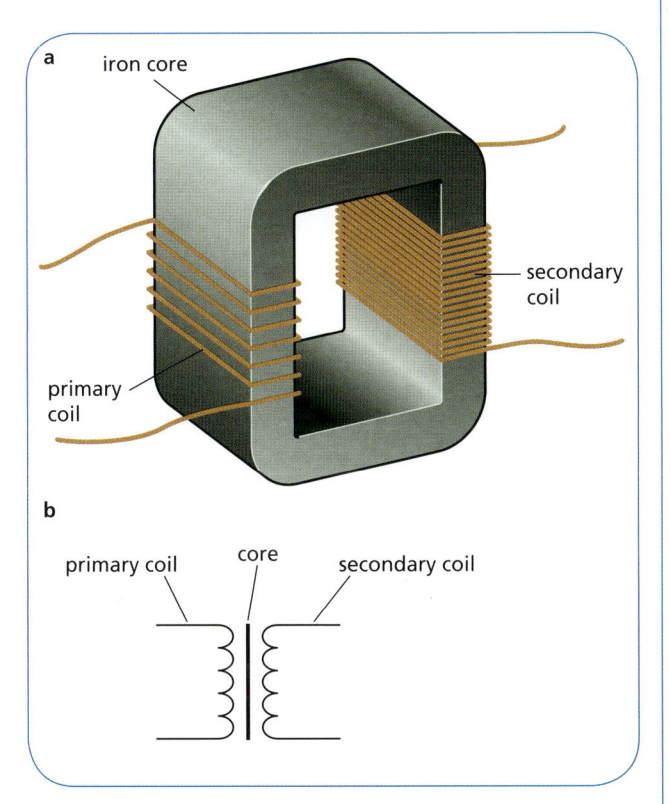

Figure 6e.3 a The structure of a transformer. This is a step-up transformer because there are more turns on the secondary coil than on the primary. If the connections to it were reversed, it would be a step-down transformer. **b** The circuit symbol for a transformer shows the two coils with the core between them.

continued on next page

Worked example 2 - continued

Step 3: substitute values from the question.

$$\frac{230\,V}{6\,V} = \frac{1000}{N_s}$$

Step 4: rearrange and solve for N_s.

$$N_s = 1000 \times \frac{6\,V}{230\,V} = 26.1 \text{ turns}$$

So the secondary coil must have 26 turns.

Is this a reasonable answer? The voltage has to be reduced, so the number of turns on the secondary coil must be much less than 1000. Mental arithmetic shows that the voltage has to be reduced by a factor of about 40 (from 230 V to 6 V), so the number of turns must be reduced by the same factor. $N_s \approx 1000 \div 40 = 25$. (This is an *approximate* answer.)

SAQ

3 A transformer has 100 turns on the primary coil and 1000 on the secondary. Is it a step-up or a step-down transformer?

4 A step-up transformer has 2000 turns on one coil and 5000 on the other. Calculate the turns ratio N_p/N_s for this transformer.

5 A transformer is designed to provide 20 V from a 240 V supply. If the primary coil has 1200 turns, how many must the secondary have?

Isolating transformers

Figure 6e.6 shows a bathroom shaver socket. Most bathrooms in the UK do not have electrical sockets in them because of the hazard of using electricity in a wet place. However, a special shaver socket like this can be used safely. It contains a small transformer called an **isolation transformer**. This doesn't change the voltage but it does mean that, when a shaver is plugged in, it is not connected directly to the mains. (Recall that the primary and secondary coils are not electrically connected.)

Isolation transformers are used in many other places. For example, at a rock concert, all of the amplifiers and other sound equipment will be connected to the mains supply via an isolation transformer, to reduce the risk of any performer or technician coming into direct contact with

Practical transformers

Figure 6e.5 shows two very different transformers – the world's first transformer, made by Michael Faraday, and a modern high-power transformer at a power station. Transformers are found in many different situations.

- Transformers step up the voltage at a power station and step it down again at the substations that supply local users.
- A portable radio might run off the mains or from batteries. The mains voltage is reduced to a suitable low value by a built-in transformer.
- Mains adapters are often used with computer games consoles and personal stereos. Adapters are heavy because of the weight of the transformer core and coils they contain.

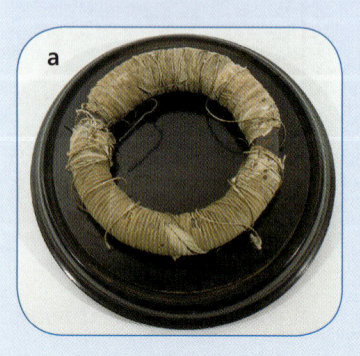

Figure 6e.5 Transformers old and new.
a This transformer was made by Michael Faraday in 1831. The core is an iron ring. It is hard to see, but there are two separate coils of wire wrapped around the ring.
b Transformers at a modern power station. The coils and core are entirely enclosed. A small proportion of the power passing through the transformer is wasted as heat. The tanks contain cooling fluid that is pumped around to remove the heat.

the supply. Similarly, builders and electricians connect their drills and other equipment to the mains supply through an isolation transformer.

Figure 6e.6 This shaver socket, installed in a bathroom, contains an isolating transformer. This greatly reduces the risk of danger to the user in a damp place where electricity would otherwise be a serious hazard.

Safety first

An isolating transformer has the same number of turns on the primary and secondary coils, so the voltage does not change. Because the two coils are not electrically connected, only AC can pass through the transformer.

If someone is working on a piece of electrical equipment, they are less likely to get an accidental shock. If they touch a live part, it is impossible for a large direct current (DC) to flow into them from the earth connection, from which they are separated by the transformer.

SAQ

6 If you had two identical transformers with a turns ratio of 10:1, how could you connect them together to make an isolating transformer? Draw a diagram to show your answer.

How transformers work

Transformers only work with alternating current (AC). To understand why this is, we need to look at how a transformer works (Figure 6e.7). It makes use of electromagnetic induction.

- The primary coil has alternating current flowing through it. It is thus an electromagnet and produces an alternating magnetic field.
- The core transports this alternating field around to the secondary coil.
- Now the secondary coil is a conductor in a changing magnetic field. A current is induced in the coil. (This is another example of electromagnetic induction at work.)

If the secondary coil has only a few turns, the p.d. induced across it is small. If it has a lot of turns, the p.d. will be large. This means that, to get a high voltage out, we need a secondary coil with a lot of turns compared with the primary.

If direct current is connected to a transformer, there is no output voltage. This is because the magnetic field produced by the primary coil is unchanging. With an *unchanging* field passing through the secondary coil, no voltage is induced in it.

Notice in Figure 6e.7 that the magnetic field links the primary and secondary coils. The energy being brought by the current in the primary coil is transferred to the secondary by the magnetic field. This means that the core must be very good at transferring magnetic energy. A 'soft' magnetic material must be used – usually an alloy of iron with a small amount of silicon. ('Soft' magnetic materials are ones that can be magnetised and demagnetised easily.) Even in a well-designed transformer, some energy is lost because of the resistance of the wires and because the core 'resists the flow' of the changing magnetic field.

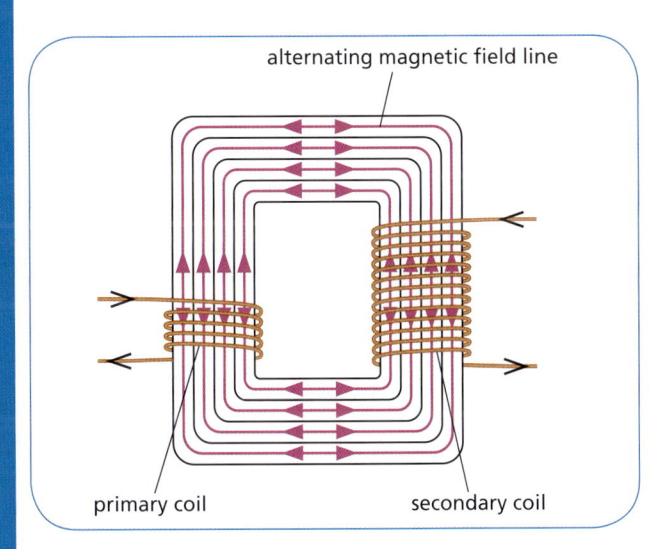

Figure 6e.7 The AC in the primary coil produces a varying magnetic field in the core. This induces a varying current in the secondary coil. The core of a transformer is often made in sheets (laminated) so that the magnetic field lines follow the sheets around from the primary coil to the secondary.

Thinking about power

If a transformer is 100% efficient then no power is lost in its coils or core. This is a reasonable approximation – well-designed transformers like the one shown in Figure 6e.5b waste only about 0.1% of the power transferred through them. This allows us to write an equation relating the primary and secondary voltages to the currents I_p and I_s flowing in the primary and secondary coils, using $P = IV$:

power into primary coil
$$= \text{power out of secondary coil}$$
$$I_p \times V_p = I_s \times V_s$$

Worked example 3

The primary coil of a transformer is connected to a 12 V alternating supply and carries a current of 5 A. If the output voltage is 240 V, what current flows in the secondary circuit? Assume that the transformer is 100% efficient.

Step 1: draw a transformer symbol and mark on it the information from the question (Figure 6e.8).

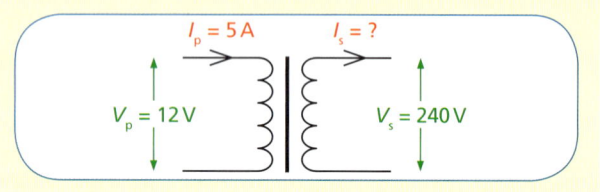

$I_p = 5\text{ A}$, $I_s = ?$, $V_p = 12\text{ V}$, $V_s = 240\text{ V}$

Figure 6e.8

Step 2: think about what a reasonable answer might be.

The voltage is being stepped up by a factor of 20 (from 12 V to 240 V). So the current will be stepped *down* by the same factor. You can probably see that the secondary current will be 1/20 of 5 A = 0.25 A.

(This is the correct answer, but we will press on with the formal calculation.)

Step 3: write down the transformer power equation.

$$I_p \times V_p = I_s \times V_s$$

Step 4: substitute values from the question.

$$5\text{ A} \times 12\text{ V} = I_s \times 240\text{ V}$$

Step 5: rearrange and solve for I_s.

$$I_s = \frac{5\text{ A} \times 12\text{ V}}{240\text{ V}}$$
$$= 0.25\text{ A}$$

The current supplied by the secondary coil is therefore 0.25 A. So, in stepping up the voltage, the transformer has stepped down the current. If both had been stepped up, we would be getting something for nothing – impossible!

Summary

You should be able to:

- state that electrical power is transmitted at high voltages, which allows the current to be relatively low so that heat losses caused the resistance of the cables are low and thinner cables can be used
- state that a transformer changes the voltage of an alternating supply
- understand that a transformer consists of primary and secondary coils linked by an iron core

continued on next page

Summary - *continued*

- **[H]** ♦ understand that the changing magnetic field produced by the primary coil induces an alternating current in the secondary coil

- ♦ state that a step-up transformer increases the voltage of the supply

- ♦ calculate the voltages and the numbers of turns using the transformer equation:

$$\frac{V_p}{V_s} = \frac{N_p}{N_s}$$

- **[H]** ♦ understand that, for a 100% efficient transformer (in which no power is wasted):

  power in = power out

  $$I_p \times V_p = I_s \times V_s$$

Questions

1. Why is electrical power transmitted across the grid at high voltage?

2. What are the three essential parts of any transformer?

3. A portable radio has a built-in transformer so that it can work from the mains instead of batteries. Is this a step-up or step-down transformer?

4. a A transformer has 200 turns on the primary coil and 5000 on the secondary. Is this a step-up or step-down transformer?

   b The transformer is to be used to supply 6 kV to an electrical machine. What voltage must be connected to the primary coil?

5. A transformer is used to increase the voltage of a 230 V mains supply to 1200 V. If the primary coil has 800 turns, how many turns must the secondary coil have?

6. **[H]** What is the function of the core of a transformer? Why must it be made of a soft magnetic material?

7. Explain why a transformer will not work with direct current.

8. In a step-up transformer, is the current in the secondary coil greater or less than the current in the primary coil?

9. A transformer is used to reduce a 230 V mains supply to 6 V, to power a radio.

   a If the primary coil has 6000 turns, how many turns must the secondary have?

   b If, in normal use, a current of 0.04 A flows in the primary coil, what current flows in the secondary?

   c What assumption must be made to answer part **b**?

P6f Charging

Selling electricity

In recent years, more and more householders have started fitting their own electricity generators to their homes. These may be wind turbines or photovoltaic panels (Figure 6f.1). Why do they do this?

First, these generators use renewable sources of energy. Using them decreases the production of carbon dioxide, the greenhouse gas that is largely responsible for climate change.

Second, they save money in the long run. Although the house is still connected to the National Grid, users have to buy less electricity from the main generating companies. They can even have the pleasure of selling back any surplus electricity to the grid.

However, there is a problem. Photovoltaic cells produce direct current (DC) but the grid is an alternating (AC) supply. A device called an 'inverter' is needed to convert the DC from the cells to AC of the correct frequency, in step with the mains supply.

National electricity supply systems are becoming increasingly complex. Generators produce both AC and DC, at different voltages and frequencies. Some consumers require AC, others DC. Engineers need to be sure that the system can cope with conversions back and forth between AC and DC.

Figure 6f.1 These Dutch houses have 200 m^2 of solar panels on their roofs that meet the occupants' electricity needs. Surplus electricity is stored in batteries or sold to the grid.

Diodes

In Item P6e *Transforming*, we saw how transformers can be used to change the voltage of an alternating supply. Many devices, such as radios and computers, are fitted with transformers to reduce the 230 V mains supply to, say, 6 V. However, the devices usually work from a DC supply such as batteries, so the AC must also be converted to DC. How can this be done?

Alternating current flows back and forth, first one way and then the other. A **diode** can be used to convert it to direct current. This is a component that allows electric current to flow in one direction only. Its circuit symbol (Figure 6f.2, page 307) represents this by showing an arrow to indicate the direction in which current can flow. This is the *forward* direction. The bar shows that current is stopped if it tries to flow in the opposite direction (the *reverse* direction). It can help to think of a diode as being a 'waterfall' in the circuit. Charge can flow over the waterfall but it cannot flow in the opposite direction, which would be uphill.

Some diodes give out light when a current flows through them. These are light-emitting diodes (LEDs). As the charge flows over the 'waterfall', some of the energy is given out as light.

SAQ

1 Draw a circuit showing a diode connected to a cell so that a large current will flow through the diode. Mark the positive terminal of the cell and add an arrow to show the direction in which the current flows round the circuit.

Figure 6f.4 shows the current–voltage graph for a diode. When the p.d. across the diode is positive, a current flows. The bigger the p.d., the greater the current. However, the current doesn't increase in proportion to the p.d. – you can tell this because the graph is not a straight line through the origin. Instead, it increases slowly at first and then curves upwards.

When the p.d. across the diode is negative, it is trying to push a current through the diode in the reverse (forbidden) direction. In fact, a very small current does flow, perhaps a few microamperes.

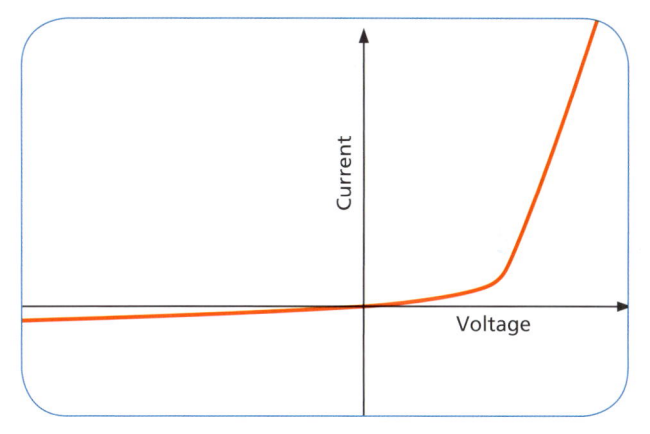

Figure 6f.4 The current–voltage characteristic graph for a diode. A large current flows only when the p.d. across the diode is positive. For negative p.d.s, a tiny current flows. The graph is curved, which tells us that the resistance of a diode decreases as the voltage across it increases – it does not obey Ohm's law.

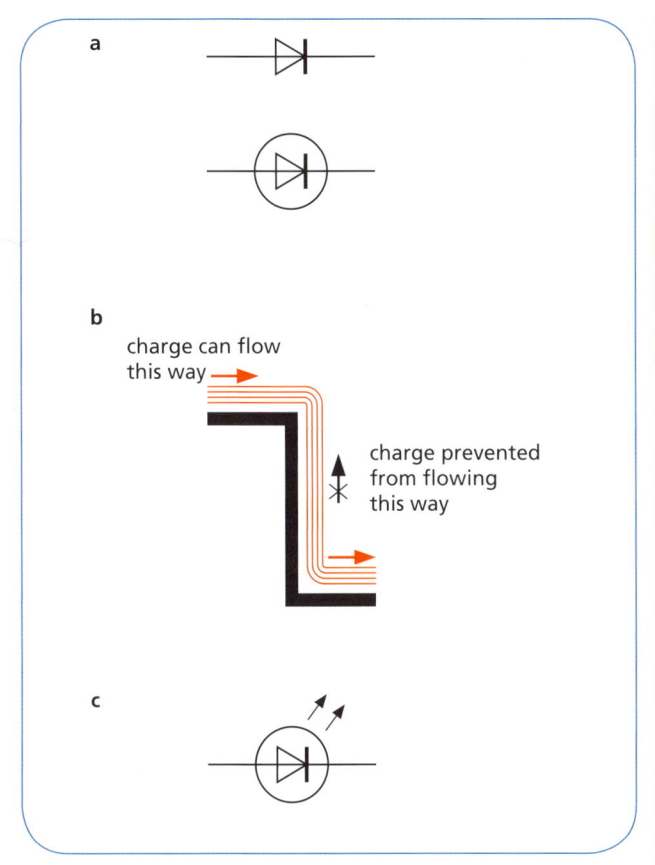

Figure 6f.2 a Two alternative circuit symbols for a diode (the circle is optional). Diodes allow current to flow in one direction only – in the direction of the arrow. **b** A diode is rather like a waterfall, in that charge can flow downhill but is prevented from flowing back uphill. **c** The arrows on the circuit symbol for a light-emitting diode represent the light that is emitted when a current flows through it.

Current-voltage characteristics

To understand how a diode works, we need to know how the current through it depends on the voltage across it. In other words, we need to use a circuit like the one shown in Figure 6f.3 to investigate its current–voltage characteristics.

SAQ

2 Describe how you would use the circuit shown in Figure 6f.3 to determine the current–voltage characteristic graph of a diode. What would you vary? What would you record? How would you obtain both positive and negative parts of the graph? (Look back to Item P6a *Resisting* if you need to remind yourself about this.)

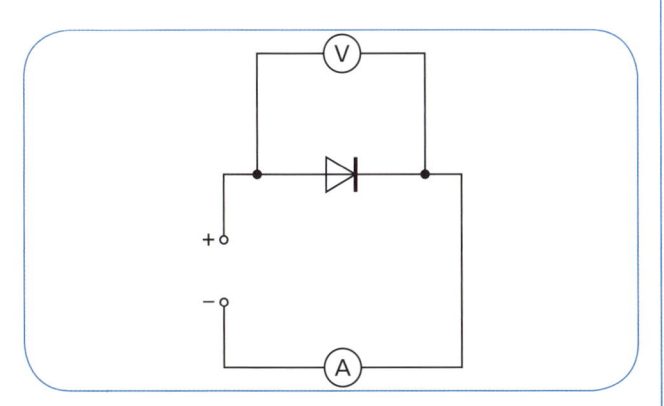

Figure 6f.3 A circuit for measuring the current–voltage characteristics of a diode. The diode is shown connected in the 'forward' direction.

Resistance - high and low

The current–voltage characteristic graph of a diode is not a straight line through the origin. We can therefore say that the current flowing is not proportional to the p.d. pushing it. In other words, the resistance of a diode does not obey Ohm's law.

- A diode has a low resistance in the forward direction.

- A diode has a high resistance in the reverse direction.

What is going on inside a diode to make it behave like this? Many diodes are made of two forms of silicon, called p-type and n-type. As shown in Figure 6f.5a, these are connected end to end so that the current has to flow through one type and then the other.

The n-type silicon has extra **electrons** that can carry a current. By contrast, p-type silicon has positively charged **holes** that can also move through the silicon to carry current. (A hole is really a place where an electron is missing from the silicon.) When a current flows through the diode in the forward direction (Figure 6f.5b), electrons in the n-type material meet holes in the p-type material. The electrons fall into the holes and their opposite charges cancel out. In this way the current flows right through the diode, meeting only a small resistance.

What happens in the reverse direction? Electrons and holes move the other way in the diode, away from the junction between the two materials (Figure 6f.5c). Soon, there are no mobile charged particles to carry the current and so no current flows. The diode has a very high resistance.

SAQ

3 Name the charged particles that carry electric current through a diode. What are their charges?

Rectification

The process of converting alternating current to direct current is known as **rectification**. A single diode can be used for this (Figure 6f.6a).

The supply voltage varies up and down, with both positive and negative values (Figure 6f.6b). The diode allows current to flow in one direction only, so the voltage across the resistor can never be negative (Figure 6f.6c). (We are ignoring the tiny reverse current.) This means that the voltage–time graph across the resistor is a series of positive 'bumps'. This is known as **half-wave rectification**, because only half of the alternating voltage contributes to the final voltage.

We can describe the half-wave-rectified voltage as 'direct' because, although it is varying up and down, it is always positive and never negative. The current through the resistor always flows in the same direction.

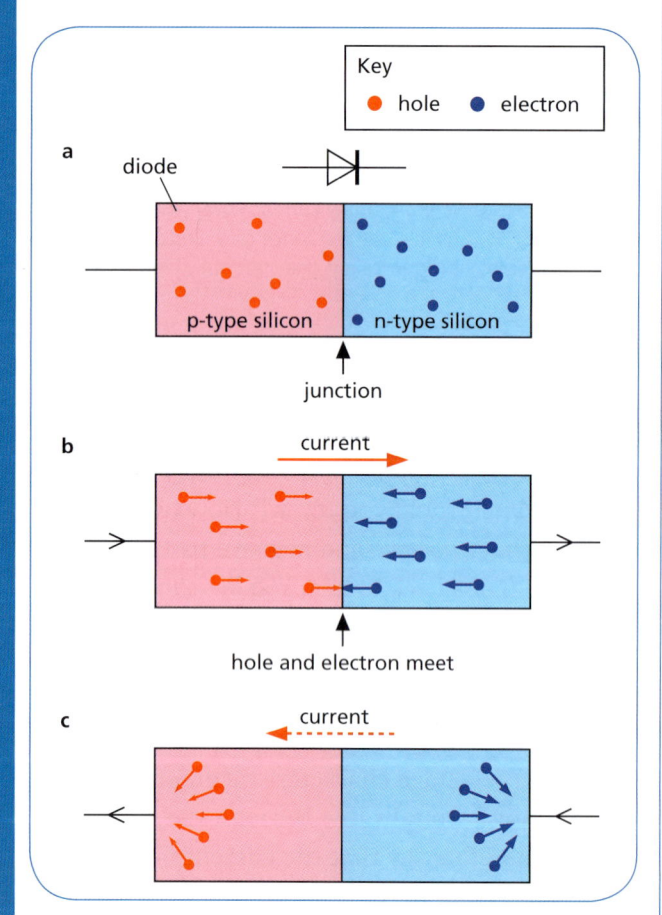

Figure 6f.5 a A diode is made of two forms of silicon. Current is carried by holes in p-type silicon and by electrons in n-type silicon. **b** When a current flows in the forward direction, electrons and holes meet at the junction and cancel each other out. **c** In the reverse direction, there are no charges to carry current.

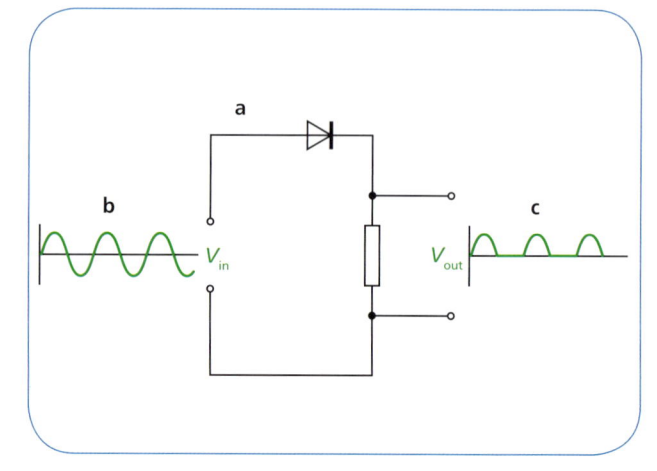

Figure 6f.6 a A circuit used to produce half-wave rectification of an alternating supply using a diode. The voltage–time graphs (**b**,**c**) show the effect of rectification.

SAQ

4 A light bulb is connected to an AC supply and shines brightly. A diode is added to the circuit in series with the bulb. Now the current in the circuit has been half-wave rectified. Explain why the bulb shines less brightly.

SAQ

5 Sketch voltage–time graphs to show:
  a an alternating voltage
  b the same voltage, half-wave rectified
  c the same voltage, full-wave rectified.

Full-wave rectification

Half-wave rectification doesn't make good use of an alternating voltage. Half of the supply voltage (the negative part) has been removed. For half of the time, no current flows. To overcome this disadvantage, a circuit using four diodes is used. This is called a **diode-bridge circuit** and an example is shown in Figure 6f.7.

The diode bridge has several important features that you should notice.
- The four diodes are connected in the shape of a diamond.
- The arrows of all four diodes point to the right.
- The supply voltage, on the left, is connected across the top and bottom corners of the diamond.
- The output resistor is connected between the left- and right-hand corners of the diamond.

You can see from the small graph on the right of the diagram that the output voltage is a series of positive 'bumps', with no gaps between them. The negative bumps of the supply have been upturned so that none of the supply voltage is wasted. This is called **full-wave rectification**.

H How a diode bridge works

In one cycle, an alternating voltage varies back and forth between positive and negative values. To understand how a diode bridge works, we need to consider the two halves of the cycle.

When point A is positive relative to point B (Figure 6f.8a, page 000), current wants to flow from A to B. What path can it follow?
- At point A, the current is blocked by diode 1 but it can flow through diode 2 to point D.
- Here it is blocked by diode 4, but it can flow downwards through resistor R.
- The current reaches point C, where it can flow down through diode 3 to point B.
- From there, it returns to the supply.

During the other half of the cycle, when point B is positive relative to point A (Figure 6f.8b), current flows from B to A through the other two diodes.

The important thing to realise is that, during both halves of the cycle, the current flows in the same direction (downwards) through resistor R. Hence the top of the resistor is always positive with respect to the bottom. This shows that the alternating voltage has been rectified in both halves of its cycle.

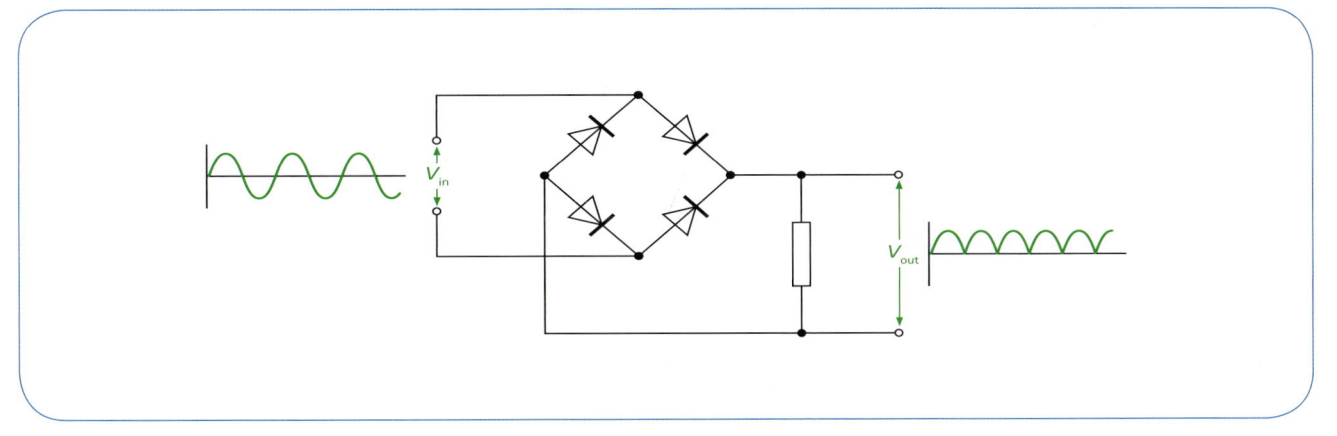

Figure 6f.7 This diode-bridge circuit is used to produce full-wave rectification. The current through the resistor always flows in the same direction.

310 P6f Charging

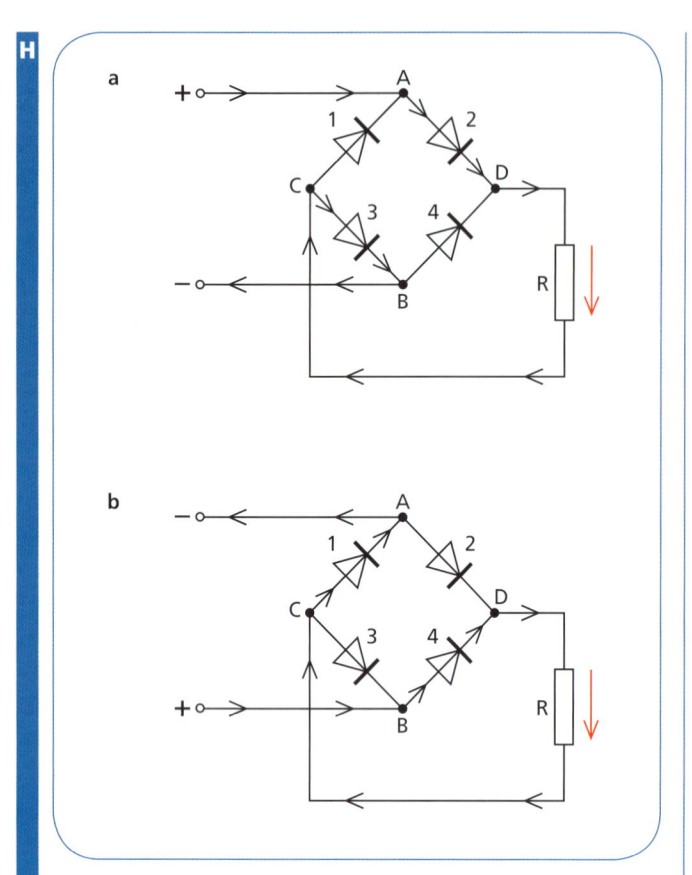

Figure 6f.8 The current through a diode bridge follows different paths in the two halves of the cycle, but it always flows in the same direction through resistor R.

SAQ

6 Explain the path followed by the current in the second half of the cycle, as shown in Figure 6f.8b.

Capacitors

Full-wave-rectified current is a form of direct current, but it is rather unsatisfactory because it varies all the time. Most electronic devices need a voltage supply that is more constant than this. The p.d. can be smoothed using a further type of electrical component called a **capacitor** (Figure 6f.9). Before looking at how a capacitor can smooth the bumps of a rectified AC supply, we need to look at how capacitors behave in circuits.

When a capacitor is connected in an electric circuit, a current flows and the capacitor stores some of the charge flowing. We say that the capacitor is now *charged*, and there is a potential difference across it. Figure 6f.10a shows this happening.

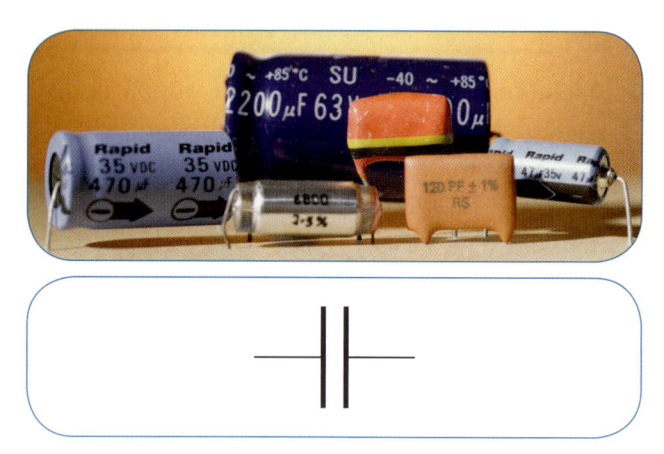

Figure 6f.9 A selection of capacitors of different sizes and the circuit symbol for a capacitor. Like resistors and diodes, capacitors have two connections.

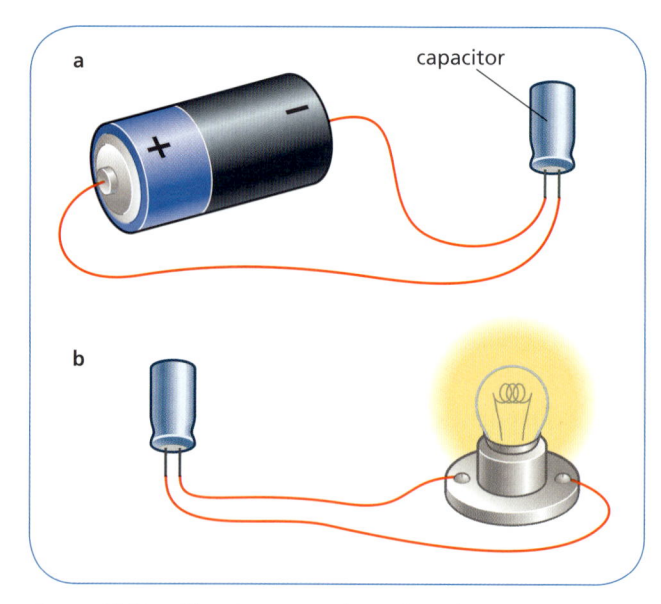

Figure 6f.10 a Charging a capacitor using a battery. **b** Discharging a capacitor though a lamp.

If the charged capacitor is now connected in a separate circuit, with no battery or power supply present, the charge that it has been storing runs out of it. The charge flows round the circuit (in the opposite direction to the current that charged up the capacitor) until the capacitor is discharged. Figure 6f.10b shows how a charged capacitor can be connected across a lamp – the lamp will shine as the current flows through it.

(It might help to think of a capacitor as a small rechargeable battery. In Figure 6f.10a, it is being charged up and, in Figure 6f.10b, it is being used to light the lamp.)

As the capacitor discharges, the p.d. across it decreases and so the current in the circuit gets smaller.

SAQ

7 a What does a capacitor store?

b How can a capacitor be charged up?

c A charged capacitor is connected in a circuit so that it gradually discharges. What happens to the p.d. across the capacitor?

Explaining charging and discharging

To understand in more detail what happens when a capacitor discharges, it helps to understand the structure of the device. Inside it are two thin sheets of metal called *plates*, with a narrow gap between them. (The circuit symbol for a capacitor represents the two plates and the gap between them.)

When a capacitor has been charged up (Figure 6f.11), one plate stores positive charge and the other negative charge. This is an example of static electricity. Just as static electricity causes a potential difference between, say, a positively charged plastic rod and a negatively charged cloth, so there is a potential difference across the charged capacitor.

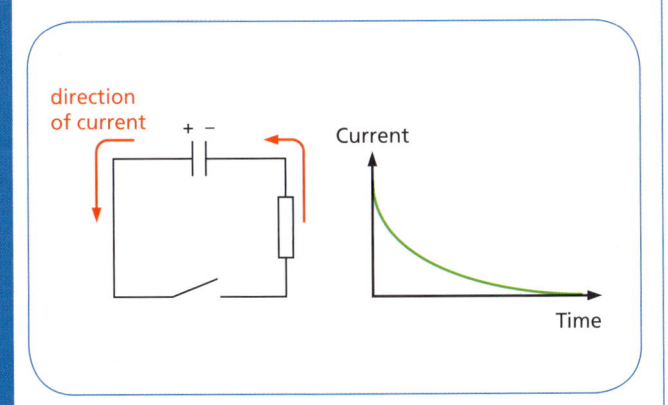

Figure 6f.11 The flow of charge when a capacitor discharges through a resistor.

If the two terminals of the charged capacitor are touched together, there will be a small flash as it discharges. A current has flowed. If the switch in Figure 6f.11 is closed, the capacitor will discharge in a more controlled way. It discharges through the resistor as positive charge flows round to cancel out the negative charge. The positive and negative charges decrease as it discharges; so the p.d. between the plates decreases and the current pushed by the p.d. also decreases. The graph in Figure 6f.11 shows the typical pattern of the decreasing current.

SAQ

8 Sketch graphs to show:

a how the charge stored by a capacitor decreases as it discharges though a resistor

b how the p.d. across it decreases.

Smoothing

Now we can combine the charge-storing ability of a capacitor with the rectification produced by diodes to see how an alternating voltage can be rectified and smoothed to produce a constant DC voltage supply.

Figure 6f.12 shows a capacitor connected across (in parallel with) the output of a diode-bridge circuit. The graph shows the result. The capacitor has removed the 'bumps' in the output voltage. There is a slight 'ripple' on the output voltage, but this is unlikely to be a problem for an appliance such as a radio or CD player.

We say that adding a capacitor to the output of the power supply produces a more constant ('smoothed') output.

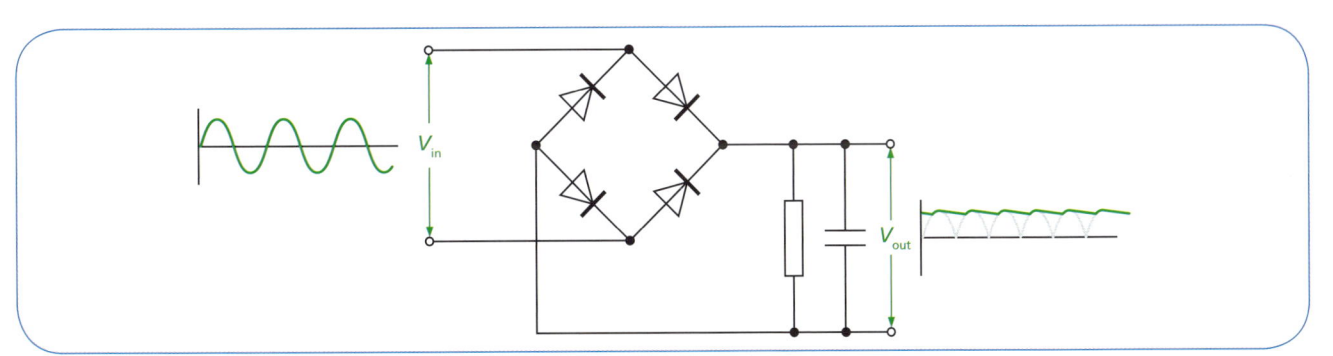

Figure 6f.12 A capacitor is connected across (in parallel with) the output of a diode bridge to produce a smoothed, almost constant voltage.

How smoothing works

How does a smoothing capacitor work? The capacitor is charged up by the full-wave-rectified voltage. As the voltage starts to drop, the current from the supply should get smaller. However, the capacitor has been storing charge, which it slowly releases, keeping the current and voltage at a high level. Once the rectified voltage returns to its high value, the capacitor charges up again, ready to discharge again as the voltage drops.

SAQ

9 The mains supply in the UK has a frequency of 50 Hz. What will be the frequency of the ripple on a full-wave-rectified supply? Explain your answer.

10 A single diode produces a half-wave-rectified voltage. A capacitor can be used to smooth this. Sketch a graph to show a half-wave-rectified voltage and superimpose the result you would expect to see if a smoothing capacitor was added to the circuit.

Summary

You should be able to:

- state that the current–voltage characteristics of a diode show that it allows current to flow in one direction only, as shown by the arrow of its circuit symbol
- state that the resistance of a diode is low in the forward direction and high in the reverse direction
- state that current in a diode is a flow of holes (in p-type silicon) and electrons (in n-type silicon)
- understand that a single diode results in half-wave rectification of alternating current
- understand that four diodes in a bridge circuit produce full-wave rectification
- understand that a capacitor stores charge when there is a p.d. across it
- understand that a capacitor discharges when a conductor is connected across it, so that a current flows
- understand that the p.d. across a capacitor decreases as it discharges, and hence the current also decreases
- understand that a capacitor connected across a rectified voltage smoothes it so that it is more constant

Questions

1. Figure 6f.13 shows four varying currents.

   a. Which of these are direct currents (DC)?

   b. What term is used to describe currents that are not direct?

   c. Which graph shows half-wave rectification?

   d. Which graph shows a smoothed, rectified current?

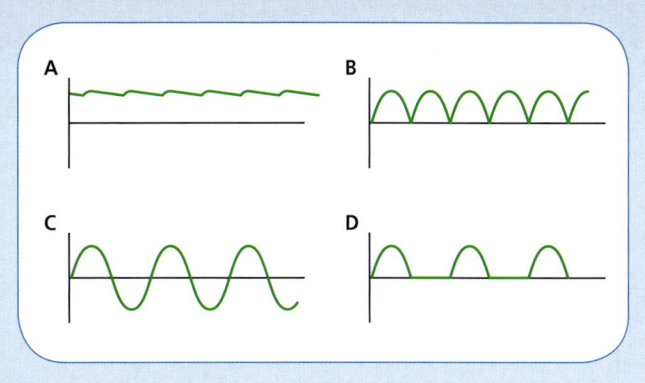

   Figure 6f.13 Graphs representing four varying currents.

2. Figure 6f.14 shows a diode-bridge circuit.

   a. Draw the circuit symbol for a single diode and add an arrow to show the direction in which current can flow through it.

   b. In the diagram, which one of the diodes is shown incorrectly connected?

   c. What is a diode bridge circuit used for?

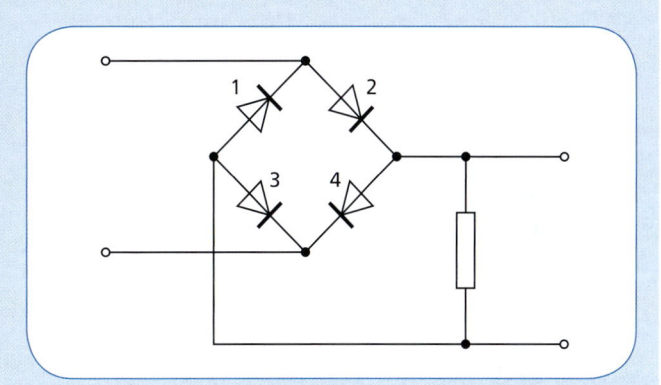

   Figure 6f.14

3. a. What does a capacitor store?

   b. A capacitor is connected to a 6V battery. It is disconnected and then connected to a resistor. Sketch a graph to show how the current that flows in the resistor changes.

   c. A capacitor may be used to smooth a varying DC supply. Why is this useful?

4. a. Draw a circuit that could be used to investigate the current–voltage characteristic graph for a silicon diode.

   b. Sketch the graph you would expect to find using this circuit.

   c. How does the resistance of a diode differ between the forward and reverse directions?

5. What charged particles carry an electric current in a silicon diode? Draw a diagram to show how they move when a current flows through a diode in the forward direction.

P6g It's logical

Getting automated

Today, many things are manufactured by robots. Fifty years ago, the same work was done by people. In a modern car factory (Figure 6g.1), robotic machines position parts with great accuracy, weld them together, paint them and so on. Computers control the robots so that they select the correct parts and fix them in the right places. On the production line, various different types of car must be constructed – they have different engine capacities, different fixtures and fittings, and are painted in different colours. The computers make sure that each car comes off the line with all the right features.

Each robot is fitted with sensors. These detect the positions of its arms and send the information back to the controlling computers. The computers then send signals to the motors that move the robot arms so that they are in the correct positions for the required task.

All of this automated control is carried out by electronic circuits. When it works well, it can produce goods of very high quality, relatively cheaply. Electronic systems have replaced many workers; those who are lucky enough to have a job can enjoy the products of modern manufacturing industries.

Automatic control systems have arrived in many aspects of life. Farming, for example, is increasingly automated. The giant greenhouse shown in Figure 6g.2 covers the area of several football pitches and is operated by one person. Sensors detect the levels of light, humidity, temperature and carbon dioxide. A computer uses this information to decide when to water the plants, open the windows, switch on the heating or burn gas to produce more carbon dioxide. If anything goes wrong at night, it can even telephone the operator at home and deliver an alarm call.

Figure 6g.1 This car production line uses robots to perform many different operations. Robotic tools have built-in sensors that detect their positions. This information is used by a computer to control the tool.

Figure 6g.2 Most of the time, only one person is needed to manage this giant glasshouse, in which tens of millions of lettuces are grown each year. Environmental sensors measure the temperature, humidity etc., and this allows the watering and heating systems to be operated automatically. Only picking the lettuces requires much human labour.

temperature sensor	processor	motors
detects the temperature	decides whether the temperature is high enough to need the windows open	open the windows

Figure 6g.3 A block diagram for a simple control system. This system automatically opens windows in the glasshouse when the temperature rises above a certain level.

Electronic processors

In an automatically run glasshouse like that in Figure 6g.2, the roof windows open automatically when the temperature rises. This requires a control system that detects when the temperature reaches a certain level and then switches on the motors that open the windows. We can describe this system as shown in Figure 6g.3.

In the past, someone might have looked at a thermometer and then turned a handle that opened the window. This person has now been replaced by the **processor**. A simple processor might be a single electronic component but, in this glasshouse, a computer is used. A single computer can perform many operations like this each second; a person takes much longer.

Figure 6g.4 shows a general block diagram that we can use to think about any control system. An **input device** such as a temperature sensor sends information to a processor, which then sends the appropriate signal to an **output device**, such as the motors that open the windows. Electronic control systems use different levels of voltage to control circuits. So we need to think about input sensors that produce voltages and processors that can respond by supplying voltages to output devices.

In earlier Items, we have looked at devices whose electrical resistance changes with temperature (thermistors), light level (LDRs) and so on. These can form the basis of electronic sensors. We have also looked at some possible output devices that produce motion (motors), light (lamps, LEDs) etc. In this Item, we will look at some electronic components that can work as processors. They have to receive a voltage from an input device and provide a voltage to operate an output device.

SAQ

1. In some situations, a human being can be thought of in the same terms as an electronic system. For example, a driver notices a pedestrian on the road ahead and applies the brakes. What are the input and output devices here? What is the processor?
2. Name a possible output device that produces sound.

Logic gates

A **logic gate** is a device that receives one or more electrical input signals and produces an output signal that depends on those input signals. These signals are voltages, typically about 0 V and 5 V:

- A high voltage (5 V) is referred to as ON and is represented by the symbol 1.
- A low voltage (0 V) is referred to as OFF and is represented by the symbol 0.

It is easiest to understand this by looking at three specific examples – the AND, OR and NOT gates, whose circuit symbols are shown in Figure 6g.5. Each symbol has inputs on the left and a single output on the right.

| input device | processor | output device |

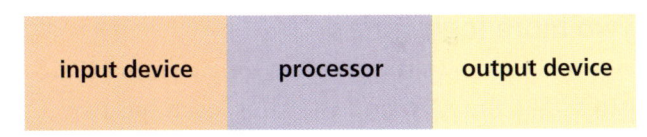

Figure 6g.4 A block diagram for a simple electronic control system. (In fact, this diagram could represent any control system, including mechanical ones. The processor might be a person.)

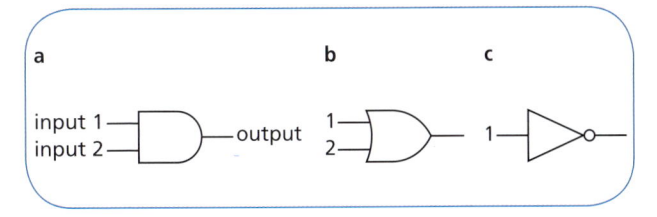

Figure 6g.5 Circuit symbols for three logic gates. (The symbol for an AND gate looks rather like the letter D in AND.)

- An OR gate functions like this: its output is ON if *either* input 1 *or* input 2 is ON.

Where might this be useful? Think of a domestic heating system. Most heating systems have only one thermostat to detect the temperature. A better system might have temperature sensors in *two* rooms. If a room was cold, the sensor would send an ON signal to the OR gate. If *either* room was cold, the output of the gate would be ON and this would switch on the heaters.

- An AND gate functions like this: its output is ON if input 1 *and* input 2 are *both* ON.
- A NOT gate functions like this: its output is ON if its input is *not* ON.

SAQ

3 Suggest an alternative use for an OR gate to the one described in the paragraph above.

4 A factory has a sprinkler system that operates when the temperature rises above 50 °C and smoke is detected. Which type of logic gate would be suitable for this?

Truth tables

The names of these three gates indicate how they operate. Another way to remember how they operate is by learning their **truth tables**, shown in Figure 6g.6.

A truth table shows all the possible combinations of inputs, and the output that results from each combination. A NOT gate has only one input, which can be ON or OFF, so this is the simplest table. AND and OR gates both have two inputs; there are four possible combinations of inputs and a corresponding output for each. For example, you can see from the last line in the AND gate table that two input 1s give an output 1; otherwise, the output is 0.

You should check that you understand how these truth tables represent the same information as in the sentences above that describe these gates.

a AND gate

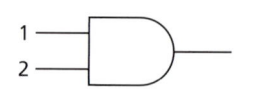

input 1	input 2	output
0	0	0
1	0	0
0	1	0
1	1	1

b OR gate

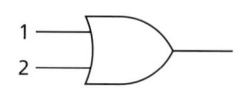

input 1	input 2	output
0	0	0
1	0	1
0	1	1
1	1	1

c NOT gate

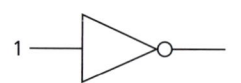

input	output
0	1
1	0

Figure 6g.6 Truth tables for three logic gates. **a** AND gate; **b** OR gate; **c** NOT gate. In a truth table, 0 stands for OFF or a low voltage; 1 stands for ON or a high voltage.

SAQ

5 The courtesy light in a car comes on when the driver's door is opened or when the passenger's door is opened. Which truth table could be used to represent this situation? What happens when both doors are opened?

Two more logic gates

Figure 6g.7 shows the symbols for two more logic gates. Like the AND and OR gates, each has two inputs. Their truth tables are also shown. From the tables, you should see that these gates can be described as follows:

- NAND gate: its output is ON if input 1 *and* input 2 are *not* both ON.
- NOR gate: its output is ON if *neither* input 1 *nor* input 2 is ON.

You could construct a NAND gate by connecting a NOT gate to the output of an AND gate. This combination has many uses, which is why NAND gates were designed in the first place.

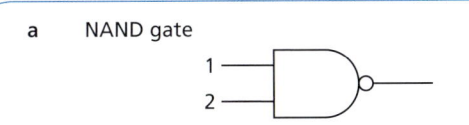

a NAND gate

input 1	input 2	output
0	0	1
1	0	1
0	1	1
1	1	0

b NOR gate

input 1	input 2	output
0	0	1
1	0	0
0	1	0
1	1	0

Figure 6g.7 Symbols for NAND and NOR gates, together with their truth tables. The little circle on each symbol is like the circle on the NOT gate symbol.

SAQ

6 Which two of the following logic gates could be combined to form a NOR gate: AND, OR, NOT? How should they be connected together?

Input devices

Figure 6g.8 shows a security light. It comes on at night if it detects anything moving in the vicinity. To do this, it incorporates two sensors. At the bottom is a movement detector and at the top is a light detector. A sensor like this is also known as a **transducer**. A transducer is any device that produces a voltage in response to some change in the environment. Similarly, an output device can be a transducer. A buzzer changes an electrical signal into a sound signal.

Figure 6g.8 A security light with two built-in sensors. You can see the movement sensor at the bottom and there is a light sensor in the top.

The security light comes on if both the movement sensor *and* the light sensor tell it to. An AND gate is part of the circuit, to combine the two inputs. The lamp comes on if the AND gate output is ON.

In Item P6b *Sharing*, we saw some devices that are suitable as input sensors:
- thermistors, whose resistance decreases (usually) over a narrow range of temperatures
- light-dependent resistors (LDRs), whose resistance decreases when light shines on them
- variable resistors, whose resistance depends on the position of the control knob or slider.

To this list we can add:
- microphones, which produce a voltage output in response to a sound
- moisture detectors, whose resistance decreases when they get damp
- a range of switches, which respond to pressure, tilting or magnetic fields.

Using a switch

Figure 6g.9 (page 000) shows how to connect two switches to an AND gate so that they act as input devices. When either of the switches is closed, it connects the input to +5V so that the input is high (ON).

In this circuit, a light-emitting diode (LED) is connected to the output of the AND gate. When the output is high, there will be a high voltage across the LED and it will light up. So the LED acts as an indicator to tell us whether the output of the AND gate is high or low.

If both switches are closed, both inputs will be high and the LED will light up.

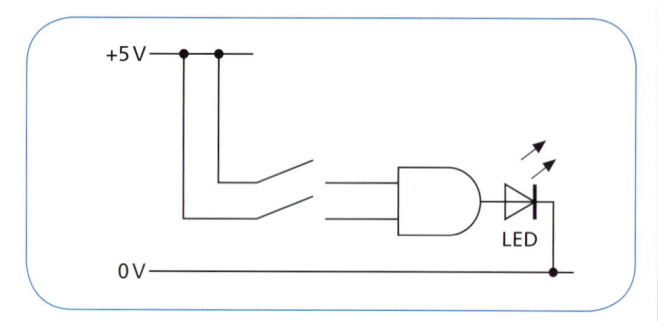

Figure 6g.9 Using switches to control the inputs to an AND gate. The LED lights up to indicate when the output is ON.

SAQ

7 Draw a similar circuit to that shown in Figure 6g.9, with an OR gate in place of the AND gate. How must the switches be set if the LED is to be unlit (OFF)?

Using a potential divider

In Item P6b *Sharing*, we saw how to connect two resistors together to make a potential divider. Such a circuit is often used with one of the above devices to make a sensor that can supply the necessary voltage to a logic gate. Figure 6g.10 shows an example of how this can be done using a thermistor. The LED indicates whether the output of the NOT gate is high or low.

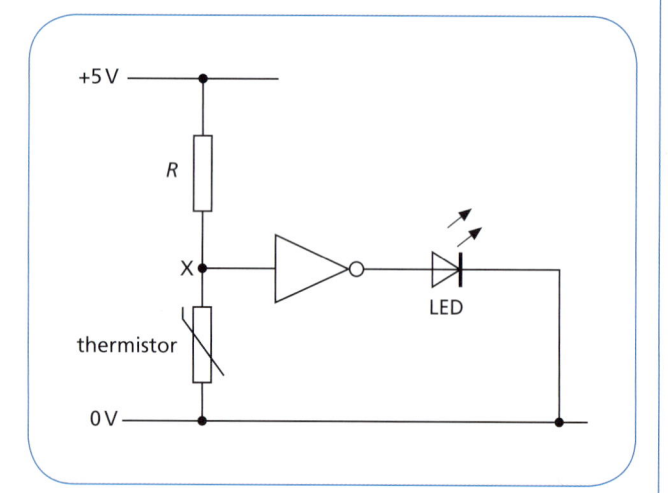

Figure 6g.10 Using a thermistor in a potential divider circuit to act as a temperature sensor. This might be used, for example, to switch off a computer if it is in danger of overheating.

This could be used to provide a warning light when an electrical appliance overheats. When the thermistor is cold, its resistance is high. It therefore has a large share of the supply voltage across it, so the voltage at point X is high. The input to the NOT gate is therefore ON, so its output is OFF. The p.d. across the LED is low, so the LED is unlit.

When the thermistor becomes hot, its resistance drops. The p.d. across it therefore decreases and the voltage at point X becomes low. The input to the NOT gate is now OFF, so its output is ON. This high voltage pushes a current through the LED and it lights up.

SAQ

8 A light-dependent resistor (LDR) could be connected in the circuit of Figure 6g.10 in place of the thermistor. Draw this circuit and describe how it can be used to turn the LED on and off.

Using a variable resistor

If the fixed resistor R in Figure 6g.10 is replaced with a variable resistor (Figure 6g.11), we can set the temperature at which the LED switches on and off.

The supply voltage (5 V) is shared between the two resistors that form the potential divider. If the thermistor is cold, it has a bigger resistance and so its share of the supply voltage will be large. Point X will be close to 5 V and the NOT gate output will be OFF.

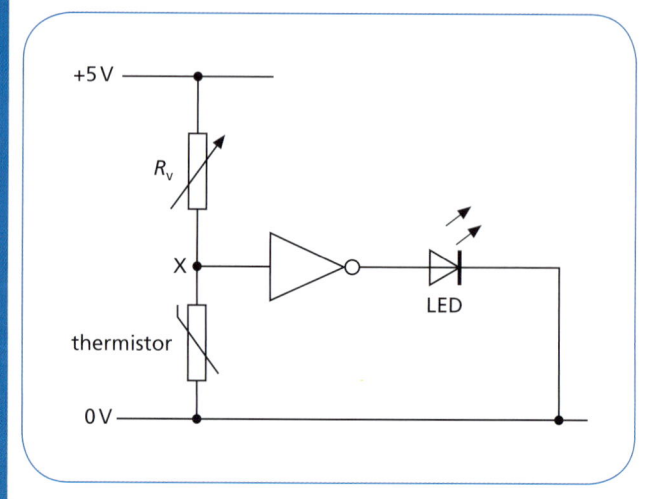

Figure 6g.11 The potential divider provides an input signal to the logic gate. The variable resistor can be altered to set the temperature at which the LED lights up.

As the thermistor is heated up, its resistance drops and its share of the supply voltage decreases. At some point, the LED will light up. Adjusting the variable resistor will alter the precise temperature at which this happens.

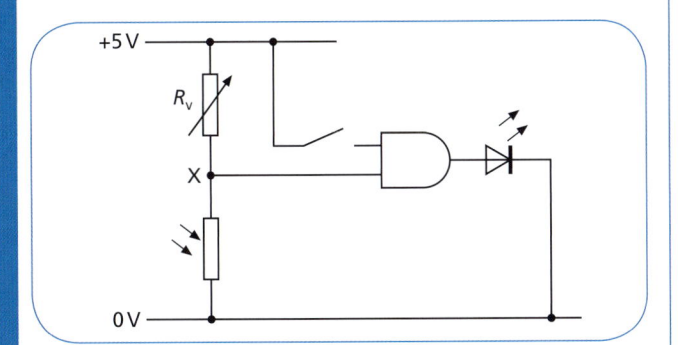

Figure 6g.12 An AND gate with two input signals, one from a switch and the other from a potential divider.

Potential dividers and switches can be combined in different ways to provide input signals to logic gates. Figure 6g.12 shows an AND gate with two controllable inputs.

SAQ

9. Draw a circuit in which an LDR is used in a potential divider to provide the input to a NOT gate. The output of the NOT gate must be high when the LDR is in the dark. Explain how to set the light level at which the NOT gate switches on.

10. Look at the circuit in Figure 6g.12 and explain its function. (Think of the possible combinations: switch open or closed; LDR illuminated or not.)

Summary

You should be able to:

- understand that the output of a logic gate can be low or high, depending on the input signals it receives

- understand that the output of a NOT gate is low if the input is high, and vice versa

- understand that truth tables are used to show the outcomes of different combinations of inputs to a logic gate

- understand that some gates can be built by combining other gates: NAND = AND + NOT; NOR = OR + NOT

- understand that switches, LDRs, thermistors etc. can supply input signals to logic gates

- understand that a potential divider consisting of a thermistor and a fixed resistor can create an input signal that depends on the temperature, and that this can be fed to a logic gate

- understand that a potential divider consisting of an LDR and a fixed resistor can create an input signal that depends on the light level, and that this can be fed to a logic gate

- understand that replacing the fixed resistor with a variable resistor allows the temperature or light level at which switching occurs to be set

Questions

1. Copy and complete the truth table for a NOT gate (Table 6g.1).

input	output
low	

Table 6g.1

2. What logic gate is represented by the truth table shown in Table 6g.2?

input 1	input 2	output
0	0	0
1	0	0
0	1	0
1	1	1

0 = OFF
1 = ON

Table 6g.2

3. Draw a truth table for an OR gate.

4. a Draw a circuit to show how a switch can be used to control the input to a NOT gate. Include an LED to act as an indicator of the output of the gate.

   b Describe what will happen when the switch is open and when it is closed.

H 5. Draw a truth table for a NAND gate.

6. The output of a logic gate called an 'exclusive OR' gate is ON if only one of its two inputs is ON; otherwise it is OFF. Draw a truth table to represent this.

7. Draw a circuit diagram showing how a potential divider circuit consisting of a thermistor and a fixed resistor can provide a signal to a NOT gate. The output of the gate should be indicated by an LED, with the LED being unlit at high temperatures.

6h Even more logical

Computers everywhere

Most people in developed countries use computers every day. There are more computers in the UK than there are people. Some of these computers are hidden away and so we might not realise they are there – in games consoles, television sets, digital radios and MP3 players. A modern car usually has a computer to monitor the engine's performance and to warn the driver that the car has been left with its lights on.

A computer makes use of combinations of logic gates (Figure 6h.1). This means that it takes some input information, processes it logically according to the software with which it is programmed and produces some outputs. Of course, computers can manage this much faster than human beings, although they are not as good as us at jumping quickly to conclusions based on judgements – should I tell my parents I have lost my bus pass? What will happen if I don't hand in my coursework on time?

The basic building block of a computer microchip is the transistor. This acts rather like a NOT gate – it can be switched on and off. Transistors can be combined to make AND gates, OR gates and so on.

In 1965, Gordon Moore (one of the founders of Intel, a major microchip manufacturer) noticed that the number of transistors in a microchip seemed to double every 24 months. But could this go on into the future? From the graph of Figure 6h.2 you can see how this has worked out – the 'law' suggested by Moore has proved to be roughly correct. This means that microchips have become increasingly complex and more powerful as the years have gone by.

No-one can be sure if Moore's law will continue into the future. New technologies for squeezing more and more computing power into the same volume will be needed if the trend is to continue.

Figure 6h.1 A computer uses many millions of logic gates, linked together. In this impression, a circuit board is shown with several gates; in reality, millions of gates are built into a single silicon chip.

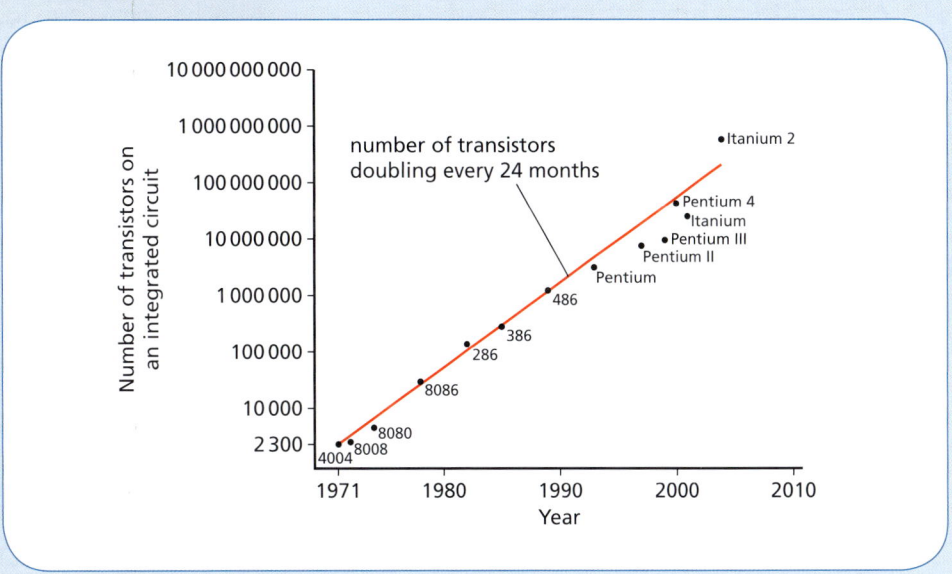

Figure 6h.2 A graph representing Moore's law, showing the numbers of transistors in the most powerful chips used in personal computers. Moore himself didn't describe his idea as a 'law'.

Combining logic gates

How can we work out what happens if we connect two or more logic gates together? This is where truth tables prove useful.

Figure 6h.3a shows an AND gate with a NOT gate connected to its output. We can work out the truth table for this combination by realising that, at point 3, the *output* of the AND gate becomes the *input* of the NOT gate; when the AND gate output is 1, the NOT gate turns this into a 0. This combination of gates might be used in a car: when the driver and the passenger have put on their seat belts, the warning light goes out.

Figure 6h.3b shows the same gates but differently connected, together with the resulting truth table.

a AND NOT

input 1	input 2	output
0	0	1
1	0	1
0	1	1
1	1	0

b NOT AND

input 1	input 2	output
0	0	0
1	0	0
0	1	1
1	1	0

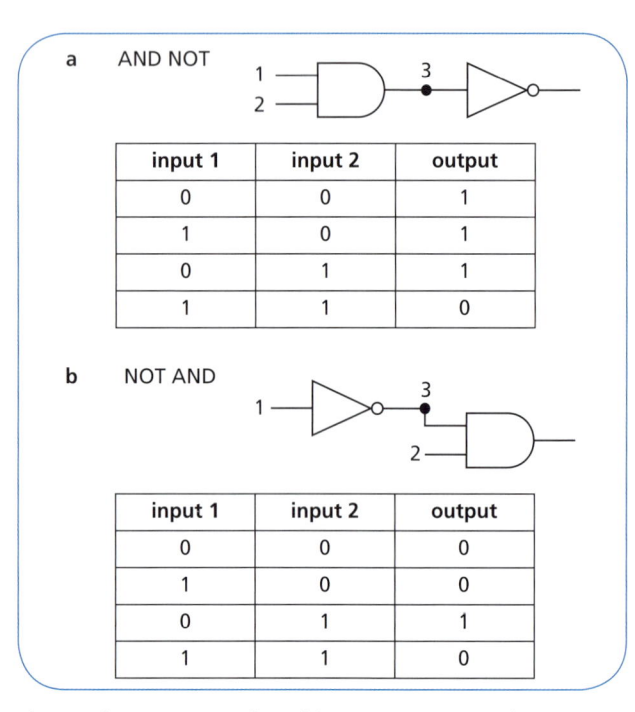

Figure 6h.3 Two ways of combining a NOT gate and an AND gate, together with the resulting truth tables.

SAQ

1 Look at the truth tables shown in Figure 6h.3.
   a What combination of inputs will cause the output of the AND NOT combination to be low?
   b What combination of inputs will cause the output of the NOT AND combination to be high?
   c Suggest an application for the NOT AND combination.

Three or more inputs

Figure 6h.4 shows an AND gate connected to one of the inputs of an OR gate. Now we have three inputs, and there are eight possible combinations of input signals.

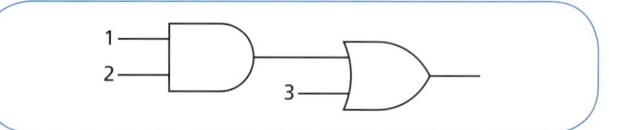

Figure 6h.4 A combination of gates with three inputs.

Worked example

If we have an AND gate connected to one of the inputs of an OR gate (Figure 6h.4), what are the output signals for each possible combination of input signals?

Step 1: draw up the truth table for the AND gate and show this twice, one below the other (see columns 1–3 of Table 6h.1). (A blank row has been left to show the two halves more clearly.)

Step 2: the AND output acts as one of the inputs to the OR gate. We can work out column 5 (the final output signal) from columns 3 and 4, using the truth table for the OR gate.

Step 3: to show the final result, delete column 3 (which is only there to help us with our working).

column 1	column 2	*column 3*	column 4	column 5
input 1	input 2	*AND output*	input 3	output
0	0	*0*	0	0
1	0	*0*	0	0
0	1	*0*	0	0
1	1	*1*	0	1
0	0	*0*	1	1
1	0	*0*	1	1
0	1	*0*	1	1
1	1	*1*	1	1

Table 6h.1 Working out the truth table for the combination of gates shown in Figure 6h.4.

Where might a combination like this be useful? Imagine a security system in a bank vault.
- If both doors in the corridor leading to the vault have been left unlocked, the alarm must sound. The door sensors are connected to the AND gate.
- The alarm must also sound if movement is detected in the vault at night. The movement sensor is connected to the OR gate.

Now the alarm will go off if movement is detected, or if both doors are left unlocked.

SAQ

2 Work out the truth table that would result if the two logic gates shown in Figure 6h.4 were swapped over. (The OR gate would be connected to one of the inputs of the AND gate.)

OR OR OR

Figure 6h.5 shows a combination of three OR gates. This might function as follows.
- A building has smoke detectors in four different places.
- Their outputs are connected via this combination of gates to a single alarm siren.
- If any detector gives an ON signal, the siren will be switched ON.

This saves the expense of a separate siren for each detector.

The combination has four inputs, so there are 16 different possible combinations of input signals. If one or more inputs is ON, the output is ON.

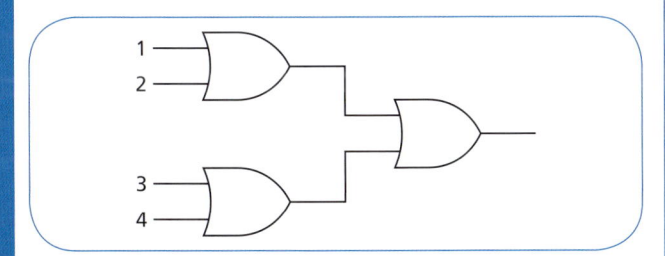

Figure 6h.5 Three OR gates connected together.

SAQ

3 Draw up a truth table for the combination of gates shown in Figure 6h.5. Under what circumstances would the output be OFF?

Latching on

A car alarm might have an alarm circuit that switches on when the door is opened without using the correct key. The siren sounds to indicate that a thief might be at work. However, there is a problem with this. Suppose the thief grabs a bag and slams the door shut. The siren will stop sounding. What is needed is a way of keeping the output of the logic system ON even when the inputs change.

There is a device that can do this, called a **latch** or **bistable** (Figure 6h.6). A latch has two inputs, SET and RESET. Here is how it works.
- When SET becomes ON, the output switches ON.
- Now, even if SET becomes OFF, the output remains ON. It is said to be *latched*.
- To return the output to OFF, the RESET must become ON.

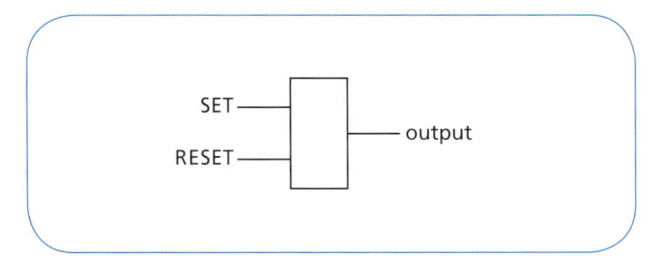

Figure 6h.6 The inputs and outputs of a latch.

Constructing a latch

Figure 6h.7a (page 324) shows how to construct a latch from two NOR gates. The output of each gate is connected back to one of the inputs of the other gate. (This is an example of *feedback* in an electronic circuit.) In what follows, S (SET) and R (RESET) are the two inputs to the latch (Figure 6h.7). Point X is the latch's output; point Y is important only for the inner workings of the latch – it is not an external connection.
- When S and R are OFF, both inputs to gate 1 are OFF and so point Y is ON.
- This feeds back to the other input of gate 2 (one input to gate 2 is ON), which means that point X (the latch's output) is OFF.
- Suppose that S turns ON. Now, one of the inputs to gate 1 is ON and so point Y turns OFF.
- This means that both inputs to gate 2 are now OFF and so point X is ON.

- This feeds back to the other input of gate 1, so gate 1 remains OFF. It will remain even if S goes OFF. We say that the output (point X) is latched ON.
- To reset the latch, R is switched ON; one input to gate 2 is now ON so its output, X, goes OFF.
- This feeds back to gate 1: both its inputs are OFF and point Y switches ON. The latch has been reset – its output (point X) is latched OFF. Even if R is switched OFF the latch will remain in its reset state. It will only change if S is turned ON again.

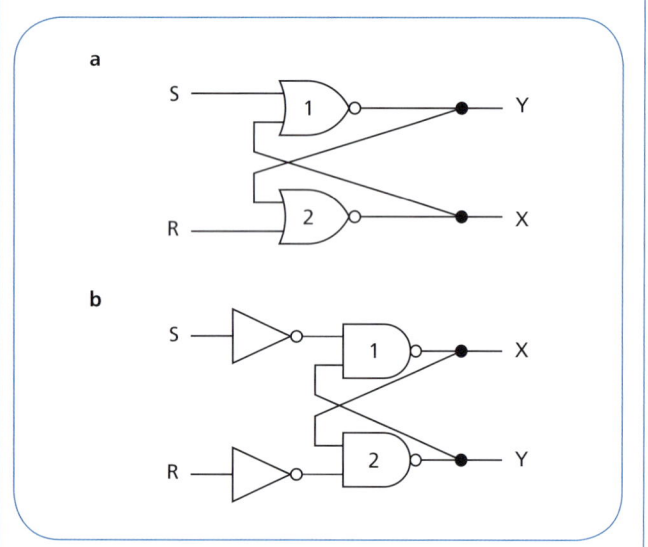

Figure 6h.7 A latch can be constructed using two NOR gates (**a**) or two NAND gates and two NOT gates (**b**). In both cases, the output of each gate is fed back to the input of the other. A combination of gates like this is sometimes known as a bistable, because it has two stable states.

SAQ

4 Figure 6h.7b shows how a latch can be constructed using two NAND gates. Explain how this circuit can act as a latch.

Output devices

Figure 6h.8 shows a circuit board used to investigate the behaviour of logic gates. You can see that it has NOT, AND and OR gates. Their inputs are on the left and their outputs on the right. The output of each gate is indicated by a small red LED. (LED bright = output ON.) An LED is suitable for this because it requires only a very small current to light it up.

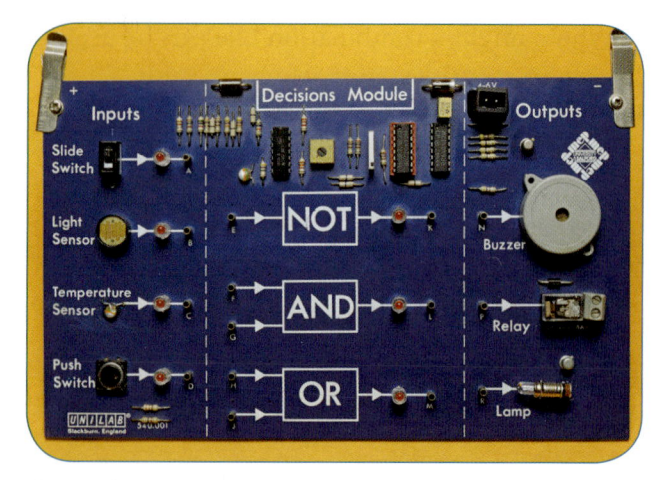

Figure 6h.8 A circuit board for investigating logic gates.

Unfortunately, a logic gate cannot provide enough current to operate, say, an electric motor or a heater. An electromagnetic switch called a **relay** (Figure 6h.9) is used to overcome this. A relay is a device that allows a small current to switch on a much larger current. A small current flows from the logic gate and operates the relay, switching on a second circuit with the bigger current needed to operate the motor or heater.

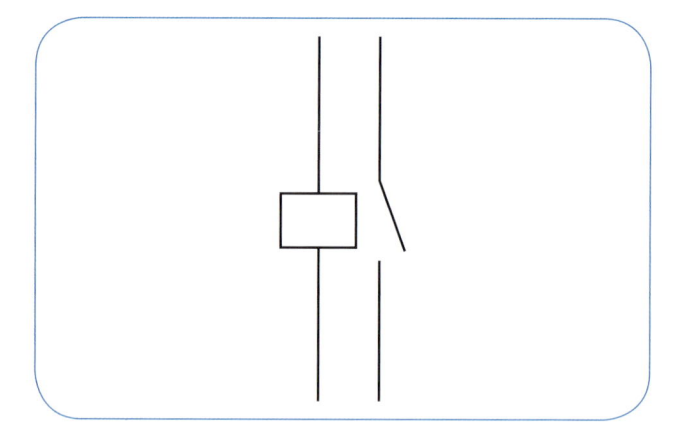

Figure 6h.9 The circuit symbol for a relay.

SAQ

5 Look at the circuit board shown in Figure 6h.8.
   a Name the four input devices on the board.
   b Name the four output devices. (Don't be fooled – there are four!)

Currents large and small

Although LEDs are used to show the output of a logic gate, the current from the gate might be too high for the LED. To prevent the LED being overloaded and damaged, a resistor is usually connected in series with the LED to ensure that only a small current flows through it.

A relay is used to overcome the fact that a logic gate can only provide a small current. It also has an additional benefit, however. A logic gate operates from a 5V supply but a heater or motor might need to use the 230V mains. The relay *isolates* the low-voltage circuit of the logic gate from the high-voltage mains supply.

In Figure 6h.9, the low-voltage circuit is on the left and the high-voltage circuit on the right. You can see that the two circuits have no electrical connection.

Summary

You should be able to:

- understand that truth tables can show the function of two or more logic gates connected together
- understand that a latch can be used to keep a circuit switched ON after it has been triggered
- understand that a bistable latch circuit can be constructed from two NAND or NOR gates
- understand that an LED can indicate the output of a logic gate
- understand that a relay uses a small current to switch a larger one
- understand that the output power of a logic gate is small, so it can only operate a high-power component via a relay

Questions

1. Draw a circuit diagram for an OR gate that has a single NOT gate connected to each of its inputs. Write down the truth table for this circuit.

2. What electronic components or devices are being described here?

    a If one of its inputs becomes high and then returns to being low, its output remains high.

    b It emits light to indicate when the output of a logic gate is high.

    c It uses a small current to switch a larger one on and off.

3. Draw the circuit symbol for a relay.

4. a Draw a diagram to show how two NAND gates must be connected if they are to act as a bistable latch.

    b Explain how a brief signal at one input results in a permanent high signal at the output.

    c How can the latch be reset?

5. Figure 6h.10 shows a combination of logic gates with three inputs. Deduce its truth table.

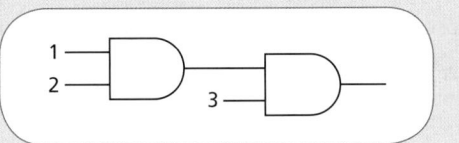

Figure 6h.10

Answers to SAQs: Biology

Item B5a In good shape

1. Internal.

2. A structure that supports an organism and holds it in shape.

3. An external skeleton can form a tough, protective covering over an animal's body. (Note that some vertebrates have also developed such coverings, for example tortoises, armadillos and pangolins.)

4. a The leg can only get longer when the skeleton has been shed.
   b Five.

5. The spinal cord lies inside the vertebral column. If the vertebral column is broken, then movement could snap or damage the spinal cord. This could cause paralysis.

6. *Advantages:* it allows pain-free movement; the person can become self-reliant again and not need constant help.
   Disadvantages: it is a large, expensive operation, and there is always a risk that something may go seriously wrong, such as in infection; the person could develop dangerous blood clots during the recovery period; it is difficult to ensure that the replacement joint leaves the leg exactly the same length as before the surgery.

7. Relax.

8. $10 \times 30 = F \times 5$

   So $F = \dfrac{10 \times 30}{5}$ $\quad F = 60$ N

Item B5b The vital pump

1. Both have arteries, veins and a pump. The closed circulatory system has capillaries, but the open one does not.

2. The closed circulatory system can move blood more rapidly around the body, as it will not pool in the tissues. An open circulatory system takes blood closer to the cells, because there are no capillary walls between them, and this could increase the speed of transfer of substances by diffusion.

3. The ventricle is at the top of the heart (as shown in the diagram) and the atrium below.

4. Having four chambers means that the oxygenated and deoxygenated blood are kept completely separate. Two chambers are used to pump deoxygenated blood to the lungs, while the other two chambers pump oxygenated blood around the body. So the blood that travels to the body tissues is fully oxygenated and can provide the maximum amount of oxygen to them.

5. It would still be better, because returning to the heart means that the blood can be given another push to make it travel more rapidly to the tissues. So, even though oxygenated blood is mixed with deoxygenated blood, oxygen could be supplied to the tissues at a greater rate than with a single circulatory system.

6. a Just under 120 mm Hg.
   b The arteries receive the blood as it is pushed out of the heart. The pressure rises with each contraction of the ventricles, and then falls as the ventricles relax.
   c The veins have valves that prevent the backflow of blood. Skeletal muscles around the veins – for example, the muscles in the legs – squeeze inwards on the veins when they contract, pushing the blood along and through the valves.

7. a All the heart muscle would contract at the same time. The blood would be squeezed in all directions, and would not flow through the heart.
   b This helps to squeeze the blood upwards, into the arteries.

8. a Three.
   b There are three complete beats in around 2.2 seconds. So in 60 seconds there will be approximately 80 beats.

Item B5c Running repairs

1. a The heart beat is irregular. There is no clear P section.
   b In the atria, because it is the P section, which relates to the contraction of the atria, that is missing.
   c The artificial pacemaker would send pulses of electric current over the walls of the atria, making them contract. It would also help to make the heart beat more regular.

Answers to SAQs: Biology 327

2. Where there is a hole in the heart, blood can flow between the right and left atria, as well as going from the atria to the ventricles. So the blood that goes to the lungs and the blood that goes to the rest of the body is carrying about the same reduced amount of oxygen, rather than one being deoxygenated and the other oxygenated.

3. Coronary heart disease means that part of the heart muscle cannot work, because it is not getting enough oxygen. The bypass provides a new route for oxygenated blood to get to this muscle, so it can start working normally again.

4. *Advantages:* it can provide a long-term solution to the person's heart problems; if their heart had several different things wrong with it, the transplant can solve all of them at once.
   Disadvantages: it is difficult to find a heart that has the same tissue type and is the same size as the recipient's; the recipient will need to take immunosupressant drugs for the rest of his or her life; the operation is a big one, and carries a significant risk.

5. a *The annotations should include:* small size to squeeze through the smallest capillaries and get close to the cells in the tissues; contains haemoglobin which combines reversibly with oxygen to transport it around the body; no nucleus so more room for haemoglobin; biconcave disc to increase surface area so more oxygen can diffuse in and out at the same time.
   b It is very flexible, so it can squeeze out of capillaries and between the cells in the tissues, enabling it to come into contact with pathogens in every part of the body.

6. If their blood was transfused into another person, any disease or infection would be transmitted to them.

7. Her blood contained anti-A and anti-B antibodies. The blood she was given contained red cells with group A agglutinins. Her anti-A antibodies locked onto these, making the red cells clump together.

8.

Group of donor	Group or groups of recipients who can safely receive the blood
A	A and AB
B	B and AB
AB	AB
O	A, B, AB and O

9. As there are no cells present, the blood will contain no agglutinins, so it can be given to any patient no matter what their blood group.

Item B5d Breath of life

1. *Large surface area* – more oxygen and carbon dioxide can diffuse across the surface at the same time.
   Moist surface – if the cells were covered with a waterproof covering, they would be too thick to allow diffusion to take place. So they have to be uncovered, and therefore kept moist, so that they do not dry out.
   Thin lining – this reduces the diffusion distance for oxygen and carbon dioxide, so they can cross more quickly.
   Good blood supply – this carries away the blood that has been oxygenated, and brings blood that has little oxygen in it; this maintains a diffusion gradient for oxygen from the alveoli into the blood. The same is true for carbon dioxide, in reverse.

2. a This increases the surface area across which diffusion can take place.
   b This reduces the diffusion distance for oxygen and carbon dioxide.
   c This provides a large surface area for gases to diffuse into and out of the blood; the haemoglobin combines with oxygen and transports it around the body.

3. a Inspiration is breathing in – the external intercostal muscles and diaphragm muscles are both contracted. Expiration is breathing out – the external intercostal muscles and diaphragm muscles are both relaxed.
   b Breathing means making movements that move air into and out of the lungs. Respiration is a metabolic reaction that takes place inside every living cell, in which glucose is oxidised to release energy.

4. a Up.
   b About $0.3\,dm^3$.
   c $3.5\,dm^3$.
   d The patient was asked to take the biggest possible breath out, followed by the biggest possible breath in.
   e There is always some air that remains in the lungs, and there is no way of measuring it because it can't be breathed out into the spirometer.

Answers to SAQs: Biology

5. Goblet cells produce mucus, in which particles get trapped. Ciliated cells have cilia which sweep the mucus up to the back of the throat, where it is swallowed.

Item B5e Waste disposal

1. a

Waste substance	Where is it produced?	Which organ removes it from the body?
undigested food	alimentary canal	alimentary canal/rectum/anus
carbon dioxide	all cells (because all cells respire)	lungs
urea	liver	kidneys (and also skin)

  b Undigested food.
  c Carbon dioxide and urea.

2. a About $0.55\,dm^3$.
   b About 11.25 breaths per minute.
   c The tidal volume increased immediately, continued increasing over the next few breaths, and then settled down to a regular level.
   d The breathing rate also showed an immediate increase, continued to increase for a few breaths and then settled down to a regular rate.
   e As she exercised, her muscle cells respired more rapidly. They therefore released more carbon dioxide into the blood. This dissolved in the plasma to form a weak acid, lowering the blood pH. The brain sensed this change, and sent nerve impulses to the diaphragm and intercostal muscles, making them contract more often and more strongly.
   f The faster, deeper breathing would move air containing oxygen into her lungs more rapidly, and remove air containing some of the carbon dioxide she was producing. This would help to make sure that her muscle cells continued to be supplied with oxygen for respiration, and that carbon dioxide was taken away from them. If this did not happen, her cells might be damaged by the low pH caused by the dissolved carbon dioxide.
   g It would go back to the resting rate, but not immediately. Her muscles would have been respiring anaerobically for some of the time, producing lactic acid. This is carried to the liver in the blood. Here it is broken down, using extra oxygen that she breathes in once exercise is finished.

3. a This provides a larger surface area through which diffusion can take place, so it speeds up the process.
   b The higher pressure helps to push substances through the membrane, again speeding up the dialysis process.
   c The blood arriving in a glomerulus in a kidney is also kept at a high pressure, by having a narrower blood vessel taking the blood away than the one that brings it.
   d Both of them allow small molecules to filter out of the blood but keep large molecules and blood cells in the blood.
   e There is no selective reabsorption in a dialysis machine. None of the substances move against their concentration gradient in a dialysis machine, whereas in a kidney tubule energy is used to move some substances out of the filtrate and into the blood.

Item B5f Life goes on

1. tubule in testis, epididymis, sperm duct, urethra

2. An egg can only be fertilised between 8 and 24 hours after release from the ovary. It will not have reached the uterus by then.

3. Sperm and egg are haploid, and the zygote is diploid.

4. If the mother and baby have different blood groups, there is a danger of the red cells of the mother or the baby clumping together (see page 22).

5. Progesterone and oestrogen inhibit the release of FSH and LH from the pituitary gland, which would normally stimulate the ovary to produce and release eggs. The pill stops egg production and ovulation.

6. The twins are genetically identical, so the cells of one twin will not be seen as 'foreign' in the other twin's body.

7. Normally, a DNA test would be able to determine who was the biological mother. But the twins' DNA is identical.

Item B5g New for old

1. Valves and pacemaker.

2. The drugs reduce the effectiveness of the body's defence system, making it more likely that infectious diseases can take hold.

3. The operation to remove the organ carries some risk, including the possibility of infection. If the donor's other kidney is later damaged, they could end up with no working kidney at all.

Item B5h Size matters

1.
   a. 16 kg is right at the upper limit of normal weight range for this age. Her parents should try to ensure that she does not get any fatter.
   b. Up until about 4 months old.
   c. Average weight at birth is 3.5 kg. Average weight at 3 years old is 13.7 kg. Growth is therefore 13.7 − 3.5 = 10.2 kg in 3 years. So the average growth per year is 10.2 ÷ 3 = about 3.4 kg per year.

2.
   a. Sub-Saharan Africa.
   b. This is where AIDS is most common, and many people die young from HIV/AIDS. Also, several countries in this region suffer from drought, which limits food supplies. War causes many deaths in some countries in this region.

Item B6a Understanding bacteria

1. They have chlorophyll, even though they have no chloroplasts. (Bacteria also use other pigments in photosynthesis.)

2. Magnification = about 200 000.

3. Rod shaped.

4.
   a. Spherical.
   b. Cilia.
   c. Mucus.
   d. Cells are making mucus to trap bacteria in the air that is being breathed down into the lungs. The cilia make sweeping movements that move the mucus, with its trapped bacteria, up to the back of the throat where it can be swallowed.

5. The milk is pasteurised to destroy any unwanted bacteria already in the milk. If it was pasteurised after the starter culture had been added, then the *L. bulgaricus* and
S. thermophilus would be killed and no yoghurt would form.

Item B6b Harmful microorganisms

1.
   a. In 1986, there were slightly more cases of *Campylobacter* than of *Salmonella*. Since then, cases of *Campylobacter* have risen, whilst cases of *Salmonella* stayed relatively steady between 1988 and 1997, and have fallen considerably since then.
   b. The longer food is stored, the more time there is for bacterial populations to increase in it. Ready meals may not always be heated to a high enough temperature to kill any microorganisms in them.
   c. Cases of *Salmonella* have dropped since the late 1990s, and it is possible that this is due to the measures taken to control the bacterium in chickens. Support for this idea comes from the fact that cases of *Campylobacter* have remained fairly steady over the same period. However, the drop in *Salmonella* cases could be from other causes that we do not know about.
   d. If electricity supplies are disrupted, then fridges and freezers will not work, and people cannot store food at low temperatures to slow down bacterial growth. They may not be able to cook food, and so have to eat food that may contain bacteria that should be destroyed by cooking.

2. You should try to obtain water from somewhere that is unlikely to be contaminated by sewage. The water could be boiled or have chlorine added to it, to kill any cholera bacteria.

3. Make sure you only eat really hot food, not food that has been kept warm for some time. Only eat salads if you know they have been washed in clean water. Wash your own hands carefully before eating, and never put your fingers into your mouth without washing them first. Drink liquids that have been sealed in bottles, or that have been made from boiling water.

4. The bacterium is present in the blood, so it is transported all over the body.

5. Tuberculosis, food poisoning and septicaemia are caused by bacteria and so can be treated using antibiotics. None of the other diseases in the table would be affected by antibiotics.

Item B6c Microorganisms – factories for the future?

1. $$\text{magnification} = \frac{\text{length in diagram}}{\text{actual length}}$$
   $$= 58\,\text{mm} \times \frac{1000}{10\,\mu\text{m}} = 5800$$

2. a Enzymes and other proteins begin to be denatured (lose their shape) above this temperature.
   b The particles involved in the reactions have more kinetic energy at higher temperatures, so they collide with each other more often and with greater energy. This means that the metabolic reactions in the yeast cells will take place faster at higher temperatures, so the population can grow faster.

3. Alcohol and carbon dioxide are made in anaerobic respiration (fermentation) in yeast, whereas in anaerobic respiration in human cells only lactic acid is made.

4. Microorganisms are killed when the sugary liquid is boiled, and when the beer is pasteurised.

5. The yeast might then respire (at least partly) aerobically. So some of the sugar would be wasted in making water and carbon dioxide, rather than alcohol and carbon dioxide.

6. Grapes are crushed to release the sugar, and then mixed with water. Yeast is added to the sugary liquid.

Item B6d Biofuels

1. methane and hydrogen

2. Biogas will be produced less quickly in winter, because bacterial enzymes will act more slowly at lower temperatures. It will contain a lower proportion of methane, because most methane is produced at temperatures between 32 °C and 39 °C, which will not be reached in winter.

3. This is described on page 40 in *Gateway Science*. The particulates irritate the lungs, and white blood cells try to remove them. This produces inflammation that can damage tissues.

Item B6e Life in soil

1. rock particles, humus, living organisms, air, water

2. One set of nematodes eats plant roots, so they are definitely herbivores. The others eat bacteria and fungi, which are not plants. If 'herbivore' means 'plant-eating', then this is perhaps not a good term. A better one would be 'primary consumers'.

3. Many mineral ions need to be taken up against their concentration gradient, by active transport. This requires an energy input by the plant. The energy is supplied by ATP, which is produced by aerobic respiration – which uses oxygen.

Item B6f Microscopic life in water

1. Water provides much more support than air, so jellyfish have no need of supportive skeletons. Once out of water, they simply collapse into a heap.

2. Sea water has a similar concentration to cytoplasm, so water will no longer enter the cell by osmosis. The rate of filling and emptying of the contractile vacuole will slow down and perhaps stop completely.

3. They need to be in the light so that they can photosynthesise. Light only penetrates a few metres into water.

Item B6g Enzymes in action

1. a Add Benedict's solution to the urine sample, and heat it. If glucose is present, the urine will turn greenish yellow, then orange, then brick red.
   b You can use the test strips with only a tiny sample of urine. You don't need any apparatus, just the test strip. It is much faster.

2. Lactose is a disaccharide, and glucose and galactose are monosaccharides.

3. Enzymes are not changed during a reaction, so they can be re-used over and over again.

Item B6h Genetic engineering

1. a Restriction enzymes are used at stages 2 and 4.
   b Ligase is used at stage 5.

2. Is it toxic to humans? Does it harm beneficial insects that visit the flowers and pollinate them? Does its pollen easily transfer to other plants nearby – this could spread the gene into other species.

Answers to SAQs: Chemistry

Item C5a Moles and empirical formulae

1. hydrogen: 1
   carbon: 12
   sulfur: 32
   zinc: 65
   lead: 207

2. a 1 g
   b 12 g
   c 32 g
   d 65 g
   e 207 g

3. one mole of atoms
   or, 600 000 000 000 000 000 000 000 atoms
   or, 6×10^{23} atoms.

4. a 40, 40 g
   b 160, 160 g
   c 100, 100 g
   d 107, 107 g
   e 164, 164 g

5. a 44, 44 g
   b 17, 17 g
   c 16, 16 g
   d 64, 64 g
   e 180, 180 g

6. a 360 g ÷ 120 g = 3 moles
   b 15 g ÷ 120 g = 0.125 moles
   c 4.92 g ÷ 120 g = 0.041 moles

7. There is 80 g of bromine in one mole of sodium bromide.
   a 6 × 80 g = 480 g
   b 25 × 80 g = 2000 g
   c 0.72 × 80 = 57.6 g

8. a 37 g − 28 g = 9 g
   b 222 g − 168 g = 54 g
   c 7.4 g − 5.6 g = 1.8 g

9. a 3.0 g of lithium oxide, 1.6 g of oxygen
   b 150 g of lithium oxide, 80 g of oxygen
   c 6.0 g of lithium oxide, 3.2 g of oxygen

10. a 3.0 g of Ca is 0.075 moles (3.0 ÷ 40 = 0.075).
    Therefore, 0.075 moles of CaO are formed. 0.075 moles of CaO have a mass of 0.075 × 56 g = 4.2 g.
    b 7.6 g of Ca is 0.19 moles (7.6 ÷ 40 = 0.19).
    Therefore, 0.19 moles of CaO are formed. 0.19 moles of CaO have a mass of 0.19 × 56 g = 10.64 g.
    c 6.4 g of Ca is 0.16 moles (6.4 ÷ 40 = 0.16).
    Therefore, 0.16 moles of CaO are formed. 0.16 moles of CaO have a mass of 0.16 × 56 g = 8.96 g.

11. a $C_2H_4O_2$
    b CH_2O

12. a HO
    b CO_2
    c C_2H_6O
    d CH_2O

13. ethanoic acid and glucose

14. a Sulphur: 40 ÷ 32 = 1.25
    Oxygen: 60 ÷ 16 = 3.75
    Simplest ratio is 1:3.
    So the empirical formula is SO_3.
    b Sulphur: 50 ÷ 32 = 1.5625
    Oxygen: 50 ÷ 16 = 3.125
    Simplest ratio is 1:2.
    So the empirical formula is SO_2.

15. a Carbon: 96 ÷ 12 = 8
    Hydrogen: 24 ÷ 1 = 24
    Simplest ratio is 1:3.
    So the empirical formula of ethane is CH_3.
    b Carbon: 360 ÷ 12 = 30
    Hydrogen: 30 ÷ 1 = 30
    Simplest ratio is 1:1.
    So the empirical formula of benzene is CH.

Item C5b Electrolysis

1. *Flow of charge* means an electric current.
   Discharge of ions means the ions leave the electrolyte.

2. a The ions are bound in position by ionic bonds and can't move.
   b Potassium ions are positive and are attracted by the negative cathode.
   c Chloride ions are negative and are attracted by the positive anode.

3. a

   DC power supply

   anode — cathode

   bubbles of chlorine gas are forming at anode

   electrolyte is molten lead bromide

   molten lead metal is forming at cathode

b Lead forms at the cathode and bromine forms at the anode.

4 at the cathode

$Al^{3+} + 3e^- \rightarrow Al$

$Pb^{2+} + 2e^- \rightarrow Pb$

$Pb^{2+} + 2e^- \rightarrow Pb$

at the anode

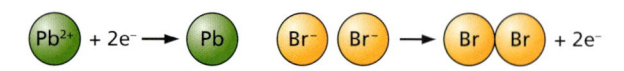

$O^{2-}\ O^{2-} \rightarrow O\ O + 4e^-$

$Br^-\ Br^- \rightarrow Br\ Br + 2e^-$

$I^-\ I^- \rightarrow I\ I + 2e^-$

5 a KNO_3
   b hydrogen gas
   c oxygen gas
   d the discharge of ions

6

Diagram: DC power supply connected to anode (+) and cathode (−) in a beaker; electrolyte is potassium sulfate solution, K_2SO_4(aq); bubbles of oxygen gas at anode; bubbles of hydrogen gas at cathode.

7 a hydroxide ions and nitrate ions
   b hydroxide ions
   c $4OH^- - 4e^- \rightarrow O_2 + 2H_2O$
   d hydrogen ions and nitrate ions
   e hydrogen ions
   f $2H^+ + 2e^- \rightarrow H_2$

8 a Hydrogen ions gain electrons more easily than potassium ions.
   b Hydroxide ions lose electrons more easily than sulfate ions.

9 a The ammeter in the circuit would be used to measure the current.
   b The cathode will gain 0.2 g of copper.
   c By using a current of more than 2 A for 5 minutes, or by using a current of 2 A for more than 5 minutes, or by increasing both the current and the time.

10 These electrons are moved from the anode to the cathode. This requires energy from the DC power supply. When the electrons reach the cathode they are gained by copper ions, which become copper atoms.

11 a $Q = It$, $Q = 0.3 \times 20 \times 60 = 360\,C$
       $45\,cm^3$ hydrogen and $22.5\,cm^3$ oxygen would be collected.
    b $Q = It$, $Q = 6 \times 80 \times 60 = 28800\,C$
       $3600\,cm^3$ hydrogen and $1800\,cm^3$ oxygen would be collected.
    c $Q = It$, $Q = 1 \times 60 \times 60 = 3600\,C$
       $450\,cm^3$ hydrogen and $225\,cm^3$ oxygen would be collected.

Item C5c Quantitative analysis

1 a 54%
   b 2 bowls

2 12 bowls, as each bowl of cereal would contain 0.5 g of salt.

3 a 0.2 g   b 0.5 g   c yes

4 a The solution shown in part **a** is more concentrated.
   b The solution shown in part **b** is more dilute.

5 a sodium nitrate
   b water
   c There are 10 g of sodium nitrate dissolved in each dm^3 of solution.
   d B
   e A
   f B

6 a $2000\,cm^3$ b $12000\,cm^3$
   c $500\,cm^3$ d $9\,cm^3$

7 a $8\,dm^3$ b $1.7\,dm^3$
   c $0.4\,dm^3$ d $0.023\,dm^3$

8 a fruit juice, medicines, baby milk
   b Over-diluting an antibiotic medicine may render it useless. Over-diluted baby milk will not nourish a child sufficiently.

9 a $4\,mol/dm^3$ b $0.4\,mol/dm^3$
   c $0.08\,mol/dm^3$

10 a $15\,g/dm^3$ b $2\,g/dm^3$
    c $40\,g/dm^3$

Answers to SAQs: Chemistry

11 a 20 g/dm³, which is 0.5 mol/dm³
    b 10 g/dm³, which is 0.25 mol/dm³
    c 80 g/dm³, which is 2 mol/dm³

12 a 6 moles
    b (0.065 dm³ × 4 mol/dm³) = 0.26 moles
    c 117 g/dm³ is 2 mol/dm³
       (0.2 dm³ × 2 mol/dm³) = 0.4 moles

Item C5d Titrations

1 Fill a burette with acid up to the zero line using a funnel.
   Use a graduated pipette and pipette filler to put 25 cm³ of alkali into a conical flask.
   Add a suitable indicator to the alkali.
   Slowly add acid from the burette to the alkali.
   Stop adding acid when the indicator changes colour.
   Note the burette reading.
   Repeat the titration until at least three results are obtained that are in good agreement with each other.

2 a 14.4 cm³ and 14.5 cm³
   b To ensure that the first titre value obtained was accurate.
   c Reasonably successful. Three identical results would have been more reassuring but 14.6 cm³, 14.4 cm³ and 14.5 cm³ are in quite good agreement. Remember, titration should be a *highly* accurate technique.
   d 14.5 cm³

3 a colourless in acid and purpley-pink in alkali
   b purple in acid and green in alkali
   c red in acid and blue in alkali

4 UI is a mixture of indicators and does not give a sharp colour change at the end-point.

5 The pH of the alkali decreases because the acid neutralises it.

6 acid + alkali → salt + water

7 a 20–21 cm³   b 1.0–1.1
   c 25 cm³

8 a $N = V \times C$
      All volumes must be converted into dm³.
      number of moles of sodium hydroxide
      = 0.0250 dm³ × 0.40 mol/dm³
      = 0.010 moles
   b the same number, 0.010 moles of hydrochloric acid

   c $$C = \frac{N}{V}$$
      concentration of hydrochloric acid = $\frac{0.010}{0.0322}$
      = 0.31 mol/dm³ (to 2 s.f.)

Item C5e Gas volumes

1 a inverted measuring cylinder, inverted burette, gas syringe
   b

Method	Advantage	Disadvantage
inverted measuring cylinder	easy to use	inaccurate
inverted burette	accurate	graduations not convenient for this use
gas syringe	accurate	plunger may stick

2 The solution has mass. If it sprays out, the flask will lose extra mass. The experimenter will not know how much of the mass loss is due to solution spray, and how much is due to gas given off.

3 a 105 cm³. Table 5e.1 shows this, because the volume recorded stops rising when it reaches 105 cm³. Figure 5e.5 shows this, because the line stops rising when it reaches 105 cm³.
   b 200 seconds
   c 73 or 74 seconds
   d 100 or 101 cm³
   e The graph in Figure 5e.5 is probably easier, but such decisions can also be made by looking at the table.

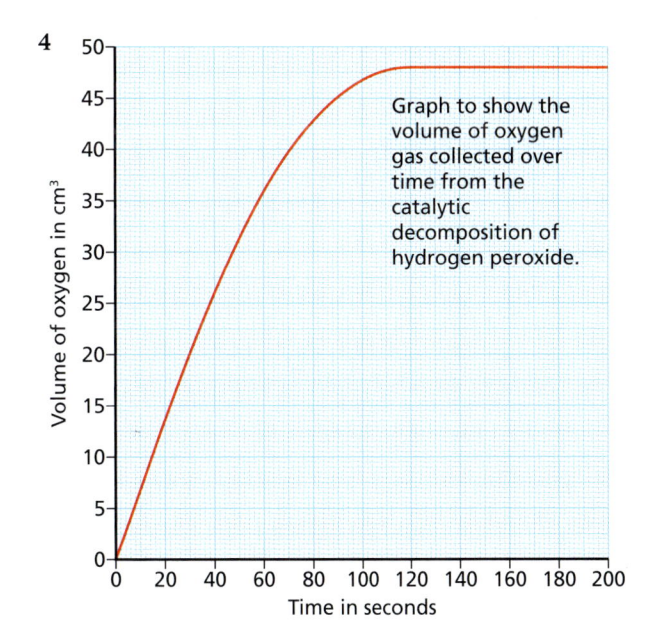

4 Graph to show the volume of oxygen gas collected over time from the catalytic decomposition of hydrogen peroxide.

5 The reactant that is in shortest supply. The limiting reactant is the only reactant that affects the amount of product formed.

6 (100 cm³ of acid and 10 g of marble give 120 cm³ of gas.)
   a 360 cm³ of gas – since the amount of the limiting reactant has been tripled, the amount of product is tripled too.
   b 60 cm³ of gas – since the amount of the limiting reactant has been halved, the amount of product is halved too.
   c 120 cm³ of gas – the acid is the limiting reactant, so increasing the amount of marble has no affect on the amount of product.
   d 240 cm³ of gas – the amount of the limiting reactant has been doubled, so the amount of product is doubled too. Increasing the amount of marble has no affect on the amount of product.
   e See above for the reasoning behind each answer.

7 a 10 moles, so 240 dm³
   b 0.0125 moles, so 0.30 dm³ or 300 cm³
   c 0.03125 moles, so 0.75 dm³ or 750 cm³
   d 0.005 moles, so 0.12 dm³ or 120 cm³

8 a 56 g is one mole of iron so one mole of hydrogen forms. This has a volume of 24 dm³ or 24 000 cm³.
   b 2.8 g is 0.05 moles of iron so 0.05 moles of hydrogen forms. This has a volume of (0.05 × 24) = 1.2 dm³ or 1200 cm³.
   c 1.0 g is (1.0 ÷ 56) = 0.018 moles of iron, to 2 s.f. Therefore, 0.018 moles of hydrogen forms. This has a volume of (0.018 × 24) = 0.43 dm³ or 430 cm³ to 2 s.f.
   d The iron is the limiting reactant.

Item C5f Equilibria

1 a In a reversible reaction, reactants can react to form products and products can react to reform reactants.
   b Dynamic equilibrium is reached when the reaction that makes the products and the reaction that reforms the reactants are occurring at the same rate. The amount of product does not change.

2 Forward reaction – the reaction in which reactants become products. In the Haber process, this is: hydrogen + nitrogen → ammonia.
   Backward reaction – the reaction in which products become reactants. In the Haber process, this is: ammonia → hydrogen + nitrogen.

3 *When the Haber process begins, the concentration of hydrogen and nitrogen is high, and the concentration of ammonia is zero. Because of this the forward reaction has a high rate, and the backward reaction has no rate. As hydrogen and nitrogen are used up, their concentrations decrease and the rate of the forward reaction decreases. As ammonia is formed, its concentration increases and the rate of the backward reaction increases.*
   There comes a time when the forward and backward reactions have equal rates. Hydrogen and nitrogen are becoming ammonia, and ammonia is becoming hydrogen and nitrogen, at the same rates. Dynamic equilibrium has now been reached.

4 a It is increased.
   b It is decreased.

5 a 74–75%
   b 440–450 atm
   c Yes, the graph confirms that higher pressure causes an increased yield.

6 In a closed system, reactants and products cannot be added or removed.

7 a The position of equilibrium shifts to the right, producing more ammonia.
   b The position of equilibrium shifts to the left, producing more hydrogen and nitrogen.
   c The position of equilibrium shifts to the left.
   d The position of equilibrium shifts to the right, producing more ammonia.

8 a There are fewer gas molecules on the right of the ⇌ arrow, so the position of equilibrium shifts to the right when the pressure is increased.
   b There are more gas molecules on the left of the ⇌ arrow, so the position of equilibrium shifts to the left when the pressure is decreased.

9 The raw materials used to make sulfuric acid are sulfur, oxygen and water.
   a Sulfur is obtained from underground by mining, oxygen from the air, water from a tap.

b The sulfur is used in step one when it reacts with oxygen from the air (it is burnt in air). The oxygen is used in step one when it reacts with the sulfur, and in step two when it reacts with the sulfur dioxide formed in step one. The water is used in step three when it reacts with the sulfur trioxide formed in step two.

c step one: sulfur + oxygen → sulfur dioxide
step two: sulfur dioxide + oxygen ⇌ sulfur trioxide
step three: sulfur trioxide + water → sulfuric acid

d step one: $S + O_2 \rightarrow SO_2$
step two: $2SO_2 + O_2 \rightleftharpoons 2SO_3$
step three: $SO_3 + H_2O \rightarrow H_2SO_4$

Item C5g Strong and weak acids

1 When HCl is dissolved in water, it ionises producing H⁺ ions (and Cl⁻ ions, but these do not account for the acidic property).

2 a Hydrochloric acid ionises fully in water giving a high concentration of H+ ions, while ethanoic acid ionises partially in water giving a lower concentration of H⁺ ions.
b Add some universal indicator to solutions of each acid. The acid solutions must be of equal concentration.
c In the hydrochloric acid solution, the universal indicator would turn red, while in the ethanoic acid solution it would turn orange. The hydrochloric acid solution has a lower pH than the ethanoic acid solution.

3 a The nitric and sulfuric acid in acid rain are strong acids, which means they are completely ionised, but their concentrations are low (*very dilute* means very *low concentration*) – there are only a small number of moles per dm³.
b The methanoic acid from red ants is a weak acid, which means it is only partially ionised, but its concentration is high – there are a large number of moles per dm³.

4 Figure 5g.3 shows a greater rate of fizzing (producing carbon dioxide gas) with the hydrochloric acid.

5 a The strength of the acid affects the rate of the reaction.

b The amount of acid and the amount of magnesium affect the amount of hydrogen produced at the end.

6 The hydrogen ions are positive and are attracted to the negative cathode.

7 The ethanoic acid has the greater electrical resistance.

8 In the experiment involving hydrochloric acid solution, there is a greater rate of gas production, and the bulb is brighter. Both observations mean that a larger current is flowing.

Item C5h Ionic equations

1 a The unidentified solution contained iodide ions.
b The unidentified solution contained chloride ions.
c The unidentified solution contained bromide ions.
d The unidentified solution did not contain chloride, bromide or iodide ions.

2 a sodium iodide (aq) + silver nitrate (aq)
→ sodium nitrate (aq) + silver iodide (s)
b sodium bromide (aq) + silver nitrate (aq)
→ sodium nitrate (aq) + silver bromide (s)
c sodium sulfate (aq) + barium chloride (aq)
→ sodium chloride (aq) + barium sulfate (s)

3 a The ions can't move and collide.
b Dissolving enables the ions to move.
c sodium chloride (aq) + silver nitrate (aq)
→ sodium nitrate (aq) + silver chloride (s)

4 a $Ag^+ (aq) + Br^- (aq) \rightarrow AgBr (s)$
b $Ag^+ (aq) + I^- (aq) \rightarrow AgI (s)$

5 $Ba^{2+} (aq) + SO_4^{2-} (aq) \rightarrow BaSO_4 (s)$

6 Mix lead nitrate solution and sodium iodide solution.
Filter out lead iodide precipitate.
Wash precipitate with distilled water.
Dry the precipitate.

Item C6a Energy transfers – fuel cells

1 a exothermic
b hydrogen + oxygen → water

2 a The glowing splint would re-light.
   b The glowing splint would not re-light.
   c The glowing splint would not re-light.
   d The flame would burn very brightly, but there would be no 'pop'.
   e There would be no 'pop'.
   f You would hear a squeaky 'pop'.

3 Look at Figure 6a.3.
   a hydrogen and oxygen
   b water
   c The product (water) is at a lower energy level than the reactants (hydrogen and oxygen), which means that energy is given out in going from reactants to products.

4 a heat and sound (for example, the 'pop' in the 'pop test')
   b electrical energy
   c hydrogen + oxygen → water

5 a $H_2 \rightarrow 2H^+ + 2e^-$
   b Hydrogen atoms lose electrons at the anode.
   c $O_2 + 4H^+ + 4e^- \rightarrow 2H_2O$
   d Oxygen atoms gain electrons at the cathode.
   e $2H_2 + O_2 \rightarrow 2H_2O$

Item C6b Redox reactions

1 a iron + oxygen + water → hydrated iron(III) oxide
   b oxygen
   c iron
   d hydrated iron(III) oxide

2 The iron is oxidised because iron atoms lose electrons: $Fe \rightarrow Fe^{3+} + 3e^-$
   Oxygen is reduced because oxygen atoms gain electrons: $O + 2e^- \rightarrow O^{2-}$

3 An *oxidising agent* takes electrons from another substance.
   When iron rusts, the oxygen atoms take electrons from the iron atoms.
   Therefore, oxygen is acting as an oxidising agent.
   A *reducing agent* gives electrons to another substance.
   When iron rusts, the iron atoms give electrons to the oxygen atoms.
   Therefore, iron is acting as a reducing agent.

4 a oiling/greasing, painting, galvanising, sacrificial protection, alloying, coating with tin
   b They stop oxygen and water reaching the surface of the iron.

   c Another metal is put in contact with the iron or steel. The other metal corrodes instead of the iron or steel. The other metal's atoms lose electrons more readily than the iron atoms. Zinc and magnesium can be used to provide sacrificial protection.
   d Because iron atoms lose electrons more readily than tin atoms.

5 a Magnesium, zinc, iron, tin.
   b i   Nothing happens (magnesium is more reactive than zinc).
       ii Iron is displaced.
       iii Nothing happens (zinc is more reactive than tin).
       iv Tin is displaced.

6 ii zinc + iron(II) sulfate → zinc sulfate + iron
   iv iron + tin(II) sulfate → iron(II) sulfate + tin

7 a $Mg + ZnSO_4 \rightarrow MgSO_4 + Zn$
   b $Mg + SnSO_4 \rightarrow MgSO_4 + Sn$
   c $Zn + SnSO_4 \rightarrow ZnSO_4 + Sn$

8 a • Magnesium is oxidised: $Mg \rightarrow Mg^{2+} + 2e^-$
        Zinc is reduced: $Zn^{2+} + 2e^- \rightarrow Zn$
      • Magnesium is oxidised: $Mg \rightarrow Mg^{2+} + 2e^-$
        Tin is reduced: $Sn^{2+} + 2e^- \rightarrow Sn$
      • Zinc is oxidised: $Zn \rightarrow Zn^{2+} + 2e^-$
        Tin is reduced: $Sn^{2+} + 2e^- \rightarrow Sn$
   b • Oxidising agent: zinc ions (or zinc sulfate)
        Reducing agent: magnesium
      • Oxidising agent: tin ions (or tin sulfate)
        Reducing agent: magnesium
      • Oxidising agent: tin ions (or tin sulfate)
        Reducing agent: zinc

Item C6c Alcohols

1 a carbon, hydrogen and oxygen
   b two carbon atoms, six hydrogen atoms and one oxygen atom
   c by covalent bonds (since carbon, hydrogen and oxygen are all non-metals)

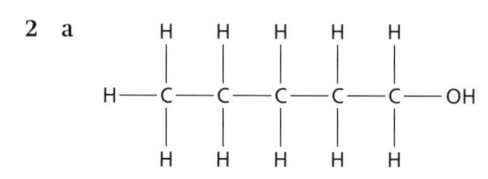

2 a

   H—C(H)(H)—C(H)(H)—C(H)(H)—C(H)(H)—C(H)(H)—OH

   b $C_5H_{11}OH$

3 a glucose → carbon dioxide + ethanol
   b The enzymes act as catalysts, and speed up the reaction (increase the rate of the reaction). They are not used up in the reaction.
   c 25–50 °C, presence of water, presence of yeast enzymes, presence of sugar, absence of oxygen
   d by distilling the wine

4 a Ethanol can be made by reacting ethene with steam, using a heated phosphoric acid catalyst.
   b ethene + steam → ethanol
   c Industrial ethanol is used as a fuel and as a solvent.

5 a Ethanol made by fermentation:
      • is produced in dilute solutions that are suitable for drinking.
      Ethanol made by hydration:
      • is produced at very high concentration, unsuitable for drinking.
   b Ethanol made by hydration:
      • is produced at a concentration that can be used directly as a fuel or solvent.
      • is made by a continuous process – continuous processes are often cheaper.
      Ethanol made by fermentation:
      • is produced as a dilute solution that has to be distilled if the ethanol is to be used as a solvent.
   c By fermentation, as crude oil may well have run out by 2050.

6 a By passing ethanol vapour over heated aluminium oxide (for a diagram, see Figure 6c.7).
   b ethanol → ethene + water
   c making polythene (polyethene)

Item C6d Chemistry of sodium chloride (NaCl)

1 a Digging the salt out and bringing it to the surface as a solid – mining.
      Dissolving the salt in water underground and pumping up the resulting solution – solution mining.
   b Mining and solution mining can cause subsidence.
   c salt solution

2 a Electrolysis is the decomposition of a molten ionic substance, or of an aqueous solution of an ionic substance, when an electric current passes through it.
   b unreactive
   c DC

3 a the anode, the positive electrode
   b treating drinking water, making household bleach, making plastics such as PVC, making solvents such as those used for dry cleaning

4 a the cathode, the negative electrode
   b making margarine

5 a The sodium hydroxide comes out as a solution from a pipe at the bottom of the container.
   b making household bleach and soap

6 a With the tube of gas from the anode, put some moist litmus paper in the gas – it will be bleached, showing that the gas is chlorine. With the tube of gas from the cathode, put a lighted splint in the gas – it will make a popping noise, showing that the gas is hydrogen.
   b Chlorine gas is poisonous, so a fume cupboard should be used. Sodium hydroxide solution is produced, which will damage skin and eyes if it comes in contact with them. Avoid skin contact with the solution and wear safety glasses.

7 a OH⁻ ions
   b H⁺ ions
   c Na⁺ ions and Cl⁻ ions

Item C6e Depletion of the ozone layer

1 a Ozone is a form of the element oxygen with three atoms per molecule, O_3.
   b The ozone layer is in the stratosphere, 15–40 km above the Earth's surface.
   c Ozone is not good for human health because it irritates the eyes and airways.

2 more sunburn, faster skin ageing, more skin cancer, eye damage

3 a CFCs are compounds of chlorine, fluorine and carbon.
   b unreactive, low boiling points, insoluble in water
   c refrigerants and aerosol propellants
   d If a refrigerant leaks into the fridge, it might contaminate food in the fridge. When someone uses an aerosol, they are likely to breathe in some of the spray. If CFCs were poisonous, both of these situations would be dangerous.

338 Answers to SAQs: Chemistry

4   a   This happens when CFC molecules are hit by UV.
    b   The single chlorine atoms cause ozone depletion.
    c   chlorine free radicals

5   a   If the bond breaks and each atom gets one of the shared electrons, free radicals are formed.
    b   If the bond breaks and one atom gets both of the bonding electrons while the other atom gets none, ions are formed.
    c   It speeds up the reaction but does not get used up itself.

Item C6f Hardness of water

1   Hard water contains dissolved substances that make it difficult for the water to form a lather with soap. Soft water contains much lower levels of dissolved substances – it is easy for the water to form a lather with soap.

2   An *aquifer* is an underground water source consisting of water soaked into porous rocks. A *reservoir* is an artificial lake formed by damming a river valley.

3   a   Take two test tubes. Put some of the first water sample into one test tube, and some of the second water sample into the second test tube. Add five drops of soap solution to each one, stopper the tubes and shake them.
    b   The hard water will not form a persistent lather. Scum will form. The soft water will form a persistent lather. Scum will not form.

4   a   Temporary hardness can be softened by boiling, while permanent hardness cannot be softened by boiling.
    b   calcium sulfate

5   Ion exchange resin – removes dissolved Ca^{2+} ions from water, putting dissolved Na^+ ions into the water in their place. The Ca^{2+} ions become bound onto the resin.
    Washing soda – sodium carbonate, which reacts with the dissolved Ca^{2+} ions, forming a precipitate of calcium carbonate.

6   a   Distilled water has no dissolved calcium ions.
    b   Yes, A is temporary hard water. It is initially hard but gets much softer when boiled.
    c   Yes, B is permanent hard water. It is initially hard and does not get softer when boiled.
    d   Adding sodium carbonate removes dissolved calcium ions by forming a precipitate of calcium carbonate.

7   a   Limescale forms when calcium hydrogencarbonate in temporary hard water is heated. This causes it to decompose, forming limescale.
    b   calcium carbonate
    c   Limescale removers contain a weak acid that reacts with limescale to form a soluble salt (and water and carbon dioxide).

8   a   limestone, chalk, marble
    b   It is a weak acid.
    c   calcium hydrogencarbonate
    d   calcium carbonate + carbon dioxide + water
        $\rightarrow$ calcium hydrogencarbonate

Item C6g Natural fats and oils

1   a   A fat is solid at room temperature, while an oil is liquid at room temperature.
    b   butter
    c   sunflower oil

2   a   An ester linkage is a linkage holding together two different sections of a molecule. It consists of one carbon atom and two oxygen atoms.
    b   Fats and oils are compounds with molecules that have two or more distinct portions, held together by ester linkages.

3   a   An unsaturated fat has at least one double bond between two carbon atoms.
    b   A saturated fat has no double bonds between carbon atoms.
    c   Mash or shake the fat with a little bromine water – if the bromine water loses its colour, the fat is unsaturated; if the colour remains, the fat is saturated.

4   a   An unsaturated fat has one or more carbon–carbon double bonds.
    b   The number of double bonds can be reduced by reacting the fat with hydrogen in the presence of a nickel catalyst.

5   a   Soap is made by boiling a fat or oil with sodium hydroxide solution.
    b $C_3H_8O_3$
    c   fat (or oil) + sodium hydroxide
        $\rightarrow$ glycerol + soap

Answers to SAQs: Chemistry 339

Item C6h Analgesics

1. a A drug is a substance that is externally administered to the body. It then modifies or affects one or more chemical reactions in the body.
   b Drugs can be swallowed, injected, inhaled, taken by suppository, or absorbed through the skin from a patch, for example.
   c A drug modifies or affects one or more chemical reactions in the body.

2. A pharmacist is a trained worker who supplies medicines.

3. a They use extremely pure, high quality starting materials. The processes they use to make the drugs are carefully and precisely controlled.
   b They are often sold in restricted amounts, such as a maximum of two to four packets, and are not sold to people under 16 years old.
   c An overdose it taking too much of a drug at one time, or during a particular time period. Always read the packet, or take advice from a pharmacist!

4. a $C_{13}H_{18}O_2$
   b $C_8H_9O_2N$

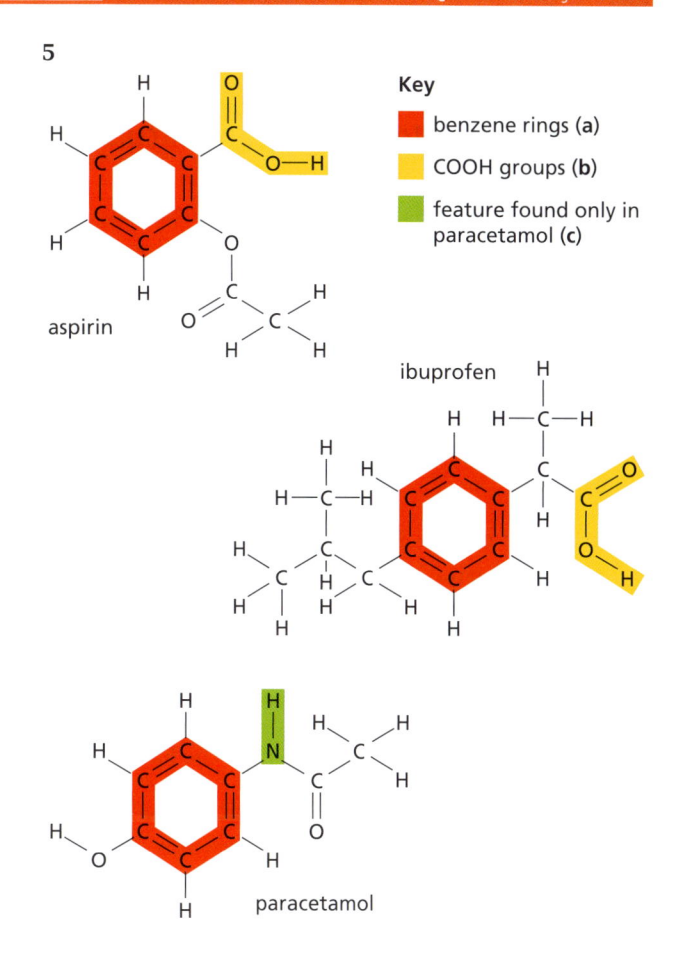

5. [Structural diagrams of aspirin, ibuprofen, and paracetamol with key: benzene rings (a), COOH groups (b), feature found only in paracetamol (c)]

6. $C_4H_6O_3$

7. a pain relief, reducing fever (lowering body temperature), thinning blood
   b able to dissolve in a solvent (in this case water)
   c easier to swallow, faster acting

8. The electric charges on the positive and negative ions in the soluble aspirin make it more soluble in water than aspirin is.

Answers to SAQs: Physics

Item P5a Satellites, gravity and circular motion

1. **a** Mars: natural satellite of Sun; ISS: artificial satellite of Earth; Moon: natural satellite of Earth
   b Other examples of natural satellites: another planet or moon of a planet; Other examples of artificial satellites: another orbiting spacecraft such as a geostationary satellite

2. mass

3. Distance has been doubled and then doubled again, so gravitational attraction is quartered and then quartered again, giving one-sixteenth of the previous value.

4. 16 orbits

5. the gravitational pull of the Sun on the Earth

6. The satellite will be slowed by friction; it will thus lose energy and move downwards towards the Earth. Its orbit will spiral in towards the Earth's surface.

7. At a height of a few hundred kilometres, a satellite can only see a small patch of the Earth's surface; being closer gives a more detailed view. From 36 000 km, a satellite can see most of one hemisphere of the Earth, so it can give a more general view (but with less detail).

Item P5b Vectors and equations of motion

1. scalars: mass, speed
   vectors: weight, force, velocity

2. 60 m/s

3. Resultant force = 450 N upwards; it will slow her down.

4. [diagram: right-angled triangle with vertical side 4 m/s, horizontal side 3 m/s, hypotenuse labelled resultant velocity from start to end]

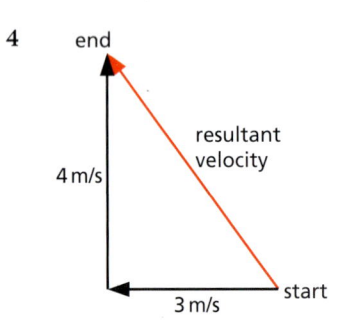

   This is again a right-angled triangle with sides of length 3 and 4 units, so the hypotenuse is 5 units. This represents the resultant of 5 m/s.

5. **a** 17 m/s
   b 330 m

6. 29.4 m

Item P5c Projectile motion

1. It is not a projectile while its fuel is still burning; after it has burnt out, it behaves as a projectile.

2. (Assuming no air resistance) the forces are unbalanced; it accelerates downwards.

3. [diagram: right-angled triangle with horizontal side 5 m/s, vertical side 4 m/s, hypotenuse labelled resultant velocity, angle 39°]

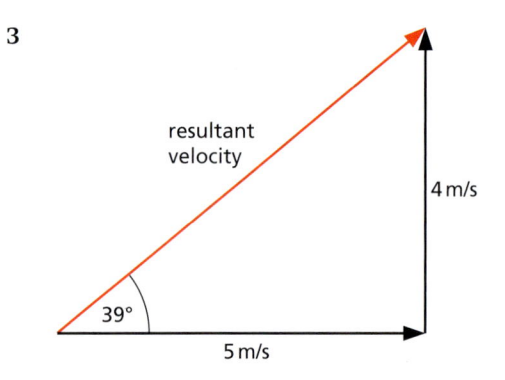

   resultant velocity = 6.4 m/s at 39° above horizontal

4. 45 m

Item P5d Momentum

1. 600 kg m/s

2. 1 kg m/s

3. before: 2 kg m/s; after: 2 kg m/s; momentum conserved

4. 1.5 m/s

5. **a** 5 kg m/s to the left
   b change = 10 kg m/s
   c 10 kg m/s

6. The high-jumper must reduce her momentum to zero; by bending her legs, she increases the time in which this happens, and so the force on them is reduced.

7. When a car bounces back, the change in momentum is twice as big as when the car stops dead. Crumpling stops the car bouncing back, so it halves the change in momentum. This reduces the force on the car.

   Also, because force = change in momentum ÷ time, a longer time of impact reduces the force.

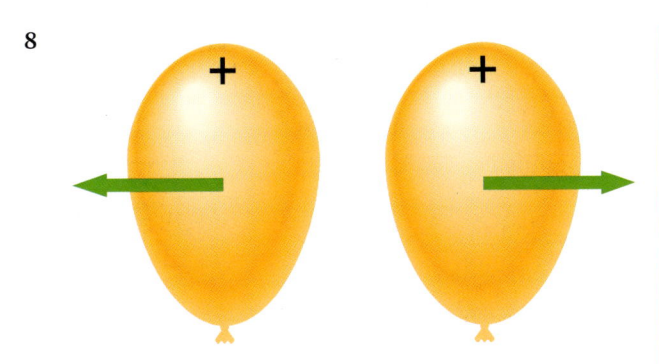

8 The forces are equal in size and opposite in direction.

Item P5e Satellite communication

1. There are many geostationary satellites in a small band across the sky; the dish must be aligned accurately so that it points exactly at the desired satellite.

2. a reflected back towards Earth's surface by the ionosphere
   b, c pass out into space
   d absorbed by atmosphere

3. They would be reflected back into space by the ionosphere.

4. The more distant listener is at the point where the reflected waves reach the ground; closer to the transmitter there is a region where neither direct nor reflected signals can be received.

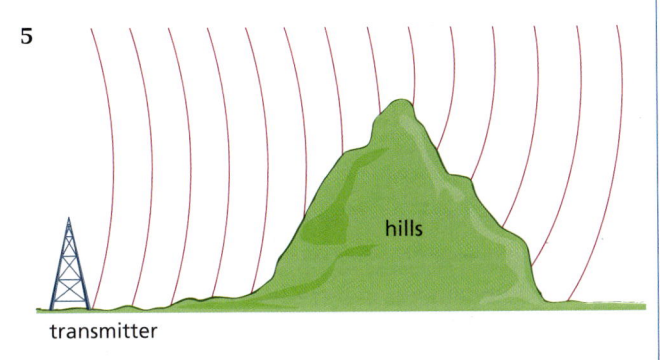

5.

6. a No
   b The size of the gap is slightly less than the wavelength.

Item P5f The nature of waves

1. a loud points
   b soft points

2. a They are produced by the same vibrating motor.
   b They are produced by loudspeakers connected to the same signal generator.

3. destructive interference

4. Constructive interference will be observed (because the path difference is one wavelength).

5. diffraction, interference

6. a The two slits act as two sources of light waves.
   b The slits must be narrow so that the light waves diffract out into the space beyond, where waves from the two slits overlap and interfere.

7. a transverse
   b Light is a transverse wave.

8. You will see the reflected light from the pond's surface when the sunglasses have been rotated through 90°. The reflected light is horizontally polarised; the Polaroid lenses are arranged vertically in normal use, to block the reflected light. Turning them through 90° allows the reflected light to pass through.

Item P5g Refraction of waves

1. Angle of incidence greater than angle of refraction; $i > r$.

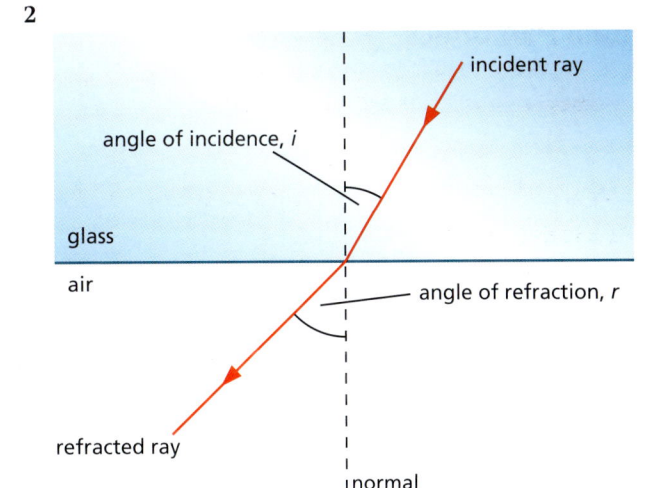

2.

3. a, b, c   Glass A

4. refractive index

5. 1.67

6. 1.84

7. 47.8°

342 Answers to SAQs: Physics

8 a Violet light travels more slowly in glass than red light.
  b violet

9 violet

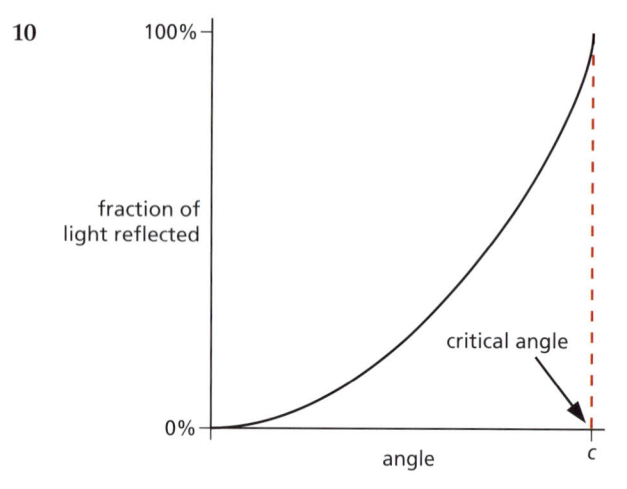

10

When $i = 0$, 0% is reflected, 100% is transmitted.

Note: the shape of the graph is not important, but it must start at 0 and end at 100% at the critical angle.

11 $\sin c = \dfrac{n_r}{n_i} = \dfrac{1}{1.33}$

$c = \sin^{-1}\left(\dfrac{1}{1.33}\right) = 48.8°$

Item P5h Optics

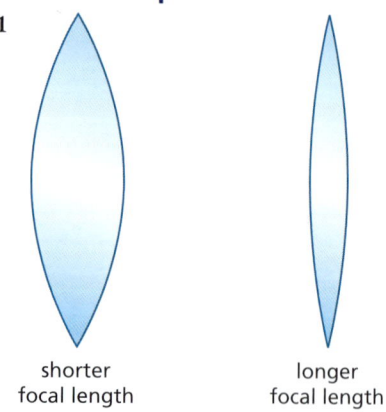

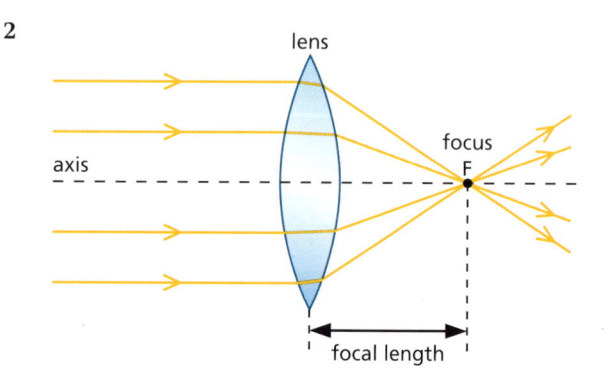

1 shorter focal length / longer focal length

2

3 A data projector uses a converging (convex) lens. It must be adjustable in order to focus the image on the screen.

4 magnified

5 real: virtual; inverted: upright (or erect); reduced: magnified (or enlarged)

6 Ray 2 starts off parallel to the axis of the lens, so it is refracted in such a way that it passes through F. (All rays parallel to the axis are refracted through F.)

7 A ray passing through a parallel-sided block is refracted first one way and then the other so that it ends up travelling in the same direction. (It is displaced slightly to one side.)

8 a 3 squares
  b 2 squares
  c 0.67 (or $\dfrac{2}{3}$)

Item P6a Resisting

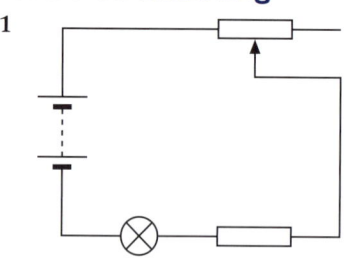

1

2 a Left and right
  b Middle and right

3 a It would include the incorrect result at 6 V.
  b The data points are scattered around the line; drawing a line of best fit allows you to eliminate outlying points; using the gradient of the line reduces the effect of the scatter.

4 Resistance = 40 Ω in both cases. However, we cannot be sure that the resistor is ohmic because we only have two values – the resistance might be different at a voltage higher than 8 V, for example.

5 Yes: the graph is a straight line through the origin at low voltages and currents.

Item P6b Sharing

1 For example: in glasshouses where artificial lighting is used on dark days; controlling the brightness of a car dashboard display to ensure that it is visible in different lighting conditions.

2   a   45 °C to 65 °C (roughly)
    b   1200 Ω to 300 Ω (roughly)
    c   0.75

3   They are equal.

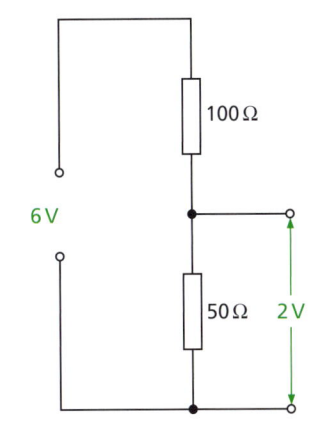

4   [Circuit diagram: 6 V supply connected to 100 Ω and 50 Ω resistors in series; 2 V measured across 50 Ω]

5   2.5 V

6   It will increase.

Item P6c Motoring

1   *Use electric motors:* washing machine, CD player, electric drill, electric lawnmower.
    Do not use electric motors: television set.
    May use electric motors: electric heater (if fan fitted); MP3 player (if uses hard drive).

2   Clockwise

3   The lines of force get farther apart.

4   The copper rod will be pushed to the right (because the two reversals cancel each other out).

5   First finger (field) points upwards; second finger (current) points to the right; so thumb (force) points towards you, i.e. the copper rod is pushed towards the power supply.

6   If the current flowed the other way, the left-hand side of the coil would become a south magnetic pole and so would be attracted to the permanent north pole.

7   A south pole would be facing a north pole, so there would be no turning effect.

8   The current arrows would be reversed, and so the force arrows would also be reversed. The motor would turn the other way.

Item P6d Generating

1   Dynamo

2   In a dynamo, the coil is fixed but the magnet moves. It works as follows: the bicycle wheel turns the knurled wheel; this causes the axle to rotate so that the permanent magnet turns inside the fixed coil; this relative movement induces a current in the coil.

3   Pull the magnet upwards out of the coil. Turn the magnet over and move the south pole down into the magnet.

4   If the wire is stationary, the magnet's lines of force are not being cut by the wire. If no lines of force are cut, no current is induced.

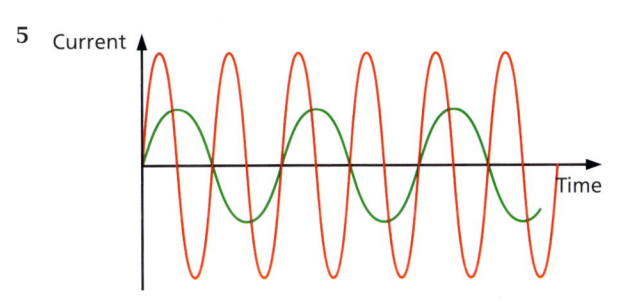

5   [Graph of Current vs Time showing two sinusoidal waves of different amplitudes]

6   60 Hz

Item P6e Transforming

1   a   200 kW
    b   12.5 kW

2   a   400 kV
    b   250 A
    c   1.5 MW

3   Step-up.

4   0.4

5   100 turns.

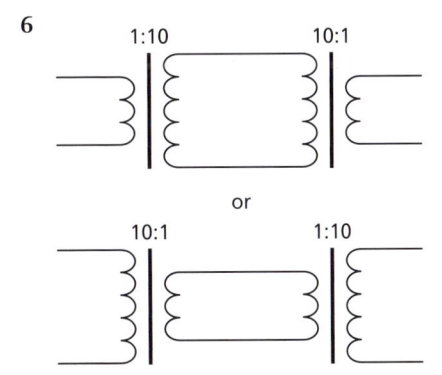

6   [Transformer diagrams: 1:10 then 10:1, or 10:1 then 1:10]

Item P6f Charging

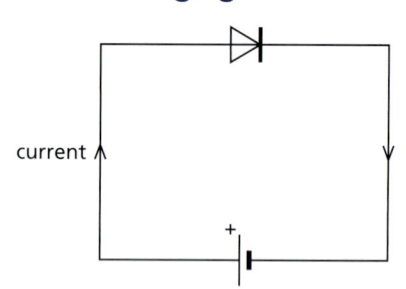

1. [circuit diagram showing diode, current direction, and battery]

2. Vary the voltage of the power supply; record voltage and current; reverse connections to diode to get negative values of voltage and current.

3. Holes (positive) and electrons (negative).

4. For half of the time, no current flows through the lamp, so it is dimmer.

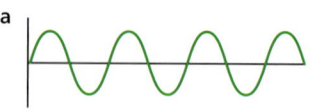

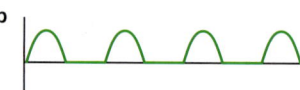

5. a [sine wave]
   b [half-wave rectified]
   c [full-wave rectified]

6. - At point B, the current is blocked by diode 3 but it can flow through diode 4 to point D.
   - Here it is blocked by diode 2, but it can flow downwards through resistor R.
   - The current reaches point C, where it can flow up through diode 1 to point A.
   - From there, it returns to the supply.

7. a Charge
   b Connect it to a battery or power supply.
   c The p.d. gradually decreases.

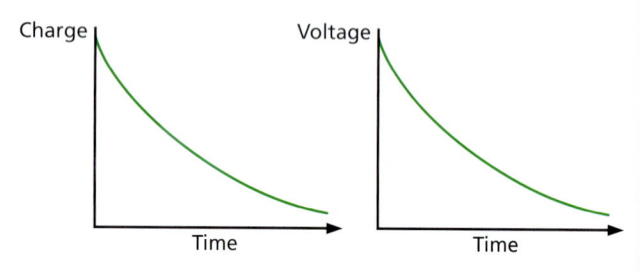

8. [Charge vs Time graph: exponential decay; Voltage vs Time graph: exponential decay]

9. 100 Hz: one ripple for each half cycle.

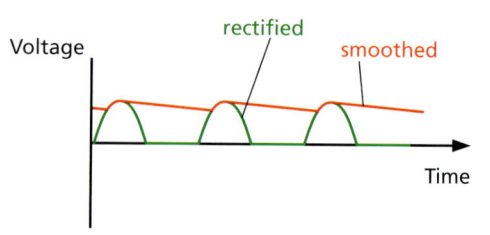

10. [Voltage vs Time graph showing rectified and smoothed waveforms]

Item P6g It's logical

1. Input: eyes. Processor: brain (central nervous system). Output: foot.

2. Examples include loudspeaker, electric bell, buzzer.

3. Examples include an art gallery or museum at night: if movement or light is detected, the alarm is triggered.

4. An AND gate.

5. The OR truth table. The light is on when both doors are open.

6. A NOT gate connected to the output of an OR gate.

7. [Circuit diagram with +5V, two switches, OR gate, LED, 0V]

   Both switches must be open if the output is to be OFF.

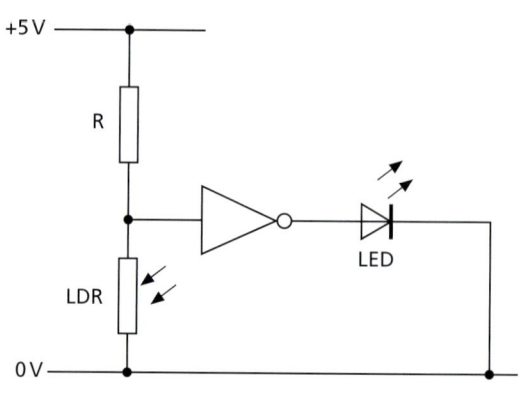

8. [Circuit diagram with +5V, resistor R, LDR, NOT gate, LED, 0V]

   When light falls on LDR, its resistance drops. The input to the NOT gate becomes low, so its output becomes high and the LED lights up.

9

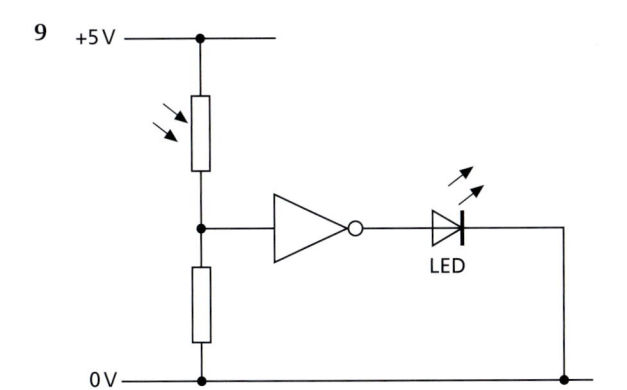

Adjust the variable resistor to set the light level at which the NOT gate switches on.

10
- Switch open: output is OFF.
- Switch closed: output depends on light level.
- In the dark: LDR has high resistance so input is high and output is ON. LDR lights up.
- In the light: LDR has low resistance so input is low and output is OFF. LDR is off.

Item P6h Even more logical

1 a Both inputs high (ON).
  b Only input 1 OFF and input 2 ON.
  c Examples include a fridge motor that switches on when the door is not open and the inside temperature is high.

2

column 1	column 2	*column 3*	column 4	column 5
input 1	input 2	OR output	input 3	output
0	0	*0*	0	0
1	0	*1*	0	0
0	1	*1*	0	0
1	1	*1*	0	0
0	0	*0*	1	0
1	0	*1*	1	1
0	1	*1*	1	1
1	1	*1*	1	1

3 Truth table has 16 rows; the output is 0 only when all four inputs are 0.

4
- When S turns ON, its signal is inverted to OFF by the NOT gate.
- The inputs to gate 1 are not both ON and so point X is ON.
- This feeds back to the other input of gate 2 – both its inputs are now ON (the OFF signal at R being turned into ON by the NOT gate) and so point Y is OFF.
- This feeds back to the other input of gate 1, so gate 1 remains ON even if S goes OFF. The output (point X) is latched ON.
- To reset the latch, R is switched ON; the inputs to gate 2 are now not both ON so its output, Y, goes ON.
- This feeds back to gate 1: both its inputs are ON (the OFF signal from S being inverted to ON by the NOT gate) and point X switches OFF. The latch has been reset.

5 a Slide and push switches; light sensor (LDR); temperature sensor (thermistor).
  b Buzzer; relay; lamp; LED.

Glossary: Biology

ADH (antidiuretic hormone) a **hormone** secreted by the **pituitary gland**, which stimulates the **kidneys** to reabsorb water from the **urine** and keep it in the body

adrenaline a **hormone** secreted by the adrenal glands, which prepares the body for fight or flight

aerobic organisms organisms that respire using oxygen

agglutination clumping together – of red blood cells, for example

agglutinins molecules on the surface of red blood cells, which determine your blood group

algal bloom a rapid and extensive growth of algae in water

alveoli air sacs in the lungs

amnion a tough membrane surrounding a developing fetus

amniotic fluid fluid secreted by the **amnion**, in which a developing fetus floats

amylase an enzyme that catalyses the conversion of starch to maltose

anaerobic bacteria bacteria that are able to survive without oxygen

angina pain caused by coronary heart disease

antagonistic muscles pairs of muscles that act together at a **joint**, one **contracting** to bend the joint and the other contracting to straighten it

antibiotic a substance that kills bacteria in the body but does not harm human cells

antibodies molecules secreted by white blood cells, which attach to specific antigens, or to **agglutinins** on the surface of red blood cells

arterioles small arteries

artificial insemination collecting sperm and inserting them into the **vagina**

asbestosis a disease caused by asbestos fibres in the lungs

aspirin a drug that relieves pain, reduces inflammation and reduces the tendency of blood to clot

AVN (atrio-ventricular node) a patch of tissue in the septum of the heart at which the electric impulses spreading through the heart muscle are slightly delayed

ball and socket joint a **joint** at which a ball at the end of one bone fits into a socket on the other, allowing circular movement

biceps a muscle that bends the arm at the elbow when it **contracts**

binary fission dividing into two

bioaccumulation the increase in concentration of a substance in organisms as you move up a food chain

biodegradable able to be broken down by the action of **microorganisms**

bioethanol ethanol produced by the action of **microorganisms** on sugars or other substances

biofuel fuel that includes substances made by living organisms

biogas a mixture of methane and carbon dioxide, made by **microorganisms** breaking down biological waste

biogas digester a structure used for the production of **biogas**

bio-inert not affected by fluids in living organisms

biosensors instruments that use enzymes to detect, or measure the concentration of, particular substances

bladder an organ where **urine** is stored

blood poisoning (septicaemia) a bacterial **infection** of the blood

bone a hard material, containing living cells embedded in **calcium phosphate**

bone marrow a tissue found in the centre of many bones, where red and white blood cells are produced

Bowman's capsule the first part of a **kidney tubule**, where the blood is filtered

bronchitis infection an inflammation in the bronchi

bypass surgery the replacement of a damaged **coronary artery** with a healthy blood vessel taken from another part of the body

calcium phosphate a mineral that gives **bone** its hardness and strength
capsule the tough tissues surrounding a **joint**
carbonic acid a weak acid formed when carbon dioxide dissolves in water
cardiac muscle the type of muscle found in the heart
cartilage a smooth, slightly flexible material found in some parts of the **skeleton** – covering the surfaces of bones at synovial joints, for example
cervix a ring of muscle at the entrance to the **uterus**
chemotrophic using inorganic chemicals as an energy source
cholesterol a lipid-like substance needed for the formation of cell membranes
chromosomes long molecules of DNA, containing **genes**
ciliated cells cells with many cilia on one surface; the cilia make sweeping movements that push fluids past them
circulatory system the heart and blood vessels
clay a substance made up of very small rock and mineral particles; clay soils are heavy and hold a lot of water
cloning producing many genetically identical organisms
closed circulatory system a **circulatory system** in which the blood is always enclosed in vessels, as in humans
clotting factors chemicals in the blood that are needed for blood clotting to take place
cocaine a highly addictive, illegal drug, which can sometimes cause a heart attack
collecting duct the final part of a **kidney tubule**
compound fracture a bone break where the surrounding tissues are also significantly damaged
continuous flow the production of a substance – for example, in a **fermenter** – where nutrients are steadily fed in, and the product is continuously removed
contractile vacuole a structure found in the cells of single-celled organisms living in fresh water, which removes excess water from the cell
contraction the shortening of a muscle
coronary artery a blood vessel supplying oxygenated blood to the heart muscle

DDT an insecticide, now banned in Europe, that does not break down in the environment and builds up to dangerous levels in organisms at high levels of food chains
deamination the breakdown of excess amino acids in the **liver**, forming **urea**
decomposers organisms that feed by breaking down dead bodies and waste from other organisms
defecation the removal of **faeces** through the anus
denatured destroyed; enzymes are denatured by high temperatures, which cause them to lose their shape and therefore catalytic activity
dialysis the use of a partially permeable membrane to allow some substances but not others to diffuse out of one fluid and into another
disaccharide a substance whose molecules are made of two sugar units combined – sucrose and maltose are examples of disaccharides
distillation heating of a mixture of liquids to separate them; those with low boiling points evaporate first
DNA ligase an enzyme that links DNA together
double circulatory system a system in which blood travels through the heart twice on one complete journey around the body
DVT (deep vein thrombosis) the formation of a blood clot in a deep-lying vein, such as a leg vein

echocardiogram an image of the heart made using ultrasound
electrocardiogram a recording of the electrical activity of the heart
embryo a developing organism; in humans, a developing child from fertilisation up to about 11 weeks

environment the surroundings of an organism; anything that affects it
epidemic a situation where many people are infected by the same disease
epididymis the part of a **testis** where sperm are stored
eutrophication the excessive growth of plants and loss of oxygen in a body of water, usually caused by pollution with material from which bacteria or plants can obtain nutrients
excretion the removal of waste products of metabolism, some of which are **toxic**, from the body
expiration breathing out
external skeleton a protective covering on the outside of the body, as in insects

factor 8 a **clotting factor**
faeces waste produced in the alimentary canal, containing undigested food
femur the thigh bone
fermentation the breakdown of substances by **microorganisms** to make a product; often used to refer to anaerobic respiration of yeast, in which sugars are converted to alcohol
fermenter apparatus in which **microorganisms** are grown, in order to produce a particular product
fetus a developing child from 11 weeks after fertilisation until birth
fibrin an insoluble, fibrous protein which forms a mesh of fibres that trap **platelets** and red cells in a blood clot
fixed joint a **joint** where the bones are not able to move – in the cranium, for example
flagellum a long, whip-like extension of a cell, which can lash against the liquid in which the cell is living and cause movement
flex bend
food web a diagram showing the feeding relationships between organisms in a habitat
FSH (follicle-stimulating hormone) a **hormone** secreted by the **pituitary gland** which encourages egg production in the **ovaries**

gaseous exchange the movement of gases across a body surface, between an organism and its environment
genes lengths of DNA that code for a particular protein or characteristic
genetic code the sequence of bases on DNA, which determines the sequence of amino acids in the proteins made in a cell
genetic engineering altering the **genes** in an organism
genetically modified organism (GMO) an organism whose **genes** have been altered – by the addition of genes from another species, for example
gills organs used for **gaseous exchange** in aquatic organisms
glomerulus a knot of blood capillaries associated with a Bowman's capsule
greenstick fracture a break in a young, growing bone, usually less serious than a break in an adult's bones
growth plates the parts of a **bone** where growth takes place

haemophilia a genetic disease in which a blood clotting factor is missing, so blood does not clot normally
heart assist device a machine that helps the heart to work while a person is waiting for a suitable heart to be found for a transplant
heart-lung machine a machine that pumps blood and oxygenates it, taking the place of the heart during open-heart surgery
heparin a substance that stops blood clotting
heroin an addictive, opiate, illegal drug, which can damage the heart
hGH (human growth hormone) a **hormone** secreted by the **pituitary gland** which controls growth
hinge joint a **joint** at which the bones can move against each other in one plane, like a door hinge
hormones chemicals secreted by endocrine glands, which travel through the body in the blood and affect target organs

humus partly broken down remains of living material, found in soil

hydrostatic skeleton a supporting system made up of fluid-filled cavities, such as in an earthworm

hypothalamus part of the brain that is involved in regulating several factors in the body, including temperature and water content

ICSI (intra-cytoplasmic sperm injection) injecting a sperm nucleus into an egg

immobilised enzymes enzymes that are trapped in something, such as beads of alginate

immunosuppressant stopping the immune system from working

implantation the attachment of an **embryo** to the **uterus** wall

incubation period the time between a **pathogen** entering the body and symptoms of disease appearing

indicator species organisms whose presence or absence can be used to estimate the degree of pollution

industrial disease a disease caused by activities of industry – because workers are exposed to harmful substances, for example

infection the entry of a **pathogen** into the body

internal skeleton a skeleton inside the body, as in humans

invertase an enzyme that catalyses the breakdown of **sucrose** to glucose and fructose

iron lung a machine that encloses a person's body and rhythmically changes the pressure around them, causing air to move in and out of the lungs

IVF *in vitro* fertilisation; adding sperm to eggs in a Petri dish

joint a place where two bones meet

kidneys organs where the blood is filtered and **urine** is made

lactase an enzyme that catalyses the breakdown of lactose to glucose and galactose

legumes plants belonging to the pea and bean family; they usually have **nitrogen-fixing bacteria** living in nodules in their roots

LH (luteinising hormone) a **hormone** made by the **pituitary gland** which encourages ovulation to take place

ligament a strong, slightly stretchy tissue that holds the two bones together at a joint

lipase an enzyme that catalyses the breakdown of lipids to fatty acids and glycerol

liver the largest organ in the body, where many different metabolic reactions occur

long bone one of the bones in the leg or arm

lubricate reduce friction by covering with a slippery liquid

lung cancer a disease in which cells in the lungs divide uncontrollably and form tumours

lung capacity the volume of air that can be held in the lungs

menstrual cycle an approximately monthly cycle of development and breakdown of the uterus lining

meristems parts of a plant where cell division takes place

mesophilic bacteria bacteria that survive at 'normal' temperatures

metamorphose change from one form to another – a caterpillar into a butterfly, or a tadpole into a frog, for example

microorganisms organisms that can only be seen using a microscope

mucus a slippery fluid that keeps surfaces of cells inside the gaseous exchange system, or other parts of the body, moist

nephrons microscopic **tubules** in the **kidneys**, where **urine** is made

nitrifying bacteria bacteria that convert ammonia to nitrites and nitrates

nitrogen-fixing bacteria bacteria that convert nitrogen gas to ammonium ions

nitrogenous excretory product a waste product of metabolism that contains nitrogen – **urea**, for example

oestrogen   a **hormone** secreted by the **ovaries**, which stimulates the production of secondary sexual characteristics, and also the thickening of the **uterus** lining during the **menstrual cycle**
open circulatory system   a circulatory system in which the blood is not always enclosed in vessels
ossification   becoming hard and bony
osteomyelitis   an **infection** of the bones
osteoporosis   a disease of the bones in which they become brittle and easily broken
ovaries   organs where eggs are produced
oviducts   tubes leading from the **ovaries** to the **uterus**; also known as Fallopian tubes

pacemaker   a small patch of muscle in the right atrium which sets the pace for the contraction of the heart muscle
pandemic   a worldwide **epidemic**
particulates   tiny particles of, for example, soot, which can cause harm if breathed in
pasteurised   heated to a high temperature for a short period of time, to kill harmful **microorganisms**
pathogen   a **microorganism** that causes disease
PCBs   chemicals, now banned, that can cause serious harm to aquatic organisms
phytoplankton   tiny plants that float in water
pituitary gland   an endocrine gland in the head, closely associated with the **hypothalamus**
placenta   the organ where the mother's blood and her developing fetus's blood are brought close together, allowing exchange of substances between them
plankton   tiny organisms that float in water
plaques   patches of **cholesterol** in artery walls
plasma   the liquid part of blood
platelets   tiny pieces of cells found in the blood, that help with blood clotting
pneumonia   an **infection** in the lungs or the tissues around them
progesterone   a **hormone** secreted by the **ovaries** and by the **placenta**, which maintains the thickness of the **uterus** lining
prostate gland   a gland near the top of the **urethra** in men, which secretes substances in which sperm can swim
protease   an enzyme that catalyses the breakdown of proteins to amino acids
pulmonary circulation   the blood vessels that carry blood between the heart and the lungs
pulse rate   the rate at which arteries expand and recoil, caused by the beating of the heart

relax   stop **contracting**; muscles can actively shorten themselves by contracting, but relaxing does not make them longer, unless they are pulled by another muscle contracting
residual air   the air that is still left in the lungs when you have breathed out as hard as possible
restriction enzyme   an enzyme that cuts DNA molecules
Rhesus   an antigen found on red blood cells; a person with these antigens is Rh positive, while a person without them is Rh negative

SAN (sino-atrial node)   the **pacemaker**
saprotrophism (saprophytism)   feeding by secreting enzymes onto the food material and then absorbing it into the body
scrotum   the skin surrounding the **testes**
selective reabsorption   taking back into the blood some of the substances filtered into a **kidney nephron**
semen   a mixture of sperm and fluids secreted by the **prostate gland** and **seminal vesicles**
seminal vesicles   glands in men that secrete fluid in which sperm can swim
septic shock   a dangerous condition caused by rapid growth of bacteria in the bloodstream
septic tank   a large container in which sewage is broken down by **microorganisms**
septicaemia   a bacterial **infection** of the blood

serum   blood plasma without the **clotting factors**
simple fracture   a clean break in a bone
single circulatory system   a **circulatory system** in which the blood travels to the **gaseous exchange** organs and then directly to the rest of the body, without going back to the heart first
skeleton   a structure that supports the body
spores   tiny structures formed by bacteria and fungi, which are very resistant to heat
starter culture   a mixture of **microorganisms** that is added to a substrate to start off a **fermentation**
sucrase   an enzyme that catalyses the breakdown of **sucrose** to glucose and fructose
sucrose   a **disaccharide** sugar whose molecules are made of glucose and fructose molecules combined together
surrogate   stand-in; a surrogate mother is one who carries an **embryo** belonging to another mother
suture   a join between bones at an immoveable **joint**, such as in the cranium
sweat   a fluid made from blood plasma and secreted onto the surface of the skin when the body is too hot; evaporation of water in sweat cools the skin
symptoms   changes caused by a **pathogen** that has infected the body
synovial fluid   the thick, slippery fluid found between the bones at a **joint**
synovial joint   a joint where the two bones are able to move relative to each other
synovial membrane   a tissue that encloses a **joint** and secretes **synovial fluid**
systemic circulation   the blood vessels that carry blood between all organs except the lungs and the heart

tendon   a strong, inelastic cord that joins a muscle to a bone
testes   organs where sperm are made
thermophilic bacteria   bacteria that survive at unusually high temperatures
thrombosis   a blood clot inside a blood vessel
tidal air or tidal volume   the volume of air that is breathed in or out in one breath
tissue type   the collection of antigens that a person has on their cells
toxic   poisonous
transgenic organism   an organism that has been given **genes** from another species
triceps   a muscle in the arm that straightens the elbow joint when it **contracts**
tubules   tiny tubes
tumour   a mass of cells caused by uncontrolled division

ultrasound scan   an image made using ultrasound
umbilical cord   the cord that links a fetus to the **placenta**
urea   a **nitrogenous excretory product** made in the liver from excess amino acids
urethra   the tube leading from the **bladder** to the outside of the body
urine   a liquid containing **urea**, made in the **kidneys**
uterus   the womb; the organ in which a fetus develops

vagina   a passageway from the **uterus** to the outside of the body
vital capacity   the biggest volume of air that you can move in and out of your lungs with one breath

warfarin   a drug that reduces the tendency of blood to clot
waterlogged   waterlogged soil is full of water and has no air in the spaces between the soil particles

yoghurt   a food made from milk, by the action of bacteria that produce lactic acid from lactose

zooplankton   tiny animals that float in water
zygote   the cell formed when two gametes fuse together

Glossary: Chemistry

alcohol a member of a particular family of chemicals, all of which are compounds of carbon, hydrogen and oxygen; one member of the alcohol family is **ethanol**, which is found in beer, wine and spirits

alloy a mixture of a metal with one or more other substances (usually one or more other metals)

amount number of particles; the amount of a substance is measured in **moles**

analgesic a **drug** that reduces pain; a painkiller

aqueous solution a solution in which the **solvent** is water

aspirin an **analgesic drug**

backward reaction the reaction in which the products, on the right of the ⇌ arrow in the equation for a **reversible reaction**, react to make the reactants on the left of the ⇌ arrow

benzene ring six carbon atoms held together in a ring, or hexagon shape, by three double bonds and three single bonds, in an alternating sequence

burette a graduated tube used to measure a volume of liquid accurately when **titrating**

CFCs compounds of chlorine, fluorine and carbon; used as **refrigerants** and aerosol **propellants** before they were found to be causing **ozone** depletion

chain reaction a continually repeating reaction; a chain reaction occurs when one of the reactants is re-formed as a product

closed system a situation where reactants and products *cannot* be added or removed; a closed system is essential if a **reversible reaction** is to reach **dynamic equilibrium**

concentration a measure of how much solute there is dissolved in each dm^3 (litre) of solution; units can be g per dm^3 or mol per dm^3

conserved neither increased or decreased, not created or destroyed; see **principle of conservation of mass**

Contact process a multi-step process for the manufacture of sulfuric acid

dehydration a chemical reaction (or other process) in which water is removed from another substance

denaturing causing protein molecules to lose their shape, usually by heating; **enzymes** do not work when denatured

discharge of ions this occurs during **electrolysis** when ions either gain or lose electrons, forming electrically neutral atoms, and leave the **electrolyte**

drug a substance that is externally administered to the body and which modifies or affects one or more chemical reactions in the body

dynamic equilibrium used to describe the situation reached in a **reversible reaction** when the **forward reaction** and the **backward reaction** are occurring at the same speed

electrolysis the decomposition of a molten ionic substance, or of an **aqueous solution** of an ionic substance, caused by the passing of an electric current (DC)

electrolyte a solution or melt that conducts an electric current (DC) and undergoes a permanent chemical change (usually decomposition)

empirical formula the formula of a covalent compound written so that it identifies the atoms in one molecule of the compound in their simplest ratios

emulsifier an additive that enables oil and water to remain mixed as an emulsion

emulsion one liquid finely dispersed in (but not dissolved in) another; milk and butter are emulsions

end-point when **titrating**, the end-point occurs when *just* enough acid (or alkali) has been added to change the colour of the indicator

enzyme a protein that acts as a biological catalyst

ester a compound of carbon, hydrogen and oxygen, having at least two distinct sections held together by one or more **ester linkages**

ester linkage a linkage holding together two different sections of a molecule; it consists of one carbon atom and two oxygen atoms

ethanol an **alcohol** found in beer, wine and spirits; its **molecular formula** is C_2H_5OH

fermentation a process in which the **enzymes** in yeast are used to catalyse the reaction that changes a sugar such as glucose into carbon dioxide and **ethanol**

forward reaction the reaction in which the reactants, on the left of the ⇌ arrow in the equation for a **reversible reaction**, react to make the products on the right of the ⇌ arrow

free radical a single atom, such as a chlorine atom, is a free radical

fuel cell a device in which the energy of an exothermic chemical reaction, between a fuel and oxygen, is released as electrical energy

gas syringe a very accurate device for measuring a volume of a gas

general formula a formula that can be used to work out the **molecular formula** of a carbon compound with a particular number of carbon atoms per molecule; the general formula of **alcohols** is $C_nH_{2n+1}OH$

glycerol a chemical with the formula $CH_2OH-CHOH-CH_2OH$; glycerol is produced from fats and oils during soap making

graduated with lines and numbers on so that measurements can be taken; each line is called a **graduation**

graduated pipette a very accurate device for measuring a fixed volume of a liquid, used when performing a **titration**; many graduated pipettes have only a single **graduation**

graduation a line marked on a measuring instrument such as a **burette**, a **graduated pipette**, or a **gas syringe**, so that measurements can be made

halide a singly charged negative ion formed by a group 7 element, such as Cl^-, or a compound containing a singly charged negative ion formed by a group 7 element, such as NaCl

halogen a group 7 element, such as chlorine

hard water water that will not lather easily when shaken with soap

hydration a chemical reaction (or other process) in which water is added to another substance

hydrolysis the breaking of a bond when it reacts with water or a similar substance – for example, the hydrolysis of the **ester linkages** in an oil when it reacts with sodium hydroxide during soap making

ibuprofen an **analgesic drug**

industrial methylated spirits impure **ethanol** made by reacting ethene with steam using a phosphoric acid catalyst; one of the impurities in the ethanol is methanol, which is highly poisonous

inert unreactive

ion-exchange resin a method of softening water; the ion-exchange resin removes dissolved Ca^{2+} ions from the water and replaces them with Na^+ ions

ionic equation an equation that omits substances whose state or electric charge does not change

limescale solid calcium carbonate formed when temporary hard water is boiled – in a kettle, for example

limiting reactant the reactant that is in shortest supply, the one that is used up first in the reactant; changing the amount of the limiting reactant will affect the amount of product formed

medicine a **drug** that benefits the person taking it

molar mass the mass in grams of one **mole** of atoms of an element or one mole of formula units of a compound

mole the unit used to measure the **amount** of a substance; one mole of any substance consists of 6×10^{23} particles (atoms, molecules or formula units)

molecular formula   the formula of a covalent compound written so that it identifies the number and type of every atom in one molecule of the compound

optimum   best possible – for example, an optimum temperature is neither too hot nor too cold
overdose   a dangerously large amount of a **drug**
oxidation   a substance is oxidised if it loses electrons
oxidising agent   a substance that takes electrons from another substance during a chemical reaction, causing the other substance to be oxidised; the oxidising agent gains electrons, and is therefore reduced itself
ozone   oxygen molecules with three atoms per molecule, formula O_3
ozone layer   a layer of the higher atmosphere which contains raised levels of **ozone**; the ozone layer protects the Earth's surface from UV radiation from the Sun

paracetamol   an **analgesic drug**
permanent hardness   hardness in water that cannot be removed by boiling
pharmacist   a trained worker who supplies medicines in a chemist shop, a dispensary or a hospital
precipitate   a solid product formed in a solution
precipitation   a reaction that forms a **precipitate**
principle of conservation of mass   the total mass of reactants in a chemical reaction is the same as the total mass of products
propellant   a liquid used in an aerosol can; the propellant turns into a gas which sprays out of the can when the nozzle on top is pressed

quantitative analysis   methods used for accurate measurement of the **amounts** of the various elements or compounds found in a substance or a mixture of substances

recommended daily allowance (RDA)   the amount of a vitamin or mineral that a person needs each day in order to stay healthy (also called 'recommended daily amount')
redox reaction   a chemical reaction in which **oxidation** and **reduction** take place
reducing agent   a substance that gives electrons to another substance during a chemical reaction, causing the other substance to be reduced; the reducing agent loses electrons, and is therefore oxidised itself
reduction   a substance is reduced if it gains electrons
refrigerant   the liquid coolant used in a refrigerator or freezer
relative atomic mass   the average mass of one atom of an element, relative to a carbon-12 atom
relative formula mass   the average mass of one formula unit of a compound, relative to a carbon-12 atom
reversible reaction   one in which reactants react to make products, and products can also react to re-form reactants

saponification   the main reaction in soap making in which a vegetable oil reacts with sodium hydroxide to make **glycerol** and soap
saturated   when used to describe a fat, this term means there are no double bonds between the carbon atoms in the hydrocarbon chains
scum   a solid formed when hardness in water reacts with soap
soft water   water that lathers easily with soap
softening water   removing the hardness from a water supply
soluble aspirin   a sodium salt made from **aspirin**, which is more soluble in water than aspirin itself
solute   a substance that can dissolve in a liquid called a solvent, for example when stirring salt into water the salt is the solute

solvent a liquid that other substances may dissolve in, for example when stirring salt into water the water is the solvent

state symbols (aq), (g), (l) and (s), meaning aqueous, gas, liquid and solid respectively

steel an alloy of iron; one of the other components is usually a low percentage of carbon, giving a strong alloy

stratosphere the Earth's upper atmosphere; starts 10 km above the Earth's surface

temporary hardness hardness in water that can be removed by boiling

titration a technique in which we measure exactly what volume of two solutions will react together – for example, measuring the exact volume of acid needed to neutralise a measured volume of alkali

titre the volume of liquid which must be added from a **burette** in order to reach the end-point of a **titration**

unsaturated when used to describe a fat, this term means there is at least one double bond between two carbon atoms in the hydrocarbon chains

washing soda sodium carbonate; washing soda is used to soften hard water when washing clothes

Glossary: Physics

acceleration due to gravity the acceleration of an object falling freely under gravity
air resistance the frictional force on an object moving through air
alternating current (AC) electric current which flows back and forth, regularly reversing its direction
amplitude modulation transmitting information by varying the amplitude of a carrier wave
angle of incidence the angle between an **incident ray** and the normal at the point where it meets a surface
angle of refraction the angle between a **refracted ray** and the normal at the point where it passes from one medium to another
artificial satellites man-made objects such as spacecraft in orbit around a larger body
axis the line passing through the centre of a lens, perpendicular to its surface

bistable an electronic circuit with two stable states
brushes devices used for sending current into and out of an electric motor or generator without the need for fixed connections

capacitor an electrical component which stores charge and energy
centripetal force any force which holds an object in a circular **orbit**, acting towards the centre of the orbit
constructive interference when two waves of the same wavelength add together because they are in step
converging lenses lenses which cause rays of light parallel to the axis to converge at the focus
corkscrew rule a rule used to determine the poles of a current-carrying coil
critical angle the minimum **angle of incidence** at which **total internal reflection** occurs
current–voltage characteristic graph a graph showing how the current through a component and the voltage across it are related

destructive interference when two waves of the same wavelength cancel because they are out of step
diffraction the spreading out of light waves as they pass through a gap or around the edge of an object
diode an electrical component which allows current to flow in one direction only
diode-bridge circuit a circuit used to produce **full-wave rectification** of **alternating current**
dispersion the separation of different wavelengths of light when they are **refracted** through different angles
diverging lenses lenses which cause rays of light parallel to the **axis** to diverge from the focus
dynamo a type of **generator**
dynamo effect the production of a voltage when a conductor and a magnetic field move relative to one another

electrons subatomic particles with negative charge
equations of motion the set of equations used in calculating the motion of a uniformly accelerated object

Fleming's left-hand rule the rule used to determine the direction of the force produced when a current flows across a magnetic field
focal length the distance from the centre of a lens to its principal **focus**
focus or focal point the point at which a **converging lens** causes rays parallel to the axis to converge
frequency the number of vibrations or waves in a second
full-wave rectification when both halves of a cycle of **alternating current** (AC) contribute to a rectified current (DC) or voltage

generator a device used to produce a voltage by moving a conductor relative to a magnetic field

geostationary   describes an **orbit** in which an **artificial satellite** will remain above a fixed point on the Earth's equator
gravity   the force that exists between any two objects with mass

half-wave rectification   conversion of AC to DC by the removal of negative voltages
holes   points in a semiconductor where an **electron** is absent
horizontal velocity   the horizontal component of the velocity of a projectile

image   what we see when we view an object by means of reflected or **refracted rays**
incident ray   a ray of light striking a surface
induced   produced
induced current   the current produced when a conductor moves relative to a magnetic field
induced voltage   the voltage produced when a conductor moves relative to a magnetic field
input device   a device used to provide a signal to the input of an electronic circuit
interference   the adding or cancelling which occurs when two waves of the same **frequency** meet
iron core   the iron or other magnetic material put inside a **solenoid** to increase the magnetic field
isolation transformer   a **transformer** used to isolate an appliance from its **power supply**

latch   an electronic circuit with two stable states
light-dependent resistor   a **resistor** whose resistance depends on the intensity of the light falling on it
logic gate   an electronic device whose output depends on the voltages applied to its inputs
longitudinal waves   waves in which the particles' motion is back and forth along the direction of transmission

magnets   devices that are permanently magnetised
magnification   the ratio of the sizes of image and object
medium   the material through which waves pass
momentum   the product of an object's mass and **velocity**
motor effect   the production of a force when a current flows across a magnetic field

natural satellite   a naturally occurring object that orbits a larger object
Newton's third law of motion   when two objects interact, they exert equal and opposite forces on each other

Ohm's law   the current through a metal wire is directly proportional to the potential difference across it, provided the temperature remains constant
ohmic conductor   a conductor which obeys Ohm's law – that is, its resistance is independent of the potential difference across it
orbit   the path of a satellite around a central object
orbital period   the time taken to travel once around an **orbit**
output device   a device whose behaviour depends on the output of an electronic circuit

parabola   the curved path followed by an object moving freely under **gravity**
path difference   the difference in distances travelled by two waves before they meet
polarisation   when a **transverse wave** vibrates in a specific plane
potential divider   a part of an electric circuit that splits a voltage into two parts
power supply   a device used to provide a variable voltage
primary coil   the input coil of a **transformer**
principle of conservation of momentum   the total **momentum** of interacting objects is constant, provided no external force acts
processor   the part of an electronic circuit that produces an output dependent on the input

projectile an object moving freely under **gravity**

ray diagram a diagram showing the paths of typical rays of light
real image an **image** that can be formed on a screen
rectification the conversion of **alternating current** to direct current
refracted ray a ray of light that has changed direction on passing from one **medium** to another
refraction the bending of light rays on passing from one **medium** to another
refractive index the property of a material that determines the extent to which it causes light rays to be refracted
relative speed the speed of one object relative to another, moving object
relay an electromagnetically operated switch
resistor a device that adds resistance to a circuit
resultant force the single force that has the same effect on a body as two or more forces
resultant velocity the combined **velocity** of an object that has two or more velocities
right-hand-grip rule a rule used to determine the direction of the magnetic field around a current
ripples small, uniform waves on the surface of a liquid

scalar quantity a quantity that has only magnitude
secondary coil the output coil of a **transformer**
slip rings devices used to transfer current into and out of an AC motor or **generator**
Snell's law the law that relates the **angles of incidence** and **refraction: refractive index** = $\sin i / \sin r$
solenoid a coil of wire, used as an electromagnet
spectrum waves separated out in order according to their wavelengths
speed the distance travelled by an object in unit time
split-ring commutator a device used to transfer current into and out of a DC motor or generator
step-down transformer a transformer used to decrease an alternating voltage
step-up transformer a transformer used to increase an alternating voltage

thermistor a resistor whose resistance changes rapidly over a narrow temperature range
total internal reflection when a ray of light strikes the inner surface of a material and is reflected back inside it
trajectory the path of a moving object
transducer any device which produces a voltage that depends on an external factor such as temperature
transformer an electrical device used to alter an alternating voltage
transformer equation the equation relating voltages to the numbers of turns on the coils of a transformer
transverse waves waves in which the particles' motion is at right angles to the direction of transmission
truth tables tables used to determine the output of a system of **logic gates**
turns ratio the ratio of the numbers of turns on the **primary** and **secondary coils** of a **transformer**

variable resistor a resistor whose value can be altered by the user
vector quantity a quantity that has both magnitude and direction
vector sum the result of adding two or more **vector quantities**, taking account of their directions
vector triangle a method for finding the vector sum of two vector quantities
velocity the speed of an object in a stated direction
vertical velocity the vertical component of the **velocity** of a **projectile**
virtual image an image which cannot be formed on a screen; formed when rays of light appear to be spreading out from a point

Periodic Table

Key

relative atomic mass
atomic symbol
name
atomic (proton) number

1	2												3	4	5	6	7	0
						1 **H** hydrogen 1												4 **He** helium 2
7 **Li** lithium 3	9 **Be** beryllium 4												11 **B** boron 5	12 **C** carbon 6	14 **N** nitrogen 7	16 **O** oxygen 8	19 **F** fluorine 9	20 **Ne** neon 10
23 **Na** sodium 11	24 **Mg** magnesium 12												27 **Al** aluminium 13	28 **Si** silicon 14	31 **P** phosphorus 15	32 **S** sulfur 16	35.5 **Cl** chlorine 17	40 **Ar** argon 18
39 **K** potassium 19	40 **Ca** calcium 20	45 **Sc** scandium 21	48 **Ti** titanium 22	51 **V** vanadium 23	52 **Cr** chromium 24	55 **Mn** manganese 25	56 **Fe** iron 26	59 **Co** cobalt 27	59 **Ni** nickel 28	63.5 **Cu** copper 29	65 **Zn** zinc 30		70 **Ga** gallium 31	73 **Ge** germanium 32	75 **As** arsenic 33	79 **Se** selenium 34	80 **Br** bromine 35	84 **Kr** krypton 36
85 **Rb** rubidium 37	88 **Sr** strontium 38	89 **Y** yttrium 39	91 **Zr** zirconium 40	93 **Nb** niobium 41	96 **Mo** molybdenum 42	[98] **Tc** technetium 43	101 **Ru** ruthenium 44	103 **Rh** rhodium 45	106 **Pd** palladium 46	108 **Ag** silver 47	112 **Cd** cadmium 48		115 **In** indium 49	119 **Sn** tin 50	122 **Sb** antimony 51	128 **Te** tellurium 52	127 **I** iodine 53	131 **Xe** xenon 54
133 **Cs** caesium 55	137 **Ba** barium 56	139 **La*** lanthanum 57	178 **Hf** hafnium 72	181 **Ta** tantalum 73	184 **W** tungsten 74	186 **Re** rhenium 75	190 **Os** osmium 76	192 **Ir** iridium 77	195 **Pt** platinum 78	197 **Au** gold 79	201 **Hg** mercury 80		204 **Tl** thallium 81	207 **Pb** lead 82	209 **Bi** bismuth 83	[209] **Po** polonium 84	[210] **At** astatine 85	[222] **Rn** radon 86
[223] **Fr** francium 87	[226] **Ra** radium 88	[227] **Ac*** actinium 89	[261] **Rf** rutherfordium 104	[262] **Db** dubnium 105	[266] **Sg** seaborgium 106	[264] **Bh** bohrium 107	[277] **Hs** hassium 108	[268] **Mt** meitnerium 109	[271] **Ds** darmstadtium 110	[272] **Rg** roentgenium 111								

Elements with atomic numbers 112-116 have been reported but not fully authenticated

*The Lanthanides (atomic numbers 58-71) and the Actinides (atomic numbers 90-103) have been omitted
Cu and Cl have not been rounded to the nearest whole number

Physics formulae

Module P5 Space for reflection

You should be able to state and use the following equations:

- The four equations of motion:

  $v = u + at$ (page 223)

  $s = \dfrac{(v + u)}{2} \times t$ (page 223)

  H $s = ut + \tfrac{1}{2}at^2$ (page 224)

  $v^2 = u^2 + 2as$ (page 224)

- Momentum and force:

  momentum = mass × velocity = mv (page 234)

  H force = $\dfrac{\text{change in momentum}}{\text{time}}$ (page 237)

- Refraction and lenses:

  Snell's law: $\dfrac{\sin i}{\sin r} = n$ (page 260)

  $\sin c = \dfrac{n_r}{n_i}$ (page 264)

  Magnification = $\dfrac{\text{image size}}{\text{object size}}$ (page 271)

Module P6 Electricity for gadgets

You should be able to state and use the following equations:

- Electrical circuits:

  Resistance (Ω) = $\dfrac{\text{voltage (V)}}{\text{current (A)}}$

  $R = \dfrac{V}{I}$ (page 274)

  H For a potential divider $V_{out} = \dfrac{V_{in} \times R_2}{(R_1 + R_2)}$ (page 283)

- Transformers:

  $\dfrac{V_p}{V_s} = \dfrac{N_p}{N_s}$ (page 301)

  H For 100% efficiency: power in = power out

  $I_p \times V_p = I_s \times V_s$ (page 304)

- Logic gates:

  NAND = AND + NOT; NOR = OR + NOT (page 317)

Index

Bold page references are to Higher material.

abortion, **49**
acceleration, **215**, 221, 222-3, **224-5**, 228, 229
acceleration due to gravity, 229
acids, 135-6, 137, **138-9**, 155-9, **156**
adrenaline, **13**
aerials, 241-3, **245**
aerobic bacteria, **83**
aerobic respiration, 26, 77, 175
aerosol propellants, 189, 192
agglutinins, **22**
air resistance, 227, 228
alcohol consumption, **20**, 37
alcoholic drinks, 65, 73, 78-80, **79**, **80**, 179, **179**
alcohols, 178-81, **178**, **179**, **180**, **181**
algae, 94, 98
alimentary canal, 33, 34, 70
alkalis, 135-6, 137, **138-9**
allergic reactions, 209
alternating current (AC), 289, 293-4, 296-7, **297**, 301, **303**, 306, 308, **309**, 311
alveoli, 26, 34
amino acids, 34, 37, **91**, 106
ammonia, 83, **91-2**, 148-50, **150**, 151
amniocentesis, 48-9, **49**
amniotic fluid, 43, 49
amoebas, 10, 26, 94, **95-6**
amphibians, 26, 94-5
amplitude modulation, **245**
anaerobic bacteria, **83**, **84**
anaerobic respiration, 77-8
analgesics, 206-9, **207**, **208**
analogue signals, **245**
angle of incidence, 257-8, **260**, 262-3
angle of refraction, 257-8, **260**, 262-3
animal fats, **20**, 201, 203
antibiotics, 74
antibodies, **22**
anti-coagulants, 24
antidiuretic hormone (ADH), **38**
antiseptics, 73-4
aorta, 11, 12, 18, 19, 20
aquatic organisms, 94-5, **95-6**, 96-9, **99**
aqueous solutions, 122-6, 161, 164
arteries, 10, 11, 12, **13**, 23, 35
arterioles, 11, 12, 23
artificial insemination, 45
asbestosis, 30, 59
asexual reproduction, 64, 77
aspirin, 23-4, 206-9, **207**, **208-9**
asthma, 31
astronomy, 256, 267
atmosphere, 242-3
atmospheric pressure, 152, **152**
atria, **13**, 15, **15**, 18
atrio-ventricular node (AVN), **13-14**

bacteria, 62-7, **63-4**
  biogas production, 82, **83**, **84**
  infection, 3-4, 7, 29-30, 31, 69-70, 72, 74, **79**
  insulin production, 64, 107-8, **107**
  microscope photograph, 267
  in soil, 87, 90, **91-2**
balanced symbol equations, 162, **163**, **174**
barium meal, 165

benzene ring, **207**
binary fission, 64, **64**
bio-diesel, 203
bioaccumulation, **99**
biodegradation, 82, **83**, **85**
bioethanol, 82, 84-5, 180
biofuels, 82-3, **83-4**, 84-5, **85**, 180, 203
biogas, 82-3, **83-4**, 84-5, **85**
biological washing powders, 102
birds, 37
bistables, 323, **324**
bladder, 35, 36, 41
bleach, 183, 184, 185-6
blood, 10-11, **11**, 21-2, **22**, 35, **36**, 42, 43
blood clotting, 19, 21, 23-4, **23**
blood groups, 22, **22**
blood plasma, 21, **22**, 33-4, 38-9
blood pressure, 12, **13**, 36, 37
blood thinning, 207, 209
blood transfusion, 21, 22, 53
blood vessels, 10, 11, 12, 20, 23, **23**
boiling, 195, 196-7, **197**
bone growth, 2, **3-4**, 59
bone marrow, 3, 53
bones, 1, 2, 3, **3-4**, 4-7, **6**, **7-8**
Bowman's capsule, **36**
breathing, 27-31, **28-9**, 33-4, **34**
brine, 183, **186**
broadcasting, 243-4, 245, **245**
bromine water, 202, **202**
brushes, 289, **289**, **297**
burettes, 135, 136, 137, 138, 141, **143**
butter, 203, 204

cables, 299-300, **300**
calcium, 5
calcium carbonate, 196, 197-8, **197**, **198**
calcium hydrogencarbonate, 196-8, **197**, **198**
calcium phosphate, 3, **3**
calcium sulfate, 195-6, 198
cameras, 269
cancer, 29, 30, 188
capacitors, 310, 311, **311**, 312
capillaries, 10, 11, 12, 26, 34, 35, 36
car fuels, 84, 124, 169, 170, 180, **180**, 203
carbon dioxide
  in biogas, 82, 84, **84**
  in climate change, 306
  from combustion, 84-5
  in fermentation, 77-8, **79**, 179, 180
  hardness of water, 198, **198**
  in respiration, 26, 33-4, **34**, 77-8
cardiac cycle, **13**
cardiac muscle, 11, 12
cartilage, 1, 3, **3**, 6
catalysts, 102, 151, 152, **152**, 180, 181, **191**, 202
cells (biology), 62, 77, 94, **95-6**, 96
centripetal force, 214
CFCs, 188-90, 191-2
chain reactions, 23, **191**
charging, 310, **311**
chemotrophic bacteria, **63-4**
chlorine, 183-4, 185-6, **186**, 190, **190-1**
chlorofluorocarbon compounds (CFCs), 188-9
cholera, 70, **71**, 73, 79
cholesterol, 20, 203
chorionic villus sampling, 49, **49**

chromosomes, 49, 106
ciliated cells, 29
circuit diagrams, 274
circular motion, 214, **215**
circulatory systems, 10-14, **11**, **13-14**, **15**, 20
climate change, 82, 84, 306
cloning, 41
closed systems, **150**
cocaine, **20**
coils of wire, 286, 289-91, **289-90**, 293-6, **296**, **297**, 301, **303**
collecting duct, **36**, 38
collisions, 232-3, **234-5**, 237
colours, 261-2
combustion, 115
comets, **215**
communications satellites, 215-17, 241-5, **245-6**
commutators, 289, **289**, **297**
computers, 314-15, 321
concentration, 129-32, **132-3**, 135, 138, **138-9**
constructive interference, **249**, **250**, **253**
contact force, 238-9
Contact process, 152, **152**
contractile vacuoles, **95**, **96**
control systems, 314-15
converging lenses, 268-9, **270-1**
copper purification, 124-6, **125**, **126**, 175
copper(II) sulfate electrolysis, 124-5, **125**, **126**
corkscrew rule, 287
coronary arteries, 11, 19, **20**, 24
coronary heart disease, 19, **20**
covalent bonds, **190-1**
covalent compounds, 117, 207, **207**
critical angle, 263, **264**
crumple zones, 236, 237
current
  in diodes, 307, **307-8**, 308-9, **309-10**
  in fuel cells, 168
  in generators, 293-6, **296**, **297**
  in logic gates, 324, **324-5**
  in motors, 286-91, **288**, **289**
  Ohm's law, 273, 276, **307**
  in resistors, 274-7, **274**
  in transformers, 300, **300**, 301, **304**
current-voltage characteristics, 275, 276-7, 307, **307**
cystic fibrosis, 31, **48**, 49, **49**, 53, 106

data presentation, 142-3
DDT, **99**
deamination, 34
decomposers, 88, **89**, **91**
decomposition, 121-2
deep vein thrombosis (DVT), 23
defecation, 33
dehydration, 181, **181**
denaturation, 102, **179**
destructive interference, **249**, **250**, **253**
detergents, 195
detritivores, 88, **89**, **91**
diabetes, 36, 64, 102, **103**
dialysis, 36-7, 52
diaphragm, 27, 28, **34**
diet, **20**, 58, 60
diffraction of waves, 243-5, **245**, 251-2
digestive system, 33, 71
digital signals, **245**

dilution, 129, 131
diode-bridge circuit, 309, **309-10**, 311
diodes, 306-7, **307-8**, 308-9, **309-10**
direct current (DC), 289, 296, **297**, 306, 308, 311
discharge of ions, 121, 122-3
discharging, **311**, **312**
disinfectants, 73, 74
dispersion of light, 261-2
displacement reactions, 174, **174**, 175
displayed formulae, 207, **208**, **209**
distance travelled, 223-4, **224-5**, 230
distillation, 80, **80**
DNA, 62, 106, 107, **107**, 109
DNA ligase, **107**
double bonds, 202, **202**, 207
Down's syndrome, 49, **49**
drugs, **20**, 21, 31, 206, **207**
dynamic equilibrium, 148-51, **149**, **150**
dynamos, 294, 295

E. coli, 62, 64, 69, 267
Earth, 211-12, **212-13**, 213-14, 215-17, **215**, **229**, 238, 243-4
earthworms, 1, 10, 26, 87, 90, **90**
echocardiograms, 15, 18
egg donation, 46, 47, **48**
eggs (ova), 41, 42, 43-4, 45
elbow joints, 6, 7, **7-8**
electric charge, **126**, **208**, 310, 311, **311**, **312**
electric motors, 186, 288-91, **289-90**, 293
electrical conduction, 157, 273
electrical impulses, 13, **13-14**, 14, 15, 18
electricity generation, 168, 169, **169**, 170, **170**, 293-7, **296**, **297**, 299, 306
electrocardiograms (ECGs), 14, **15**, 18
electrode processes, **122**, **123-4**, **125**, **168-9**, **186**
electrolysis, 121-7, 157, 175, 183-6, **186**
electrolytes, 121, **123**, 157
electromagnetic induction, 293-6, **296**, 303
electromagnetic spectrum, 188, 242
electromagnets, 273, 286, 289, **290**, 290, 293-4
electron gain/loss, **122**, **123**, **125**, **168-9**, **172**, **174**, **186**, **190-1**
electronic systems, 315-16
electrons (current carrying), 308
embryos, 42, 43, 46, 47, **48**, **49**, 106
empirical formulae, 117, **118**
emulsions, 204
end-points, 135-6, 137, **138-9**
energy level diagrams, 168
energy losses, 300, **301**, 303-4
enzymes, 34, **64**, 82, **91**, 102-4, **105**, 107, 179
equations of motion, 222-4, **224-5**, **229-30**
equilibria, 148-52, **149**, **150**, **152**
ester linkages, 201, **204**
ethanoic acid, 83, 155, **156**, **179**, **208**
ethanoic anhydride, 208
ethanol, 82, 84-5, 178, 179, **179**, 180, **180**, 181
ethene, 180, 181
ethics, **48**, **49**, 54
eutrophication, 98
excretory system, 33-9, **34**, **36-7**
exercise, 5, 13, **34**, 38, 58
exothermic reactions, 92, 167, **168**, 174
expiration, 27-8, 33, 34
explosions, **84**, 85, **235-6**
external skeleton, 1, 2

face transplants, 55
faeces, 33
Fallopian tubes, 41
Faraday, Michael, 295, 296, **296**, 302
fats, 201-4, **202**, **204**
female reproductive organs, 41-2
femur, **3**, 6
fermentation, 65, 73, 77-9, 80, 179, **179**, **180**
fertilisation, 42, 43, 57, **99**
fertility treatment, 45-8, **48**
fetal screening, 48-9, **49**
fetus, 43, 48, 57
fibrin, 23, **23**
filament lamps, 277
filtration, 35, **36**, 79, **80**, 164
fish, 10, 11, 27, **27**, 96, **99**
Fleming, Alexander, 74
Fleming's left-hand rule, **288**, 290
fluorescent genes, 106, **107**
focal length, 268-9, **271**
focal point, 268, 269, **270**
follicle stimulating hormone (FSH), **44**, 45
food hygiene, **64**, 66-7, **66**, 69-70
food labels, 128, 138
food poisoning, 66-7, 69-70
food webs, 88, 90, 97, **99**
force, 220, 221, 227, 228, 236-9, **237**, **288**, 290
formula units, 112-13, 117, **138**
fossil fuels, 84-5, 167, 170
fractures, 4-5
free radicals, 190, **190-1**
frequency, 241, 242-3, 259-60, 296-7, **297**
friction, 6, 214, 221, 238, 239
fructose, 103, **103**
fuel cells, 124, 168-70, **168-9**, **170**, 175
full-wave rectification, 309, 310, **312**
fungi, 69, 74, 77, **91**

gametes, 41
gardening, **89-90**, 92
gas collection, 141, **143**
gas syringes, 141
gas volumes, 141-4, **143**, **144-5**
gaseous exchange, 26-7
gasohol, 180
general formula, **179**
generators, 293-7, **296**, **297**, 306
genes, 31, 49, 58, **58**, 104
genetic engineering, 64, 65, 106-9, **107**
geostationary satellites, 215-16, **215**, 241
gills, 10, 11, 27, **27**, 96
Global Positioning System (GPS), 217
glomerulus, **36**
glucose, 26, 35, **36**, 36-7, 77, 102-3, **103**, 175, 179, **179**
glycerol, 201, 203
graphs, **15**, 137, 142, **143**, 275, 307
gravity, 5, 212, **212-13**, 213-14, **215**, 227, 228, 238, 239
growth, 2, 3-4, 57-9, **58**

Haber process, 148-9, **150**, 151
haemoglobin, 21
half-wave rectification, 308-9
halide ion test, 161, **163**
hardness of water, 194-7, **197**, 198, **198**
heart, 10, 11-14, **11**, **13-14**, 15, **15**
heart assist devices, 19-20
heart problems, 14, 18-20, **20**
heart transplants, 19, 51-3
heart valves, 12, **13**, 19

heroin, **20**
HFCs, 192
high-voltage electricity, 299-302, **300**, 325
hip replacement, 6-7, **7**, 51
holes (current carrying), 308
horizontal velocity, 228, **228-30**
hormones, 5, 13, **13**, 38, 43, **44-5**, 45-6, 47, 58, **58**, 59
human growth hormone (hGH), 58, **58**, 59
humus, 87, **89**, 90, **90**
hydration, 180, **180**, **181**
hydrocarbons, 192, 201-2
hydrochloric acid, 155, **156**, 197
hydrogen, 184, 202
hydrogen fuel, 124, 167, 168, **168-9**, 169-70, **170**
hydrogen ions, 155, **156**, 157, 158
hydrogen test, 167
hydrolysis, **204**
hydrostatic skeleton, 1
hypothalamus, 38

ibuprofen, 206-7, **207**, 209
images, 269, **270-1**
immobilising enzymes, 103-4
immune system, 69, 73
immunosuppressant drugs, **54**
implantation, 42, 43
in vitro fertilisation (IVF), 45-6, **48**, **49**, 106
indicator species, 98
indicators, 135, 136-7, 138
induced current, 293-5, **296**, **297**, 303
induced voltage, 294, 295, 296, **296**, **297**
industrial methylated spirits, 180
inertness, 183, 189, 191
infection, 3-4, **7**, 20, 23, 31, 36, **64**, 69, 73
infectious diseases, 64, 69-74, **71**, 267
influenza, 71-2
injection, **20**, 21
input devices, 315, 317-18, **318-19**
input voltage, 282, **283**, 301, 309
insects, 1, 2, 10, 95, 98, **99**
inspiration, 27-8
insulin, 64, 107-8, **107**
intercostal muscles, 27, 28, **34**
interference of waves, 248-9, **249**, 251, 252, **252-3**
internal skeleton, 1, 2
intra-cytoplasmic sperm injection (ICSI), 46
invertase, 103
inverted images, 269, **270**, **271**
inverters, 306
ion-exchange resins, 196
ionic equations, **163**, **172**
ionic salts, 162, 164, **208**
ionisation, 155, **156**, 157-8
ionosphere, 243, 244
ions, 121-7, 155, **156**, 157-8, **186**, 196
in solution, 161-5, **163**
iron, 172, **172**, 173
iron core, 301, **303**
isolating transformers, 302-3, **303**

joint replacements, 6-7, **7**, 51
joints, 2, 6-7, **6**, **7-8**, 37

kidney transplants, **36**, 51, 53, 54
kidneys, 33, 34-6, **36-7**, 38, 96

lactase, 103-4
lactose tolerance, 104
landfill sites, **83-4**, 85
laser light, 252, **252-3**, 262

latches, 323, **323-4**
lather, 194-5
lenses, 267-9, **270-1**
levers, **7-8**
life expectancy, 59-60, **60**
lifestyle, **20**, 30
ligaments, 6, 7
light, 250-2, **252-3**, 253-4, **254**, 256-9, **259-61**, 261-4, **262**
light-dependent resistors (LDRs), 280-1, **283-4**, 315, 317
light-emitting diodes (LEDs), 306, 317-18, **318-19**, **324-5**
lightning strikes, 14
limescale, 197-8
limiting reactants, 143-4, **145**
Lister, Joseph, 73-4
liver, 34
logic gates, 315-18, **316-17**, 321-4, **323-4**
longitudinal waves, 253
lubrication, **6**
lung capacities, 28, **28-9**
lung diseases, 29-31
lungs, 26, 27-8, 33-4, 53
luteinising hormone (LH), **44**

magnetic fields, 273, 286-91, **288**, **290**, 293, **296**, **297**, 303
magnetic force, 238, 239, 289, 290-1
magnification, **269**, **270**, **271**
magnifying glasses, 267, 268, 269, **271**
mains supply, 299, 302-3, 306
male reproductive organs, 41
margarine, 202-3
mass, 114-16, 142, 212, 220-1, 232-3, **234-5**
measuring cylinders, 141
mechanical replacements, 6-7, 7, 51-2
medicines, 59, 107, 206
menstrual cycle, 43-4, **44-5**, **99**
metamorphosis, 95
methane, 82-3, **83-4**, 84, 85, **85**
methanol, 180
microorganisms, 69-74, **71**, 77-9, **77**, **79**, 82, 87, 90
microscopes, 267
microwaves, 241, 242, 243, **245**, 248
milk, 103-4, 204
minerals, 87, **89-90**, 90, 194, 198
mitochondria, **36**
mitosis, 42, 57
molar mass, 111-13, **113**, **116**, **132-3**, **144-5**
molecular formulae, 117, 178, **178**, 188, **188**, 189, 190, 201-2, 204, 207, **207**, **208**
molecular mass, 112-13
moles, 111-13, **113-14**, **115-16**, 130, 132, **132-3**, **138-9**, **144-5**
molten ionic compounds, 121-2, 185, **186**
momentum, 232-3, **234-6**, 236-9, **237-8**
Moon, 211, 214, 215
Moore's law, 321
motor effect, 287-8, 295
motors, 286-9, **288**, **289-90**, 290-1
mucus, 26, 29, 31
multiple pregnancies, 45, 48, **48**
muscle contraction/relaxation, 7, 12, 13, **13-14**, 14, 15, **15**, 27-8
muscles, 2, 6, 7, 11, 59

nephrons, 35, 37
nervousness, 13, **13**
neutralisation, 135, 136, 137, **138**
Newton, Isaac, 213, **215**, 227, 238, 250-1, 261-2

Newton's third law of motion, 239
nitrifying bacteria, **91**
nitrogen cycle, **91-2**
nitrogen-fixing bacteria, **92**
nitrogenous excretory products, 34, 37
nutrient cycles, 91, **91-2**

oestrogen, 5, **44-5**
Ohm's law, 273, 276, **307**
oils, 201-4, **202**, **204**
optics, 267-9, **270-1**
optimum temperature, 77, **152**, 179, **179**
orbits, 211, 212, 213-15, 215-17, **215**
organ transplants, 19, **36**, 47-8, 51-5, **54**
osmosis, 94, **95-6**
ossification, 3
osteomyelitis, **3-4**
osteoporosis, 5
output devices, 315, 324
output voltage, 282, **283-4**, 309, 311
ova (eggs), 41, 42, 43-4, 45
ovaries, 41, 42, 44, **44**, 46
ovary transplants, 47-8, 52
overdoses, 206, 209
oviducts, 41, 42, 43, 45
ovulation, 42, 43, 44, 45, 47
oxidation, 168, 172, **172**, **173**, **174**, 175
oxygen
    in fuel cells, 168, **168-9**, **170**
    in respiration, 26, **34**
    in soil, 87, **89**, 90, **91**
    volume, 141-3, **143**, **144-5**
    in water, 94, 98
oxygen test, 167
oxygenated blood, 11, 13, 18, 21
ozone layer depletion, 188-90, **188**, **190-1**, 191-2

pacemaker, 13, **13-14**, 14, 18
painkillers, 206-9, **207**, **208**
pandemics, 71-2
parabola, 227, 228
paracetamol, 206-7, **207**, 209
particle theory of light, 250-1
particulates, 29, 84
Pasteur, Louis, 73, **79**
pasteurisation, 65, 78, **79**
path difference, **249-50**, 253
pathogens, **20**, 23, 69, 70, 74
PCBs (polychlorinated biphenyls), 98, **99**
penicillin, 74
Periodic Table, **114**
permanent hardness of water, 195-6, 198
permanent magnets, 287-8, **289-90**, 290, 293, 295-6, **296**, **297**
pH, 34, **34**, 64, 65, 77, **89**, 102, 137, 155, 159
photoelectric effect, 252
photosynthesis, 62, **63**, 96, 97, 98, 175
phytoplankton, 96-8
pipettes, 135, 138
pituitary gland, **38**, **44-5**, **58**
pivots, **7-8**
placenta, 42, 43, 49
planets, **215**, 267
plankton, 96-8, 99
plant growth, 57, 87, 88, **89-90**
plaques, 20, 203
platelets, 21, **23**
pneumonia, 29-30
polar-orbiting satellites, **215**, 216-17
polarisation, 253-4, **254**
pollution, 98, **99**

potential difference (p.d.), 168, 274-6, 282, **283-4**, **303**, 307, **307**, 310, **311**, 318
potential divider circuits, 282-3, **283-4**, 318
power stations, **170**, 293-4, 299, 300-1, **300**
precipitation, 161-3, **163**, 164-5
pressure effects, 149, **150**, 151, **152**
primary coils, 301, **303**, **304**
principle of conservation of mass, 114-16
principle of conservation of momentum, **234-5**, 329
processors, 315
products, 143-4, 148-51, **149**, **150**, 168
progesterone, 43, **44-5**
projectile motion, 227-8, **228-30**
projectors, 269
propellants, 189, 191, 192
prostate gland, 41
proteins, 34, 35, 58, **91**, 102, 106, **107**
protozoa, 69, 71
pulmonary artery, 12, 19
pulmonary circulation, 10
pulmonary thrombosis, **7**
pulse rate, 12

quantitative analysis, 128-34, 135, 138

radio waves, 242, 243-4, 245, **245**
rate of reaction, 149, 151, **152**, 156-7
ray diagrams, **270-1**
reabsorption, 35, **36**
reactants, 143-4, **145**, 148-51, **149**, **150**, 168
reactivity, 174
real images, **270**, **271**
receivers, 241, 242, **245**
recoil, **236**
recommended daily amounts (RDA), 128-9, **128**
rectification, 308-9, **309-10**, 310, 311, **312**
red blood cells, 21, **23**
redox reactions, 172-5, **172-3**, **174-5**
reduction, 116-17, **169**, 172, **172**, **174**, 175
reflection of waves, 250, 262-4, **264**
refraction, 250-1, 256-9, **259-61**, 261-4, **262**, **264**, 268
refractive index, 258, **259**, **260**, 264
refrigerants, 189
relative atomic mass, 111-12, **114**, **118**, **128-9**
relative formula mass, 112-13, **128**, 132
relative velocities, 220
relays, 324, **325**
residual air, 28
resistance, 273-7, **274**, 280-4, **283-4**, 300, **307-8**
resistors, 274-7, **274**, 280-3, **283-4**, 309, **309-10**, 318, **318-19**, **325**
respiration, 26-31, **27**, **28-9**, 33-4, **34**, 77, 175
respiratory diseases, 29-31
restriction enzymes, **107**
resultant force, 221
resultant velocity, **221-2**, **229**, **230**
reversible reactions, 148-9, **149**, 151, 152, **152**, 155, **156**
right-hand-grip rule, 287
ripples, 244, **245**, 248, **249**, 250, 256
road traffic accidents, 222-3
robotic machines, 314
rusting, 172, **172-3**, 173

sacrificial protection, **173**
salicylic acid, **208**
Salmonella, 69, 70

salt
  in food, **128–9**
  in soil, 109
  see also sodium chloride
salt mines, 183
salt water regulation, **95–6**
saponification, 203–4, **204**
saprotrophs, **91**
satellite telephone systems, 241–2
satellites, 211–17, **212–13**, 215
saturated fats, 201–3, **202**
saturated solutions, **186**
scalar quantities, 220
scale in kettles, 158
scum, 194–5
secondary coils, 301–2, **303**, **304**
selective reabsorption, **36**
semen, 41, 42
sensors, 280, 314–16, 317, 318
septicaemia, 72, 73
septum, **11**, 12, **13**, 18
sexual reproduction, 41–3, **44–5**, 45–8, **48**
side-effects, 47, 209
silicon, 308
single bonds, 201, **207**
sino-atrial node (SAN), **13**, **14**
skeletons, 1–3, **3–4**, 4–5, 94
skin, 33, 38, **38–9**, 188
slip rings, **297**
smallpox, 72
smelting, 116–17
smoking, **20**, 29, 30
smoothing, 311, **312**
Snell's law, **260**
soap manufacture, 184, 185, 203, **204**
soap solution, 194–5
sodium carbonate, 196
sodium chloride, 183–5, **186**
  see also salt
sodium hydroxide, 184, **186**, 203, **204**, **208–9**
sodium in salt, **128–9**
soft water, 194–5
soil, 87–8, **89–90**, 90, 91, **91–2**, 92
solar cells, 241
solenoids, 286–7
solutes, 122, 129–30, 132, **132–3**
solution mining, 183
solutions, 129–33
solvents, 129–30, 131, 132–3, 180, **180**
sound waves, 248, 249, **249–50**, 250, 253
spacecraft, 5, 169, 170, 211–17, **212–13**, 215
spectrum, 261–2
speed, 220, 296, **297**
speed of light, 251, 258–9, **259–60**, 262
sperm cells, 41, 42, 43, 45
spirometers, 28, **28–9**
split-ring commutators, 289, **289**, **297**
stalactites, 198

state symbols, 162, **163**
static electricity, **311**
steel, 173
stem cells, **48**
step-down transformers, 301
step-up transformers, 301
stratosphere, 188, **188**, 190, 191
stress, **20**
strong acids, 155–8, **156**
sucrose, 103, **103**
sugars, 103–4, 103, 179, **179**
sulfate ion test, 162
sulfur trioxide, 152, **152**
sulfuric acid production, 152, **152**, 175
Sun, 188, **213**, **215**, 262, 270
surgery, 19, 21, 24, 59
surrogacy, 46, **48**
sweat, 33, 38, **38–9**
switches, 317–18, **319**, 324
synovial joints, 6, **6**
systemic circulation, 10

telescopes, 256, 267
temperature effects, 64, **77**, **84**, 102, 149, 151, **152**, 152, 179, **179**
temperature regulation, 38–9
temporary hardness of water, 195, 196–8, **197**
thermistors, 280, 281, **284**, 315, 317, 318, **318–19**
thrombosis, 23–4
thrust, **237**
tidal air, 28
time, **237**
tissues, 3, **3**, 11, 53–4, **54**
titrations, 135–8, **138–9**
torque, 290
total internal reflection (TIR), 262–4, **264**
toxins, 66–7, 69, 108
trajectories, 227, 228, **229**
transducers, 317
transformer equation, 301
transformers, 300–3, **303–4**
transgenic organisms, 106–8
transistors, 321
transmitters, 241, 242, 243–4, **245**
transverse waves, 253–4
truth tables, 316, **316–17**, 322
tubules, 35, 36, 36
turbines, 293–4
turns ratio, 301

ultrasound, 15, 19, 37, 45, 48
ultraviolet light (UV), 188, **188**, 190, 191, **191**
umbilical cord, 43
units of concentration, 130, 132, **132–3**
Universal indicator, 137
unsaturated fats, 202, **202**, 203

urea, 33, 34, 35–6, **36**, **37**, 39
ureters, 35, 36, 37
urethra, 35, 36, 41, 42
uric acid, 37
urine, 34, 35–6, **36**, 38, **38**, **96**, 102
uterus, 41–2, 43, 43–4, 46

vagina, 42, 43
variable resistors, 274, **274**, 280–1, 282, **283**, **284**, 317, **318–19**
vectors, 220–1, **221–2**, **228–30**, **235–6**
vegetable oils, 201, 202, 203
veins, 10, 11, 12, **13**, 19, 21, 35
velocity, 220, **221–2**, 222–4, **224–5**, 228, 232–3, **234–6**
vena cava, 12, 19
ventricles, **13**, 15, **15**, 19, 20
vertical velocity, 228, **228–30**
virtual images, **270**
viruses, 69, 71, 72, 79, **107**
vital capacity, 28
vitamins, 128, **128**, 129, 138
voltage
  in fuel cells, 168, **170**
  in generators, 293, 294
  high voltage, 299–302, **300**, **325**
  in logic gates, 315, 317–18, **318–19**
  Ohm's law, 273, 276, **307**
  in resistors, 274–7, **274**, 280, 282–3, **283–4**
volume
  of air breathed, 28, **28–9**
  of gases, 141–4, **143**, **144–5**
  of solutions, 129–30, 132, **132–3**
  in titrations, 135–6, 137, **139**

washing soda, 196
water, in soil, 87, **89**, 90
water hardness, 194–7, **197**, 198, **198**
water life, 94–5, **95–6**, 96–9, **99**
water regulation, 38–9, **38**
wavelength, 245, **245**, 251–2, **259–60**
waves, 244–5, **245–6**, 248–54, **249–50**, 252–3, **254**
weak acids, 155–9, **156**, 197, **197**
weather-forecasting, 216, 243, 280
weight, 220, 238
weightlessness, 5
white blood cells, 21, 30, 31, 106
word equations, 162

yeast, 77–8, **79**, 80, **80**, 179
yield, 149–50, **152**
yoghurt, 65–7, **66**

zooplankton, 96, 97
zygotes, 41, 42, 43, 49, 57